כתבי האקדמיה הלאומית הישראלית למדעים

PUBLICATIONS OF THE ISRAEL ACADEMY OF SCIENCES AND HUMANITIES

SECTION OF SCIENCES

———

FLORA PALAESTINA

EQUISETACEAE TO UMBELLIFERAE

by

MICHAEL ZOHARY

ERICACEAE TO ORCHIDACEAE

by

NAOMI FEINBRUN–DOTHAN

FLORA PALAESTINA

PART ONE · TEXT

EQUISETACEAE TO MORINGACEAE

BY

MICHAEL ZOHARY

JERUSALEM 1966

THE ISRAEL ACADEMY OF SCIENCES AND HUMANITIES

ISBN 965–208–000–4
ISBN 965–208–001–2

First Published in 1966
Second Printing, 1981
Printed in Israel
Jerusalem Academic Press

PREFACE

FLORISTIC EXPLORATION IN PALESTINE has a long history. Many botanists have contributed to the knowledge of the local flora, mainly in the course of the last hundred years. Research by scholars resident in this country began early in the current century.

The first of them was A. Aaronsohn who, among other botanical activities, collected extensively both in Palestine and in the adjacent countries. His collections were revised and published by H. R. Oppenheimer in 1931 and H. R. Oppenheimer and M. Evenari in 1941.

J. E. Dinsmore, American by birth and a resident of Jerusalem, was another pioneer in the research of the local flora. He published the *Jerusalem Catalogue of Palestine Plants* in 1912 and later (1932–1933) revised Post's *Flora of Syria, Palestine and Sinai*.

In the early twenties, A. Eig and E. Faktorovsky began their floristic studies which were continued in the Institute of Natural History headed by O. Warburg.

Since the foundation of the Hebrew University of Jerusalem, four decades ago, floristic investigations have been undertaken by the staff of its Botany Department, and extensive plant collections from Palestine and from neighbouring countries have been brought together in the herbarium of the University. A. Eig, an outstanding taxonomist and phytogeographer, who devoted himself to the study of the local flora until his untimely death in 1938, published, together with the present authors, the *Analytical Key to the Plants of Palestine* (1931, Hebrew) which was the basis for the *Analytical Flora of Palestine* (1948, Hebrew).

The present is the first Flora confined to Palestine, that is, to Cis- and Transjordan, comprising Israel, Jordan and the Gaza Strip. It includes many species and varieties newly described, recorded for the first time or renamed since the publication of the second edition of Post's Flora (1932–1933). Certain groups have been taxonomically revised. Habitats and distribution in and outside Palestine have been reexamined.

The flora treats about 2,400 species and consists of four parts, each comprising one volume of text and one volume of plates.

Parts One and Two, written by Michael Zohary, include the families of the Pteridophyta, Gymnospermae, Apetalae and Dialypetalae.

Parts Three and Four, written by Naomi Feinbrun-Dothan, include the families of the Sympetalae and the Monocotyledoneae.

With minor deviations, the authors followed the sequence of families presented in Engler's *Syllabus der Pflanzenfamilien*, 12th edition (1964).

Keys to families, genera and species are provided; to avoid repetition, keys to varieties are given only in species comprising more than two varieties.

Data on flowering times refer to the main flowering season and not to the out-of-season flowering sometimes observed in single individuals. In general, a given species flowers much earlier in the Coastal Plain and in the Jordan and Arava Valleys than in the mountainous areas.

The descriptions, with few exceptions, are based on plants deposited in the Herbarium of the Hebrew University. The local distribution of each species is recorded for whole districts as demarcated on Map 1. The data are primarily based on specimens from the above herbarium; some are taken from literature. General distribution is presented in terms of plant geographical regions (Map 2) instead of single countries, thus indicating the phytogeographical character of the taxa. The four phytogeographical territories of Palestine are shown on Map 1.

The brief notes on the uses of local plants are in part founded on personal observation and in part taken from reliable literary sources. Data on chromosome numbers are based almost exclusively on counts made on local plants. Of Biblical plant names only those identified with some certainty are quoted. Modern Hebrew plant names are given on the plates together with the Latin names; they are indexed at the end of each volume of plates.

The drawings were made by artists under botanical supervision, mostly from living specimens; the plants are portrayed in natural size, with enlargement of diagnostically important parts. Many of the plates represent species not hitherto illustrated. The explanations are given at the end of each volume of plates.

For general reference, a selected bibliography is provided at the beginning of the first part. It is limited to basic taxonomic works, standard Floras of the regions concerned, and selected taxonomic and floristic publications on plants of Palestine.

The authors are aware of shortcomings in their work, due partly to the fact that many difficult groups in the flora of Palestine still await critical revision.

The authors

ACKNOWLEDGEMENTS

BY THE AUTHOR OF PARTS ONE AND TWO

I wish to express my sincere thanks to my collaborators and assistants who, with great devotion, helped me to accomplish this work with their advice and criticism, by checking, by reading the text and by elucidation of certain taxonomic and nomenclatural problems.

I am much indebted to Mrs. Irene Gruenberg-Fertig for her critical reading; her expert knowledge of technical literature and her sense of exactness were exceedingly valuable. To Mrs. Stephanie Wachtel, for her services in typing, checking and indexing, I am also deeply grateful.

My appreciation is due to Miss Augusta Horovitz, who read a great part of the manuscript very thoroughly. I am also indebted to Mrs. Ziva Altman, Mrs. Rivka Frank-Winter and Mrs. Edna Ben-Shlomo for checking many descriptions. Thanks are due, as well, to Miss Sara Leinkram, who checked and prepared descriptions for many species, especially of Umbelliferae and Papilionaceae. To Mr. David Heller I am grateful for scientific and technical help rendered to me in a variety of ways.

To the young taxonomists, Mr. Uzi Plitmann and Mr. Baruch Baum, I am much obliged for their share in clarifying some intricate groups, such as *Silene, Rumex, Amaranthus, Hammada, Spergularia, Reseda, Fagonia, Tamarix, Erodium, Vicia, Lathyrus, Daucus,* etc.

I am also grateful to Dr. Chaia C. Heyn for enabling me to use her knowledge of the genera *Medicago* and *Lotus* as well as to quote her unpublished data on chromosome numbers of many species.

The plates were drawn by Mrs. Esther Huber, Mrs. Ruth Koppel and Mrs. Katty Torn; a few were made by Mrs. Dulic Amsler. I would like to compliment all these artists on their precise and meticulous draftsmanship, and I trust that users of the Flora will rightly appreciate their contribution to it.

My sincere acknowledgements are extended to the authorities of the Hebrew University, the Ford Foundation and the United States Department of Agriculture; all of them, directly or indirectly, provided financial support for research on some families.

And, finally, I record my especial gratitude to the Israel Academy of Sciences and Humanities, which spared no effort to publish this work in a distinguished form, and to its technical editor, Mr. R. Eshel, for his knowledgeable and unremitting editorial assistance. The author is indebted to Dr. M. Spitzer for his advice and guidance in the lay-out and typography of this work. To the Goldberg Press I owe my warmest thanks for its careful and precise work in the printing of this book, and to Photo-offset Ziv for the printing of the plates.

M. Z.

CONTENTS

PREFACE   v

ACKNOWLEDGEMENTS   vii

LITERATURE   xi

ABBREVIATIONS   xvi

GLOSSARY   xvii

KEY TO THE FAMILIES   xxviii

SYNOPSIS OF FAMILIES IN PART ONE   xxxix

TEXT   1

APPENDIX: DIAGNOSES PLANTARUM NOVARUM   341

ADDENDA AND ERRATA   344

INDEX   345

MAPS   *at end*

LITERATURE

1. Basic Taxonomic Works

Bentham G. & Hooker J. D., *Genera Plantarum,* 3 vols., London 1862–1883.

Candolle A. L. P. P. de & Candolle A. C. P. de, *Monographiae Phanerogamarum,* 9 vols., Paris 1878–1896.

Candolle A. P. de, *Prodromus Systematis Naturalis Regni Vegetabilis,* 17 vols. & index, Paris 1824–1874 (Vols. 8–17 by A. L. P. P. de Candolle).

Engler H. G. A., *Syllabus der Pflanzenfamilien,* ed. 12, 2 vols., Berlin 1954–1964 (Vol. 1 by H. Melchior & E. Wedermann, Vol. 2 by H. Melchior).

Engler H. G. A. & Prantl K. A. E., *Die natürlichen Pflanzenfamilien,* ed. 1, 4 vols., Gesamtregister & Nachtrag 1–4, Leipzig 1887–1915; ed. 2, Leipzig–Berlin 1925.

Gmelin J. F., see Linné C. von, *Systema Naturae,* ed. 13.

Jackson B. D. et al. (ed.), *Index Kewensis Plantarum Phanerogamarum,* 2 vols. & 12 suppls.; Vols. 1–2, Oxford 1895; Suppl. 1, Brussels 1901–1906; Suppls. 2–12→, Oxford 1904–1959→.

Lamarck J. B. A. P. M. de, *Encyclopédie méthodique; Botanique,* 8 vols. & 5 suppls., Paris 1783–1817 (Vols. 5–13 by J. L. M. Poiret).

Lamarck J. B. A. P. M. de, *Tableau encyclopédique et méthodique; Botanique,* 3 vols. & suppl., Paris 1791–1823 (Vol. 2, Part 5, 2 & Vol. 3 by J. L. M. Poiret).

Linné C. von, *Amoenitates Academicae,* 7 vols., Stockholm 1749–1769.

Linné C. von, *Demonstrationes Plantarum in Horto Upsaliensi 1753,* Uppsala 1753.

Linné C. von, *Species Plantarum,* ed. 1, Stockholm 1753; ed. 2, Stockholm 1762–1763.

Linné C. von, *Systema Naturae,* ed. 10, Vol. 2, Stockholm 1759; ed. 11, Vols. 1–2, Leipzig 1762; ed. 12, Vols. 2–3, Stockholm 1767–1768; ed. 13 by J. F. Gmelin, Vol. 2, Leipzig 1791–1792.

Linné C. von, *Mantissa Plantarum,* Stockholm 1767.

Linné C. von, *Mantissa Plantarum Altera,* Stockholm 1771.

Linné C. von, *Systema Vegetabilium,* ed. 13 (*Systema Naturae,* ed. 12, botanical part) by J. A. Murray, Göttingen–Gotha 1774; ed. 15 by J. J. Roemer & J. A. Schultes, 7 vols., Stuttgart 1817–1830 (Vol. 7 by J. A. & J. H. Schultes); ed. 16 by C. P. J. Sprengel, 4 vols. & suppl., Göttingen 1824–1828.

Miller P., *The Gardeners Dictionary,* ed. 8, London 1768.

Murray J. A., see Linné C. von, *Systema Vegetabilium,* ed. 13.

Pritzel G. A., *Thesaurus Literaturae Botanicae,* ed. 2, Leipzig 1872–1877 (partly by F. W. Jessen).

Roemer J. J. & Schultes J. A., see Linné C. von, *Systema Vegetabilium,* ed. 15.

Sprengel C. P. J., see Linné C. von, *Systema Vegetabilium,* ed. 16.

Stapf O., *Index Londinensis to Illustrations,* 6 vols. & 2 suppls., Oxford 1929–1941 (suppls. by W. C. Worsdell, under the direction of A. W. Hill).

2. Selected Standard Floras

Ascherson P. F. A. & Graebner K. O. P. P., *Synopsis der mitteleuropäischen Flora,* ed. 1, Vols. 1–7, 12, Leipzig 1896–1938; ed. 2, Vol. 2, Leipzig 1912–1913.

Ascherson P. F. A. & Schweinfurth G. A., *Illustration de la Flore d'Égypte,* in: *Mém. Inst. Eg.,* 2 : 25–260 (1887); suppl., ibid., 2 : 745–821 (1889).

Battandier J. A. & Trabut L., *Flore de l'Algérie,* 2 vols., Algiers–Paris 1888–1895 (Dicotylédones by J. A. Battandier; Monocotylédones by J. A. Battandier & L. Trabut).

Boissier P. E., *Diagnoses Plantarum Orientalium Novarum,* Ser. 1, 13 fascs.; Ser. 2, 6 fascs., Geneva–Leipzig–Paris–Neuchâtel 1843–1859.

Boissier P. E., *Flora Orientalis,* 5 vols. & suppl., Basel–Geneva–Leiden 1067–1000.

Bonnet E. & Barratte G., *Catalogue raisonné des plantes vasculaires de la Tunisie,* Paris 1896.

Bouloumoy L., *Flore du Liban et de la Syrie,* 1 vol. & atlas, Paris 1930.

Briquet J. I., *Prodrome de la Flore corse,* 3 vols., Geneva–Basel–Lyons–Paris 1910–1955 (Vol. 2, Part 2 & Vol. 3 by R. de Litardière).

Clapham A. R., Tutin T. G. & Warburg E. F., *Flora of the British Isles,* ed. 1, Cambridge 1962.

Cufodontis G., Enumeratio Plantarum Aethiopiae; Spermatophyta, *Bull. Jard. Bot. État Bruxelles,* Suppl., 23–35→ (1953–1965→).

Davis P. H., *Flora of Turkey and the East Aegean Islands,* Vol. 1→, Edinburgh 1965→.

Delile A. R., *Florae Aegyptiacae Illustratio & Flore d'Égypte; Explication des planches,* in: *Description de l'Egypte,* Vol. 2 : 49–82, 145–320, Pls. 1–62, Paris 1813.

Desfontaines R. L., *Flora Atlantica,* 2 vols. & atlas, Paris 1798–1799.

Durand E. & Barratte G., *Florae Libycae Prodromus,* 2 vols., Geneva 1910.

Eig A., Zohary M. & Feinbrun Naomi, *Analytical Flora of Palestine,* ed. 1, Jerusalem 1948 (Hebrew).

Fiori A., *Nuova Flora Analitica d'Italia,* 2 vols., Florence 1923–1929.

Fiori A. & Paoletti G., *Iconographia Florae Italicae,* ed. 3, San-Casciano Val di Pesa 1933.

Forsskål P., *Flora Aegyptiaco-Arabica,* Copenhagen 1775.

Grenier J. C. M. & Godron D. A., *Flore de France,* 3 vols., Paris 1847–1856.

Halacsy E. von, *Conspectus Florae Graecae,* 3 vols. & 2 suppls., Leipzig–Budapest 1900–1912; Vols. 1–3, Leipzig 1900–1904; Suppl. 1, Leipzig 1908; Suppl. 2, in: *Magyar Bot. Lapok,* 11 : 114–215 (1912).

Hayek A. von, *Prodromus Florae Peninsulae Balcanicae,* 3 vols., in: *Repert. Sp. Nov. Reg. Veg. Beih.,* 30 (1924–1933) (Vols. 2–3 by F. Markgraf).

Hegi G., *Illustrierte Flora von Mittel-Europa,* ed. 1, 7 vols., Munich 1906–1931; ed. 2, Vols. 1–3, Vol. 4, Parts 1–2→, Munich 1936–1963→.

Holmboe J., *Studies on the Vegetation of Cyprus,* in: *Bergens Mus. Skrifter* (N.S.), 1, No. 2 (1914).

Jahandiez E. & Maire R. C. J. E., *Catalogue des plantes du Maroc (Spermatophytes et Ptéridophytes*), 3 vols. & suppl., Algiers 1931–1941 (suppl. by L. Emberger & R. C. J. E. Maire).

Jaubert H. F. & Spach E., *Illustrationes Plantarum Orientalium,* 5 vols., Paris 1842–1857.

Komarov V. L. et al. (ed.), *Flora URSS,* 30 vols., Leningrad–Moscow 1934–1964.

Maire R. C. J. E., *Flore de l'Afrique du Nord,* Vols. 1–12→, Paris 1952–1965→ (Vols. 1–3 by M. Guinochet & L. Faurel, Vol. 4 by M. Guinochet, Vol. 5 by M. Guinochet & P. Quézel, Vols. 6–12 by P. Quézel).

Mouterde P., *La Flore du Djebel Druze,* Beirut 1953.

Muschler R., *A Manual Flora of Egypt,* Berlin 1912.

Pampanini R., *Plantae Tripolitanae,* Florence 1914.

Pampanini R., *Prodromo della Flora Cirenaica,* Forli 1931.

Parsa A., *Flore de l'Iran (Le Perse),* 5 vols. & 2 suppls., Teheran 1943–1959.

Post G. E., *Flora of Syria, Palestine and Sinai,* Beirut 1883–1896; ed. 2 by J. E. Dinsmore, 2 vols., Beirut 1932–1933.

Quézel P. & Santa S., *Nouvelle Flore de l'Algérie et des régions désertiques méridionales,* 2 vols., Paris–Gap 1962–1963.

Rechinger K. H., *Flora Aegaea,* in: *Denkschr. Akad. Wiss. Math.-Nat. Kl. Wien,* 105, No. 1 (1943); *Florae Aegaeae Supplementum,* in: *Phyton* (Austria), 1: 194–228 (1949).

Rechinger K. H. (ed.), *Flora Iranica,* Contrib. 1–10→, Graz 1963–1965→.

Rechinger K. H., *Flora of Lowland Iraq,* Weinheim 1964.

Rouy G. C. C., *Flore de France,* 14 vols., Asnières–Rochefort–Paris 1893–1913 (Vols. 1–3 in collab. with J. Foucaud, Vol. 6 in collab. with E. G. Camus & N. Boulay, Vol. 7 in collab. with E. G. Camus).

Rouy G. C. C., *Conspectus de la Flore de France,* Paris 1927.

Sauvage C. & Vindt J., *Flore du Maroc,* Vols. 1–2→; Vol. 1, in: *Trav. Inst. Sci. Chérif.,* No. 4 (1952); Vol. 2, ibid., Ser. Bot., No. 3 (1954).

Savulescu T. (ed.), *Flora Republicii Populare Române,* Vols. 1–10→, Bucharest 1952–1965→.

Schwartz O., *Flora des tropischen Arabien,* in: *Mitt. Inst. Allg. Bot. Hamburg,* 10 (1939).

Sibthorp J. & Smith J. E., *Florae Graecae Prodromus,* 2 vols., London 1806–1816.

Sibthorp J. & Smith J. E., *Flora Graeca,* 10 vols., London 1806–1840 (Vols. 1–7 by J. E. Smith; Vols. 8–10 by J. Lindley).

Täckholm Vivi, *Students' Flora of Egypt,* Cairo 1956 (in collab. with M. Drar & A. A. Abdel Fadeel).

Täckholm Vivi & Täckholm G., *Flora of Egypt,* Vols. 1–3→, in: *Bull. Fac. Sci. Fouad Univ. Cairo,* 17, 28, 30 (1941–1954→) (in collab. with M. Drar).

Thiébaut J., *Flore Libano-Syrienne,* 3 vols.; Vols. 1–2, in: *Mém. Inst. Eg.,* 31 (1936), 40 (1940); Vol. 3, Paris 1953.

Tutin T. G. & Heywood V. H. (ed.), *Flora Europaea,* Vol. 1→, Cambridge 1964→.

Willkomm H. M. & Lange J., *Prodromus Florae Hispanicae,* 3 vols. & suppl., Stuttgart 1861–1893 (suppl. by H. M. Willkomm).

Zohary M., *The Flora of Iraq and its Phytogeographical Subdivision*, in : *Bull. Dir. Gen. Agr. Baghdad*, 31 (1950).

3. SELECTED PAPERS ON THE FLORA OF PALESTINE

Barbey C. & Barbey W., *Herborisations au Levant; Égypte, Syrie et Méditerranée*, Lausanne 1882.

Bornmüller J., Ein Beitrag zur Kenntniss der Flora von Syrien und Palästina, *Verh. Zool.-Bot. Ges. Wien*, 48 : 544–653 (1898).

Bornmüller J., Zur Flora von Palästina, *Beih. Bot. Centralbl.*, 29, Part 2 : 12–15 (1912).

Bornmüller J., Zur Flora von Palästina, *Magyar Bot. Lapok*, 11 : 3–12 (1912).

Bornmüller J., Weitere Beiträge zur Flora von Palästina, *Mitt. Thür. Bot. Ver.* (N. F.), 30 : 73–86, Pl. I (1913).

Decaisne J., Enumération des plantes recueillies par M. Bové dans les deux Arabies, la Palestine, la Syrie et l'Égypte; Florula Sinaica, *Ann. Sci. Nat. Bot.*, Ser. 2, 2 : 5–18, 239–270 (1834), 3 : 257–291, Pl. VII (1835), Liste des plantes recueillies par M. Bové dans la Palestine et la Syrie, 4 : 343–360 (1835).

Dinsmore J. E., *The Jerusalem Catalogue of Palestine Plants*, ed. 3, Jerusalem 1912.

Eig A., A Contribution to the Knowledge of the Flora of Palestine, *Bull. Inst. Agr. Nat. Hist. Tel-Aviv*, 4 (1926).

Eig A., A Second Contribution to the Knowledge of the Flora of Palestine, *Bull. Inst. Agr. Nat. Hist. Tel-Aviv*, 6 (1927).

Feinbrun Naomi & Zohary M., *Flora of the Land of Israel; Iconography* (plates by Ruth Koppel), 3 vols.; Vol. 1, ed. 1, Jerusalem 1949, ed. 3, Tel-Aviv 1959; Vol. 2, Jerusalem 1952; Vol. 3, Jerusalem 1958.

Hart H. C., *Some Account of the Fauna and Flora of Sinai, Petra and Wâdy 'Arabah*, London 1891.

Lowne B. T., On the Vegetation of the Western and Southern Shores of the Dead Sea, *Journ. Linn. Soc. Lond. Bot.*, 9 : 201–208 (1867).

Nabelek F., Iter turcico-persicum, *Publ. Fac. Sci. Univ. Masaryk*, 35 (1923), 52 (1925), 70 (1926), 105 (1929), 111 (1929).

Oppenheimer H. R., *Florula Transiordanica* (Reliquiae Aaronsohnianae, I), *Bull. Soc. Bot. Genève*, Ser. 2, 22 : 126–409 (1931), 23 : 510–519 (1931).

Oppenheimer H. R. & Evenari M., *Florula Cisiordanica* (Reliquiae Aaronsohnianae, II), *Bull. Soc. Bot. Genève*, Ser. 2, 31 : 1–432, Pls. 1–12 (1941).

Paine J. A., A List of Plants Collected between the Two Zarqas Eastern Palestine in the Spring of 1873, *Palest. Explor. Soc. Statement*, 3 : 91–130 (1875).

Post G. E., *Plantae Postianae*, Fascs. 1–4, Lausanne 1890–1892; Fasc. 5, *Bull. Herb. Boiss.*, 1 : 1–18 (1893); Fascs. 6–10 (in collab. with E. Autran), *Bull. Herb. Boiss.*, 1 : 393–411 (1893), 3 : 150–167 (1895), 5 : 755–761 (1897), 7 : 146–161 (1899), 18 : 89–102 (1900).

Range P., Die Flora der Isthmuswüste, *Veröffentl. Ges. Palaestina–Forsch. Berlin*, 7 (1921).

Rechinger K. H., Zur Flora von Palästina und Transjordanien (Reliquiae Samuelsonianae, V), *Ark. Bot.*, Ser. 2, 2, No. 5 : 271–455, Pls. I–XI (1952).

Tristram H. B., *The Survey of Western Palestine; The Fauna and Flora of Palestine*, London 1884.

4. ON THE PLANT GEOGRAPHY OF PALESTINE

Eig A., *Les Eléments et les groupes phytogéographiques auxiliaires dans la flore palestinienne*, 2 parts, in : *Repert. Sp. Nov. Reg. Veg. Beih.*, 63 (1931–1932).

Feinbrun Naomi & Zohary M., A Geobotanical Survey of Transjordan, *Bull. Res. Counc. Israel*, 5D : 5–35 (1955).

Hart H. C., *Some Account of the Fauna and Flora of Sinai, Petra and Wâdy 'Arabah*, London 1891.

Oppenheimer H. R., Esquisse de géographie botanique de la Transjordanie, *Bull. Soc. Bot. Genève*, Ser. 2, 22 : 410–438 (1931).

Post G. E., *The Botanical Geography of Syria and Palestine*, London 1889.

Zohary M., Vegetation Map of Western Palestine, *Journ. Ecol.*, 34 : 1–19 (1947).

Zohary M., *Plant Life of Palestine (Israel and Jordan)*, New York 1962.

5. PUBLICATIONS ON USES OF LOCAL PLANTS

Crowfoot Grace M. & Baldensperger Louise, *From Cedar to Hyssop*, London 1932.

Dalman G., *Arbeit und Sitte in Palästina*, 7 vols., Gütersloh 1921–1942.

Zohary M. & Feinbrun Naomi, Useful Plants Growing Wild in Palestine, *Hassadeh*, 10 : 298–301, 363–368, 433–437, 608–612 (1930) (Hebrew).

6. ON PLANTS OF THE BIBLE

Feliks J., *Plant World of the Bible*, Tel-Aviv 1957 (Hebrew).

Löw I., *Die Flora der Juden*, 4 vols., Vienna–Leipzig 1924–1934.

Moldenke H. N. & Moldenke Alma L., *Plants of the Bible*, Waltham, Mass. 1952.

Zohary M., The Flora of the Bible, *Biblical Encyclopaedia*, Jerusalem 1952→ (Hebrew).

Zohary M., Flora of the Bible, *The Interpreter's Dictionary of the Bible*, New York 1962.

ABBREVIATIONS

C. central
comb. combination
diam. diameter
E. east, eastern
env. environs
excl. excluding
f. forma
Fl. flowering
Fr. fruiting
Hab. habitat
HUJ Hebrew University Herbarium,
   Jerusalem
incl. including
l.c. loco citato
Mt(s). Mount(ains)

N. north, northern
nov. novus
p.p. pro parte
2n somatic chromosome number
S. south, southern
Sect. section
Subfam. subfamily
Subgen. subgenus
Subsp., ssp. subspecies
sp., spp. species
stat. status
Tab. tabula
Trib. tribe
Var. variety
W. west, western

Measurements given without qualification refer to length. Measurements connected by ×
indicate length followed by width. Figures in parentheses indicate uncommon dimensions.

GLOSSARY

Acaulescent Stemless or almost stemless.

Accrescent Increasing in size with age.

Accumbent (cotyledons) With their faces turned towards the axis.

Achene A small, dry, indehiscent, 1-seeded fruit.

Acicular Needle-shaped and very slender.

Actinomorphic Radially symmetrical; with parts similar in shape and size.

Aculeate Armed with prickles.

Acuminate Sides somewhat concave and tapering to a protracted point.

Acute Ending in a point, with sides essentially straight.

Adelphous (stamens) In sets or bundles.

Adherent Touching an organ or part of organ connivently without being fused with it.

Adnate Fused with another part.

Alar (flower) Axillary between dichotomously branching organs.

Alternate Not opposite or whorled; placed singly at nodes.

Alternation of generations Regular succession of sexual and asexual phases.

Amplexicaul Clasping the stem.

Anastomosing (veins) Joining each other and forming a network.

Anatropous (ovule) Inverted along the funicle and adnate to it, with micropyle close to the point of attachment of the funicle.

Androecium The stamens collectively.

Androphore A stalk bearing the androecium

Anemophilous Wind-pollinated.

Annual A plant that lives only for 1 season or 1 year.

Annulus Ring; in ferns special thick-walled cells forming part of the opening mechanism of the sporangium.

Anterior On the front side; away from the axis (abaxial).

Anther The pollen-bearing part of the stamen.

Antheridium An organ containing male gametes.

Anthesis The process or time of expansion of a flower.

Antrorse Directed forwards or upwards.

Apetalous Without petals.

Aphyllous Leafless.

Apiculate With short and abrupt point.

Apocarpium A fruit composed of free (not connate) carpels.

Apocarpous Of separate (not connate) carpels.

Appressed Closely pressed against another organ.

Approximate(d) Close to each other.

Archegonium An organ containing female gametes.

Arcuate Arched or curved.

Aril An outer, sometimes pulpy covering of a seed growing out from the hilum or funicle of seed.

Aristate Bearing a bristle-like awn or seta.

Articulate Jointed; with segments and nodes at which separation may naturally take place.

Ascending Rising upwards somewhat obliquely.

Asperous Rough to the touch.

Attenuate Tapering gradually.

Auricle An ear-shaped part or appendage at the base of a leaf or petal.

Auriculate Ear-shaped.

Awl-shaped Tapering upwards to a point very gradually.

Awn A stiff bristle on the top of an organ.

Axil Angle formed between axis and any organ arising from it, especially leaf.

Axile (placentation) At or near the centre of a compound ovary.

Axillary Rising from or located in the axil of leaf or other organ.

Baccate Berry-like.

Barbate Bearded; having long weak hairs in tufts; barbulate — finely bearded.

Basifixed Attached at the base.

Batha Mediterranean dwarf-shrub formation.

Berry Fleshy fruit containing several (rarely 1) seeds.

Biennial A plant completing its life cycle within 2 years.

Blade The expanded part of a leaf.

Bract A small, scale-like leaf subtending a flower cluster or a flower.

Bracteole A bract borne on a pedicel or very close to the flower (diminutive of bract).

Bulb A thickened shoot serving as a storage organ, composed of a much shortened axis, fleshy leaf bases and 1 or more renovation buds.

Caducous Falling off early.

Caespitose Growing in tufts or mats.

Calcarate Spurred.

Callosity A hard or tough protuberance or swollen area; callus.

Calyx The outer, usually green whorl of a floral perianth composed of free or united sepals.

Campanulate Bell-shaped.

Campylotropous (ovule) Curved in a way that the micropyle is near the funicle (stalk) with the side of the ovule not being adnate to funicle.

Canaliculate Deeply grooved or channelled longitudinally.

Canescent Grey-pubescent.

Capensis Floristic kingdom of S. Africa.

Capillary Hair-like.

Capitate Head-like; in a dense, rounded cluster.

Capitulum A head; an inflorescence with a shortened axis and short-pedicelled or sessile flowers.

Capsule A dry fruit (of 2 or more carpels), usually opening at maturity by teeth, valves, a lid, etc.

Carinate Keeled.

Carpel Foliar unit forming a simple pistil; part of a compound pistil or ovary or fruit.

Carpophore Stalk of the ovary or of the fruit; a prolongation of the receptacle into a stalk-like structure; a thin stalk that supports each mericarp in Umbelliferae.

Caruncle An excrescence or appendage at or near the hilum of the seed.

Caryopsis The 1-seeded fruit of the grasses with pericarp adnate to the seed.

Casual An alien, not established plant.

Catkin A flexuous spike or spike-like inflorescence, usually falling as a whole.

Caudate Bearing a tail-like appendage.

Caulescent Having a well developed stem above ground.

Cauline Belonging to or arising from a stem or axis.

Cell A compartment within an ovary, a fruit or an anther.

Chalaza The basal part of an ovule.

Chamaephyte Dwarf-shrub, the top parts of which die back in the dry season.

Chartaceous Of papery texture.

Ciliate Bearing hairs on the margin; ciliolate — minutely ciliate.

Circinate Rolled downwards or inwards in one plane, with apex in centre.

Circumscissile Opening or dehiscing around a transversal, circular line, the top usually coming off as a lid.

Clasping Partly or wholly surrounding the stem.

Clathrate Latticed or pierced with holes.

Clavate Club-shaped; thickened towards the top.

Claw Long narrow base of petal or sepal.

Cochleate, cochleariform Coiled like a snail shell; spoon-shaped.

Coherent (similar parts) Growing close to each other but not fused.

Column Tube formed by coalescent filaments or adnate filaments and styles.

Commissure The surface by which one carpel adheres to another, as in the Umbelliferae; the edge along which sepals or other parts are joined.

Compound (leaf) Divided into leaflets.

Concrescent Growing together.

Conduplicate Folded lengthwise.

Cone A structure composed of an axis bearing sporophylls or other seed- or pollen-bearing organs.

Connate (mostly similar organs) United or joined.

Connective Tissue connecting the 2 cells of an anther.

Connivent Converging, coming together.

Cordate Heart-shaped.

Coriaceous Leathery.

Corm A thickened, rounded, subterranean stem serving as a storage organ and composed of 1 or few internodes and renovation buds.

Corolla Inner circle or whorl of floral perianth, often coloured.

Corona Any appendage between corolla and stamens or on the corolla or an outgrowth of the staminal whorl.

Corymb A flat-topped raceme in which pedicels of the lower flowers are much longer than those of the upper.

Cotyledon The primary leaf or leaves in the embryo.

Crenate Round-toothed; crenulate — finely crenate.

Crisp Curled, undulate, ruffled.

Crozier-shaped With a coiled end.

Crustaceous Crust-like; hard, thin and brittle.

Culm Stem of grasses.

Cuneate Wedge-shaped; with the narrow end at point of attachment.

Cupule Cup-like structure.

Cuspidate With an apex abruptly constricted into an elongated tip.

Cyme A determinate inflorescence; a cluster of flowers with central flower opening first; see also dichasium, monochasium and pleiochasium.

Deciduous Falling off at the end of the growing season, as the leaves of summer-green trees; falling off.

Decumbent Lying on the ground, with apical parts ascending.

Decurrent Extending downwards from point of insertion and adnate to stem (leaf) or to style (stigma).

Decussate In opposite pairs orientated at right angles to each other.

Deflexed Bent or turned downwards.

Dehiscent (fruit or anther) Splitting open along definite lines.

Dentate With projecting teeth along margin; denticulate — with small teeth.

Depressed Somewhat flattened from above.

Diachenium A fruit composed of 2, 1-seeded carpels, as in Umbelliferae, Rubiaceae.

Diadelphous (stamens) Occurring in 2 sets or groups.

Dichasium A cyme, in which each axis is terminated by a flower or head, below which 2 opposite branches arise.

Dichotomous Forked in 1 or more pairs.

Diffuse Loosely branching or spreading in all directions.

Digitate (leaf) Hand-like; compound with the leaflets arising from one point.

Dimorphic Of two forms.

Dioecious Having staminate and pistillate or ovulate flowers on different plants.

Diploid Having 2 sets of chromosomes.

Discoid Resembling a disk; tubular floret; head with tubular florets only.

Disk Central portion of a receptacle; central portion of a fruit or seed.

Dissected Cut into segments.

Distichous 2-ranked.

Divaricate Widely divergent.

Dominant The most abundant constituent of a plant community.

Dorsal Relating to the back or outer surface of an organ.

Dorsifixed Attached by the back.

Drooping Inclining downwards but not quite pendent.

Drupe A fleshy, 1-seeded indehiscent fruit with a stony endocarp; stone-fruit.

E- Prefix denoting without.

Echinate With prickles.

Ellipsoidal Said of 3-dimensional body,

elliptical in longitudinal section.

Elliptical Narrowed at ends and widest at or near the middle.

Emarginate With a shallow notch at the apex.

Embryo The young plantlet enclosed in the seed.

Endemic Restricted in occurrence to a particular geographical area.

Endocarp The innermost layer of the pericarp.

Endosperm Nutritive tissue accompanying the embryo.

Ensiform Sword-shaped; linear with an acute point.

Entire Without toothed or otherwise divided margin.

Epicalyx A series of bracts close to and outside the main calyx, sometimes resembling an additional (outer) calyx or involucre.

Epicarp, exocarp Outer layer of the pericarp.

Epipetalous, -sepalous Opposite petals or sepals respectively.

Epiphytic Growing upon another plant but not parasiting on it.

Erose (margin) As if gnawed.

Exserted Projecting beyond an envelope, as stamens from a perianth.

Exstipulate Without stipules.

Extrorse Facing outwards; anthers dehiscing outwards.

Falcate Sickle-shaped.

Farinose Covered with meal-like powder.

Fascicle A condensed cluster of flowers or leaves.

Fibrous (roots) A root system with several major about equal roots arising from approximately the same point.

-fid Cut about halfway to the midrib.

Filament Thread; stalk of the stamen.

Filiform Thread-like.

Fimbriate Fringed; with margin bordered by long filiform processes.

Flabellate Fan-like.

Flexuous (stem) Curved alternately in opposite directions.

Floret Individual small flower of a dense inflorescence, usually of the Compositae and Gramineae.

-foliate Having leaves; foliolate — having leaflets.

Follicle Dry, dehiscent fruit or part of fruit, made of 1, 2- or many-seeded carpel opening by 1 suture.

Foveolate Pitted.

Free Not adnate to nor connate with other organs.

Fruticose Shrub-like; shrubby.

Funicle The stalk by which an ovule is attached to the ovary wall or placenta.

Fusiform Spindle-shaped.

Gamopetalous, -sepalous With petals or sepals respectively united at least at base.

Garigue Mediterranean half-shrub formation somewhat taller than batha.

Geniculate Bent like a knee.

Gibbous Swollen or distended at base or on one side.

Glabrous Not hairy; glabrescent — nearly glabrous or becoming glabrous with age.

Gland A secreting part or appendage, often used for any protuberance having the appearance of such an organ; fruit (without cupule) of the oak.

Glandular-hairy With hairs, bearing glands on top.

Glaucous Bluish-green; covered with a bloom; glaucescent — slightly glaucous.

Glomerate In clusters.

Glomerule Cluster.

Glume One of the 2 sterile bracts at the base of the grass spikelet.

Glutinous Sticky.

Gynoecium The female part or parts of the flower collectively.

Gynophore Stipe of an ovary.

Gynostemium Column in the flower resulting from the union of the stamens and style.

Halophyte Plant growing in saline soils.

Hammada Desert with stony or gravelly cover.

Haploid Having 1 set of chromosomes.

Hastate Having the shape of an arrow-head but with diverging basal lobes.

Helicoid Curved or spiral like a snail-shell.

Hemicryptophyte Perennial herb all overground parts of which die back in the dry season.

Hemiparasite Parasitic plant with green leaves.

Herbaceous Not woody or membranous; leaf-like in texture.

Hermaphrodite Bisexual; with functional stamens and pistils in the same flower.

Heterophyllous With 2 kinds of leaves.

Heterosporous Producing 2 kinds of spores.

Heterostylous With style differing in length in flowers of different plants.

Hilum Scar on the seed marking the point of attachment of the funicle.

Hirsute With rather coarse hairs.

Hirtellous (hirtulous) Softly or minutely hairy.

Hispid With stiff or bristly hairs.

Holarctis Temperate regions of northern hemisphere.

Holoparasite Parasitic plant lacking green leaves.

Homosporous Producing 1 kind of spores.

Hyaline Thin and translucent.

Hydrophyte Plant growing in moist habitats.

Hypanthium Cup-like structure enclosing ovary and derived from fusion of the receptacle and outer floral parts.

Hypocotyl The portion of the axis below the cotyledons and above the radicle or root.

Hypogaeous Below ground.

Hypogynous Inserted beneath the base of ovary.

Imbricate(d) Overlapping, as shingles on a roof.

Imparipinnate Odd-pinnate, with a single terminal leaflet.

Incised More or less deeply and sharply cut.

Included Not projecting from the surrounding structure.

Incumbent (cotyledons) Lying face to face and with the back of one against the axis.

Indehiscent (fruit) Not opening.

Indeterminate (inflorescence) Inflorescence with lowest or peripheral flowers opening first.

Indumentum Hairy covering.

Indurate(d) Hardened.

Indusium Tissue (epidermal outgrowth) covering or enclosing a group of sporangia (sorus) in ferns.

Inferior (ovary) Adnate to the receptacle or to the floral cup and appearing below the other floral parts.

Inflexed Bent inwards.

Inflorescence A flower cluster or a flower-system; part of a shoot bearing flowers.

Infraspecific Referring to any unit of classification (taxon) below the species level.

Infructescence A cluster of fruits.

Infundibuliform Funnel-shaped.

Integument The envelope of an ovule turning later into the seed coat.

Internode Part of stem between two nodes.

Introrse Turned inwards, as the anther whose line of dehiscence faces towards the centre of flower.

Involucel Secondary or small involucre.

Involucre One or more whorls of small leaves or bracts underneath a flower cluster (especially underneath a head or an umbel).

Involute (petals in bud) Rolled inwards lengthwise.

Irregular (flower) Unequal in size, form or union of its similar parts; zygomorphic.

Keel A prominent ridge along the outside of a leaf or bract, etc.; the innermost pair of petals in a papilionaceous flower.

Labellum Lip; particularly the lip of an orchid flower.

Labiate Lipped; often referring to corolla or calyx.

Lacerate Torn; irregularly divided by deep incisions.

Laciniate Slashed into narrow lobes.

Lamina Blade; expanded portion of leaf.

Lanceolate Lance-shaped; widening towards base and tapering to apex.

Latex Milky juice.

Leaflet A leaf-like unit of a compound leaf.

Legume Fruit derived from a single carpel and generally dehiscing along both sutures; pod.

Lemma The lower of the 2 bracts enclosing the grass flower.

Lenticular Lens-shaped, biconvex.

Lepidote Covered with small scurfy scales.

Ligulate Strap-shaped.

Ligule A variously built projection at the junction of sheath and blade in the grass leaf; the strap-shaped corolla in the Compositae.

Limb The expanded part of a petal; the expanded portion of calyx or corolla.

Linear Long and narrow with parallel or almost parallel sides.

Lingulate Tongue-shaped.

Lip One of the parts of an unequally divided corolla or calyx.

Lobe Any segment of an organ; one of the free parts of a gamosepalous or gamopetalous calyx or corolla; lobule — secondary lobe.

Locule (loculus) Compartment or cell of an ovary, anther or fruit.

Loculicidal (fruit) Dehiscing along midrib of the carpels.

Lodicule One of the 2 or 3(6) scales near the base of the ovary in the grass flower.

Loment Fruit separating transversely into 1-seeded segments.

Lunate Crescent- or half-moon-shaped.

Lyrate (leaf) Pinnatifid with an enlarged terminal lobe and smaller lateral lobes.

Macrosporangium, macrospore, macrosporophyll. See megasporangium, megaspore, megasporophyll.

Mammillate Nipple-shaped; having nipple-shaped processes.

Maquis A formation of Mediterranean evergreen, mostly hard-leaved low trees and tall shrubs.

Marcescent Persisting on the plant after withering.

Megasporangium Sporangium producing the larger type of spores.

Megaspore The larger of the 2 types of spores, mainly referring to ferns.

Megasporophyll Carpel or sporophyll which bears megaspores.

Membranous Of parchment-like texture, thin and transparent.

Mericarp One of the carpels into which a schizocarp splits at maturity.

-merous (flower) Referring to the number of parts of the flower whorls.

Micropyle An opening between the integuments in an ovule, becoming a pit-like mark on the mature seed.

Microsporangium Sporangium producing the smaller type of spores; anther sac.

Microspore The smaller type of spore; pollen grain.

Microsporophyll A leaf-like organ bearing one or more microsporangia; anther.

Monadelphous (stamens) All united, usually by their filaments.

Monochasium A cymose, determinate inflorescence composed of several consecutive axes, each terminated by a flower.

Monoecious With separate staminate and pistillate flowers on the same plant.

Monopodial (stem) Of a single main axis which continues to grow from the same apical growing point.

Monotypic (genus) Composed of a single species.

Mucro A short and abrupt sharp tip.

Muricate Rough due to the presence of many, sometimes lightly prickly points.

Muticous Without a point, blunt.

Nectary A nectar-secreting gland, scale or pit.

Neophyte Plant which has recently arrived in the area.

Neotropis Tropics of the New World.

Nerve See vein.

Netted-veined With the veins forming a net-work.

Node A joint where a leaf is borne.

Nucellus The tissue in the ovule enclosing the embryo sac.

Nut An indehiscent, mostly 1-celled and 1-seeded hard fruit.

Nutlet A small nut.

Ob- Prefix denoting reversely.

Obconical Conical but with the pointed end below.

Obdiplostemonous Stamens in 2 whorls with those of the outer whorl opposite petals.

Oblanceolate Broader in upper third than in the middle and tapering towards the base.

Oblique With the sides slightly unequal or slanting.

Oblong Longer than broad with sides nearly parallel most of their length.

Obovate Ovate but with the terminal half broader than the basal.

Obovoid Said of a 3-dimensional body, obovate in longitudinal section.

Obtuse Blunt, not acute.

Ochrea A nodal sheath formed by the fusion of 2 stipules (as in Polygonaceae).

Ochreole A sheath formed by the fusion of 2 bracts.

Odd-pinnate A pinnate leaf with a single terminal leaflet.

Operculum A lid or cover of a circumscissile fruit.

Orbicular Circular, round.

Orthotropous (ovule) Erect and straight, with the funicle at the base and the micropyle at the apex.

Ovary Part of a pistil bearing ovules.

Ovate With the outline of an egg having the broader end at base.

Ovoid Said of a 3-dimensional body, ovate in longitudinal section.

Ovule The body (mostly within the ovary), which after fertilization becomes the seed.

Palea The upper of the 2 bracts enclosing the grass flower.

Paleotropis Tropics of the Old World.

Palmate Lobed, divided or ribbed in a palm-like or hand-like fashion.

Panicle A compound indeterminate inflorescence in which the primary axis branches once or repeatedly.

Papilionaceous (corolla) With standard, wings and keel, as in pea flower.

Papilla Minute, soft protuberance or thickened hair.

Papillose With protuberances or thick short hairs.

Pappus Tuft of hairs on fruit; sometimes modified outer perianth series in the Compositae.

Paraphysis Sterile filament among sporangia in some ferns.

Parietal Borne on the inner walls of the ovary.

Paripinnate Pinnate without terminal leaflet.

Parted, partite Divided not quite to the base, or to the leaf rhachis.

Patulous Spreading.

Pectinate Comb-like, pinnatifid with many or few close narrow divisions.

Pedate (leaf) Palmately lobed or divided with the 2 side lobes 2-cleft in turn.

Pedicel Stalk of an individual flower or fruit.

Peduncle Stalk of a cluster of flowers, or of a solitary terminal flower.

Pellucid Transparent.

Peltate (leaf) Attached to the petiole more or less at the middle of its lower surface; usually shield-shaped.

Pentagonous (3-dimensional) 5-angled in cross section.

Perennate, perennating Becoming perennial occasionally.

Perennial Of more than 2 seasons' or 2 years' duration.

Perfoliate Sessile, with base completely surrounding the stem.

Perianth Calyx or corolla or both; collective term for a floral envelope not differentiated into calyx and corolla.

Pericarp The wall of a fruit.

Perigynous Borne or arising from around the ovary.

Persistent Not falling off.

Petal One of the separate parts of a corolla.

Petiole Leaf stalk; petiolule — stalk of a leaflet.

Phylloclade Flattened or cylindrical green branch functioning as a leaf.

Phyllode Broadened, leaf-like petiole without a blade.

Pilose With rather long and soft hairs.

Pinna A primary segment or primary leaflet of a pinnate leaf, generally composed of pinnules or secondary leaflets.

Pinnate (leaf) With leaflets placed on either side of the rhachis.

Pistil The female organ of the flower, comprising ovary, style and stigma.

Placenta The tissue of the ovary which bears ovules.

Placentation The arrangement of ovules within the ovary.

Pleiochasium A cymose, determinate inflorescence in which each axis is terminated by a flower beneath which 3 or more lateral branches arise.

Plicate Folded, as in a fan, or approaching this condition.

Plumose Feather-like; of hairs or bristles which have finer hairs along each side.

Plumule Leaf-bud of an embryo.

Pod A dehiscent dry fruit; generally referring to fruit of the Leguminosae.

Pollen Microspores produced in the anther.

Pollen sac One of the microsporangia contained in the anther.

Pollinium A coherent mass of pollen grains.

Polygamous With unisexual and bisexual flowers on the same plant.

Polymorphic Of 2 or more forms.

Pome Fruit of apple, pear, etc.

Posterior On the back side; facing the axis (adaxial).

Praemorse (leaf) As if bitten off.

Prickle A small and weak spine-like structure on branches, fruits, leaves, etc.

Procumbent Trailing or lying loosely on the ground.

Prostrate Lying flat on the ground.

Prothallium The gametophyte in pteridophytes, usually a flattened expansion, bearing rhizoids and sexual organs.

Pruinose With a bloom on the surface.

Pubescent Downy or hairy; often covered with short soft hairs; puberulent — minutely pubescent.

Pulverulent Powdery.

Pyrene Kernel or stone in a drupe.

Pyriform Pear-shaped.

Quinate Growing together in fives.

Raceme A simple, indeterminate inflorescence with an elongated axis and pedicellate flowers.

Radical Basal, referring to base of stem.

Radicle The embryonic root of a seed.

Ray Outer floret of some Compositae, with an extended part of the corolla; a branch of an umbel bearing an umbellet or a flower.

Receptacle Dilated or elongated axis on which flower parts or flowers are borne.

Recurved Curved backwards or downwards.

Reflexed Bent down- or backwards.

Regular (flower) A flower with the parts so arranged as to be vertically divisible into equal halves by 2 or more planes; actinomorphic.

Relic (plant) Remnant of a former flora.

Remote Placed at some distance from each other.

Rendzina Greyish-white calcareous soil.

Reniform Kidney-shaped.

Repand Weakly sinuate.

Replum Frame-like structure between the 2 valves of a fruit, such as found in the Cruciferae.

Retrorse Directed backwards or downwards.

Retuse   Notched slightly at an obtuse apex.

Revolute   Rolled backwards.

Rhachilla   Axis of a grass spikelet.

Rhachis   Axis of an inflorescence or of a divided or compound leaf.

Rhizome   Underground stem; rootstock.

Rib   A prominent nerve.

Rosette   Basal, circular cluster of leaves often present in stemless plants.

Rosulate   Arranged in a rosette.

Rotate   Wheel-shaped; gamopetalous corolla with broad circular limb and very short tube.

Rotundate   Rounded.

Ruderal   Growing in waste places or on roadsides.

Rudiment, rudimentary   Imperfectly developed and non-functional or vestigial organ.

Rugose   Wrinkled; rugulose — minutely wrinkled.

Runcinate   Coarsely serrate to sharply incised with the teeth pointing towards the base.

Runner   A slim horizontal overground shoot, usually with elongated internodes and with scale-like leaves, rooting at nodes.

Saccate   Pouch-shaped.

Sagittate   Triangular, with elongated basal lobes pointing downwards.

Samara   An indehiscent winged fruit.

Scabrous   Rough; roughish to the touch; scabridulous — minutely scabrous.

Scale   A small often dry leaf or bract.

Scandent   Climbing.

Scape   Flowering stem of a plant, all the leaves of which are radical.

Scarious   Thin, dry, membranous, not green.

Schizocarp   A dry fruit that splits at maturity into 2 or more 1-seeded parts (mericarps).

Scorpioid   Circinately coiled determinate inflorescence.

Scrobiculate (seeds)   Pitted.

Scurfy   Covered with minute scales.

-sect   Divided down to the midrib.

Segetal   Growing as a weed among crops.

Segment   One of the parts of a divided leaf or of a gamosepalous or gamopetalous perianth.

Semiterete   Semicircular in cross-section; flat on one side, rounded on the other.

Sepal   One of the separate parts of a calyx.

Sepaloid (involucre or perianth)   Sepal-like, green.

Septate   Partitioned.

Septicidal (fruit)   Dehiscent along the partitions (septa), or along the sutures of carpels.

Septum   A partition wall.

Sericeous   Silky.

Serrate   Saw-toothed, with teeth pointing forwards; serrulate — minutely serrate.

Sessile   Not stalked.

Seta   Bristle, a stiff hair.

Setaceous   Bristle-like.

Setose   Covered with bristles; setulose — with minute bristles.

Sheath   A tubular or rolled structure surrounding an organ.

Shrub   A comparatively low woody plant without main trunk but with several main branches.

Silicle, silicule   Fruit of Cruciferae, the length of which is less than 3 times its width.

Siliqua   Fruit of Cruciferae, the length of which is more than 3 times its width.

Silky   Covered with fine, closely appressed hairs.

Simple   (Stem) not or scarcely divided or branched; (leaf) not compound; (perianth) 1-whorled, not differentiated into calyx and corolla.

Sinuate   With a wavy margin.

Sinus   Space between 2 lobes or divisions.

Sorus (pl. sori)   Cluster of sporangia.

Spadix   A thick or fleshy spike.

Spartoid (shrub)   Broom-like, with green stems shedding leaves very early.

Spathe   A large bract or leaf surrounding or subtending a flower cluster or a spadix.

Spatulate With shape of a spatule.

Sperm A male gamete or reproductive cell.

Spicate, spiked With spikes; of spike-like structure; arranged in spikes.

Spike A simple indeterminate inflorescence with an elongated axis and sessile flowers.

Spikelet The unit of a grass inflorescence comprising 1 or more flowers and usually 2 subtending outer bracts (glumes).

Spine Stiff, sharp-pointed woody or horny body.

Spinescent Terminating with a sharp or spine-like tip.

Spinose Spine-like; with spines.

Spontaneous Propagating freely.

Sporangiophore Organ or stalk bearing sporangia.

Sporangium A sac or body bearing spores.

Spore A single detached cell capable of developing into a new individual.

Sporocarp Hard or leathery structure containing sporangia.

Sporophyll A sporangium-bearing leaf; leaf-like organ bearing reproductive parts.

Spreading Expanding outwards, almost horizontally.

Spur A tubular or sac-like projection of a petal or sepal, often containing a nectar-secreting gland.

Squamose Covered with scales.

Stamen The pollen-bearing organ of a flower.

Staminate Having stamens only; male.

Staminode A sterile stamen or a structure borne in the staminal part of the flower.

Standard The upper and broad petal of the papilionaceous flower.

Stellate Star-like.

Sterile (Flowers) lacking functional sex organs; (bracts) not supporting fertile flowers; (branches) not bearing flowers.

Stigma Top part of pistil that receives the pollen.

Stipe Stalk (within the perianth) bearing a pistil or a fruit; stipitate — stalked.

Stipule One of a pair of appendages of a petiole.

Stolon Horizontal stem below or above soil surface with elongated internodes and with scale-like leaves, usually rooting at nodes and giving rise to new shoots.

Stone The bony endocarp of a drupe with the enclosed seed.

Striate With longitudinal lines or ridges.

Strict Growing upwards; close to the main axis.

Strigose With sharp, straight and stiff hairs, often swollen at base; strigulose — minutely strigose.

Style The part of the pistil between the ovary and the stigma.

Stylopodium A disk-like structure at the base of the style, as in some Umbelliferae.

Sub- Prefix denoting somewhat, slightly or rather.

Subulate See awl-shaped.

Succulent Fleshy, or juicy; thick and soft.

Suffruticose Perennial with overground woody base the annual shoots of which decay in the dry season.

Sulcate Grooved or furrowed.

Superior (ovary) Free from and above the insertion of the perianth and stamens.

Suture A line of joining; a seam or other mark along a natural division or union of a fruit.

Sympetalous With petals united at least at the base.

Synangium United sporangia.

Synaptospermic Dispersing by whole many-seeded fruits or infructescences and not by single seeds.

Syncarpous Having carpels united.

Taxon (pl. taxa) A general term applied to any unit of classification irrespective of its rank.

Tepal Segment of perianth differentiated into calyx and corolla.

Terete More or less cylindrical, often of varying diameter.

Ternate Growing or divided in threes.

Testa Outer seed coat (developed from the integument).

Tetragonous (3-dimensional) 4-angled in cross section.

Thalloid Resembling a thallus.

Thallus A flat, leaf-like organ without differentiation into stem and foliage.

Throat (perianth) Narrow opening at limit between tube and limb.

Tomentose Woolly, with soft wool-like indumentum; tomentulose, tomentellous — delicately tomentose.

Tortuous Bent or twisted in different directions.

Torulose (fruit) Swollen at close intervals; constricted between seeds.

Torus Receptacle of a flower.

Tri- Three- or 3 times.

Trifoliate With 3 leaves or leaflets.

Trigonous (3-dimensional) 3-angled in cross section.

Triquetrous (3-dimensional) More or less acutely 3-angled in cross section, often with projecting angles.

Truncate Appearing as if cut off at the end.

Tube The united, more or less cylindrical hollow part of the perianth.

Tuber A thickened portion of an underground stem or root.

Tubercle Small rounded process on a surface.

Tunicated Covered with membranous coats as in onions.

Twig A young, woody shoot; a current year's branch of a woody plant.

Twining Winding spirally.

Umbel A simple or compound indeterminate inflorescence with pedicels or rays arising from a common point; umbellet — umbel borne on a ray.

Umbilicate Depressed in the centre, navel-like.

Umbo (umbonate) A conical projection arising from the surface; projection on the apophysis of a cone scale.

Uncinate Hooked.

Unguiculate Narrowed into a claw.

Urceolate Urn-shaped.

Utricle A 1-seeded, usually indehiscent fruit; a small bladder or bladdery part of the corolla tube in *Aristolochia*.

Vaginate Sheathed.

Vallecular Situated in the grooves between the ridges as in fruits of Umbelliferae.

Valvate Opening by valves; with contiguous but not overlapping edges of sepals or petals in the bud.

Valve A separable part of a pod or capsule.

Vascular Relating to conducting elements.

Vein Thread of conducting tissue in a leaf, flower or fruit.

Velutinous Velvety.

Venation Arrangement of veins.

Ventral Relating to the inner surface of an organ; opposite the back or dorsal part; adaxial.

Ventricose Swelling unequally or on one side.

Verrucose Warty.

Versatile Attached near the middle.

Verticillate Whorled.

Vesicle A small bladdery sac or cavity filled with air or fluid.

Villose With long and soft hairs.

Viscid Sticky.

Wadi Bed of water course which is dry in summer.

Whorl Three or more leaves or flower parts arranged in a circular row.

Wing(s) Dry or membranous expansion or appendage of an organ; the lateral petals of a papilionaceous flower.

Zygomorphic Bilaterally symmetrical; corolla divisible into equal halves in one plane only.

Zygote Fertilized egg from which the embryo develops.

KEY TO THE FAMILIES

1. Plants not producing seeds but propagating by spores contained in sporangia. Stems with well developed vascular tissue. Leaves of several cell layers   2
- Plants producing seeds   12
2. Stems jointed; nodes green, sheathed, with numerous small tooth-like leaflets. Sporangia arranged in a terminal spike-like body.   **1. Equisetaceae**
- Stems not jointed. Leaves large or fairly large   3
3. Leaves made of 4 entire leaflets. Sporangia enclosed in bean-like bodies near base of leaves.   **11. Marsileaceae**
- Plants not as above   4
4. Sori of sporangia arranged in a spike.   **2. Ophioglossaceae**
- Sori on lower surface of leaves   5
5. Sori linear, arranged along the leaf veins and not along margin; indusium elongated or absent but then leaves covered beneath with leathery scales.   **8. Aspleniaceae**
- Plants not as above   6
6. Small plants (3–15 cm.). Leaves procumbent, 2-pinnate, with almost orbicular leaflets. Sori covered by marginal, fringed indusia or concealed in dense wool.   **3. Sinopteridaceae**
- Plants with erect or ascending leaves. Sori not concealed in wool nor covered by fringed indusia   7
7. Sori arranged along margin of leaf segment (pseudo-indusium) and covered by it   8
- Sori not marginal, either naked or covered by a membranous indusium   9
8. Segments linear-lanceolate. Sori forming a continuous line along leaf margin. Plants at least 40 cm. high.   **5. Pteridaceae**
- Segments cuneate or fan-shaped. Sori not forming a continuous line along leaf margin.   **4. Adiantaceae**
9 (7). Sori naked (without indusia)   10
- Sori with indusia, at least when young   11
10. Sori linear or irregularly scattered along the secondary veins, appearing confluent at maturity. Leaves dimorphic, 2–3-pinnate.   **6. Gymnogrammaceae**
— Sori orbicular or short-ovate, in distinct rows. Leaves all alike, 1-pinnatifid.   **10. Polypodiaceae**
11 (9). Leaves 1-pinnatisect; lobes with revolute margin.   **7. Thelypteridaceae**
— Leaves 2-pinnatisect; leaflet margin dentate or lobulate, not revolute.   **9. Aspidiaceae**
12 (1). Seeds not enclosed in a fruit but borne on scales of a cone or between bract-like, sometimes fleshy, scales   13
— Seeds enclosed within a fruit developing from a pistil   15
13. Leaves needle-shaped, 5–10 cm. long, borne on short branches. Cones ovoid.   **12. Pinaceae**
— Leaves small, scale-like or spiny or almost 0   14
14. Trees or high shrubs with opposite or whorled leaves. Seeds contained in globular, woody or fleshy cones.   **13. Cupressaceae**
— Leafless or almost leafless shrubs or climbers. Seeds surrounded by fleshy or winged bracts.   **14. Ephedraceae**
15 (12). Plants free-floating on surface of water, lanceolate, lenticular or globular in shape and not exceeding 1 cm.   **Lemnaceae**

— Plants not as above, often much larger, with leaves or stems 16
16. Aquatic submerged plants with whorled leaves cut into filiform or linear lobes or leaflets. Flowers inconspicuous, unisexual 17
— Aquatic or terrestrial plants. Leaves or flowers not as in above 18
17. Leaves dichotomously divided. Flowers axillary, solitary. **39. Ceratophyllaceae**
— Leaves pinnatipartite. Flowers arranged in terminal spikes. **80. Haloragaceae**
18(16). Flowers unisexual, crowded on a fleshy spadix surrounded by a leaf-like, often coloured, spathe. Perennial herbs often with netted-veined, not linear leaves.
Araceae
— Plants not as above 19
19. Trees with pinnate or fan-shaped leaves about 0.6–3 m. broad or long. **Palmae**
— Not as above 20
20. Aquatic, submerged or partially floating plants. Leaves grass-like or not. Stamens 1–4. Flowers small, green or rarely coloured but then plants dioecious and corolla actinomorphic 21
— Terrestrial or aquatic plants not as above 27
21. Leaves rigid, spiny-dentate. **Najadaceae**
— Leaves neither rigid nor spiny-dentate 22
22. Dioecious plants. Leaves hairy — or glabrous but then leaves reniform or ligulate or pistillate flowers with a long, spirally coiled pedicel. **Hydrocharitaceae**
— Plants not dioecious. Leaves and/or flowers not as in above 23
23. Flowers in spikes 24
— Flowers not in spikes, mostly solitary or in small clusters in leaf axils 25
24. Fresh-water plants. Flowers hermaphrodite, arranged around a terete rhachis.
Potamogetonaceae (Potamogetoneae)
— Marine plants. Flowers unisexual, arranged on one side of a flat rhachis.
Potamogetonaceae (Zostereae)
25(23). Fruit of 1 or of several united carpels. **61. Callitrichaceae**
— Fruit of 2 or more free carpels 26
26. Carpels sessile or subsessile. **Zannichelliaceae**
— Carpels borne on a stalk several times their length. **Potamogetonaceae**
27(20). Dioecious vines or half-shrubs. Leaves ovate or elliptical and spiny-tipped, or needle-shaped, scale-like or filiform. Flowers small, 3- or 6-merous. Fruit a berry 28
— Plants not as above 29
28. Small shrubs with leaf-like phylloclades in the axils of scale-like leaves. **Liliaceae**
— Plants with large, cordate leaves, and without phylloclades. **Dioscoreaceae**
29(27). Annual or perennial herbs. Leaves usually narrow, not fleshy, parallel-nerved, sheathing at base, alternate or rosulate, entire. Flowers conspicuous, with a 3- or 6-merous perianth or flowers minute, green, arranged in spikes or spikelets, rarely in other inflorescences. Cotyledon 1 (rarely 2) 30
— Annual or perennial herbs or shrubs or trees. Leaves (if present) usually not parallel-nerved, alternate, opposite or whorled, entire, dentate or variously divided. Flowers coloured or not. Perianth (if present) only very rarely 3- or 6-merous. Cotyledons 2 (rarely 1) 40
30. Ovary superior 31
— Ovary inferior 38
31. Flowers green, small 32
— Flowers coloured, conspicuous 36
32. Aquatic monoecious plants. Flowers minute, forming thick spikes on upper part of

thick, long, leafless stalk; the staminate spike above the pistillate. Perianth composed of several bristles.   **Typhaceae**

— Plants not as above   33

33. Fruit a many-seeded capsule. Flowers hermaphrodite. Perianth made of 6 green segments. Stamens 6.   **Juncaceae**

— Fruit a 1-seeded achene or caryopsis. Flowers unisexual or hermaphrodite, with a perianth of 1–6 green scales. Stamens 1–5 (rarely 6)   34

34. Flowers unisexual, arranged in globular heads, the pistillate ones with a perianth of 3–6 narrow scales. Fruit beaked.   **Sparganiaceae**

— Plants not as above   35

35. Stems mostly hollow with more or less thickened nodes. Leaves with an open sheath, a ligule and a blade. Flowers with 2 scale-like leaflets and usually 3(–6) stamens. Fruit a caryopsis.   **Gramineae**

— Stems not as in above. Leaf sheath tubular, closed or leaves without a sheath. Flowers subtended by a single scale or pale. Fruit an achene.   **Cyperaceae**

36(31). Ovary single, of 3 connate carpels.   **Liliaceae**

— Ovaries 3 or numerous, separate   37

37. Stamens 6. Carpels many. Flowers in whorls, panicles or spikes.   **Alismataceae**

— Stamens 9. Carpels 6. Flowers in umbels, showy.   **Butomaceae**

38(30). Stamens 6.   **Amaryllidaceae**

— Stamens 1–3   39

39. Stamens 3, with distinct filaments. All leaves of perianth more or less equal in size and shape.   **Iridaceae**

— Stamens 1–2, without filaments. Lower leaf of perianth differing from others in shape and size.   **Orchidaceae**

40(29). Perianth simple or 0 (at least in pistillate flowers)   41

— Perianth consisting of a calyx and a corolla, differing from each other in shape and/ or in colour   93

41. Parasitic plants without chlorophyll. Leaves scale-like, colourless or yellow or red   42

— Plants with chlorophyll   43

42. Flowers about 1 cm., white or yellow, arranged in a short dense inflorescence surrounded by large red bracts.   **23. Rafflesiaceae**

— Flowers minute, purple, arranged in a long fleshy spike; scales minute, not concealing the flowers.   **24. Cynomoriaceae**

43(41). Flowers unisexual, staminate or pistillate   44

— Flowers hermaphrodite, rarely part of flowers unisexual, very rarely all plants unisexual but then fruit a 1-seeded capsule, dehiscing by a lid   65

44. Trees, shrubs, half-shrubs with woody base or woody vines   45

— Annuals or perennials not woody at base   60

45. Fruit an acorn, i.e. composed of a cupule and a gland. Pistillate flowers solitary or few on a very short axis.   **16. Fagaceae**

— Fruit not an acorn   46

46. Fruits aggregate, i.e. consisting of numerous minute nutlets, enclosed in a fleshy, fig-like receptacle.   **18. Moraceae**

— Fruits not aggregate   47

47. Fruit a capsule   48

— Fruit a drupe, berry, samara, nutlet or dry achene   49

48. Flowers arranged in catkins. Capsule many-seeded. Seeds minute, with long, silky hairs. Trees.   **15. Salicaceae**

— Flowers not in catkins. Capsule 3-seeded. Seeds not hairy.   **60. Euphorbiaceae**

49 (47). Leaves simple, entire, dentate or lobed   50

— Leaves pinnate   57

50. Leaves covered with minute, silvery scales. Fruit a drupe.   **71. Elaeagnaceae**

— Leaves without silvery scales   51

51. Leaves opposite, green, somewhat fleshy. Hemiparasites living on trees or shrubs.

    21. Loranthaceae

— Plants not as above   52

52. Anthers opening by valves. Evergreen dioecious trees with 2–5 cm. broad, entire, fragrant leaves.   **34. Lauraceae**

— Anthers opening by longitudinal slits. Plants not as above   53

53. Ovary inferior. Fruit a drupe. Perianth 3-lobed.   **20. Santalaceae**

— Ovary superior   54

54. Large trees. Leaves large (10 cm. or more in diam.), deeply lobed. Flowers sessile in globular (1–2 cm. wide) heads, borne on long, pendulous stalks.   **48. Platanaceae**

— Not as above   55

55. Fruit a white, many-seeded berry. Leaves linear-oblong, not fleshy, glabrous. Dioecious desert shrubs.   **46. Resedaceae**

— Fruit a 1-seeded nutlet or achene, enclosed in perianth   56

56. Leaves conspicuous, succulent or covered with mealy or vesiculous hairs. Stamens 5 or less.   **32. Chenopodiaceae**

— Leaves scale-like, not succulent. Stamens 8.   **70. Thymelaeaceae**

57 (49). Fruit a samara. Leaves opposite.   **Oleaceae**

— Fruit a drupe, berry or fleshy pod. Leaves alternate   58

58. Thorny half-shrubs. Leaves small. Ovary inferior.   **51. Rosaceae**

— Unarmed trees or tall shrubs. Ovary superior   59

59. Fruit a linear, fleshy pod.   **54. Caesalpiniaceae**

— Fruit a globular berry or drupe.   **64. Anacardiaceae**

60 (44). Staminate flowers with 1 or 4 stamens. Fruit an achene.   **19. Urticaceae**

— Flowers not as in above   61

61. Plants with milky juice. Fruit a 3-valved and 3-seeded capsule.   **60. Euphorbiaceae**

— Plants without milky juice. Fruit not as in above   62

62. Pistillate flowers and fruits enclosed within a prickly or bladdery or tuberculate involucre.   **Compositae**

— Pistillate flowers not as in above   63

63. Ovary inferior. Perianth of pistillate flowers funnel-shaped. Fruit naked. Small, prostrate herbs.   **81. Thelygonaceae**

— Perianth not funnel-shaped. Ovary superior. Fruit mostly enclosed within perianth or between bracts   64

64. Mature fruits enclosed within a winged or prickly perianth. Plants not stinging.   **25. Polygonaceae**

— Mature fruits not enclosed as in above. Plants stinging or densely covered with sticky hairs.   **19. Urticaceae**

65 (43). Plants with milky juice. Fruit a 3-valved and 3-seeded capsule.   **60. Euphorbiaceae**

— Plants without milky juice. Fruit not as in above   66

66. Ovary superior   67

— Ovary inferior or half-inferior   85

67. Trees or shrubs, thorny or unarmed. Leaves alternate, undivided. Fruit a samara, drupe or berry   68

— Plants not as above   70

68. Unarmed, deciduous trees. Leaves oblique, i.e. composed of 2 unequal parts with a long, attenuate apex. Fruit a samara.   **17. Ulmaceae**

— Leaves and fruit not as in above   69

69. Fruit a berry borne on a long gynophore.   **44. Capparaceae**

— Fruit a berry or drupe, not borne on a gynophore.   **67. Rhamnaceae**

70(67). Perianth 3- or 6-merous; segments yellow. Stamens 6; anthers opening by valves. Perennial herbs with a large, subterranean tuber and pinnately or ternately divided leaves.   **36. Berberidaceae**

— Perianth and anthers not as in above   71

71. Climbing or creeping perennials. Leaves opposite. Flowers funnel-shaped. Fruit a cylindrical or obconical, tuberculate achene.   **27. Nyctaginaceae**

— Plants with a different set of characters   72

72. Small, rare annuals. Leaves small, fan-shaped. Flowers minute. Perianth 8-lobed. Stamens 1–4. Fruit of 1–2 achenes included in calyx.   **51. Rosaceae**

— Plants not as above   73

73. Fruit a group of several free carpels, a many-seeded capsule or a berry   74

— Fruit a 1-seeded achene or nutlet   81

74. Leaves opposite, entire, not succulent. Perianth with 4–5 segments or teeth. Styles 2–3. Fruit a 1-celled, many-seeded capsule. Small annuals.   **31. Caryophyllaceae**

— Plants differing from above in one or more characters   75

75. Stamens 5. Leaves divided into 3 leaflets.   **58. Zygophyllaceae**

— Stamens 6 to many   76

76. Fruit a berry   77

— Fruit a capsule or a group of free carpels   78

77. Berry composed of 5–25 cells, usually black.   **26. Phytolaccaceae**

— Berry 1-celled, white. Desert shrubs.   **46. Resedaceae**

78(76). Style 1. Lobes of perianth 4–8, inconspicuous. Leaves opposite or whorled. Tiny annuals of damp sites.   **77. Lythraceae**

— Plants not as above   79

79. Fruit a group of several free carpels.   **35. Ranunculaceae**

— Fruit a capsule   80

80. Leaves succulent, not tomentose. Desert plants.   **29. Aizoaceae**

— Leaves not succulent, tomentose. Non-desert plants.   **28. Molluginaceae**

81(73). Stamens 6 to many or if less then leaves with ochreae   82

— Stamens 1–5. Leaves not ochreate   83

82. Leaves with stipules united into an ochrea. Styles or stigmas 2–3.   **25. Polygonaceae**

— Leaves without stipules or ochreae. Style and stigma 1.   **70. Thymelaeaceae**

83(81). Leaves opposite, entire, stipulate.   **31. Caryophyllaceae**

— Leaves alternate, rarely opposite, exstipulate   84

84. Floral bracts usually herbaceous or 0. Flowers mostly axillary. Perianth mostly herbaceous or fleshy, sometimes winged in fruit. Fruit a nutlet enclosed in the perianth and becoming detached together with it. Plants often succulent or hairy, or sometimes with mealy indumentum. Mostly halophytes or ruderal plants.
32. Chenopodiaceae

— Flowers subtended by 2–3 scarious, persistent bracts and forming long, dense, spike-like inflorescences or axillary glomerules. Perianth often scarious, neither winged nor mealy-hairy. Fruit an utricle, readily dehiscing and/or separating from perianth. Plants neither succulent nor covered with a mealy indumentum.   **33. Amaranthaceae**

85 (66). Leaves covered with silvery scales. Fruit a berry.   **71. Elaeagnaceae**
— Plants not as above   86
86. Perianth large, zygomorphic, with an ear-shaped limb. Stamens 6. Fruit a many-
  seeded capsule.   **22. Aristolochiaceae**
— Perianth and stamens not as in above   87
87. Flowers crowded into heads subtended by an involucre of small leaflets or bracts;
  all flowers tubular or ligulate, or part of them tubular and part ligulate. Anthers mostly
  united into a tube. Stigmas or styles 2.   **Compositae**
— Flowers not as in above or anthers not connate   88
88. Flowers arranged in a compound or simple umbel or rarely in heads. Umbels or
  umbellets or both often accompanied by an involucre of bracts. Perianth 5-merous.
  Ovary of 2 carpels. Fruit mostly of 2, 1-seeded carpels, often separating from each
  other at maturity.   **83. Umbelliferae**
— Plants with another set of characters   89
89. Leaves fleshy. Fruit a dry berry with 4–8 horns.   **29. Aizoaceae**
— Leaves not fleshy. Fruit not horned   90
90. Leaves alternate. Flowers solitary or few together.   **20. Santalaceae**
— Leaves opposite or whorled. Flowers often many, arranged in heads or groups   91
91. Flowers in heads subtended by an involucre, each flower accompanied by an outer
  involucel. Stamens 5, free. Style 1, stigma 1.   **Dipsaceae**
— Flowers without involucre, or individual flowers not subtended by an involucel. Stamens 1–4, rarely more but then styles 2   92
92. Stamens 1 or 3. Style 1. Fruit 1-seeded. Leaves opposite.   **Valerianaceae**
— Stamens 4–5. Styles 2. Leaves whorled, rarely opposite.   **Rubiaceae**
93 (40). Petals free, rarely united at apex only and sometimes slightly united at base but then
  fruit a schizocarp of 10 or more carpels or fruit a 5-celled capsule with a false partition
  in each cell   94
— Petals united at least in lower part   145
94. Ovary superior   95
— Ovary inferior or half-inferior   135
95. All or part of petals divided or dissected into few or many lobes. Leaves alternate.
  Flowers white or yellow, zygomorphic, 2–8-merous, in terminal spikes. Calyx parted
  to base. Fruit a capsule of 1 or 6 cells, tipped with 3–4 teeth (stigmas).   **46. Resedaceae**
— Not as above   96
96. Green, hemiparasitic shrubs. Leaves opposite. Petals 6.   **21. Loranthaceae**
— Plants not as above   97
97. Sepals 2 or 4. Petals 4, as long as or longer than calyx. Stamens 2, 4, 6 or
  more than 10; seeds without a tuft of hairs   98
— Number of sepals and/or petals and/or stamens not as in above (rarely all flower
  parts 4 but then petals much shorter than sepals or seeds with a hair tuft) 101
98. Sepals 4. Petals 4   99
— Sepals 2. Petals 4   100
99. Stamens numerous, rarely 4 or 6 and then fruit a siliqua with glandular valves and
  without septum.   **44. Capparaceae**
— Stamens usually 6. Fruit a dehiscent or indehiscent silicle or a loment or a siliqua,
  the latter always with septum.   **45. Cruciferae**
100 (98). Stamens numerous. Fruit a many-seeded capsule or siliqua.   **42. Papaveraceae**
— Stamens 2 or 4. Fruit a nutlet or a 1–2-seeded capsule or a loment.   **43. Fumariaceae**
101 (97). Leaves succulent, rarely not succulent and then opposite; margin of leaves entire

or denticulate but never divided into lobes. Fruit of 2 to many free carpels. Herbs.
49. Crassulaceae
— Plants not as above   102
102. Stamens numerous, united into a tubular column. Style divided into many branches and surrounded by the staminal column.   **69. Malvaceae**
— Stamens not united into a tubular column and/or styles not as in above   103
103. Stamens 10–15. Pistil of 5 carpels adnate to a long or short central axis; each mature carpel 1-seeded, furnished with a long beak separating from axis; rarely carpels not beaked and not separating but then plants perennial, glandular-hairy, with 2–3-pinnatisect leaves.   **57. Geraniaceae**
Not as above   104
104. Flowers actinomorphic   105
— Flowers zygomorphic   130
105. Flowers 3- or 6-merous   106
— Flowers not as above   107
106. Perennial herbs with large, divided leaves and subterranean tubers. Fruit an inflated capsule.   **36. Berberidaceae**
— Woody vines with entire, subsessile, small leaves. Fruit a drupe.   **37. Menispermaceae**
107 (105). Calyx of united sepals with 8 or 12 (rarely 4) teeth or lobes. Fruit a capsule enclosed in the persistent calyx.   **77. Lythraceae**
— Calyx and fruit not as in above   108
108. Fruit 1-seeded. Styles 5, long.   **Plumbaginaceae**
— Fruit and/or styles not as in above   109
109. Fruit a 1-seeded drupe or a group of many 1-seeded druplets, rarely of many 1-seeded achenes but then plants woody at base and with 10 sepals. Leaves alternate, neither succulent nor cuneate.   **51. Rosaceae**
— Plants not as above   110
110. Flowers about 10 cm. or more in diam. Fruit of 2–5 densely tomentose, few-seeded, large follicles.   **40. Paeoniaceae**
— Flowers and fruit not as in above   111
111. Fruit of many free, 1-seeded carpels — or of 2–5 connate or partly connate carpels and then petals with a 2-lipped limb bearing a nectar pore.   **35. Ranunculaceae**
— Fruit and/or petals not as in above   112
112. Fruit a double samara. Trees or shrubs with opposite, lobed leaves.   **65. Aceraceae**
— Not as above   113
113. Plants with a heavy odour. Leaves alternate, glandular-dotted, not succulent. Flowers yellow. Petals hooded. Stamens 8–10. Fruit a 3–5-celled capsule.   **62. Rutaceae**
— Not as above   114
114. Leaves undivided, entire or toothed   115
— Leaves pinnately, ternately or otherwise divided   127
115. Seeds provided with a tuft of long hairs. Trees or shrubs. Fruit a 1-celled capsule.
74. Tamaricaceae
— Not as above   116
116. Succulent desert plants. Leaves linear or wedge-shaped. Fruit a drupe or a capsule.
58. Zygophyllaceae
— Not as above   117
117. Sepals unequal, the outer 2 much smaller, often bract-like. Anthers numerous. Style 1, not divided. Fruit a 1-celled, many seeded capsule, loculicidally dehiscing by 3 valves.   **73. Cistaceae**

— Sepals and other characters not as in above   118

118. Leaves alternate, rarely slightly whorled   119

— Leaves opposite   123

119. Stamens 5. Styles 5. Capsule of 5 cells, each falsely and partly septate by an intrusion arising from midrib of carpel.   **59. Linaceae**

— Not as above   120

120. Fruit a 3–4 cm. long capsule, rarely a 4-lobed drupe but then leaves orbicular or cordate.   **68. Tiliaceae**

— Fruit not as in above   121

121. Flowers green or greenish-yellow. Fruit a fleshy or dry drupe. Trees or shrubs, spiny or spinescent.   **67. Rhamnaceae**

— Not as above   122

122. Leaves stipulate. Glabrous herbs.   **31. Caryophyllaceae**

— Leaves exstipulate. Tomentose herbs.   **28. Molluginaceae**

123(118). Trees or shrubs with leathery leaves. Fruit a 1-seeded berry. Flowers unisexual, green, in panicles.   **66. Salvadoraceae**

— Herbs or dwarf-shrubs. Fruit a capsule or achene. Flowers hermaphrodite   124

124. Stamens mostly 3–5-adelphous at base. Leaves usually glandular-dotted. Flowers yellow.   **41. Hypericaceae**

— Stamens free, not 3–5-adelphous. Leaves and flowers not as in above   125

125. Minute swamp plants with glabrous stipulate leaves. Fruit a 2–5-celled capsule.
                                                                              76. Elatinaceae

— Not as above   126

126. Placentae central or axile. Styles 2–5. Gynophore mostly present.
                                                                          31. Caryophyllaceae

— Placentae parietal. Style 1, 3–4-parted. Gynophore 0.   **75. Frankeniaceae**

127(114). Thorny or prickly trees or shrubs. Leaves 2–3-pinnate. Flowers small, yellow or whitish, in heads or spikes. Fruit a 1-celled pod.   **53. Mimosaceae**

— Not as above   128

128. Herbs with trifoliate leaves. Flowers yellow. Stamens 10. Styles 5.   **56. Oxalidaceae**

 Not as above   129

129. Styles 3. Fruit a drupe. Trees or shrubs.   **64. Anacardiaceae**

— Style 1. Fruit a capsule (sometimes spiny), rarely fruit a large drupe but then plants thorny with 2-foliolate leaves.   **58. Zygophyllaceae**

130(104). Stamens 6–10 or more   131

— Stamens 4–5. Fruit a 1-celled, many-seeded capsule, dehiscing by 3 valves.
                                                                               72. Violaceae

131. Flowers with a long spur. Fruit composed of 1 or several, 1- or more-seeded follicles.
                                                                             35. Ranunculaceae

— Flowers without spur. Fruit not composed of follicles   132

132. Petals 3, connate below and adnate to the stamen tube. Ovary 2-celled. Seeds hirsute.
                                                                             63. Polygalaceae

— Petals 4–5. Ovary 1-celled. Seeds glabrous   133

133. Fruit a 10–20 cm. long capsule, dehiscing by 3 valves. Almost leafless trees.
                                                                             47. Moringaceae

–- Fruit not as in above   134

134. Corolla papilionaceous, the posterior petal (standard) outermost. Stamens 10; all or 9 filaments fused, rarely free. Fruit a legume, dehiscing by 2 valves.   **55. Papilionaceae**

— Corolla not or rarely papilionaceous but then the posterior petal the innermost.

Filaments mostly free.   **54. Caesalpiniaceae**

135 (94). Flowers in simple or compound umbels; the umbel or umbellet or both accompanied by a whorl of bracts (involucre); rarely flowers in head-like umbels. Ovary of 2 carpels. Fruit mostly a diachenium. Leaves mostly sheathed at base.

83. Umbelliferae

— Plants with another set of characters   136

136. Woody climbers with umbellate flowers. Ovary 5-celled. Fruit a berry. Branches with brush-like rootlets.   **82. Araliaceae**

— Plants not as above   137

137. Aquatic plants with large, ovate or peltate floating leaves borne on long petioles. Flowers large, long-pedicelled. Petals 4 to many. Stamens numerous.

38. Nymphaeaceae

— Plants not as above   138

138. Fruit a berry. Flowers white. Leaves opposite or whorled. Evergreen shrubs.

78. Myrtaceae

— Fruit a capsule, a nutlet, a drupe or a hypanthium enclosing 5–10 achenes   139

139. Petals 4. Stamens 8. Carpels 4, connate into 4-celled ovary.   **79. Onagraceae**

— Petals and/or stamens and/or carpels not as in above   140

140. Fruit a 1-celled capsule opening by a lid. Leaves succulent, glabrous.

30. Portulacaceae

— Plants not as above   141

141. Petals and stamens numerous. Succulent herbs.   **29. Aizoaceae**

— Not as above   142

142. Calyx 4-, 6- or 8–12-toothed. Fruit a capsule or nutlet.   **77. Lythraceae**

— Calyx not as in above   143

143. Fruit a 2-beaked capsule opening by a valve on each side. Rare, tiny, white-flowered annuals, growing in maquis.   **50. Saxifragaceae**

— Fruit not a capsule   144

144. Fruit a pome or a hypanthium enclosing 1–2 achenes or a drupe.   **51. Rosaceae**

— Fruits composed of 5–10 achenes enclosed in a flat, spiny, indurated calyx. Prostrate desert and coastal annuals.   **52. Neuradaceae**

145 (93). Ovary superior   146

— Ovary inferior   170

146. Parasites with scale-like leaves   147

— Green plants   148

147. Flowers zygomorphic, rather large. Root-parasites.   **Orobanchaceae**

— Flowers actinomorphic, small. Plants twining, adhering to stems or leaves of the host.

Convolvulaceae (Cuscutoideae)

148 (146). Ovary divided by a deep groove into 4 lobes, each lobe containing 1 ovule   149

— Ovary not 4-lobed as in above   150

149. Stamens 5. Leaves mostly alternate.   **Boraginaceae**

— Stamens 2 or 4. Leaves opposite.   **Labiatae**

150 (148). Stamens 2   151

— Stamens more than 2   152

151. Herbs. Corolla rotate with a very short tube, slightly zygomorphic, blue, pinkish, white or blue and white. Fruit a capsule.   **Scrophulariaceae**

— Trees or shrubs. Corolla actinomorphic, variously shaped. Fruit a drupe or a double berry.   **Oleaceae**

152 (150). Ovary of 2 free carpels developing into 2 follicles. Seeds numerous. Leaves op-

posite or whorled; sometimes succulent herbs with reduced leaves   153
— Ovary and fruit not as in above   154
153. Anthers united and adnate to the compound stigma. Pollen grains massed into pollinia. Styles 2.   **Asclepiadaceae**
— Anthers free. Pollen not massed into pollinia. Style 1.   **Apocynaceae**
154(152). Tropical trees (Dead Sea area). Style forked twice. Ovules 4. Fruit a drupe.
                                                                Boraginaceae
— Style not forked twice. Plants not as above   155
155. Stamens 6–16. Fruit a berry or drupe. Mediterranean trees or tall shrubs   156
— Plants not as above   157
156. Corolla urceolate with short lobes. Anthers with 2 pores at tip. Leaves leathery, persistent. Fruit a many-seeded berry.   **Ericaceae**
— Corolla lobes long. Anthers not as in above. Leaves deciduous. Fruit a 1–2-seeded drupe.   **Styracaceae**
157(155). Styles or stigmas 5. Fruit a 1-seeded utricle.   **Plumbaginaceae**
— Not as above   158
158. Stamens opposite the corolla lobes. Placentation central. Flowers actinomorphic. Fruit a many-seeded capsule.   **Primulaceae**
— Not as above   159
159. Flowers actinomorphic, 4-merous, small, sessile, wind-pollinated, in spikes or dense heads. Filaments long-exserted. Corolla scarious, whitish.   **Plantaginaceae**
— Flowers not as in above   160
160. Leaves and bracts spiny. Flowers rather large. Corolla 1-lipped.   **Acanthaceae**
— Plants not as above   161
161. Aquatic submerged carnivorous plants with minute bladders on the leaves. Corolla 2-lipped, spurred.   **Lentibulariaceae**
— Plants not as above   162
162. Flowers actinomorphic, white or yellow, small, numerous, in scorpioid cymes. Stigma nearly sessile, broad at base. Nutlets 1–4, 1-seeded.   **Boraginaceae**
— Flowers not as in above   163
163. Stamens 4   164
— Stamens 5 or more   166
164. Flowers in dense globular heads. Corolla blue, 1–2-lipped. Leaves alternate. Fruit a 1-seeded nutlet. Perennial herbs.   **Globulariaceae**
— Plants not as above   165
165. Leaves opposite, digitate or incised-dentate. Corolla pink, lilac or white, distinctly or slightly 2-lipped. Fruit a 1-seeded drupe or a 2–4-seeded schizocarp.   **Verbenaceae**
— Leaves all alternate or all, or the lower ones only, opposite. Corolla variously shaped. Fruit not as above.   **Scrophulariaceae**
166(163). Leaves opposite. Corolla actinomorphic, salver-shaped, pink or yellow. Herbs, mostly annual.   **Gentianaceae**
— Leaves alternate   167
167. Fruit of 1–5, many-seeded follicles. Calyx and corolla with a long spur.
                                                             35. Ranunculaceae
— Fruit not as above. Calyx not spurred   168
168. Corolla rotate, with a very short tube, yellow. Filaments partly hirsute. Flowers bracteate. Plants usually densely covered with simple or branched hairs.
                                                             Scrophulariaceae
— Plants not as above   169

169. Fruit a berry or 2-celled capsule with numerous seeds. Calyx united into a tube, 5-toothed or -lobed, often growing in fruit. Style 1.   **Solanaceae**
— Fruit a capsule with 4 seeds, rarely with 6–8 seeds but then sepals free or connate only below. Styles 2 or more or style cleft or branched.   **Convolvulaceae**
170(145). Anthers connate, surrounding the style. Fruit a 1-seeded achene. Calyx absent or transformed into hairs or scales. Flowers in heads subtended by an involucre of bracts.   **Compositae**
— Anthers free — or connate but then fruit a many-seeded capsule and calyx green   171
171. Leaves alternate   172
— Leaves opposite or whorled   173
172. Flowers unisexual. Corolla yellow. Fruit a berry.   **Cucurbitaceae**
— Flowers hermaphrodite. Corolla rarely yellow. Fruit a capsule.   **Campanulaceae**
173(171). Stamens 5. Fruit a 1-celled berry. Climbers or tall shrubs.   **Caprifoliaceae**
— Plants not as above   174
174. Leaves whorled or with interpetiolar stipules. Flowers actinomorphic, 4–5-merous. Calyx reduced to minute teeth. Fruit a dry diachenium or a 2-celled berry, rarely a many-seeded capsule.   **Rubiaceae**
— Leaves opposite, exstipulate. Flowers zygomorphic. Fruit 1-seeded. Calyx not as in above   175
175. Stamens 4–5 or more. Each flower with a calyx-like involucel, usually adnate to the fruit.   **Dipsaceae**
— Stamens 1 or 3, rarely 4, fewer than lobes of corolla. Calyx limb persistent in fruit, rarely absent.   **Valerianaceae**

SYNOPSIS OF FAMILIES IN PART ONE

PTERIDOPHYTA

SPHENOPSIDA
1 Equisetaceae

FILICOPSIDA
2 Ophioglossaceae
3 Sinopteridaceae
4 Adiantaceae

5 Pteridaceae
6 Gymnogrammaceae
7 Thelypteridaceae
8 Aspleniaceae
9 Aspidiaceae
10 Polypodiaceae
11 Marsileaceae

SPERMATOPHYTA

GYMNOSPERMAE

CONIFEROPSIDA
12 Pinaceae
13 Cupressaceae

GNETOPSIDA
14 Ephedraceae

ANGIOSPERMAE

DICOTYLEDONEAE

Salicales
15 Salicaceae

Fagales
16 Fagaceae

Urticales
17 Ulmaceae
18 Moraceae
19 Urticaceae

Santalales
20 Santalaceae
21 Loranthaceae

Aristolochiales
22 Aristolochiaceae
23 Rafflesiaceae

Balanophorales
24 Cynomoriaceae

Polygonales
25 Polygonaceae

Centrospermae
26 Phytolaccaceae
27 Nyctaginaceae
28 Molluginaceae
29 Aizoaceae
30 Portulacaceae
31 Caryophyllaceae
32 Chcnopodiaceae
33 Amaranthaceae

Magnoliales
34 Lauraceae

Ranunculales
35 Ranunculaceae
36 Berberidaceae
37 Menispermaceae
38 Nymphaeaceae
39 Ceratophyllaceae

Guttiferales
40 Paeoniaceae
41 Hypericaceae

Papaverales
42 Papaveraceae
43 Fumariaceae
44 Capparaceae
45 Cruciferae
46 Resedaceae
47 Moringaceae

PTERIDOPHYTA

PLANTS with distinct alternation of generations. Sporophyte well developed, generally perennial, provided with stem, leaves and true roots, all with vascular tissue. Sporangia borne on leaves or on stalks or at ends of stems. Gametophyte – a leaf-like or tuberous thallus (prothallium), bearing rhizoids and sexual organs : archegonia and antheridia; archegonia contain egg cells, antheridia – motile sperm cells (antherozoids). Sporophyte develops from an embryo which originates from a zygote.

The Pteridophyta are represented in the local flora by the Sphenopsida and the Filicopsida.

SPHENOPSIDA

PLANTS with whorled branches, jointed stems and very small, scale-like leaves; leaf gaps lacking. Spores many, generally uniform, produced in sporangia which are borne on peltate sporophylls arranged in terminal cones or spikes Antherozoids multiciliate.

1. EQUISETACEAE

Perennial herbs. Rhizome subterranean, jointed, creeping, branching into aerial, erect, usually annual stems. Stems hollow, grooved, simple or verticillately branched. Leaves whorled, scale-like, usually not green, united into a toothed or lobed sheath. Reproductive organs in terminal cones or spikes consisting of whorled, pedicellate, peltate, scale-like sporangiophores. Sporangia numerous on each sporangiophore, membranous, dehiscent by a longitudinal slit. Spores numerous, globular, with four hygroscopic spiral bands (elaters) attached at one point. Prothallia cushion-like at base, with green flat lobes on upper surface, generally dioecious, sometimes with archegonia and antheridia succeeding each other on the same prothallium.

One genus, almost cosmopolitan.

1. EQUISETUM L.

Aerial stems jointed, mostly whorled, of two kinds : sterile and fertile (bearing spikes). Spikes ovoid or oblong-cylindrical; sheaths below spike 1 or more. Sporangiophores in close whorls, each with (4–)5–7 (–10) sporangia hanging in a circle from its lower surface.

Twenty three species, almost cosmopolitan.

1. Stems persistent, green, sterile and fertile ones alike; sheaths of lateral branches with 6–8 short teeth.   **2. E. ramosissimum**
– Stems deciduous, whitish, the sterile ones branching into 10–30 green branches, the fertile simple; sheaths of lateral branches with 4 subulate teeth.   **1. E. telmateia**

1. Equisetum telmateia Ehrh., Hannov. Mag. 18 : 287 (1783); Boiss., Fl. 5 : 741 (1884). *E. maximum* auct. non Lam., Fl. Fr. 1 : 7 (1778). [Plate 1]

Perennial, 60–100 cm. Rhizome creeping, black, bearing tubercles. Sterile stems about 1 cm. in diam., branching, 20–40-ridged, white, their sheaths 1.5–3 cm., ending in subulate teeth of same number as ridges; lateral branches numerous, spreading to erect, simple, thin, 4–5-angled, green, with short 4-toothed sheaths; fertile stems 20–40 cm., simple, hollow, pale brown; sheaths close together, subulate, with about 30 teeth. Spike 5–15 × 1–2 cm. April–July.

Hab.: Shady river banks, both on heavy and sandy soils. Acco Plain, Philistean Plain, Upper Galilee, Dan Valley, Hula Plain.

Area : Euro-Siberian, Mediterranean and Irano-Turanian.

A rare plant on the verge of extinction owing to the drying up of swamps and the expansion of farmed areas. Rubin swamp in the Philistean Plain is the southernmost station of its distribution area.

2. Equisteum ramosissimum Desf., Fl. Atl. 2:398 (1799); Milde, Fil. Eur. Atl. 235 (1867). *E. ramosum* Schl. ex DC., Syn. Pl. Fl. Gall. 118 (1806); Boiss., Fl. 5 : 742 (1884). [Plate 2]

Perennial, 30–100 cm. Rhizome creeping, dark brown to black. Fertile and sterile stems alike, persistent, erect, ridged, greenish; lateral branches 0–15 in number, thin, whorled; sheaths of main stems 0.6–2 cm, many-toothed, teeth darkening with age, with narrow, white-membranous margins and a persistent, hair-like apex; sheaths of lateral branches with 6–8 short teeth. Spikes 0.8–3 × 0.4 cm. February–October.

Hab.: River banks and swamps. Coastal Galilee, Acco Plain, Sharon Plain, Philistean Plain, Upper Galilee, Esdraelon Plain, Dan Valley, Hula Plain, Upper and Lower Jordan Valley, Dead Sea area, Edom.

Very variable. The local forms are closest to the following varieties: (a) var. **simplex** (Döll) Milde, Sporenpfl. 118 (1865): stems without or with 1–2 lateral branches, 8–16-ridged. (b) var. **procerum** (Poll.) Aschers. in Aschers. et Graebn., Syn. 1: 140 (1896): stems with 3–8 lateral branches, 8–16-ridged. (c) var. **altissimum** A. Br. ex Milde, l.c. 117 : stems with many long branches, 14–20-ridged.

As there are transitions between these forms, their taxonomic value is questionable at present.

Area : In several temperate and subtropical regions.

Used in popular medicine; eaten by goats.

FILICOPSIDA

Mostly terrestrial plants, sometimes epiphytic or aquatic. Sporophyte mostly with root, stem and leaves. Leaves spirally arranged, usually large, with simple or compound blade, mostly circinnate in bud; leaf gaps present. Sporangia mostly borne on lower surface of leaf or rarely within thickened containers (sporocarps) or on spike-like organs; spores alike or differentiated into macro- and microspores, often grouped in sori, the latter mostly covered with a scale-like indusium, rarely naked. Gametophyte small, mostly with a small green prothallium, bearing rhizoids, archegonia and antheridia, the latter with multiciliate antherozoids.

This division is locally represented by 10 families. Eight of them were formerly included within the family Polypodiaceae.

Literature : E. G. Copeland, *Genera Filicum*. Waltham (1947). R. E. Holttum, The classification of ferns. *Biol. Rev. Cambridge Philos. Soc.* 24 : 267–296 (1949). R.E. Pichi-Sermolli, The nomenclature of some fern-genera. *Webbia* 9 : 387–454 (1954).

2. OPHIOGLOSSACEAE

Low, perennial ferns, mostly terrestrial. Rhizome short, usually vertical, not scaly. Leaves mostly solitary or few, petiolate, not circinnate in bud. Sporophyll reduced to 1 or rarely more, simple or branched stalks, with spikes of 2-rowed sporangia. Sporangia sessile, with walls of several cell layers, without annulus, opening by a slit. Spores numerous, all alike, mostly globular-tetrahedral. Prothallium monoecious, usually subterranean, more or less tuberous, usually without chlorophyll but with endotrophic mycorrhiza.

Comprises 3 genera: 2 cosmopolitan, 1 in Trop. Asia and Australasia.

1. OPHIOGLOSSUM L.

Rhizome short, fibrous. Leaves leathery, simple, rarely lobed, with reticulate veins. Sporophyll forming a pedunculate, simple, linear spike of 2-rowed, connate sporangia sunk into the rhachis and opening by a transverse slit. Prothallia cylindrical or ovoid.

About 50 species, cosmopolitan.

1. Ophioglossum lusitanicum L., Sp. Pl. 1063 (1753); Boiss., Fl. 5:720 (1884). [Plate 3]

Perennial, 5–15 cm. Rhizome short, fairly thick, simple or with long, laterally creeping branches and fasciculate roots. Leaves 1–2, 3–7 cm., long-petioled, oblanceolate, tapering towards a somewhat sheathed base, acute, entire. Sporophylls 1–2, 1–1.5 cm., long-peduncled, with a linear, acute, distichous spike, mostly exceeding leaves in length. Spores smooth. January–March.

Hab.: Sandy loams and sandy-calcareous hills. Sharon Plain, Philistean Plain.

Area: Mediterranean and W. Euro-Siberian; also in Trop. Africa.

Very rare, on verge of extinction.

3. SINOPTERIDACEAE

Small perennial and xerophytic ferns, often living on rocks or in rock crevices. Rhizome scaly, erect or creeping, often with black, shining petioles and palmately or pinnately divided, usually uniform leaves. Sori borne on the distal parts of the veins along the leaf margin, sometimes extending rather far towards midrib, often confluent; indusium formed by the deflexed leaf margin. Spores tetrahedral or globular.

About 8 genera, mostly in the tropics, but also in warm-temperate regions of both hemispheres.

1. CHEILANTHES Swartz

Low perennials. Rhizome mostly short, with rusty scales. Leaves persistent, few or many, mostly tufted, rigid, erect to spreading, 1–3-pinnate or -cut, glabrous or hairy; veins free. Sori borne on the thickened ends of veins, about 0.5 mm., nearly orbicular, distinct or almost confluent; indusia (pseudo-indusia) formed by the deflexed, membranous, entire or fimbriate or finely tomentose margin of leaf segments.

About 250 species, in tropical and warm-temperate regions of both hemispheres.

1. Pinnules, at least of adult plants, glabrous or somewhat hairy beneath.   **1. C. fragrans**
– Pinnules densely covered on both surfaces with flat or filiform scales.   **2. C. catanensis**

1. Cheilanthes fragrans (L.) Swartz, Syn. Fil. 127/325 (1806) excl. descr. et loc. *Polypodium fragrans* L., Mant. Alt. 307 (1771) non L., Sp. Pl. 1089 (1753). *C. fragrans* (L.) Webb et Berth., Phyt. Canar. 3 : 452 (1849); Boiss., Fl. 5 : 725 (1884). *C. pteridioides* (Reichard) Christens., Ind. Fil. 170 (1905). *P. pteridioides* Reichard, Syst. Pl. Nov. 404 (1780). [Plate 4]

Perennial. Rhizome short, covered with rusty scales. Leaves 5–20 cm., smelling of coumarin, densely tufted, 2 (–3)-pinnatisect; blade ovate or oblong-lanceolate in outline, glabrous or somewhat hairy beneath; segments 5–10 on each side, mostly opposite, divided into sessile, small, ovate or oblong, obtuse pinnules with revolute, mostly fimbriate and glandular margin; ultimate segments 1–3 mm., oblong to orbicular; petiole as long as or longer than blade, slender, shining, dark brown, covered with scales. Sori round, brown, almost confluent at maturity and forming a marginal interrupted band. February–July.

Hab.: Rocks. Sharon Plain, Upper and Lower Galilee, Mt. Carmel, Esdraelon Plain, Mt. Gilboa, Samaria, Judean Mts., Judean Desert, Upper and Lower Jordan Valley, Ammon, Edom.

Area: Mediterranean, extending into the W. Euro-Siberian and W. Irano-Turanian territories.

2. Cheilanthes catanensis (Cosent.) H. P. Fuchs, Brit. Fern Gaz. 9 : 45 (1961). *Notholaena vellea* (Ait.) R. Br. emend. Desv., Journ. Bot. Appl. 3 : 93 (1814). *N. vellea* (Ait.) R. Br., Prodr. Fl. Nov. Holl. 146 (1810) p. p. *Acrostichum velleum* Ait., Hort. Kew. 3 : 457 (1789). *N. lanuginosa* (Desf.) Desv. ex Poir. in Lam., Encycl. Suppl. 4 : 110 (1816); Boiss., Fl. 5 : 725 (1884). [Plate 5]

Perennial. Rhizome short, thick. Leaves 5–25 cm., densely tufted, short-petioled; blade oblong-lanceolate in outline, 1–2 (–3)-pinnatisect, dark green, covered on both surfaces with white or yellow fleece or almost glabrescent above; segments ovate or lanceolate, obtuse, cut into small, entire or dentate lobes rounded at apex; petioles yellow-brown, shiny, densely covered with narrow scales. Sori covering almost the whole surface of lobes and concealed by fleece. March–April.

Occurs in two forms: f. *lanuginosa* with lobes hairy on upper surface, and f. *cheilanthoides* with lobes green and glabrescent on upper surface.

Hab.: Rock crevices and walls. Upper and Lower Galilee, Mt. Carmel, Esdraelon Plain, Judean Mts., Upper and Lower Jordan Valley, Ammon – f. *lanuginosa*; Coastal Negev, Lower Jordan Valley, Dead Sea area, Ammon – f. *cheilanthoides*.

Area: S. Mediterranean, extending into W. Euro-Siberian, Sudanian and Irano-Turanian territories.

4. ADIANTACEAE

Perennial terrestrial ferns, various in habit. Rhizome erect or creeping, covered with scales. Leaves uniform, usually with shining petiole and rhachis; blade rarely entire, generally 1–5-pinnate or -pedate, the ultimate pinnules mostly trapeziform or cuneate-flabellate; veins very often free. Sori close to the leaf margin; sporangia borne along and sometimes also between the ends of the veins on the underside of a reflexed continuous or lobed margin (pseudo-indusium) which conceals the sori. Spores tetrahedral.

One genus, mostly tropical, in both hemispheres.

1. ADIANTUM L.

Perennials with scaly rhizome. Leaves all alike, often pinnate, segments mostly delicate, more or less cuneate or fan-shaped; veins simple or forked, free; petiole filiform, mostly glossy black. Sori at the ends of the veins, covered by the reflexed, scale-like and membranous margin of lobes (pseudo-indusium).

About 200 species, mostly in tropical, some in temperate regions, predominantly in S. America.

1. Adiantum capillus-veneris L., Sp. Pl. 1096 (1753); Boiss., Fl. 5 : 730 (1884). [Plate 6]

Perennial. Rhizome creeping, with dense scales. Leaves 5–30 cm., persistent, 2–3-pinnatisect, glabrous; blade ovate to oblong in outline, with a slender, black rhachis, ultimate segments alternate, petiolulate, obliquely cuneate-obovate, irregularly lobed or incised, often serrate at tip; veins thin, forked, free; petiole long, hairy, brown to black. Sori 2–10 on each segment, covered by rectangular or reniform recurved parts of lobes (pseudo-indusium). Almost the whole year.

Hab.: Shady and damp walls, rocks and cave mouths. Acco Plain, Sharon Plain, Philistean Plain, Upper and Lower Galilee, Mt. Carmel, Esdraelon Plain, Mt. Gilboa, Samaria, Judean Mts., Judean Desert, N. and C. Negev, Dan Valley, Upper and Lower Jordan Valley, Dead Sea area, Arava Valley, Gilead, Moav, Edom.

Area: Mediterranean, Irano-Turanian and Euro-Siberian, extending also into other temperate and tropical regions.

Was used medicinally in former times (herba capillorum veneris).

5. PTERIDACEAE

Terrestrial ferns with erect or creeping rhizome covered with hairs or scales. Leaves uniform or subdimorphic, pinnately, rarely digitately divided; veins free or anastomosing. Sori submarginal, forming a confluent band (coenosorus) along the veinlets connecting the vein ends and covered by the reflexed membranous leaf margin. Spores tetrahedral.

Twelve genera, mostly tropical, a few in warm-temperate regions.

1. Pteris L.

Erect perennials with creeping, scaly rhizome. Leaves long-petioled, of two forms: sterile and fertile, or lower segments sterile and upper fertile; blade 1–4-pinnate; veins free or anastomosing. Sori linear, continuous along margin of segments; indusium (pseudo-indusium) linear, continuous, formed of the reflexed margin of segments, more or less covering the sori.

About 280 species, mainly in tropical and subtropical regions.

1. Pteris vittata L., Sp. Pl. 1074 (1753). *P. longifolia* auct. non L., l.c.; Boiss., Fl. 5 : 727 (1884). [Plate 7]

Perennial, 30–100 cm. Rhizome creeping , covered with brown scales. Leaves densely tufted, leathery, lanceolate in outline, pinnate; pinnae 3–10 cm., numerous, generally opposite, rarely alternate, subsessile, linear-lanceolate, somewhat cordate at base, acute, segments of sterile leaves serrulate and broader than those of the fertile ones, all pinnately nerved with simple or forked lateral veins; petiole much shorter than blade, brown, sparsely covered with scales. Sori 1–1.5 mm. broad, forming a band along the reflexed margin of segments; sporangia surrounded by paraphyses. March–August.

Hab.: River banks and maquis. Upper Galilee, Mt. Carmel, Dan Valley. Very rare.

Area: Mediterranean and W. Euro-Siberian; also in Trop. America, Africa and Asia.

6. GYMNOGRAMMACEAE

Mostly terrestrial ferns, various in habit. Rhizome solenostelic, creeping or erect, hairy or scaly. Leaves various. Sporangia seriate along the veins, scattered and not in well defined sori, without indusium and often forming a confluent coenosorus. Spores tetrahedral.

Fourteen genera, mostly Neotropical.

1. Anogramma Link

Low glabrous annuals or biennials with perennial prothallium. Rhizome short, with few scales. Leaves thin, 2–3-pinnate; veins free. Sori arranged along the secondary veins, naked, often confluent at maturity and occupying almost the whole lower surface of segments.

Seven species, chiefly in tropical and south-temperate zones.

1. Anogramma leptophylla (L.) Link, Fil. Sp. 137 (1841). *Polypodium leptophyllum* L., Sp. Pl. 1092 (1753). *Gymnogramme leptophylla* (L.) Desv., Mag. Ges. Nat. Fr. Berlin 5 : 305 (1811); Boiss., Fl. 5 : 721 (1884). [Plate 8]

Annual or biennial, 10–25 cm.; prothallium perennial. Rhizome very short, sparsely scaly when young. Leaves few, sparsely hairy, later glabrous, of two forms: the outer short, 3–7 cm., ovate to orbicular in outline, deeply pinnatisect or cut into fan-shaped, lobulate segments; others much longer, 5–20 cm., oblong-lanceolate in outline, 2–3-pinnatisect, segments obovate-cuneate, incised or pinnatifid; petiole as long as or longer than blade, slender, purplish. Sori scattered along the secondary veins, sometimes becoming confluent and occupying whole surface of segment. February–April.

Hab.: Shady and damp places, walls, rock fissures and cave openings. Acco Plain, Sharon Plain, Philistean Plain, Upper and Lower Galilee, Mt. Carmel, Esdraelon Plain, Mt. Gilboa, Samaria, Judean Mts., Hula Plain, Upper Jordan Valley.

Area: Mediterranean, W. Euro-Siberian and Irano-Turanian; also in Australia, S. America, and elsewhere.

7. THELYPTERIDACEAE

Terrestrial or aquatic ferns. Rhizome creeping or ascending, dictyostelic, with sparse scales; rhachis and veins often with simple or branched hairs. Leaves mostly 2-pinnatisect; petioles with 3–7 vascular strands. Sori submarginal, mostly round, rarely elongated, with reniform indusia or without indusia.

About 20 genera in tropical and temperate regions.

1. THELYPTERIS Schmidel

Terrestrial, perennial herbs. Rhizome short or long, creeping, vertical or ascending, with few, usually hairy scales. Leaves long-petioled, all similar, with more or less pinnatipartite segments, mostly more or less pubescent; veins free. Sori near margin of segments, small, almost orbicular, sometimes confluent at maturity; indusium reniform, sometimes caducous or 0.

About 500 species, cosmopolitan.

1. Thelypteris palustris Schott, Gen. Fil. Observ. Nephrod. t. 10 (1834). *Dryopteris thelypteris* (L.) A. Gray, Man. Bot. 630 (1848). *Nephrodium thelypteris* (L.) Stremp., Fil. Berol. Syn. 32 (1822); Boiss., Fl. 5 : 737 (1884). [Plate 9]

Perennial. Rhizome slender, creeping below ground. Leaves 25–80 cm., not tufted, long-petioled, with glabrous, shining petioles; blade oblong-lanceolate in outline, not glandular, sparsely hairy when young, 1-pinnatisect; segments 10–30 on either side, oblong-lanceolate to linear, pinnatipartite into oblong, acute, entire or slightly repand lobes, those of fertile leaves with somewhat reflexed margin. Sori small, arranged half-way between main vein and margin, later becoming confluent; indusium caducous. May–June.

Hab.: Swamps. Dan Valley, Hula Plain.

Area: Almost all temperate regions of the Holarctis; also Trop. Africa, S. Africa, New Zealand.

The Hula Plain is the southernmost station of this fern in S.W. Asia. With the draining of the swamps in this district it is threatened with extinction.

8. ASPLENIACEAE

Terrestrial or epiphytic ferns with erect, creeping or climbing dictyostelic rhizomes, covered with scales. Leaves of diverse habit, simple to compound, heteromorphic to uniform; petioles with 1–2 vascular bundles, often scaly. Sori usually oblong to linear, borne either on one or on both sides of the vein, mostly with membranous, simple or double indusium. Spores bilateral.

About 7–10 genera, mostly Neo- and Paleotropical but some also in temperate regions

1. Leaves entire to slightly lobed.   **3. Phyllitis**
– Leaves 1–3-pinnatisect   2
2. Leaves covered with scales on lower surface. Indusium absent.   **2. Ceterach**
– Leaves without scales on lower surface. Indusium present.   **1. Asplenium**

1. Asplenium L.

Perennial glabrous herbs, terrestrial or epiphytic. Rhizome scaly, fibrous, mostly creeping. Leaves tufted, mostly firm, simple or compound; veins usually free. Sori borne on secondary oblique veins, not marginal, oblong to linear; indusium of same shape as sorus, arising from vein and usually opening towards midrib.

About 700 species, almost cosmopolitan.

1. Asplenium adiantum-nigrum L., Sp. Pl. 1081 (1753); Boiss., Fl. 5 : 734 (1884). [Plate 10]

Perennial. Rhizome creeping, horizontal or oblique, branching, covered with brown scales. Leaves 8–45 cm., persistent, densely tufted, 2–3-pinnatisect; blade triangular to lanceolate in outline, often acuminate, shining; primary segments 8–15 on either side, mostly alternate, the upper sessile, ultimate segments oblong-lanceolate, dentate; petiole as long as or longer than blade, rigid, blackish. Sori elongated, oblong or linear; indusium with free, entire margin.

Subsp. **onopteris** (L.) Heufl., Verh. Zool.-Bot. Ges. Wien 6 : 310 (1856); Luerss., Farnpfl. 260 (1889). *A. onopteris* L., Sp. Pl. 1081 (1753). *A. adiantum-nigrum* L. var. *virgilii* (Bory) Boiss., l.c. Blade and segments long-acuminate, lower segments often recurved; ultimate segments lanceolate or linear-lanceolate, subaristate-dentate. March–July.

Hab. : Shady rocks in maquis. Upper Galilee, Mt. Carmel.

Area : Mediterranean and Euro-Siberian, also extending into adjacent territories of the Irano-Turanian and Sudanian regions.

2. Ceterach DC.

Dwarf perennial herbs. Rhizome short, erect, covered with black scales. Leaves tufted, thick, pinnatifid to almost pinnatisect with ovate to rounded lobes, densely covered below with large, leathery, clathrate scales; veins anastomosing towards margin;

petioles very short. Sori in 2 rows on one side of the secondary veins, almost linear, without or with rudimentary indusium more or less hidden by scales.

Three species in Europe, W. and C. Asia and Africa.

1. Ceterach officinarum DC. in Lam. et DC., Fl. Fr. ed. 3, 2:566 (1805); Boiss., Fl. 5 : 722 (1884). *Asplenium ceterach* L., Sp. Pl. 1080 (1753). [Plate 11]

Perennial. Rhizome short, covered with blackish-brown, lanceolate scales. Leaves 5–25 cm., persistent, densely tufted, sinuately pinnatipartite; blade leathery, linear-lanceolate in outline, obtuse, green and glabrous above, covered beneath with brown, shining, triangular-lanceolate, overlapping scales; segments 8–12 on each side, alternate, oblong or semiorbicular with rounded apex, mostly entire, the lower distinct, the upper confluent; petiole shorter than blade. Sori about 2 mm., oblong-linear; indusium 0. Almost the whole year.

Hab. : Rock crevices and walls. Sharon Plain, Upper and Lower Galilee, Mt. Carmel, Mt. Gilboa, Judean Mts., Dan Valley, Upper and Lower Jordan Valley, Dead Sea area, Gilead, Ammon, Edom. Fairly common.

Area : Mediterranean, Euro-Siberian and Irano-Turanian, extending also into the Sudanian region and S. Africa.

Formerly used as remedy for splenitis and other diseases (herba ceterachi).

3. PHYLLITIS Hill

Perennial herbs with thick, scaly rhizome. Leaves mostly tufted, entire or slightly lobed, the fertile ones shorter; secondary veins 2–3-forked, not attaining margin of blade. Sori oblong, in pairs, arranged obliquely in parallel rows, 1 sorus of pair borne on upper fork of one vein, the other on lower fork of adjacent vein; indusia of the pair of sori opening towards each other.

Two species. North-temperate regions.

1. Phyllitis sagittata (DC.) Guinea et Heywood, Collect. Bot. 4 : 246 (1954). *Scolopendrium sagittatum* DC. in Lam. et DC., Fl. Fr. ed. 3, 5 : 238 (1815). *P. hemionitis* (Swartz) O. Ktze., Rev. Gen. 2 : 818 (1891). *S. hemionitis* Swartz in Schrad., Journ. Bot. 2 : 50 (1801) non *Asplenium hemionitis* L., Sp. Pl. 1078 (1753); Boiss., Fl. 5 : 729 (1884). [Plate 12]

Perennial. Rhizome short, densely covered with scales. Young leaves ovate, cordate at base, obtuse, adult ones firm, oblong-lanceolate, acuminate, entire or wavy-margined or slightly lobed, frequently somewhat constricted above the deeply cordate or often hastate and divaricately auriculate base; petiole sometimes longer than blade, more or less scaly. Sori 0.5–1.5 cm., oblong; indusia linear. April–July.

Hab. : Shady rocks. Upper Galilee, Samaria. Very rare.

Area : Mediterranean.

Formerly used in diarrhoea and catarrhal disturbances (herba scolopendrii).

9. ASPIDIACEAE

Terrestrial ferns with creeping or erect dictyostelic rhizomes covered with opaque scales. Leaves simple to compound, petiolate; petioles with 5–7 vascular strands. Sori dorsal, rarely marginal, orbicular, rarely elongated; indusia reniform. Spores bilateral.

About 25 genera in tropical and extratropical regions.

1. DRYOPTERIS Adans.

Perennial herbs. Rhizome short, stout, decumbent to ascending, densely covered with scales. Leaves tufted, long-petioled, all similar, 1–3-pinnate, with somewhat scaly petiole; veins forked, free. Sori borne on veins near midrib, mostly in rows, rather large, more or less orbicular; indusium reniform, attached by inner margin of umbilicate sinus.

About 150 species, in north-temperate zones, S. Africa and the tropics of both hemispheres.

1. Dryopteris villarii (Bellardi) H. Woynar ex Schinz et Thell., Viert. Naturf. Ges. Zürich 60 : 339 (1915). *Polypodium villarii* Bellardi, Mém. Acad. Turin 5 : 255 (1792). *D. rigida* (Swartz) A. Gray, Man. Bot. 631 (1848). *Nephrodium rigidum* (Swartz) Desv., Ann. Soc. Linn. Par. 6 : 261 (1827); Boiss., Fl. 5 : 738 (1884). *Aspidium rigidum* Swartz in Schrad., Journ. Bot. 2 : 37 (1801). [Plate 13]

Perennial. Rhizome thick, oblique. Leaves 20–60 cm., rigid; blade ovate-lanceolate to deltoid in outline, dull green, glabrous or sometimes glandular, 2-pinnate, pinnae lanceolate, pinnatisect into dentate or pinnatifid segments; petiole shorter than blade, green, covered with rather large, buff-coloured, scarious scales. Sori in 2 rows; indusium glabrous or sometimes glandular.

Var. **australis** (Ten.) Maire, Fl. Afr. N. 1 : 32 (1952). *Aspidium rigidum* Swartz var. *australe* Ten., Atti Ist. Incor. Napol. 5 : 144 (1832). Lower pinnae pinnatisect, short-petioled. Leaves and indusium almost glabrous, not glandular or scarcely so. March–September.

Hab.: Maquis. Upper Galilee, Dan Valley.

Area: Mediterranean, Euro-Siberian and Irano-Turanian.

10. POLYPODIACEAE

Epiphytes, rarely terrestrial ferns with creeping to ascending rhizomes, polystelic or solenostelic, mostly covered with broad scales. Leaf blade simple or compound, scaly or pilose, often leathery, netted-veined; petioles with 1–3 vascular strands. Sori dorsal, often without indusium, round or oblong. Spores bilateral.

Ten (or 50–65) genera, predominantly tropical.

1. POLYPODIUM L.

Mostly epiphytic, rarely terrestrial perennials, with a fleshy, creeping, scaly rhizome. Leaves all similar, pinnatifid or pinnatisect with segments confluent at base; veins forked or otherwise branched, usually anastomosing evenly into areoles (loops) with free endings inside loops. Sori terminal or almost so on veins, arranged in 1 (–3) rows on each side of the midrib, large, round or rarely oblong, yellow when ripe; indusium 0.

About 75 species, mainly in the northern hemisphere and in Trop. America, Asia and Polynesia.

1. Polypodium vulgare L., Sp. Pl. 1085 (1753); Boiss., Fl. 5 : 723 (1884). [Plate 14]

Perennial, 10–40 cm. Rhizome subterranean or aerial, creeping, thick, densely covered with rust-coloured scales. Leaves 5–45 cm., solitary, long-petioled, erect, oblong-lanceolate, ovate or deltoid in outline, pinnatipartite to pinnatisect, mostly glabrous on both sides; segments 5–20 on either side, opposite or alternate, oblong-lanceolate to oblong-linear, acute or obtuse, entire or serrulate, confluent at base. Sori in 2 rows along the midrib, round.

Var. **serratum** Willd., Sp. Pl. 5 : 173 (1810). Blade deltoid, tapering; segments acute or obtuse, crenate-serrate. January–July.

Hab.: Maquis, shady rocks. Upper Galilee.

Area: Throughout the regions of the Holarctis, Capensis and Antarctis.

Used in pharmacy for preparation of laxatives and expectorants (herba et rhizoma polypodii).

11. MARSILEACEAE

Aquatic or subaquatic perennials. Rhizome long, slender, creeping, hairy. Leaves radical or some of them cauline, circinnate in bud, simple, filiform-linear and without blade or with blade of 2 or 4 leaflets palmately arranged on a long petiole; veins dichotomous. Sporangia borne in leathery or membranous, globular or ovoid-oblong, 1- or many-celled sporocarps dehiscent at maturity, each sporocarp containing 1 or more sori surrounded by an indusium; sorus mostly consisting of macro- and microsporangia; macrosporangium with 1 macrospore, microsporangium with numerous small microspores; annulus rudimentary or 0.

Three genera and about 75 species, almost cosmopolitan.

1. MARSILEA L.

Perennial. Rhizome slender, rooting at nodes. Sterile leaves fasciculate, long-petioled; blade of 2 contiguous pairs of opposite, obovate-cuneate leaflets with flabellate anastomosing venation. Sporocarp leathery, sessile or stipitate, globular-ovoid to compressed, with truncate base, divided by membranous partitions (indusia) into two series of transverse compartments, each with a single sorus. Macro- and microsporangia intermixed in the same sorus; mature sporocarp dehiscent under water along the ventral suture into 2 valves and liberating a gelatinous ring which opens and turns into a long thread, bearing the sori.

About 60 species in tropical and temperate zones.

1. Marsilea minuta L., Mant. Alt. 308 (1771). *M. diffusa* auct. Fl. Palaest. non Lepr. ex A. Br., Flora 22 : 300 (1839); Boiss., Fl. 5 : 750 (1884). [Plate 15]

Perennial. Rhizome long, rooting at nodes, almost glabrous, at least when adult. Leaves long-petioled; leaflets 1–2.5 cm., obovate-cuneiform, with repand to praemorse upper margin, pubescent, later glabrescent. Sporocarps inserted at base or slightly above base of petiole in groups of 2–4, each on a free pedicel twice as long as sporocarp or longer; sporocarp recurved, ellipsoidal and compressed or lenticular, rounded at apex or more or less truncate, with 2 teeth at base, appressed-hairy when young, then glabrescent; sori 5–6 on each side of the gelatinous ring. Summer.

Hab. : Swamps and swamp edges. Hula Plain.

Area : Trop. Africa, Canary Islands, Egypt and probably elsewhere.

This species reaches the northern limit of its distribution here, but has apparently disappeared from the country since the drying up of the Hula swamps.

SPERMATOPHYTA

PLANTS with alternation of generations. Sporophyte well developed, bearing true roots, stems, leaves and flowers. Gametophyte much reduced. Male gametophyte developing from a microspore (pollen grain) and producing a few cells, among them a pollen tube and 2 or more eciliate (rarely ciliate) gametes. Female gametophyte of many or few cells, among them archegonia or egg cells retained within the macrospore. Seeds generally formed as a result of fertilization, comprising a testa and an embryo, with or without endospermous tissue.

The Spermatophyta or seed plants are divided into the Gymnospermae and Angiospermae.

GYMNOSPERMAE

TREES or shrubs with secondary thickening and often with resin ducts. Xylem without vessels (except in Gnetopsida). Leaves needle-like, scale-like or broad, sometimes 0. Flowers mostly unisexual; the staminate ones with pollen sacs (microsporangia), borne on the upper or lower side of stamens (microsporophylls); ovulate flowers consisting of ovules (macrosporangia), borne on a macrosporophyll or at the end of a flower axis but not enclosed in an ovary. Pollen reaching the ovule directly (not by means of a stigma). Female prothallium many-celled, forming an endosperm and containing reduced archegonia. Seeds consisting of a seed coat, a diploid embryo with 2 to many cotyledons and a haploid endosperm (prothallium).

The Gymnospermae are represented in the local flora by the Coniferopsida and Gnetopsida.

CONIFEROPSIDA

Monoecious, sometimes dioecious trees or shrubs. Secondary xylem without vessels. Leaves simple, usually small, narrow or scale-like, mostly evergreen. Pollen sacs borne on the under-surface of microsporophylls arranged in cones. Ovules usually borne on a scale, mostly subtended by a sterile bract; scales usually becoming woody or fleshy after fertilization, forming woody or rarely berry-like cones. Male gametes not motile. Embryo with 2–15 cotyledons.

12. PINACEAE

Monoecious trees, rarely shrubs, mostly resinous. Leaves mostly persistent, solitary or fasciculate, mostly spirally arranged, linear, semiterete, terete or compressed-triquetrous, entire or minutely serrulate. Reproductive organs in cones with spirally arranged scales. Staminate cones surrounded at base by a whorl of sterile bracts and consisting of scale-like stamens each with 2 pollen sacs (microsporangia) on the under-surface. Scales of ovulate cones mostly double : the lower one (bract) sterile, the upper (ovuliferous scale) with 2 ovules on its inner surface; mature cones usually woody, closed. Seeds mostly winged, usually 2 on each scale.

Nine genera and about 250 species, mainly in temperate regions of the northern hemisphere.

1. PINUS L.

Evergreen trees, usually with whorled branches. Buds covered with imbricated scales. Shoots of two forms : long shoots with deciduous leaves, scale-like and woody at base, and short shoots arising from axils of the scale-like leaves and bearing fascicles of 2, 3 or 5 green, needle-like leaves, surrounded at base by scarious scales forming a sheath. Staminate cones axillary, clustered at base of young shoots, catkin-like and consisting of spirally arranged scales each with 2 pollen sacs. Pollen grains with 2 air sacs. Ovulate cones lateral or subterminal in the upper part of the tree, maturing mostly in second or third year; ovuliferous scales spirally arranged, persistent, leathery or woody, with enlarged apex (apophysis), mostly bearing a dorsal or terminal protuberance (umbo); bracts distinct, very small. Seeds winged; cotyledons 4–15.

About 90 species mainly in the northern hemisphere and in mountains of the northern tropics.

1. Pinus halepensis Mill., Gard. Dict. ed. 8, no. 8 (1768); Boiss., Fl. 5 : 695 (1884). [Plate 16]

Tree (6–)10–15 m., with diffuse crown and ascending somewhat whorled branches. Leaves 8(–15) cm., in pairs, slender, bright green, with marginal resin ducts. Staminate cones clustered in heads, ovoid to cylindrical. Ovulate (mature) cones 7–12 × 4–6 cm., solitary or 2–3 in a whorl, reflexed on thick peduncles, oblong or oblong-conical, reddish-brown, maturing 15–16 months after pollination; scales oblong, exposed portion shining, flat, rhombic, transversely keeled, with small flattish umbo. Seeds 5–6 × 3 mm., oblong; wing 3–5 times as long as seed. Fl. March–April. Seed maturing July–August.

Hab.: Hills and mountains, mostly on light-coloured calcareous rendzina soils at 100–1,000 m. altitude. Coast of Galilee, Upper Galilee, Mt. Carmel, Samaria, Judean Mts., Upper Jordan Valley, Gilead, Ammon.

Area : Mediterranean.

Forms natural forests and occurs in remnant stands in several places of Mediterranean Palestine, e.g. Coastal Galilee, Mt. Carmel, Samaria and Gilead.

P. halepensis is also extensively and successfully planted as a timber tree. It is noteworthy for its relative drought resistance; it is rather variable in growth rate and dimensions. Mycorrhizally associated with *Boletus granulatus*.

Believed to be the עֵץ שֶׁמֶן of the Bible (Is. xli : 19 and elsewhere).

13. CUPRESSACEAE

Evergreen, mostly monoecious trees or shrubs with resin ducts. Buds naked. Leaves opposite or in whorls of 3–4, frequently dimorphic : the juvenile ones larger, needle-like or subulate; adult leaves small, scale-like, rarely needle-like. Reproductive organs in small cones with opposite or ternately whorled scales. Staminate cones terminal or axillary, with 4–8 whorls of scales each with 2–6 pollen sacs (microsporangia) on the under-surface; pollen grains without air sacs. Ovulate cones terminal or lateral on short branches, scales usually few, each with 1 to many erect ovules on upper surface, sterile bract united with scale; mature cone mostly woody, sometimes fleshy and berry-like. Seeds sometimes narrowly winged; cotyledons 2, rarely 5–6.

About 15 genera and about 140 species, mainly in temperate and subtropical regions of the northern hemisphere, some in tropical mountains.

1. Scales fleshy, coalescing at maturity to form a berry-like body. **2. Juniperus**
-- Scales woody, diverging at maturity and shedding their seeds. **1. Cupressus**

1. Cupressus L.

Monoecious trees or shrubs with scaly bark. Leaves opposite, imbricated when adult, scale-like. Staminate and ovulate cones on separate branches. Staminate cones oblong, with decussately arranged pairs of imbricated scales each with 2–4 sessile pollen sacs. Ovulate cones ovoid to globular, maturing in the second year; scales 6–14, free, woody, peltate, mucronate on back, the upper ones sterile; mature cones first tightly closed, then diverging; scales persistent, angular to orbicular, umbonate. Seeds 4–20 on each scale, in several rows, bony, frequently narrowly winged.

About 15 species: N.W. America, Mexico and E. Mediterranean to China.

1. Cupressus sempervirens L., Sp. Pl. 1002 (1753); Boiss., Fl. 5 : 705 (1884). [Plate 17]

Tree 10–30 m., with trunk up to 1 m. in diam. Branches very dense, erect or horizontally spreading; twigs more or less in 2 rows, obscurely tetragonous. Leaves decussate and imbricated, triangular, obtuse, convex and frequently with a sunken resin gland on back. Staminate cones terminal on short twigs, oblong-cylindrical. Ovulate mature cones 2–3 cm., solitary, ovoid or subglobular, composed of 6–14 scales with radiating grooves on outer surface; umbo flattened. Seeds 5–7 mm., furnished with a narrow wing-like margin, ripening one year after pollination. Fl. March–May.

Var. **horizontalis** (Mill.) Gordon, Pinet. 68 (1858). *C. horizontalis* Mill., Gard. Dict. ed. 8, no. 2 (1768). Crown with spreading, nearly horizontal branches.

Hab.: On light-coloured calcareous rendzina soils; in natural pine forests. Gilead. In Edom a stand of about 30 aged specimens was found above Busseira village (Chapman, Palest. Journ. Bot. Jerusalem ser. 4 : 55, 1947).

Area: E. Mediterranean (with relics in N. Iran) and Himalaya.

The above variety and also var. *pyramidalis* (Targ.-Tozz.) Nym., Consp. 675 (1881) with more or less upright appressed branches, are widely and successfully planted in this country both as ornamental trees and wind-breaks. An intermediate form, probably a cross between the two varieties, is also met with in cultivation.

Yields timber and tannins; oil derived from leaves was formerly used in perfumery and medicine (remedy for whooping-cough).

Believed by many to be the biblical תְּאַשּׁוּר (Is. xli : 19 and elsewhere).

2. Juniperus L.

Dioecious or monoecious shrubs or trees. Leaves evergreen, decussate or ternately whorled, needle-like and spreading or scale-like and appressed, with a dorsal gland. Reproductive organs on different branches; cones lateral or terminal, with decussate or ternately whorled scales. Staminate cones globular or ovoid, with imbricated, suborbicular scales each with 3–6 pollen sacs. Ovulate cones with several sterile lower bracts and 3–10 upper scales with solitary erect ovules; mature cones fleshy, berry-like, containing 1–10 free, hard, angular seeds. Cotyledons 2.

About 60 species, mostly in the northern hemisphere, some extending to the equatorial mountains of Mexico and to W. Indies, Nyasaland, Himalaya, Formosa.

1. Leaves all needle-like, prickly. Mature cones with 3–6 scales. Dioecious shrubs or trees.
1. J. oxycedrus

– Leaves (except juvenile ones) scale-like. Mature cones with 6–8 scales. Monoecious shrubs or small trees.   **2. J. phoenica**

1. Juniperus oxycedrus L., Sp. Pl. 1038 (1753); Boiss., Fl. 5 : 707 (1884). [Plate 18]

Dioecious shrub or tree up to 10 m. Young branches obscurely triangular. Leaves ternately whorled, spreading, needle-like, linear-lanceolate, prickly, obsoletely 2-sulcate on upper surface, keeled on lower; resin duct below keel. Staminate cones solitary, nearly sessile, subglobular or ovoid. Mature ovulate cones 0.6–1 cm. in diam., globular, red; scales 3–6, with scarcely prominent margin. Seeds 2–3, rarely 1, ripening in the second year. Fl. March–April.

Hab. : Calcareous soil among pines. Upper Galilee. Very rare.

Area : Mediterranean, extending far into territories of adjacent regions.

Upper Galilee is the southernmost station of this species in S.W. Asia.

Yields a dark brown tar used in skin healing; also known for other healing qualities (oleum cadinum).

2. Juniperus phoenica L., Sp. Pl. 1040 (1753); Boiss., Fl. 5 : 710 (1884). [Plate 19]

Monoecious shrubs or trees up to 8 m., with dense, erect or ascending branches, terete when young. Leaves 1–2 mm., ternately whorled or opposite and imbricated, scale-like, appressed, ovate-rhombic, somewhat obtuse, with convex back marked by a linear resin duct. Staminate cones terminal, sessile, erect, ovoid-oblong, with rounded scales. Mature ovulate cones 0.8–1.5 cm. in diam., nearly sessile, fleshy, subglobular or ovoid, brownish-red to tawny, glossy, with 6–8 scales. Seeds 3–8, free, oblong, ripening in summer of the second year. Fl. March–April.

Hab. : Steppe-forests; nubian sands and sandstone. Edom.

Area : Mediterranean, extending into N.W. Arabia.

The occurrence of this species in S. Edom, N.W. Arabia and Sinai and its absence in other parts of Palestine, Lebanon and Syria testifies to its relic nature in the E. Mediterranean.

The biblical עַרְעָר (Jer. xvii : 6) is probably identical with the Arabic name for this tree.

GNETOPSIDA

Dioecious or rarely monoecious, erect, decumbent or climbing shrubs. Xylem without resin ducts and often with vessels. Leaves opposite or whorled, connate below, mostly small or reduced to membranous sheaths or scales, rarely large to very large. Flowers in short, spicate clusters. The staminate ones with at least 2 connate bracts and stamens borne on a common column with sessile or stalked 2–3-celled anthers opening by pores. Ovulate flowers with 1 to several pairs of bracts enclosing 1–3 erect ovaries; macrospore with archegonia or egg cells or egg nuclei. Seeds with 2 cotyledons.

Comprises 3 unigeneric families: Ephedraceae, Gnetaceae and Welwitschiaceae. Only the first is represented in this area.

14. EPHEDRACEAE

Mostly dioecious, much branched, erect or climbing shrubs. Branches usually green. Leaves opposite or whorled, often reduced to membranous sheaths. Reproductive organs axillary, opposite or arranged in whorls of 3 or 4 cones, each cone composed of 2–8 decussate pairs of bracts, the lower pair or pairs sterile, the others fertile. Staminate flowers of 2 opposite scales (perianth) subtending the single column of stamens; anthers 2–8, at end of column, sessile or stalked, each with 2–3 pollen sacs dehiscing by a terminal pore. Ovulate flowers in groups of 1–3, each consisting of 2–4 connate bracts (perianth) and an ovule with 2 pairs of bracts forming the outer and inner integuments, the latter terminating in an elongated style-like micropyle (micropylar tubule). Seed enclosed in the leathery, winged or fleshy integuments; cotyledons 2.

One genus and 42 species mainly in subtropical regions of the northern and southern hemisphere.

Some species contain the alkaloid ephedrine used in medicine; all local species of *Ephedra* are browsed by goats and camels.

Literature: C. A. Meyer, Versuch einer Monographie der Gattung *Ephedra,* durch Abbildungen erläutert. *Mém. Acad. Sci. Pétersb.* 5 : 35–108 (1846). O. Stapf, Die Arten der Gattung *Ephedra. Denkschr. Akad. Wiss. Math.-Nat. Kl. Wien* 56 : 1–112 (1889). H. Riedl, Ephedraceae, in : Rech. f., *Fl. Iranica* 3 : 1–8 (1963).

1. EPHEDRA L.

Description as for family.

1. Bracts of mature ovulate cones 4–5 pairs, yellowish, with membranous, wavy-margined wings. Seeds triquetrous.   **1. E. alata**
– Bracts of mature ovulate cones in 2–3 pairs, fleshy. Seeds berry-like   2
2. Leaves up to 3 cm., mainly on sterile branches. Ovulate and staminate cones at the ends of long twigs. Anthers 2–4, sessile on slightly exserted column. Climbing shrubs with delicate, flexuous, hairy-scabrous branches.   **4. E. peduncularis**

– Leaves 0 or minute, soon deciduous. Ovulate and staminate cones mostly clustered at
   nodes. Anthers 3–7, sessile on a long-exserted column   3
3. Staminate cones with 4–6 flowers of which the terminal 2 are ovulate. Ovulate cones
   with 2 flowers each. Bracts and sheaths not ciliate.   **3. E. campylopoda**
– Staminate cones without terminal ovulate flowers. Anthers 3–5 in each flower. Ovulate
   cones often 1-flowered. Bracts and sheaths finely ciliate.   **2. E. alte**

1. Ephedra alata Decne., Ann. Sci. Nat. Bot. ser. 2, 2 : 239 (1834); Stapf, Eph. 37
(1889); Boiss., Fl. 5 : 717 (1884). [Plate 20]

Erect shrub, not climbing, much branched. Branches rigid, yellowish-green, scabrous.
Leaves very short, united into a short sheath. Staminate cones densely crowded at ends
and nodes of branches; flowers with (3–)4–6 long-stipitate or sessile anthers on a
slightly exserted column. Ovulate cones with 2 fertile flowers and with 4–5 pairs of
loosely imbricated, obtuse or notched bracts, yellowish when mature, with membranous,
wavy, wing-like margin. Seeds ovoid, triquetrous, acuminate. Fl. March–May.

Hab.: Stony deserts. S. Negev. Very rare.

Area : Saharo-Arabian (Morocco-Iran).

In the eastern part of its area (incl. Palestine) this species is represented by var. **decaisnei**
Stapf, Eph. 37 (1889). The latter is distinguished by its 4–6 stipitate anthers, the 8–9 mm.
long mature cone and the 6–6.5 mm. long seeds.

Yields d-pseudo-ephedrine; extract from branches used in eye diseases.

2. Ephedra alte C. A. Mey., Monogr. Eph. 75, t. 3, f. 4 (1846); Boiss., Fl. 5 : 715 (1884).
[Plate 21]

A much branched shrub. Branches elongated, erect or climbing, tortuous, scabrous;
twigs whorled, leafless or with very short triangular leaves when young. Staminate
cones sessile or pedunculate, mostly crowded at nodes; flowers with (3–)4–5 anthers,
sessile on a long-exserted column. Ovulate cones in clusters of (1–2–)3–8 at nodes,
each borne on an ascending or recurved peduncle with 1(–2) flowers; bracts short-
ciliate, in 3 pairs, the lower cup-shaped, the upper shorter than or as long as seed;
micropylar tubule straight. Fl. March–May.

Hab.: Steppes and deserts; also in maquis and on hedges of the coastal sandy soil
belt. Sharon Plain, Philistean Plain, Judean Desert, Negev, Upper and Lower Jordan
Valley, Dead Sea area, Arava Valley, Moav, Edom.

Area : E. Saharo-Arabian (Somalia, Cyrenaica, Egypt, Arabian Peninsula; Syrian
Desert).

A component of the sandy-soil variant of the *Ceratonia siliqua–Pistacia lentiscus* maquis
in the coastal plain of Palestine.

3. Ephedra campylopoda C. A. Mey., Monogr. Eph. 73, t. 2 (1846); Boiss., Fl. 5 : 715
(1884). [Plate 22]

Shrub, much branched, with long, tortuous, scabridulous branches. Twigs whorled,
divaricate or retrorse. Staminate cones clustered at nodes, mostly sessile, ovoid, with
4–6 flowers, the 2 terminal ovulate; anthers 4–7, sessile on a long-exserted column.
Ovulate cones in clusters of 2–6 at nodes, on reflexed or recurved peduncles generally

longer than cones, 2-flowered; bracts in 2–3 pairs, the lowermost short, truncate, the upper long, almost enclosing the seed; micropylar tubule straight. Fl. April–October.

Hab.: Hills, stony ground, mostly in batha and garigue or on bare rocks. Acco Plain, Sharon Plain, Upper and Lower Galilee, Mt. Carmel, Esdraelon Plain, Samaria, Shefela, Judean Mts., Moav, Edom.

Area: E. Mediterranean.

4. Ephedra peduncularis Boiss., Fl. 5 : 717 (1884). [Plate 23]

Climbing or prostrate, scabrous-puberulent shrubs. Branches delicate, spreading, mostly with whorled twigs. Leaves conspicuous, especially on sterile branches, up to 3 cm., opposite or in whorls of 4, narrowly subulate. Staminate cones almost sessile on top of whorled branches, unequal in length, each with 3–6 or more flowers; bracts ciliate; column slightly exserted, with 2–4 sessile anthers. Ovulate cones at the ends of young branches, ovoid, long-peduncled, with 2–3 fertile flowers and 3 pairs of sheathing ciliate bracts, length of lower sheath one third to one fifth that of the upper one, the latter shorter than seed; micropylar tubule erect. Fl. March–April.

Hab.: Hot rocky slopes and wadis. S. Negev, Arava Valley.

Area: Sudanian (Nubo-Sindian), extending into Baluchistan.

E. peduncularis is not synonymous with *E. foliata* Boiss. et Ky. ex Boiss., Diagn. ser. 1, 7 : 101 (1846) as maintained by Riedl in Rech. f., Fl. Iranica, fasc. Ephedraceae 3 (1963), and by V. and T. Täckholm and Drar, Fl. Eg. 1 : 83 (1941).

ANGIOSPERMAE

TREES, shrubs and herbs. Stems usually with vessels. Organs of sexual reproduction are the androecium and gynoecium. The androecium comprises 1 or more stamens; the stamen usually consists of a filament and an anther bearing the microsporangia (pollen sacs). The gynoecium consists of 1 or more carpels forming an ovary in which the ovules are enclosed; the ovary is usually connected with the stigma by means of a style. The androecium and gynoecium or only one of them form the flower which usually comprises also a perianth arising from the same axis. After pollination and fertilization, the ovary turns into a fruit and the ovules into seeds.

The angiosperms are the largest group of plants and the most important both in the earth's vegetation cover and in the economy of man. They comprise about 300 families and over 200,000 species. They are divided into Dicotyledoneae and Monocotyledoneae.

DICOTYLEDONEAE

Woody or herbaceous plants. Stems with open vascular bundles arranged in a single ring (as seen in cross-section); cambium present between the xylem and phloem of the bundle. Leaves mostly with netted venation. Flowers mostly 5-, rarely 4-merous. Seeds usually with 2 cotyledons.

The dicotyledons comprise about 280 families grouped in 48 orders.

Salicales

15. SALICACEAE

Dioecious trees or shrubs. Leaves generally alternate, simple; stipules free, scale-like and soon deciduous or leaf-like and persistent. Flowers in erect or pendulous catkins, each in the axil of a bract. Perianth reduced to a cupuliform disk, or to 1–2 nectar glands. Staminate flowers of 2–30 stamens; filaments free or united; anthers extrorse, longitudinally dehiscent. Pistillate flowers with 1 pistil having 1 or more short styles and 2 stigmas; ovary free, 1-celled, of 2 (rarely 3–4) carpels; placentae 2–4, parietal. Fruit a capsule, longitudinally dehiscing by 2 (rarely 3–4) valves. Seeds small, numerous, with a tuft of long hairs inserted at base.

Three genera, mainly in temperate and cold zones of both hemispheres.

1. Perianth reduced to a cup-like nectariferous disk. Bracts usually incised or laciniate. Stamens 4–30. Buds with several outer scales. **2. Populus**
– Perianth reduced to 1–2 nectar glands. Bracts entire. Stamens 2–5 (12–24). Buds with 1 or 2 outer scales. **1. Salix**

1. SALIX L.

Trees or shrubs, rarely herbaceous perennials. Branches more or less flexible. Buds some-times glutinous, with 1 outer scale woolly inside. Leaves deciduous or persistent, gene-rally alternate, mostly short-petioled, oblong, lanceolate or linear (rarely obovate), entire or serrulate-dentate. Catkins axillary and mostly solitary, erect or pendulous, sessile or on a short peduncle. Staminate catkins caducous. Bracts entire. Perianth reduced to 1 or 2 nectar glands. Stamens usually 2–5; filaments usually free, sometimes con-nat at base. Ovary sessile or stipitate, 1-celled or incompletely 2-celled; style almost 0 or well developed with 2, often 2-fid stigmas. Capsule many-seeded, dehiscing by 2 recurving valves. Seeds numerous, long, with silky hairs arising from the funicle.

About 300 species mainly in the north-temperate regions and also in higher altitudes of the tropics.

One of the most intricate genera of the plant kingdom, owing to the occurrence of fertile interspecific hybrids which obscure the limit of species.

Literature : N. J. Andersson, Monographia Salicum hucusque cognitarum. *Kungl. Svenska Vetensk.-Akad. Handl.* 6, 1 : 1–180 (1867). A. et E. G. Camus, *Classification des Saules d'Europe et Monographie des Saules de France,* 1–2. Paris (1904–1905). C. Schnei-der, Ueber die systematische Gliederung der Gattung *Salix. Oesterr. Bot. Zeitschr.* 65 : 273–278 (1915). R. Görz, *Salicaceae Asiaticae.* I. Brandenburg, 1–23 (1931); II. *Repert. Sp. Nov.* 32 : 387–398 (1933); III. *Repert. Sp. Nov.* 36 : 20–38 (1934).

1. Stipules more or less narrowly lanceolate, very small, soon deciduous. Fairly common plants   2
 – Stipules ovate to subcordate, conspicuous, more or less persistent. Rare plants   4
2. Branches erect or ascending, straight, more or less fragile. Buds often hairy. Leaves ovate-lanceolate to linear-lanceolate, less than 10 times as long as broad. Stamens twice as long as bracts   3
 – Branches slender, flexible, recurved-drooping. Buds glabrous. Leaves narrowly linear-lanceolate, 10 or more times as long as broad, rather densely serrulate. Catkins short. Stamens 2, almost 3 times as long as bracts. Capsule sessile.   **3. S. babylonica**
3. Trees or shrubs with reddish, glabrous branches. Leaves obsoletely serrulate, all glabrous. Catkins 2–3(–3.5) cm. long. Bracts hooded, ovate. Stamens 4–5. Pedicels of mature capsules about 1.5 mm. long.   **1. S. acmophylla**
 – Trees up to 10 m. with yellow-green to brown branches, appressed white-hairy when young, later more or less glabrous. Leaves minutely serrulate or glandular-serrulate, appressed-silky when young, later glabrescent or sparsely pilose. Catkins 3–6.5 cm., flexuous. Bracts lanceolate. Stamens 2. Capsule subsessile.   **2. S. alba**
4(1). Branches brown-red, pilose when young, later glabrous. Buds glabrous. Leaves leathery, glossy above, glaucescent beneath, serrate; stipules rather large, about as long as petioles, mostly longer than 1 cm., persistent. Staminate catkins up to 5 cm. Bracts obovate or oblong, persistent. Stamens 3, almost 3 times as long as bracts.   **5. S. triandra**
 – Branches brown or grey, glabrous. Buds sparsely pilose. Leaves pruinose beneath, not glossy above, denticulate; stipules shorter than 1 cm., deciduous. Catkins up to 4 cm. long. Bracts more or less ovate, partly persistent. Stamens (3–)4–7, twice as long as bracts.   **4. S. pseudo-safsaf**

1. Salix acmophylla Boiss., Diagn. ser. 1, 7 : 98 (1846); Anderss., Monogr. Sal. 7, t. 1, f. 6 (1867); Boiss., Fl. 4 : 1183 (1879). [Plate 24]

Tree or shrub, 3–5 m., with elongated, frequently reddish branches. Buds somewhat pilose. Leaves up to 14 × 0.9–1.8 cm., petiolate, oblong or narrowly ovate to linear-lanceolate, obsoletely serrulate, some (those of lower parts of branches and peduncles) small and mostly obtuse, mucronulate, others long-acuminate, all glabrous, upper surface somewhat shining with light-coloured nerves, lower mostly glaucous-pruinose with 1 prominent nerve; stipules shorter than petiole, soon deciduous, narrowly triangular, somewhat denticulate, acuminate. Catkins 2–3(–3.5) cm., erect on short peduncles bearing 2–4 leaves. Bracts pale in colour, ovate, hooded, those of staminate flowers crisp-hairy, those of pistillate flowers slightly hairy at base, soon deciduous. Staminate flowers with 4–5 stamens twice as long as bracts. Stigmas sessile. Capsule about 4 mm., ovoid-oblong, glabrous, on a pedicel of about 1.5 mm., 2–3 times as long as the nectar gland. Fl. March–May.

Hab.: By water. Coastal Galilee, Acco Plain, Sharon Plain, Philistean Plain, Upper and Lower Galilee, Esdraelon Plain, Samaria, Judean Mts., Judean Desert, Dan Valley, Hula Plain, Upper and Lower Jordan Valley, Dead Sea area, Gilead, Ammon, Moav.

Area : E. Mediterranean and Irano-Turanian.

Honey plant; tanniferous; used for wicker work.

The biblical עֲרָבִים (Is. xliv : 4) refers to the above and other species of *Salix*.

2. Salix alba L., Sp. Pl. 1021 (1753); Anderss., Monogr. Sal. 47 (1867); Boiss., Fl. 4 : 1185 (1879). [Plate 25]

Mostly a tree up to 10 m., with rigid, somewhat fragile, more or less glabrous, yellow-green or chestnut-brown branches, sometimes white- and appressed-tomentose when young. Buds small, glabrous or frequently white-pilose at apex. Leaves up to 11 (–13) × 2.5(–3.5) cm., short-petioled, lanceolate, rarely almost ovate, tapering at both ends, acuminate and sometimes oblique at apex, almost entire and more or less appressed white-silky on both sides when young; the adult ones minutely denticulate or glandular-serrulate, more or less glabrous, sometimes appressed-pilose above and pruinose beneath; stipules generally shorter than petiole, soon deciduous, narrowly lanceolate, denticulate. Staminate catkins 3–6.5 cm., on short, leafy peduncles, densely flowered, flexuous; bracts soon deciduous, small, oblong-lanceolate, green, yellow or brown, hairy. Pistillate catkins less densely flowered and slightly shorter; bracts soon deciduous, ovate-lanceolate, villose. Stamens 2 (very rarely 3), about twice as long as bracts; filaments villose at base. Style almost 0; stigmas divaricate, notched at apex. Pedicel almost 0 when young, later almost as long as or somewhat longer than nectar gland. Capsule ovoid, to short conical, glabrous. Fl. March–June.

Var. **alba.** Branches yellow-green, more or less glabrous. Leaves thin, lanceolate to oblanceolate, rarely oblong-lanceolate (=f. *latifolia*), more or less white-silky when young, later glabrescent, minutely serrulate. Leaves at base of catkins more or less entire. Bracts greenish to yellowish. Capsule sessile or on a pedicel almost as long as nectar gland.

Hab.: By water. Philistean Plain, Esdraelon Plain, Dan Valley, Hula Plain, Upper Jordan Valley, Ammon.

Var. **micans** (Anderss.) Anderss. in DC., Prodr. 16, 2 : 212 (1868); Post, Fl. Syr. Pal. Sin. ed. 2, 2 : 530 (1933). *S. micans* Anderss., Monogr. Sal. 49 (1867). Branches chestnut-brown, shining, densely pilose when young. Leaves leathery, broadly ovate-lanceolate, acute, with glossy-silvery, appressed hairs on both surfaces, coarsely serrulate; leaves at base of catkins denticulate. Bracts of pistillate catkins and ovary brownish. Fruiting pedicel twice as long as nectar gland.

Hab.: As in above. Acco Plain, Sharon Plain, Philistean Plain, Upper Galilee, Samaria, Hula Plain, Ammon.

Area of species : Mediterranean, Euro-Siberian and Irano-Turanian.

Honey plant; wood used for inferior carpentry; bark used for tanning — and also medicinally.

3. Salix babylonica L., Sp. Pl. 1017 (1753); Anderss., Monogr. Sal. 50 (1867); Boiss., Fl. 4 : 1185 (1879).

Tree, 3–6 m. Branches long, simple, flexible, loose, recurved-drooping, yellowish-green or reddish. Buds small, acute, glabrous, appressed to the branches. Leaves up to 15 × about 1 (–1.6) cm., with short (about 1 cm. long) petioles, narrowly linear-lanceolate, often obliquely tapering at apex, mostly rather densely serrulate, pruinose-glaucescent beneath, glabrous when adult; stipules much shorter than petiole, deciduous, lanceolate-subulate, denticulate. Catkins short, less than 4 cm., slender, acute, at first erect, more or less densely flowered, then recurved, overtopped by the few leaves of the short peduncle. Bracts small, soon deciduous, ovate-lanceolate, somewhat obtuse, more or less pilose at base. Staminate flowers with 2 stamens about 3 times as long as bracts. Ovary conical; style almost 0; stigmas divaricate, entire. Capsule short, sessile, ovoid-conical, glabrous. Fl. March–April.

Hab.: Grown as an ornamental tree and escaped from cultivation. Humid sites. Sharon Plain, Philistean Plain, Upper and Lower Galilee, Dan Valley.

Area : Origin unclear; according to Komarov — temperate China or Iran.

Honey plant; young leaves and branches used in popular medicine.

4. Salix pseudo-safsaf A. Camus et Gomb., Bull. Soc. Bot. Fr. 86 : 136 (1939). *S. acmophylla* Boiss. var. *pseudo-safsaf* (A. Camus et Gomb.) Thiéb., Fl. Lib.-Syr. 3 : 162 (1953). [Plate 26]

Tree, 3–10 m. Branches brown or greenish-grey, the young ones nearly glabrous. Buds somewhat pilose. Leaves up to 11 (–15) × about 2 (–3) cm., lanceolate, rarely narrowly ovate-elliptical, tapering to roundish at base, mostly acuminate at apex, sometimes obtuse, mucronate, obsoletely glandular-denticulate, pruinose beneath, slightly hairy when young, then glabrous; petiole 0.3–1.5 cm.; stipules mostly shorter than 1 cm., suborbicular or ovate, more or less denticulate, later deciduous. Staminate catkins 3–4 cm., thin, with persistent, lanceolate-ovate, pilose bracts. Pistillate catkins 1.5–4 cm., on short, leafy peduncles; rhachis white-villose; bracts small, ovate, pilose, partly persistent. Stamens (3–)4–7; filaments twice as long as bracts, hairy at base. Nectar glands one third as long as fruiting pedicel. Capsule about 5 mm., borne on a very short pedicel, ovoid, glabrous, with very short but distinct style and 2-partite, divaricate stigmas. Fl. March–May.

Hab.: River banks. Lower Jordan Valley, Dead Sea area, Moav. Very rare.

Area: Irano-Turanian (Palestine-Syria).

Description based on A. Camus et Gomb., l.c. The specimens from the cited localities in Palestine are incomplete.

5. Salix triandra L., Sp. Pl. 1016 (1753); Anderss., Monogr. Sal. 23 (1867); Boiss., Fl. 4 : 1186 (1879); A. Camus, Class. Saules d'Europe 1 : 90 (1904). [Plate 27]

Shrub or tree, up to 10 m. Branches brown-red, glabrous, pilose when young. Buds ovoid-conical, brown, glabrous. Leaves up to 11 × 3 cm., oblong-lanceolate, roundish or subcordate at base, tapering at apex, glandular-serrate, glabrous when adult, more or less glossy above and glaucescent beneath, with branching secondary veins; petioles mostly shorter than 1.5 cm., with 2–3 small glands at the top; stipules about as long as petioles, rather large, persistent, ovate or semicordate, serrate. Catkins erect on short leafy peduncles, cylindrical, with a more or less pilose, angular rhachis; the staminate ones 3–5 cm., the pistillate shorter and denser. Bracts persistent, obovate or oblong, yellowish-green, more or less pilose at base, glabrous or hairy towards apex. Stamens 3; filaments almost 3 times as long as bracts, hairy at base. Capsule on a pedicel 3–4 times as long as the single nectar gland, conical-ovoid, glabrous, with a very short style and short, somewhat notched, divaricate stigmas. Fl. April.

Hab.: River banks. Acco Plain, W. Negev, Dan Valley. Rather rare.

Area: Mediterranean, Euro-Siberian and Irano-Turanian.

Honey plant; tanniferous; used in basket-work.

Forms recorded as *S. australior* Anderss. and *S. subserrata* Willd. in the Philistean Plain (Yarkon R.) and the Dead Sea area, are probably hybrids between *S. acmophylla* Boiss. or *S. alba* L. and other species. *S. dinsmorei* Enander ex Dinsmore in Post, Fl. Syr. Pal. Sin. ed. 2, 2 : 529 (1933), recorded in the Upper Galilee, is probably one of them. We have found no species in addition to those dealt with here.

The following intermediate forms (hybrids) have been observed:

S. acmophylla × *alba* [Plate 28]: leaves more or less serrulate with slight cover of white hairs when young, later glabrous; stamens 2 or more; capsule mostly short-pedicelled. Common.

S. alba × *babylonica*: leaves linear, 6–8 times as long as broad, young ones more or less appressed-pilose, the adult ones glabrous. Rare.

S. acmophylla × *babylonica*: all leaves glabrous, linear; stamens 2 or more; capsule very short-pedicelled. Rare.

S. triandra × *pseudo-safsaf*: leaves ovate-lanceolate, more or less serrulate, roundish at base, not leathery, nearly glabrous when young, with rather large, persistent stipules; in Jordan Valley and Hula Plain.

2. Populus L.

Tall trees. Buds mostly resiniferous, covered with scales. Leaves alternate, petiolate, broad, oblong to ovate or rhombic; stipules membranous, small, narrow. Catkins loose, appearing before leaves, the staminate ones mostly pendulous. Bracts generally lobed or incised, sometimes crenate. Pedicels distinct. Perianth reduced to a cup-like necta-riferous disk. Stamens 4–30 or more; filaments free; anthers 2-celled, oblong to ovate. Ovary mostly sessile or subsessile, 1-celled, many-ovuled; style short; stigmas 2, entire or lobed. Capsule 2–3 (–4)-valved. Seeds small, ovoid or obovoid, with a tuft of long silky hairs arising from the funicle.

About 30 or more species, mainly in north-temperate regions.

1. Populus euphratica Oliv., Voy. Othoman ed. min. 6 : 319 (1807); Boiss., Fl. 4 : 1194 (1879). [Plate 29]

Tree with spreading to ascending branches. Twigs and erect buds somewhat pube-rulent. Leaves 1.5–10 × 1–8 cm., long-petioled, broadly ovate or triangular, some-times deltoid-rhombic or suborbicular, cuneate or rounded or obsoletely subcordate at base, glabrous, glaucous, irregularly acute-dentate towards apex, those of young and lower branches oblong-lanceolate to linear, almost entire. Bracts caducous, incised-dentate. Pedicels as long as or longer than flowers, spreading. Nectariferous disk cupuliform or patelliform, irregularly incised-lobed. Stamens 12–24; anthers purple, about as long as filaments. Ovary sessile; stigmas 3, 2-partite, purple. Capsule 0.9–1.3 cm., ovoid, glabrescent or villose. Fl. February–April.

Hab.: River banks and springs. C. Negev, Upper and Lower Jordan Valley, Dead Sea area, Moav. Forms riverine forests, especially on the banks of the lower course of the Jordan River; salt tolerant to some extent.

Area: Irano-Turanian and Saharo-Arabian.

Resin of bark used in popular medicine; timber soft and suitable for some kinds of car-pentry; fast growing.

Believed to be identical with the צַפְצָפָה of the Bible (Ez. xvii : 5). In post-biblical literature the Hebrew names for willow and poplar were exchanged, i.e. willow was named צַפְצָפָה and poplar עֲרָבָה. These names have survived to the present day in Arabic where *safsaf* is *Salix* and *gharab* – *Populus*.

Fagales

16. FAGACEAE

Monoecious trees or large shrubs, evergreen or deciduous, with scaly buds. Leaves alternate, simple, with caducous stipules. Flowers unisexual, usually in different inflo-rescences on the same main twig. Staminate flowers arranged in many-flowered catkins or in tassel-like heads, rarely solitary or in 3's, each flower in the axil of a bract, usually with a green, simple perianth of 4–8 divisions and with as many or twice as many stamens; filaments distinct; anthers basifixed, 2-celled; rudiments of gynoecium present or 0. Pistillate flowers solitary or in groups of 2–5 with a common involucre composed of concrescent bracts accrescent at maturity and forming a cupule; perianth usually of (4–)6 tepals arranged in 2 whorls; ovary inferior, usually 3(–6)-celled, with 2 anatropous, pendulous ovules in each cell; ovule with 2 integuments; styles as many as cells. Fruit through abortion of the other ovules a 1-celled, 1-seeded nut, surrounded or enclosed by the scaly or spiny cupule. Seed with no endosperm; cotyledons thick; coat membranous.

Six genera and over 800 species mainly in the northern hemisphere but also occurring in some areas of the southern hemisphere.

1. Quercus L.

Deciduous or evergreen trees or shrubs. Leaves usually dentate, sinuate or pinnately lobed, rarely entire. Flowers green, inconspicuous. Staminate flowers in drooping catkins; bracts minute, soon deciduous; perianth 4–7-lobed; stamens 4–12, usually with no rudiment of an ovary. Pistillate flowers few, sessile on a short or long rhachis, each enclosed by an involucre of imbricated concrescent scales; perianth urceolate, 6-lobed or -dentate, with tube adherent to ovary; ovary usually 3-celled, with 2 ovules in each cell; styles 3, rarely 4–5, erect or recurved; stigmas minute on the flat inner surface of style; rudimentary stamens sometimes present. Fruit an acorn, solitary or in small groups, composed of a large gland and a cupule; gland 1-celled, 1-seeded, generally much longer than cupule; cupule mostly with imbricated, unarmed or somewhat prickly scales. Seeds with thick and large cotyledons. Germination hypogaeous.

Over 500 species, mainly in the northern hemisphere, in temperate regions and in mountains of some tropical areas.

Literature: T. Kotschy, *Die Eichen Europas und des Orients*. Wien–Olmüz (1862). A. S. Oersted, Etudes préliminaires sur les Cupulifères de l'époque actuelle, principalement au point de vue de leurs rapports avec les espèces fossiles. *Vidensk. Selsk. Skrift. Raekke* 5, 9, 6: 331–538 (1871–1872). A. Camus, *Les Chênes: Monographie du genre Quercus*, 1–3. Paris (1934–1954). O. Schwarz, *Cavanillesia* 8: 65–100 (1936). O. Schwarz, Monographie der Eichen Europas und des Mittelmeergebietes & Atlas. *Repert. Sp. Nov. Sonderbeih.* D: 1–200, 64 pls. (1936–1939). M. Zohary, On the oak species of the Middle East. *Bull. Res. Counc. Israel* D, 9: 161–186 (1961).

1. Evergreen shrubs or trees with leathery, 2–4 cm. long leaves; margin often somewhat prickly-dentate.   **3. Q. calliprinos**
– Deciduous trees with herbaceous, 4–8(–10) cm. long leaves; margin dentate or dentate-aristate, but not prickly   2
2. Adult leaves usually glabrous on both sides, rarely slightly tomentellous beneath; teeth of margin obtuse or acute but not aristate. Scales of cupule small, strongly appressed; cupule usually about 1 cm. in diam.   **1. Q. boissieri**
– Adult leaves tomentose-hairy beneath; teeth aristate (with long bristly tip). Scales of cupule long, usually not appressed; cupule usually more than 1 cm. in diam.   **2. Q. ithaburensis**

Subgen. QUERCUS. Often summergreen trees or shrubs. Endocarp glabrous.

1. Quercus boissieri Reut. in Boiss., Diagn. ser. 1, 12 : 119 (1853); O. Schwarz, Repert. Sp. Nov. Sonderbeih. D : 185 (1936–1939). *Q. infectoria* Oliv. ssp. *boissieri* (Reut.) Gürke, Pl. Eur. 2 : 69 (1897); A. Camus, Monogr. Querc. 2 : 187 (1939). [Plate 30]

Deciduous small tree, 4–8 m.; bark grey, scaly; crown ovoid. Branches ascending and horizontal; young twigs striate, puberulent, yellowish-tomentose, soon becoming glabrescent. Buds ovoid, scales ovate, hairy at margin or glabrous. Leaves 4–7 × 2–4 cm., ovate-oblong to oblong-lanceolate, rounded or tapering at base, obtuse at apex, undulate or more or less regularly sinuate-toothed to crenate with ascending, triangular, acute, muticous or mucronulate teeth, glossy on upper surface, pubescent to fleecy on lower when young, later glabrescent or glabrous, with 7–9 lateral nerves; petiole 0.5–1 (–2) cm., glabrous or glabrescent; stipules hairy, deciduous. Staminate catkins 4–6 cm., with hairy axis; flowers loose; perianth of 4–6 ovate, hairy lobes; anthers glabrous or glabrescent. Pistillate catkins 1–2-flowered, on a very short, puberulent axis; styles 3 (–4), recurved. Acorns usually 3–4 × 1 cm., solitary or in pairs, short-pedicelled or sessile, oblong to cylindrical, maturing in the first year; cupule hemispherical or cyathiform, somewhat narrow at throat; scales of cupule appressed, triangular-ovate to short-lanceolate, obtuse or acute, more or less thickened, tomentose or glabrous on back; gland oblong-cylindrical, 3–5 times as long as cupule. Fl. March–April. Fr. October.

Hab. : Forests and maquis; usually preferring higher altitudes or sites with colder winters. Occurs in association with *Q. calliprinos* but sometimes also in pure stands. The following 2 varieties are distinguished locally.

Var. boissieri. Leaves oblong and oblong-lanceolate, glabrous on both sides, margins undulate or acutely dentate.

Upper Galilee, Mt. Carmel, Samaria, Judean Mts., Gilead.

Var. latifolia (Boiss.) Zoh., Bull. Res. Counc. Israel D, 9 : 165 (1961). *Q. boissieri* Reut. ssp. *latifolia* (Boiss.) O. Schwarz, Notizbl. 13 (116) : 17 (1936) et Repert. Sp. Nov. Sonderbeih. D : 187, Atlas t. 44 (1936–1939). *Q. infectoria* Oliv. ssp. *boissieri* (Reut.) Gürke var. *latifolia* (Boiss.) A. Camus, Monogr. Querc. 2 : 189 (1939). Leaves broadly ovate, mostly undulate at margin, obtusely crenate or almost entire, frequently tomentellous on lower surface. Glands mostly large.

Upper and Lower Galilee, Mt. Carmel, Samaria, Judean Mts., Gilead.

Area of species : E. Mediterranean and W. Irano-Turanian.

In contrast to the two other local oak species, *Q. boissieri* is rather uniform in its fruit characters. Outside Palestine it is exceedingly polymorphic as to its leaves, and crosses readily with other species.

Yields gallotannins.

Sugben. CERRIS (Sp.) Oerst. Summergreen, rarely with leaves partly persisting during winter. Endocarp glabrous.

2. Quercus ithaburensis Decne., Ann. Sci. Nat. Bot. ser. 2, 4 : 348 (1835); Ky., Eichen t. 12 (1862); Zoh., Bull. Res. Counc. Israel D, 9 : 168 (1961). *Q. aegilops* L. var. *ithaburensis* (Decne.) Boiss., Fl. 4 : 1172 (1879). [Plate 31]

Deciduous tree, 5–6(–15) m.; trunk up to 0.5–1 m. in diam., generally 4–6 m. high; bark grey. Branches divaricate-ascending, forming a globular or broadly ovoid crown; young twigs grey-tomentose with stellate hairs. Buds crowded, ovoid with appressed scales, hairy mainly at margin. Leaves usually 5–8(–10) × 3–6 cm., late in falling (rarely in warm winters partly persistent), herbaceous and slightly rigid in texture, ovate to lanceolate, rarely oblong, cordate or rounded at base, obtuse or acute at apex, varying in size and shape on the same tree, glabrescent and somewhat shining above, pale and stellately pubescent beneath, rarely pubescent also above, with 6–9(–11) lateral nerves ending in triangular, often aristate or long-mucronate teeth; teeth unequal in size and shape; petiole 1–2 cm., tomentose. Staminate catkins 5–6 cm., with pubescent axis; flowers short-pedicelled; perianth of 4–6 lanceolate lobes, hairy at margin; stamens with hairy, almost ellipsoidal anthers. Pistillate catkins whitish, densely hairy, with 1–3 flowers on short pedicels; styles 3–4, linear, recurved. Acorns solitary or in pairs, borne on a very short and thick peduncle or sessile, maturing in the second year; cupule (0.7–)1.3–2 cm. in diam., usually top-shaped, semiglobular or cyathiform, usually one third as long as 2–5 cm. long gland, rarely half as long as gland or more; scales thick, rather woody, ovate to linear, recurved, spreading, deflexed or erect, often upper scales long, recurved or coiled and the lower short, erect; gland acute, obtuse or barrel-shaped. Fl. February–April. Fr. October–November.

Hab.: Forests and forest remnants on various soils (rendzina, terra rossa, basalt, sandy soils) up to 500 m. above sea level. Coastal Galilee, Acco Plain, Sharon Plain, Upper and Lower Galilee, Samaria, Dan Valley, Hula Plain, Upper Jordan Valley, Gilead.

The leading plant of various Tabor Oak forest communities in Palestine.

Q. ithaburensis varies widely in size and shape of leaves, glands and cupules.

1. Cupule 0.7–1 cm. across; gland 2–3 times as long as cupule.   var. **calliprinoides**
   – Cupule and/or gland much larger   2
2. Upper margin of cupule scaleless.   var. **subcalva**
   – Cupule scaly all over   3
3. Cupule about two thirds as long as gland.   var. **subinclusa**
   – Cupule much shorter   4
4. Scales of cupule, at least most of them, free to base, linear or filiform, 1–2 cm. long, straight, erect or deflexed.   var. **dolicholepis**
   – Scales of cupule usually with broad base, connate below, most of them coiled or recurved.
                                                                                          var. **ithaburensis**

Var. ithaburensis. *Q. ithaburensis* Decne., l.c. *Q. aegilops* L. var. *ithaburensis* (Decne.) Boiss., l.c. *Q. aegilops* ssp. *ithaburensis* (Decne.) Eig, Beih. Bot. Centralbl. 51, 2 : 228 (1933). *Q. pyrami* Ky., Eichen t. 3 (1862). *Q. ungeri* Ky., l.c. t. 13, p.p. [Plate 31]

Fruit large; cupule generally top-shaped with thick, mostly recurved scales. In addition to the form illustrated on Pl. 31, a whole series of others, differing from the above in size and shape of cupule and gland also belong to this variety.

Sharon Plain, Upper and Lower Galilee, Samaria, Dan Valley.

Var. **subcalva** Zoh., l.c. 169. Differs from other varieties in the scaleless upper margin of its cupule.

Sharon Plain, Dan Valley.

Var. **dolicholepis** Zoh., l.c. 169. All or part of cupule scales linear or filiform, distinct, 1–2 cm., erect, patulous or deflexed, generally not recurved.

Sharon Plain, Upper and Lower Galilee, Samaria, Dan Valley.

Var. **subinclusa** Zoh., l.c. 170. Gland enclosed by two thirds or more of its length in the semiglobular or globular cupule.

Dan Valley.

Var. **calliprinoides** Zoh., l.c. 170. Cupule 0.7–1 cm. long and broad; scales minute, lanceolate, appressed, pubescent; gland 2–3 times as long as cupule.

Lower Galilee. Rare.

This variety closely resembles *Q. calliprinos* in its fruits and is perhaps a hybrid between the latter and some form of *Q. ithaburensis*.

The varieties described here are differentiated mainly by their fruits, particularly by the shape of the cupule and its scales. Fruits maintain their homogeneity on each individual tree in contrast to leaves which, to a certain extent, vary from branch to branch according to position, age, etc. From our observation it seems that fruit characters are fairly constant at least in the first generation. No correlation has been found between cupule and leaf characters.

Some of the above varieties may occur together within the same population, but not infrequently one variety – often the most widespread var. *ithaburensis* – dominates in individual populations.

Area of species : E. Mediterranean (Palestine–Turkey).

Cupules have a high content of tannins; timber used in turnery and for agricultural tools.

The biblical names אַלּוֹן and אֵילוֹן evidently refer to this and the following oak species.

Subgen. SCLEROPHYLLODRYS O. Schwarz. Evergreen trees or shrubs. Leaves dentate to entire, often with spiny teeth. Endocarp tomentose.

3. Quercus calliprinos Webb, It. Hisp. 15 (1838); A.DC. in DC., Prodr. 16, 2 : 54 (1864); A. Camus, Monogr. Querc. 1 : 451 (1938); Zohary, Bull. Res. Counc. Israel D, 9 : 161 (1961). *Q. coccifera* L. var. *calliprinos* (Webb) Boiss., Fl. 4 : 1169 (1879). [Plate 32]

Evergreen tree or shrub, 2.5–4 m., not rarely up to 15 m.; crown semiglobular; trunk in very aged trees up to 1 m. or more in diam.; bark grey. Branches divaricate, ascending, very dense; young twigs yellowish, stellate-pubescent; indumentum persistent sometimes until the second year; older twigs brownish-grey. Buds congested at ends of twigs, ovoid, reddish grey-brown, tomentose. Leaves 2–4(–6) × (0.8–)1–1.5(–2) cm., persistent for 2 years or more, leathery, rigid, oblong to elliptical, rounded or subcordate at base, obtuse or somewhat acute at apex, serrate-dentate, often spiny-toothed, rarely nearly entire or wavy-margined, glabrous or sparingly and minutely pubescent beneath; petiole short, tomentose. Staminate catkins numerous, flowers rather dense; perianth with 4–5 hairy, obtuse or rounded lobes; stamens 4–5, opposite the lobes, with acute and mucronate anthers. Pistillate catkins 1–2 cm., solitary or in pairs in leaf axils of current year's branches, erect, tomentose, 1–3-flowered; flowers mostly with glabrous scales; styles 3, linear, elongated; stigmas 5–6, recurved. Acorns short-peduncled, maturing in the second year, very variable in shape and size but uniform throughout on each individual tree; cupule 0.7–1.5 cm. in diam., hemispherical, cyathi-form or almost campanulate, enclosing one to two thirds of the gland, finely velvety within; scales variously shaped, short or long, appressed, erect, recurved or ascending, hairy, mostly prickly, the lowermost ovate, the intermediate oblong-lanceolate; gland 1–3 cm., ovoid, ellipsoidal or oblong, rarely cylindrical, acute or rounded to mucro-nate at apex. Fl. March–April. Fr. (ripe) December.

Hab.: Maquis and forest on terra rossa, rendzina, sandy clay and also on rock fissures, from sea level to 1,200 m. Sharon Plain, Upper and Lower Galilee, Mt. Car-mel, Samaria, Shefela, Judean Mts., Dan Valley, Gilead, Ammon, Moav, Edom.

Q. calliprinos can be distinguished from its W. Mediterranean relative *Q. coccifera* L. by a series of characters (see Zohary, Bull. Res. Counc. Israel D, 9 : 182, 1961). It is in itself very polymorphic. The local populations have been divided into the following varieties:

1. Cupule globular or campanulate, enclosing the entire gland or two thirds of it   2
 – Cupule shorter and not as in above   4
2. Acorns more or less globular, 1.5–1.8 cm. across; gland obtuse, truncate.
                                                                                                    var. **brachybalanos**
 – Acorns and glands not as in above   3
3. Cupule nearly globular, covering almost the whole gland; middle scales linear-lanceolate, slightly recurved, upper filiform, deflexed.   var. **subglobosa**
 – Cupule almost campanulate, covering two thirds of the gland; scales lanceolate, mid-dle ones ascending, the upper appressed.   var. **eigii**
4(1). Leaves with short, forked or stellate hairs beneath. Plants occurring in Edom.
                                                                                                      var. **puberula**
 – Leaves glabrous   5
5. Scales of cupule strongly reflexed and somewhat hooked.   var. **palaestina**
 – Scales erect or spreading   6
6. Acorns rather large, oblong-cylindrical; scales rather long, lanceolate, not tightly ap-pressed to cupule.   var. **calliprinos**
 – Acorns small, ellipsoidal-oblong; scales short, imbricated, ovate-triangular, very tightly appressed to cupule.   var. **fenzlii**

Var. calliprinos. *Q. calliprinos* Webb var. *eucalliprinos* A.DC. in DC., Prodr. 16, 2 : 55 (1864); A. Camus, Monogr. Querc. 1 : 456 (1938). *Q. calliprinos* Webb var. *dispar* Ky., It. no. 442 (1866). *Q. pseudococcifera* Labill., Ic. Pl. Syr. Dec. 5 : 9, t. 6. f. 1 (1812) non Desf., Fl. Atl. 2 : 349 (1799). Acorns rather large, oblong-cylindrical; scales of cupule 5–6 mm., erect to spreading, not tightly appressed, lanceolate; gland thick, about one and a half times as long as cupule.

Upper and Lower Galilee, Mt. Carmel, Samaria, Judean Mts., Gilead.

Var. **fenzlii** (Ky.) A. Camus, Bull. Soc. Bot. Fr. 80 : 355 (1933). *Q. fenzlii* Ky., Eichen t. 24 (1862). Leaves ovate to obovate. Acorns ellipsoidal-oblong; cupule hemispherical; scales short, imbricated, erect, very tightly appressed, ovate to triangular, acute, the uppermost shorter; gland up to twice as long as cupule.
    Judean Mts., Ammon.

Var. **palaestina** (Ky.) Zoh., Bull. Res. Counc. Israel D, 9 : 183 (1961). *Q. coccifera* L. var. *palaestina* (Ky.) Boiss., Fl. 4 : 1170 (1879). *Q. palaestina* Ky., Eichen t. 19 (1862). *Q. calliprinos* Webb var. *arcuata* (Ky.) A. DC. in DC., Prodr. 16, 2 : 56 (1864). Acorns broadly ovoid, mucronate at apex; cupule almost hemispherical; scales of cupule strongly reflexed and somewhat hooked, ovate to lanceolate and linear; gland up to twice as long as cupule. Leaves obscurely dentate, 3–4 cm.
    Lower Galilee, Mt. Carmel, Esdraelon Plain, Judean Mts.

Var. **eigii** A. Camus, Bull. Soc. Bot. Fr. 80 : 355 (1933); Zoh., l.c. Cupule almost campanulate, covering two thirds of the gland; lower and uppermost scales lanceolate, the lower and upper mostly appressed, the middle ones ascending.
    Mt. Carmel, Judean Mts.

Var. **subglobosa** Zoh., l.c. Cupule almost globular, 2.5–3 cm. in diam.; middle scales linear-lanceolate, slightly recurved, upper filiform, deflexed; gland wholly or almost wholly included in the cupule.
    Judean Mts.

Var. **brachybalanos** (Ky.) A.DC. in DC., Prodr. 16, 2 : 55 (1864); Zoh., l.c. 184. *Q. brachybalanos* Ky., It. no. 441 (1866). Acorns more or less globular, 1.5–1.8 cm. in diam.; scales of cupule thick, short, the middle ones ovate-lanceolate, erect, more or less appressed; gland obtuse, truncate, 2 cm. or less in length, longer than cupule by one third.
    Mt. Carmel, Samaria.

Var. **puberula** Zoh., l.c. 184. Leaves broadly ovate, much thicker than in other varieties, with short, forked or stellate hairs beneath.
    Edom.

Nothing is known about the genetic nature of the above varieties, and certain intermediate forms are sometimes met with.
    Area of species : E. Mediterranean.

Q. calliprinos is the most common tree or shrub of the local maquis, where it occurs as a leading and dominant component in various plant communities. It is generally shrubby as a result of heavy browsing, fire and continuous cutting for charcoal production. Lofty and aged trees are found only in cemeteries and other protected sites. This favours the opinion that the oak maquis is at least partly derived from oak forest degraded by man.

Urticales

17. ULMACEAE

Trees or shrubs without latex. Stems sympodial. Leaves alternate, simple, often asymmetrical at base; stipules soon deciduous. Flowers hermaphrodite or unisexual, fasciculate in axillary cymes or solitary. Perianth often sepaloid, (3–)4–6(–9)-lobed. Stamens (3–)4–6(–9), opposite perianth segments. Ovary superior, 2-carpelled, 1–2-celled, each cell containing a single pendulous ovule; styles short, 2-fid or 2-partite with filiform stigmas. Fruit a samara, nut or drupe. Seeds with or without endosperm; embryo straight.

About 15 genera and over 150 species, in north-temperate and tropical regions, mainly of Asia and America.

1. Ulmus L.

Trees. Leaves alternate, unequal at base; stipules scarious, soon deciduous. Flowers generally hermaphrodite or partly also unisexual, yellowish-green, arising from one-year-old twigs, arranged in fascicles and appearing before leaves. Perianth more or less campanulate, marcescent, of (3–)5(–9) segments. Stamens as many as segments; filaments inserted at base of perianth, later exserted. Ovary usually 1-celled, compressed, with short style and 2 pubescent stigmas. Fruit a compressed samara, notched at tip of wing. Seeds with straight embryo and without endosperm.

About 20 species in north-temperate regions and in mountains of Trop. Asia.

1. Ulmus canescens Melv., Kew Bull. 1957 : 499 (1958). [Plate 33]

Deciduous tree of medium height. Young shoots about 2 mm. in diam., white-pubescent, becoming glabrous in the second year. Buds 4–6 mm., with brown, orbicular-ovate scales, ciliate at apex, appressed-puberulent on back. Leaves 6–12 × 3–7 cm., broadly ovate to oblong-elliptical, crenate-dentate or simply or doubly serrate, asymmetrically subcordate at base, generally acuminate at apex, sparsely puberulent or almost glabrous or scabrous above, grey-pubescent beneath; petiole shorter than 1 cm. Flowers many, in catkins, very short-pedicelled. Perianth 2.5–3 mm., longer than the small, ciliate, caducous bracteoles, 5–8-lobed; lobes obtuse, ciliate. Stamens 3–5, longer than perianth. Stigmas ciliate, recurved in fruit. Fl. February–March.

Hab. : Shady creeks by water. Lower Galilee, Mt. Carmel, Samaria.

Area : N. and E. Mediterranean.

Celtis australis L., Sp. Pl. 1043 (1753).

Tree up to 15 m. Leaves 4–10 × 2–5 cm., ovate to ovate-lanceolate, obliquely cordate at base, long-acuminate, serrate, scabrous-pubescent. Flowers staminate and hermaphrodite. Fruit a globular, blackish drupe borne on a long pedicel.

Hab. : Planted, rarely subspontaneous. Upper Galilee, Samaria, Judean Mts., Gilead.

Very aged specimens of this tree are encountered among the forests of *Quercus ithaburensis* in Gilead.

18. MORACEAE

Trees, shrubs or woody climbers, rarely herbs, mostly with milky juice. Leaves deciduous or persistent, mostly alternate, stipulate. Inflorescences cymose but modified, head-like with fleshy axis or globular or discoid or with flowers inserted on the inner surface of a hollow globular pyriform body (fig). Flowers unisexual, small, with a persistent simple perianth or without perianth. Tepals normally 4, rarely 2–6, coalescent. Stamens generally 4, rarely 1–2, generally equalling tepals in number and opposite them. Carpels 2, united into a 1-celled ovary; styles 1–2, filiform. Fruit an achene or drupe, free or adnate to perianth and often crowded together into an aggregate body with or without receptacle.

About 60 genera and 1,550 species, mostly pantropical and only a few extratropical.

1. Ficus L.

Trees or shrubs with unisexual flowers (plants monoecious, rarely dioecious). Leaves deciduous or persistent, alternate, rarely opposite; stipules enclosing buds, caducous. Inflorescence a fig (syconium), i.e., a globular or pyriform hollow fleshy receptacle bearing the flowers on the inner surface; apex of fig with a narrow open mouth furnished with scales. Flowers subtended by bracts; the staminate ones with 3–5 tepals and (2–)3–6 stamens; the pistillate with a 5-fid perianth, a 1-celled ovary and a lateral style with (1–)2 stigmas. Fruiting fig consisting of a fleshy receptacle and minute, crustaceous, 1-seeded druplets. Seeds with membranous coat and incurved embryo.

About 1,000 species mainly in tropical regions, a few extratropical.

1. Figs crowded in grapes on the main stem or older branches. Leaves entire.
   3. F. sycomorus
- Figs solitary or few on the previous year's branches. Leaves, at least part of them, lobed or dentate   2
2. Figs 2–5 cm. in diam. Leaves mostly 3–5-lobed.   **1. F. carica**
- Figs 1–1.5 cm. in diam. Leaves repand or slightly 3-lobed. Hot desert plants.
   2. F. pseudo-sycomorus

1. Ficus carica L., Sp. Pl. 1059 (1753); Boiss., Fl. 4 : 1154 (1879).

Deciduous tree with milky juice, up to 5(–8) m., with rounded or broadly ovoid crown, sometimes shrubby and branching from base; bark grey, more or less smooth. Buds glabrous. Leaves rather variable, 7–15 cm. across, thick and scabrous, usually orbicular or broadly ovate, 3–5-lobed (rarely part of leaves undivided); lobes with dentate or dentate-crenate margin; petiole 2–5 cm., hairy. Figs usually 2–5 cm. in diam., short-peduncled, pyriform or almost globular, variously coloured, glabrous, fleshy, sweet-tasting and edible. For flowers and fruits see description of genus.

F. carica comprises several varieties considered by some as distinct species. While the fig of cultivated varieties contains only pistillate flowers, that of wild forms has both pistillate and staminate ones.

F. carica is widely cultivated in several varieties; it produces 2–3 crops during the year, but the main edible crop matures between June and September. Subspontaneous along river banks and on walls, but probably not indigenous.

The wild variety *F. carica* L. var. *caprificus* has been observed in several localities: Upper Galilee, Mt. Gilboa, Samaria. Although no longer used in this country for caprification, it may be a remnant of earlier cultures. The primary range of this species, cultivated since the Bronze Age and perhaps earlier, is still unknown. Its occurrence in the jungles of the S. Caspian coast and elsewhere in Mediterranean "primary" habitats does not necessarily prove its indigeneity in these sites. On the other hand, there are a series of taxa very close to the cultivated *F. carica* that grow in the Irano-Turanian mountains of Iran (e.g. *F. johannis*, *F. persica*), which should definitely be regarded as wild (ancestral) forms of the common fig tree. Schweinfurth (Bull. Herb. Boiss. 4, App. 2 : 127, 1896) considers *F. palmata* Forssk., Fl. Aeg.-Arab. 179 (1775) as the ancestor of the cultivated fig tree.

The fig tree is mentioned several times in the Bible under the name תְּאֵנָה.

2. Ficus pseudo-sycomorus Decne., Ann. Sci. Nat. Bot. ser. 2, 2 : 44 (1834); Boiss., Fl. 4 : 1155 (1879). *F. palmata* Schweinf. exs. non Forssk., Fl. Aeg.-Arab. 179 (1775). [Plate 34]

Shrub, mostly straggling, with long and thin branches, 1–3 m. Young twigs pruinose, somewhat tortuous. Leaves (3–)5–10 cm., deciduous, petiolate, broadly ovate, cordate or rarely truncate at base, obtuse and mucronate at apex, crenate-dentate or with 3 short, often repand lobes, gree and scabrous on upper surface, pale coloured and tomentellous beneath, later glabrescent. Figs 1–1.5 cm. across, solitary or several on the previous year's branches, subtended by a number of bracts forming a cup-like structure, short-peduncled to sessile, globular or somewhat pear-shaped, pubescent. Fl. April.

Hab.: Among rocks in dry, hot creeks. N. and C. Negev, Arava Valley, Edom. Rare.

Area: E. Sudanian (Egypt, Palestine, Arabian Peninsula).

3. Ficus sycomorus L., Sp. Pl. 1059 (1753); Boiss., Fl. 4 : 1155 (1879).

Tree, 8–20 m.; crown broadly ovoid to flat and umbrella-like. Leaves usually 10–20 cm., persistent (but deciduous in cold winters), leathery, ovate-subcordate, obtuse, undivided, glabrous above and somewhat hairy beneath, especially along nerves. Figs about 2 cm., usually depressed-globular, crowded in grapes on tortuous leafless short twigs arising from the trunks or older branches. Mature figs edible, somewhat watery and insipid as compared with the cultivated fig; several crops produced in the year. Fl. Summer.

Cultivated. Acco Plain (rare), Sharon Plain, Philistean Plain, W. Negev, Lower Jordan Valley.

Native in Ethiopia and elsewhere in Trop. E. Africa.

Once extensively grown both in the coastal plain and in the Jordan Valley. Individuals and groups of these showy and aged trees still occur in several localities and also in coastal towns and on dunes. Does not produce viable seeds but is easily propagated by cuttings.

Famed for its durable timber which was used for sarcophagi in ancient Egypt and elsewhere. This is the שִׁקְמָה of the Bible (e.g. 1 Ki. x : 27).

19. URTICACEAE

Annual or perennial herbs or shrubs, usually pubescent, frequently with stinging hairs. Leaves alternate or opposite, simple, usually stipulate. Flowers unisexual, rarely herma- phrodite (plants monoecious, dioecious or polygamous), small, crowded in cymes. Perianth sepaloid, free or united. Staminate flowers with (1–)4–5 stamens opposite the perianth segments, mostly bent inwards in bud and exploding at flowering; anthers 2-celled, opening lengthwise; rudimentary ovary often present. Pistillate flowers often with small staminodes, with a 1-celled, 1-ovuled ovary; style simple or 0. Fruit an achene. Seeds with oily endosperm; embryo usually straight.

About 40 genera and 700 species, mainly in tropical and partly in temperate zones.

1. Leaves entire. Pubescent-pilose plants.   **2. Parietaria**
– Leaves dentate. Plants with stiff hairs   2
2. Leaves opposite, with stinging hairs.   **1. Urtica**
– Leaves alternate, hispid and woolly, white at lower surface; hairs not stinging.
                                                                                    3. Forsskaolea

Trib. URTICEAE. Plants with stinging hairs. Leaves opposite or alternate. Pistillate perianth 4-merous.

1. URTICA L.

Annual or perennial herbs, monoecious or dioecious, with stinging hairs. Stems ridged or 4-angled. Leaves opposite, dentate to incised-lobed; stipules free or connate. Inflo- rescences uni- or bisexual, axillary, spike-like (catkins) or glomerate with clustered cymes. Flowers unisexual, ebracteate, with 4-merous green perianth. Staminate flowers with 4 stamens, opposite the equal tepals and with rudimentary pistil. Pistillate flowers with outer tepals much smaller than inner ones; pistil 1, with sessile stigma. Achenes somewhat compressed, oblong, enclosed in the persistent, hairy perianth.

About 40 species mainly in temperate regions; a few in mountains of the tropics.

The biblical name סִרְפָּד (e.g. Is. LV : 13) may refer to one or more local species of this genus.

1. Stipules 2 on each node (each consisting of 1 connate pair), green   2
– Stipules 4 on each node, membranous   3
2. Annuals, 20–50 cm. Tooth at leaf apex equal to other teeth; stipules small. Rhachis of staminate catkin dilated into narrow, membranous wings.   **3. U. dubia**
– Perennial herbs, 70–100 cm. Leaves up to 12 cm., with apical, lanceolate tooth (1–3 cm.), twice the length of other teeth or longer; stipules up to 1.5 cm. Rhachis of sta- minate catkins wingless.   **1. U. hulensis**
3(1). Pistillate racemes globular, long-peduncled. Leaves ovate, up to 7(–9) cm. broad, coarsely dentate.   **4. U. pilulifera**
– Bisexual racemes spike-like. Leaves generally narrower than in above, lanceolate to ovate, dentate-serrate.   **2. U. urens**

1. Urtica hulensis Feinbr., Palest. Journ. Bot. Jerusalem ser. 4 : 114 (1947). [Plate 35]

Perennial, monoecious or dioecious herb, 70–100 cm., sparingly covered with long stinging hairs. Leaves up to 12 × 8 cm., ovate, coarsely dentate, subcordate at base, acuminate at apex with 1–3 cm. long, lanceolate apical tooth; petiole almost equal to or shorter than blade; stipules 2 on each node, up to 1.5 cm., sessile, broadly ovate to orbicular, subcordate at base, 2-lobed at apex, green. Staminate racemes shorter than petioles, many-flowered, dense; rhachis wingless; flowers minute, with hispid perianth. Pistillate racemes very short, with flexuous rhachis bristly at nodes; flowers 0.75 mm. Fl. June–September.

Hab.: Swamps, among reeds. Hula Plain. Rare.

Area: E. Mediterranean (endemic in Israel).

2. Urtica urens L., Sp. Pl. 984 (1753); Boiss., Fl. 4 : 1146 (1879). [Plate 36]

Annual, monoecious, more or less setose, 20–60 cm. Stems erect, branching below. Leaves up to 8 × 4 cm., broadly lanceolate or ovate, subcordate or tapering at base, acutely dentate-serrate; petiole shorter than to almost as long as blade; stipules small, 4 at each node. Inflorescences axillary, spike-like, up to 1.5–2 cm., shorter than petioles, with few staminate and many pistillate flowers. Achene about 1.5–2 mm. Fl. December–June.

Hab.: Waste places. Coastal Galilee, Acco Plain, Sharon Plain, Philistean Plain, Upper and Lower Galilee, Esdraelon Plain, Mt. Carmel, Samaria, Judean Mts., Hula Plain, Upper and Lower Jordan Valley, Transjordan.

Area: Euro-Siberian and Mediterranean, extending into Irano-Turanian territories.

3. Urtica dubia Forssk., Fl. Aeg.-Arab. CXXI (1775). *U. membranacea* Poir. in Lam., Encycl. 4 : 638 (1798); Boiss., Fl. 4 : 1147 (1879). *U. caudata* Vahl, Symb. Bot. 2 : 96 (1791) non Burm. f., Fl. Ind. 198 (1768). [Plate 37]

Annual, monoecious or sometimes dioecious, sparsely hairy, 20–50 cm. Stems erect or ascending, simple or branching. Leaves petiolate, ovate, subcordate or rounded at base, incised-serrate, with stinging or also with simple hairs; stipules small, much shorter than petiole, 2 at each node. Inflorescences each of 2 axillary spikes; the staminate longer than petioles, with rhachis dilated into membranous, narrow wings; pistillate catkins almost as long as petioles, with no dilated rhachis, spreading or almost pendulous. Staminate flowers many, 1 mm., inserted on the upper surface of rhachis; pistillate ones about 1.5 mm., subsessile. Achenes 1–2 mm., shining, greenish-yellow. Fl. January–May.

Hab.: Waste shady places and in orchards. Sharon Plain, Philistean Plain.

Area: Mediterranean, extending into adjacent territories of the Euro-Siberian region.

4. Urtica pilulifera L., Sp. Pl. 983 (1753); Boiss., Fl. 4 : 1147 (1879). [Plate 38]

Annual, monoecious, setose, occasionally almost glabrous, 25–70 cm. Stems erect, branching. Leaves up to 9 (–12) × 7 (–9) cm., almost ovate, coarsely dentate, subcordate at base; petiole about as long as blade or longer; stipules short, not exceeding 7 mm., 4 on each node, membranous. Staminate inflorescences in the lower axils or together with pistillate ones in upper axils, long-peduncled, branching, spike-like; the pistillate ones (1–)2 on each peduncle, up to 1.3 cm. in diam., globular, later deflexed. Perianth of pistillate flowers densely setose. Fl. January–May (–July).

Hab. : Waste places, hedges and on damp roadsides or near dwellings. Sharon Plain, Philistean Plain, Upper and Lower Galilee, Mt. Carmel, Esdraelon Plain, Samaria, Shefela, Judean Mts., Judean Desert, Hula Plain, Upper and Lower Jordan Valley, Gilead, Ammon, Edom.

Area : Mediterranean, Irano-Turanian and partly Euro-Siberian.

The data by Hart (Fauna and Fl. Sin. Petr. Wad. Arab. 110, 1891) on *U. dioica* L., Sp. Pl. 984 (1753) from Jericho are doubtful.

Trib. PARIETARIEAE. Hairs not stinging. Leaves alternate, exstipulate. Pistillate perianth usually tubular. Stamens 2–5.

2. PARIETARIA L.

Annual or perennial herbs, without stinging hairs. Leaves alternate, petiolate, simple and entire; stipules mostly 0. Inflorescences small, axillary, cymose, clustered. Flowers hermaphrodite and unisexual in the same cluster (plants polygamous), bracteolate, green. Perianth of hermaphrodite and staminate flowers of 4 segments, elongating after flowering, that of pistillate flowers tubular, 4-toothed, not elongating. Stamens 4; filaments free, inserted at base of tepals, inflexed in bud, straightening elastically at flowering. Ovary 1-celled, ovoid or oblong; stigma sessile or borne on a very short style, tufted. Achenes more or less ovoid, enclosed in the persistent perianth.

About 10 species, mainly in temperate regions.

1. Leaves about 1 cm. or less, rounded at base. Bracts of pistillate flowers ovate to orbicular or cordate, membranous, 2–3 times as long as fruiting perianth.   **3. P. alsinifolia**
– Leaves generally 1.5–5 cm. or more, tapering at base. All bracts oblong-ovate or lanceolate, green, those of pistillate flowers shorter to somewhat longer than fruiting perianth   **2**
2. Annuals. Leaves generally 1.5–2.5 cm. Perianth about 1.5 mm., hardly growing in fruit. Achenes brownish.   **2. P. lusitanica**
– Perennials with stems indurated at base. Leaves 3–7 cm. Perianth growing in fruit up to 4 mm. Achenes blackish.   **1. P. diffusa**

1. Parietaria diffusa Mert. et Koch in Röhling, Deutschl. Fl. ed. 3, 1 : 827 (1823). *P. judaica* sensu Boiss. non L., Amoen. Acad. 4 : 464 (1759) nec Sp. Pl. ed. 2, 1492 (1763); Boiss., Fl. 4 : 1149 (1879). [Plate 39]

Perennial, pubescent herb, 10–50(–60) cm. Stems indurated below, ascending to erect, simple or branched. Leaves usually 3–7 × 1–4.5 cm., ovate to oblong-lanceolate, tapering at apex, cuneate-tapering at base, minutely papillose above. Cymes many, axillary, subsessile. Bracts crowded, lanceolate, somewhat connate at base. Central flower of cyme pistillate, the lateral hermaphrodite. Perianth elongating in fruit, exceeding stamens and reaching 4 mm. in length. Achenes blackish, shining. Fl. (December–) January–July (–August).

Hab. : In shady rocks, wall crevices, cave entrances, etc. Acco Plain, Sharon Plain,

Upper and Lower Galilee, Mt. Carmel, Esdraelon Plain, Samaria, Judean Mts., Judean Desert, Hula Plain, Gilead, Ammon, Edom. Very common.

Area : Mediterranean and Irano-Turanian, extending into the Sudanian and Euro-Siberian regions.

Used in popular medicine for healing wounds.

2. Parietaria lusitanica L., Sp. Pl. 1052 (1753); Boiss., Fl. 4 : 1150 (1879). [Plate 40]

Annual, puberulent, 10–30 cm. Stems thin, flaccid, prostrate to ascending. Leaves (1–)1.5–2.5(–4) cm., generally less than 1 cm. broad, ovate or ovate-lanceolate, tapering at base and apex, minutely papillose and ciliate above; petiole shorter than blade, filiform, pubescent. Cymes many, subsessile at axils, dense. Bracts small, about as long as flowers, lanceolate, green, scarcely growing in fruit. Hermaphrodite flowers few, sterile, the rest pistillate, fertile. Perianth about 1.5 mm., with 4 connivent segments enclosing the brownish, shining achene. Fl. February–May.

Hab. : Shady rocks and grassy places. Sharon Plain, Upper and Lower Galilee, Mt. Carmel, Esdraelon Plain, Mt. Gilboa, Judean Mts., Judean Desert, Hula Plain, Upper and Lower Jordan Valley, Gilead, Ammon, Edom.

Area : Mediterranean, extending into contiguous areas of the Irano-Turanian region.

A form with larger leaves (3–4 × 1–1.5 cm.) is occasionally met with; its taxonomical status has not yet been elucidated.

3. Parietaria alsinifolia Del., Fl. Eg. 137 (1813); Boiss., Fl. 4 : 1151 (1879). [Plate 41]

Annual, pilose-hairy, 5–30 cm. Stems more or less flaccid, procumbent or ascending. Leaves usually 0.5–1 × 0.3–1 cm., ovate, rounded at base, hairy, sometimes minutely papillose above; petiole about as long as blade, slender, pubescent. Cymes many, axillary, (2–)3-flowered, short; lowest flower hermaphrodite but sterile, with a small lanceolate bract, somewhat shorter than perianth; other flowers pistillate, with 3, about 3 mm. long, membranous, ovate to orbicular or cordate bracts. Perianth 1–2 mm., cleft halfway or two thirds down its length, scarcely growing in fruit, enclosing the brown, shining achene. Fl. March–June.

Hab. : Rocks, wadis and shady places in stony and mostly hot deserts. Judean Desert, W., N. and C. Negev, Upper and Lower Jordan Valley, Dead Sea area, Arava Valley, Gilead, Ammon, Moav, Edom.

Area : Mainly Saharo-Arabian, extending into contiguous regions.

Trib. FORSSKAOLEAE. Leaves alternate, hairs not stinging. Stamen 1 in each staminate flower.

3. FORSSKAOLEA L.

Perennial herbs or shrubs with rigid but not stinging hairs. Leaves alternate, simple; stipules lateral, free. Inflorescences axillary, consisting of outer staminate flowers and inner pistillate ones, all enclosed in an involucre of 3–6 bracts, fleecy on upper (inner) and often villose on lower surface (rarely pistillate flowers alone in a 2-bracted involucre). Staminate flowers numerous, with perianth tubular below and dilated above, entire or obtusely 3-dentate or 3-lobed; stamen 1 in each flower. Pistillate flowers few, naked. Ovary 1-ovuled, ellipsoidal, straight, fleecy, with linear-filiform, hairy stigma. Achenes more or less compressed, fleecy. Seeds with scanty endosperm and rounded, somewhat notched, cotyledons.

Five or 6 species in Trop. Asia and Africa.

1. Forsskaolea tenacissima L., Oppobalsam. Decl. 18 (1764) et Mant. 11 (1767); Boiss., Fl. 4 : 1151 (1879). [Plate 42]

Perennial, appressed-woolly, hispid herb or shrublet, 20–80 cm. Stems more or less erect, simple or branching. Leaves 1–4 × 0.4–3 cm., petiolate, obovate-orbicular to ovate-rhombic, cuneate at base, coarsely dentate-serrate, upper surface green, hispid, often tomentose, lower surface densely white-woolly; stipules membranous, hispid at margin. Inflorescences subsessile, axillary, 1–4 in a cluster. Involucral bracts mostly 3–6 (rarely 2), up to 8 mm., lanceolate, silky-hispid. Perianth of staminate flowers with green, rather hispid, tomentose lobes. Ovary 1 or more in each inflorescence. Fl. January–August.

Hab. : Wadis in hot deserts. Judean Desert, C. and S. Negev, Lower Jordan Valley, Dead Sea area, Arava Valley, Moav, Edom.

Area : E. Sudanian (Nubo-Sindian) and Saharo-Arabian.

Bark used as source of fibre.

Santalales

20. SANTALACEAE

Hemiparasitic herbs, shrubs or trees. Leaves alternate or rarely opposite, entire, exstipulate. Flowers hermaphrodite or unisexual. Perianth simple, sepaloid or petaloid, of 3–6 segments valvate in bud. Stamens 3–6, opposite tepals; filaments inserted at base of perianth. Ovary inferior, 1-celled, containing 1–4 pendulous ovules; style 1; stigma capitate or lobed. Fruit 1-seeded, indehiscent, nut-like or drupe-like. Seed solitary, with copious endosperm and small embryo.

About 35 genera and 400 species, mainly tropical.

Literature : R. Pilger, Santalaceae, in : Engl. u. Prantl, *Nat. Pflznfam.* ed. 2, 16b : 52–91 (1935).

1. Flowers always hermaphrodite. Fruit a nutlet. Annual or perennial low herbs.
                                                                    2. Thesium
– Flowers usually unisexual. Fruit a drupe. Perennial evergreen shrublets.   **1. Osyris**

Trib. OSYRIDEAE. Perianth either not tubular below or tubular but then provided with an epigynous disk.

1. Osyris L.

Evergreen, root-parasitic shrubs or trees. Branches angular. Leaves mostly alternate. Inflorescences axillary. Bracts and bracteoles caducous. Flowers unisexual (plants dioecious or polygamous) or rarely hermaphrodite; staminate flowers crowded on peduncles; pistillate and hermaphrodite ones solitary or few. Perianth 3–4(–5)-partite with ovate caducous segments; tube very short in staminate flowers, long-obconical in the pistillate ones. Receptacle disk-like, convex, 3–4-angled. Stamens 3–4, with short thick filaments inserted at base of tepals; anthers 2-celled. Ovary inferior, 1-celled, 3(–4)-ovuled, fleshy, with a short style; stigma 3–4-lobed. Drupe 1-seeded, ovoid-globular, bearing remnants of perianth and disk on top. Seed filling the drupe, mostly with endosperm; embryo short, straight or somewhat arcuate.

About 7 species, mainly tropical.

1. Osyris alba L., Sp. Pl. 1022 (1753); Boiss., Fl. 4 : 1058 (1879). [Plate 43]

Dwarf-shrub, dioecious, evergreen, rather glabrous, 40–80 cm. Stems green, much branching, angular, with creeping rhizomes. Leaves 0.5–2.5 × 0.2–0.4 cm., somewhat fleshy, sessile or nearly so, elliptical, oblong to linear or oblanceolate, mucronulate, mostly early deciduous. Staminate cymes short-peduncled with many pedicellate flowers; pistillate (and hermaphrodite) ones with 1 to few nearly sessile flowers. Perianth sepaloid, 3(–4)-merous with ovate segments connate below, somewhat yellowish (with rudimentary petals). Drupe 6–8 mm. in diam., more or less globular, red, borne on erect peduncle. Fl. March–August.

Hab.: Batha and garigue. Coastal Galilee, Sharon Plain, Philistean Plain, Upper and Lower Galilee, Mt. Carmel, Samaria, Judean Mts., Judean Desert, Hula Plain, Upper Jordan Valley, Gilead, Ammon. Parasitic on various batha and garigue shrubs.

Area: Mediterranean, slightly extending towards W. Europe.

Berries edible.

Trib. THESIEAE. Perianth long-tubular; epigynous disk 0.

2. Thesium L.

Annual or perennial green root-parasites. Leaves alternate, narrowly linear, sometimes scale-like. Flowers hermaphrodite, in small cymes or solitary, with 1 bract and 2 bracteoles. Perianth 4–5-merous, funnel-shaped or campanulate. Stamens 4–5; filaments inserted at base of tepals. Ovary inferior, containing 2–4 ovules; stigma more or less capitate. Fruit a dry nutlet, globular or ellipsoidal, reticulate, with persistent perianth at summit. Seed 1; embryo at centre or apex of endosperm.

About 220 species in tropical and temperate regions.

1. Annuals. Peduncles almost 0. Leaves scabrous. Nutlets up to 6 mm., abundant on each branch.   **2. T. humile**
– Perennials. Peduncles 1–4 mm., spreading. Leaves almost smooth. Nutlets up to 4 mm., sparse on each branch.   **1. T. bergeri**

1. Thesium bergeri Zucc., Abh. Bayer. Akad. Wiss. 2 : 324 (1837); Boiss., Fl. 4 : 1065 (1879). [Plate 44]

Perennial, glabrous, branching herb, 8–20 cm. Roots rather thick. Stems angular, decumbent to ascending. Leaves up to 3.5 × 0.3 cm., not fleshy, narrowly linear, acute, 1-nerved, almost smooth or slightly scabrous at margin. Peduncles 1–4 mm., ascending, bearing 1–3 flowers. Bracts leaf-like, about 4 times as long as flowers; bracteoles about as long as fruit, much longer than flower. Perianth greenish, minute. Nutlets 2–3(–4) mm., short-stipitate, ovoid-ellipsoidal, reticulate between nerves, umbonate. Fl. March–June.

Hab. : Batha and maquis. Upper Galilee, Mt. Carmel, Judean Mts. Parasitic on batha plants.

Area : E. Mediterranean.

2. Thesium humile Vahl, Symb. Bot. 3 : 43 (1794); Boiss., Fl. 4 : 1064 (1879). [Plate 45]

Annual herb, 5–25 cm. Stems few or many, ascending to erect, rarely decumbent, striate-angular. Leaves 1.5–4(–5) × 0.075–0.15 cm., fairly fleshy, narrowly linear, 1-nerved, scabrous, especially at margin. Flowering branches with many flowers. Peduncles almost 0, mostly 1-flowered. Bracts long, leaf-like, about 2–3 times as long as flowers; bracteoles much shorter than bracts but longer than flowers. Lobes of perianth short, light green. Nutlets about 6 mm., subsessile, ovoid-globular to ellipsoidal, reticulate between nerves, umbonate. Fl. February–May.

Hab. : Fields and batha. Coastal Galilee, Acco Plain, Sharon Plain, Philistean Plain, Upper and Lower Galilee, Mt. Carmel, Esdraelon Plain, Samaria, Shefela, Judean Mts., W. and N. Negev, Upper Jordan Valley, Transjordan. Parasitic on various plants. Formerly a noxious weed in grain fields of the Negev.

Area : Mainly Mediterranean.

21. LORANTHACEAE

Green, shrubby plants with adventitious haustorial roots, hemiparasitic on trees and shrubs. Leaves mostly opposite, often thick, entire, sometimes scale-like or absent; stipules 0. Bracts and bracteoles small, scale-like, sometimes rudimentary. Flowers hermaphrodite or unisexual, actinomorphic, single or in cymes, spikes or fascicles. Perianth simple or double, petaloid or sepaloid. Calyx or simple perianth 3–4-lobed. Petals (where present) 3–8, free or united, valvate. Stamens as many as perianth lobes, opposite and adnate to them, with 1 to many pollen sacs. Ovary inferior, adnate to tube of perianth, 1-celled, with 1 ovule and with integument not differentiated from placenta; stigma simple, sessile or borne on a filiform style. Fruit usually drupe- or berry-like. Seed solitary, without testa and with copious endosperm, adherent to the pericarp; embryo 1 or many.

About 40 genera and about 1,400 species, mainly in tropical regions.

1. Flowers unisexual (plants dioecious). Perianth small, green.   **2. Viscum**
– Flowers generally hermaphrodite. Perianth large, red.   **1. Loranthus**

1. Loranthus L.

Perennial, green hemiparasites. Leaves mostly opposite or nearly so, generally fleshy. Inflorescences racemose or spicate, mostly axillary, many- or few-flowered. Bracts and bracteoles present, small. Flowers hermaphrodite, rarely unisexual. Calyx tube ovoid or top-shaped, with short, truncate or dentate limb. Petals 4–6(–8), free or connate at base, valvate. Filaments adnate to base of petals; anthers ovoid, oblong or linear, 2-celled. Ovary inferior; style filiform, with capitate stigma. Fruit a viscid berry or drupe, 1-celled, 1-seeded.

About 500 species, parasitic or hemiparasitic mainly in tropical and subtropical regions of the Old World.

1. Loranthus acaciae Zucc., Abh. Akad. Münch. 3 : 249 t. 2, f. 3 (1840); Boiss., Fl. 4 : 1070 (1879). [Plate 46]

Perennial, glabrous, green hemiparasite. Branches terete, woody. Leaves up to 7 × 2.5 cm., nearly opposite, short-petioled, fleshy-coriaceous, obovate to oblong, tapering at base, obtuse, entire. Bracteoles small, truncate. Pedicels 4–7 mm. Flowers up to 4 cm., crimson-red, 2–7 in axillary fascicles, the upper ones solitary. Calyx short, about 2–3 mm., truncate. Petals 4–8, free or connate in lower part, linear, gibbous at the sinuses, later revolute. Stamens as long as corolla and style; anthers linear. Stigma simple. Fruit 0.5–1.5 cm., baccate, ovoid-oblong. Fl. December–August.

Hab.: Judean Desert, N. Negev, Lower Jordan Valley, Dead Sea area, Arava Valley, Edom. Parasitic on various trees and shrubs, mostly on *Acacia* and *Ziziphus*. Also observed on *Punica, Tamarix, Atriplex* and others.

Area : E. Sudanian.

2. Viscum L.

Perennial, green hemiparasites. Branches opposite or dichasial. Leaves opposite, fleshy, sometimes reduced. Inflorescences cymose, usually axillary. Flowers (in ours) unisexual (plants dioecious), crowded, usually (1–)2–5 in axillary fascicles. Bracts and bracteoles minute. Staminate flowers : sepals 4; petals 0; stamens 4, anthers sessile, many-celled, many-pored, concrescent with perianth lobes. Pistillate flowers : sepals reduced to 4 scale-like teeth; petals minute; ovary inferior, stigma thick, sessile or nearly so. Fruit baccate, with fleshy, viscid mesocarp. Embryos 1–2, straight, with copious endosperm.

About 70 species, mostly in tropical and subtropical regions of the Old World.

Literature : K. F. von Tubeuf, *Monographie der Mistel.* München (1923).

1. Viscum cruciatum Sieb. ex Boiss., Voy. Bot. Midi Esp. 2 : 274 (1839); Boiss., Fl. 4 : 1068 (1879). [Plate 47]

Perennial, dioecious plant. Branches opposite, woody, brittle, terete. Leaves up to 5 × 0.5–1.5 cm., short-petioled, fleshy-leathery, oblong, tapering at base, obtuse, obscurely 3-nerved. Flowers usually 2–4 at each node, short-pedicelled or subsessile, green. Staminate flowers with 4 sepals up to 9 mm. long and without corolla; stamens

4, with linear-lanceolate, many-celled anthers, concrescent with calyx lobes. Pistillate flowers 1–2 mm.; calyx obsoletely 4-dentate; corolla of 4 scaly petals. Berry 4–6 mm. in diam., globular, umbonate, red, viscid. Fl. February–May (–September).

Hab.: Philistean Plain, Upper and Lower Galilee, Samaria, Judean Mts., Gilead, Ammon, Moav, Edom. Parasitic mainly on *Olea*, but also on *Rhamnus, Crataegus, Rubus, Amygdalus, Robinia, Citrus*, etc.

Area: Mediterranean. Very disjunct: Palestine, Algeria, Morocco, S. Portugal, Spain.

Heavily damaging olive trees.

Aristolochiales

22. ARISTOLOCHIACEAE

Rhizomatous herbs or shrubs, with much branched, mostly climbing or twining stems. Leaves alternate, petiolate, simple, mostly cordate at base, entire, exstipulate. Flowers hermaphrodite, axillary, solitary or racemose, actinomorphic or zygomorphic, usually 3-merous. Perianth simple, petaloid, forming a tube with 3 lobes or with a unilateral limb. Stamens 6–36, in 1–2 whorls; the anthers and filaments free or adnate to the stylar column (gynostemium). Ovary usually inferior, 4–6-carpelled, with numerous anatropous ovules; placentation axile; styles 3 to many, free or united into a column with a 3- to many-lobed stigma. Capsule dehiscent or rarely indehiscent. Seeds many; endosperm copious; embryo minute.

Seven genera and about 380 species, mainly in tropical and warm-temperate zones of both hemispheres.

1. ARISTOLOCHIA L.

Perennial, rhizomatous herbs or shrubs with frequently twining or climbing stems. Leaves alternate, usually petiolate, cordate or hastate at base, entire. Flowers axillary, pedicellate, variously coloured, sometimes variegated. Perianth zygomorphic, lower part of tube (utricle) swollen around the style, upper part more or less cylindrical, narrowing, straight or variously curved, ending with entire, lobed or appendiculate limb. Stamens mostly 6, with more or less sessile anthers adnate to the stylar column. Ovary inferior, (4–)6-celled, many-ovuled; stylar column composed of (4–)6 or more styles and as many stigmas. Capsule septicidal, usually dehiscent into 6 valves. Seeds many in 1 row, compressed.

About 300 species, mainly in tropical regions.

Literature: P. H. Davis and M. S. Khan, *Aristolochia* in the Near East. *Not. Roy. Bot. Gard. Edinb.* 23 : 515–546 (1961).

1. Perianth limb deeply cordate and auriculate, broader than 2.5 cm.   2
   – Perianth limb lingulate, neither cordate, nor auriculate, up to 2 cm. broad   3
2. Perianth tube rusty-red, rather glabrous, limb hirtellous on upper surface. Plants more or less puberulent. Leaves with oblong auricles and oblong-ovate to lanceolate-triangular blades. Pedicels about as long as flowers.   **4. A. maurorum**
   – Perianth tube greenish-yellow, hirtellous outside, limb glabrous on upper surface. Plants

densely hirtellous. Leaves with ovate auricles and triangular to ovate blade. Pedicels
shorter than flowers.   **5. A. paecilantha**
3 (1). Limb longer than the hardly incurved tube, linear-lanceolate. Leaves 1–3.5 cm., very
    short-petioled, ovate-oblong or roundish.   **2. A. parvifolia**
– Limb as long as or shorter than the recurved tube. Leaves 2–9 cm. long, conspicuously
    petiolate, ovate   4
4. Evergreen plants. Leaves acute, glabrous, glossy. Pedicels longer than flowers. Perianth
    brown-purple. Capsule 2–3 cm.   **1. A. sempervirens**
– Deciduous plants. Leaves obtuse, papillose-hispid on both sides or glabrous above and
    puberulent beneath. Pedicels shorter than flowers. Perianth violet. Capsule about
    1–2 cm.   **3. A. billardieri**

1. Aristolochia sempervirens L., Sp. Pl. 961 (1753); Davis and Khan, Not. Roy. Bot.
Gard. Edinb. 23 : 520 (1961). *A. altissima* Desf., Fl. Atl. 2 : 324, t. 249 (1799); Boiss.,
Fl. 4 : 1075 (1879). [Plate 48]

Evergreen, perennial climber. Stems elongated, angular, often thickened. Leaves
3–9 × 1.5–6 cm., persistent, petiolate, somewhat leathery, ovate, cordate at base with
rounded auricles and more or less acute apex, almost entire, glabrous, glossy, palmately-
reticulately veined. Pedicels distinctly longer than petioles, somewhat puberulent.
Flowers 2–3.5 cm., solitary. Perianth yellow, striped with purple; limb brownish-
purple, minutely puberulent; utricle subglobular; tube more or less curved, longer
than limb. Capsule 2–3 cm., oblong, pendulous. Fl. March–July.

Hab.: Maquis. Upper and Lower Galilee, Mt. Carmel, Samaria, Dan Valley.
Uncommon.

Area: Mediterranean.

2. Aristolochia parvifolia Sibth. et Sm., Fl. Gr. Prodr. 2 : 222 (1813); Boiss., Fl. 4 :
1076 (1879). [Plate 49]

Low, perennial herb. Rhizome subterranean, fusiform. Stems slender, procumbent
or ascending. Leaves 1–2 (–3.5) × 0.7–1.5 (–2.5) cm., ovate-oblong or suborbicular,
cordate-auriculate at base with rounded auricles, obtuse at apex, more or less puberulent beneath; petiole about as long as sinus or longer. Flowers solitary in upper axils,
3–5 (–6) cm., very short-pedicelled. Perianth dark-purple, glabrous outside; tube hardly
incurved, with ovoid utricle; limb linear-lanceolate, puberulent on inner surface, much
longer than tube. Ovary hirtellous. Capsule 0.8–1.4 cm., nearly globular. Fl. December–May (–July).

Hab.: Batha and stony ground. Sharon Plain, Philistean Plain, Upper and Lower
Galilee, Mt. Carmel, Esdraelon Plain, Shefela, Judean Mts., Upper and Lower Jordan
Valley, Gilead.

Area: E. Mediterranean (Palestine–Turkey, Cyprus).

3. Aristolochia billardieri Jaub. et Sp., Ill. Pl. Or. 1 : 175, t. 100 (1844); Boiss., Fl.
4 : 1077 (1879). [Plate 50]

Perennial climber. Stems slender, flexuous, more or less puberulent. Leaves 2–
7 (–8) × 2.5 (–7) cm., ovate-cordate with rounded auricles, obtuse and slightly retuse
at apex, entire, papillose-hispid on both sides or sometimes glabrous above, green,

glaucous beneath; petiole as long as blade, sometimes shorter. Pedicels shorter than flowers, elongating in fruit, hispid-puberulent or subglabrous. Flowers 1.5–3 cm., solitary in axils. Perianth violet, hispid-pubescent or partly glabrous; utricle shorter than pedicel, almost globular; tube strongly incurved; limb as long as or shorter than tube, oblong-lingulate or oblong-triangular, pubescent or glabrous. Ovary more or less hirsute. Capsule 1–2 cm., pyriform to subglobular, densely or sparingly pubescent. Fl. March–May.

Var. **galilaea** Zoh., Palest. Journ. Bot. Jerusalem ser. 2 : 183 (1941). Leaves long-petioled, less papillose-hispid, with upper surface almost glabrous. Pedicels subglabrous. Limb oblong-triangular, shorter than tube. Ovary and capsule sparingly pubescent.

Hab. : Maquis. Upper Galilee.

Area : E. Mediterranean. Var. *galilaea* is endemic in Palestine.

4. Aristolochia maurorum L., Sp. Pl. ed. 2, 1363 (1763); Boiss., Fl. 4 : 1080 (1879). *A. maurorum* L. var. *latifolia* Boiss., l.c. [Plate 51]

Perennial, hirtellous-puberulent, 30–60 cm. Stems more or less erect, simple or branched, not climbing. Leaves 2.5–9 (–10) × 1.5–5 (–6) cm., ovate-oblong to lanceolate-triangular, hastate-auriculate at base with rather oblong or rounded auricles and a deep sinus, tapering towards the more or less acute apex; petiole much shorter than blade. Bracts small, leaf-like. Pedicels about as long as flowers, slender, deflexed. Flowers 4–6 cm., solitary. Perianth rusty-red with yellow spots; tube almost glabrous, strongly incurved, gradually broadening towards apex, with ovate-oblong limb hirtellous on upper spotted surface, cordate and auriculate at base. Capsule 4–6 cm., obovoid-oblong. Fl. March–May.

Hab. : Fields. Philistean Plain, Upper Galilee, Esdraelon Plain, Samaria, Judean Mts., N. Negev, Hula Plain, Upper Jordan Valley, Gilead, Ammon.

Area : E. Mediterranean and W. Irano-Turanian.

There are forms intergrading in width of·leaves between the "type" variety and var. *latifolia* Boiss., l.c.

Roots sometimes used for healing wounds.

5. Aristolochia paecilantha Boiss., Diagn. ser. 1, 12 : 104 (1853) et Fl. 4 : 1080 (1879). [Plate 52]

Perennial, densely hirtellous, 30–50 cm. Stems erect, simple or branched, ridged. Leaves 4–10 (–11) × 2–8 cm., petiolate, triangular to ovate, hastate-auriculate with ovate and obtuse auricles and with deep sinus at base, obtuse or acute at apex. Bracts leaf-like. Pedicels shorter than flowers, deflexed. Flowers 4–7 cm., solitary. Perianth greenish-yellow, hirtellous outside; tube strongly curved, with ovoid utricle; limb oblong-triangular, cordate with 2 auricles at base, upper surface glabrous with blackish-purple spots on yellowish ground colour. Capsule 3.5–5 cm., oblong-pyriform. Fl. March–May.

Hab. : Batha and maquis. Upper Galilee, Samaria, Judean Mts., Dan Valley.

Area : E. Mediterranean (Palestine, Lebanon, Syria).

23. RAFFLESIACEAE (CYTINACEAE)

Herbs, mostly with a thalloid or mycelium-like body, parasitic on branches or roots of other plants. Leaves scale-like. Flowers unisexual, rarely hermaphrodite, solitary or in spike-like heads. Perianth simple, petaloid, fleshy, with 4–10 segments connate at base. Stamens 8 to many; anthers sessile, 2-celled. Ovary inferior to half-inferior, 4–8-carpelled, 1-celled, with parietal placentae and many ovules; style 1 or 0; stigma discoid, capitate or lobed. Fruit baccate. Seeds many, minute, with endosperm and indistinct embryo.

Nine genera and 55 species, mainly in tropical regions of the Old World.

1. CYTINUS L.

Monoecious or dioecious parasitic herbs. Stems simple, short, fleshy. Leaves alternate, imbricated, scale-like, coloured. Flowers unisexual, in terminal spikes or heads, sessile or short-pedicelled, usually with 2 bracteoles at base. Perianth tubular-campanulate, 4–8-lobed. Stamens 8–10, in 1 row, with filaments united into a column. Ovary inferior; ovules orthotropous; style cylindrical; stigma capitate. Berry with many minute seeds, containing endosperm.

Six species in the Mediterranean region, Madagascar and S. Africa.

1. Cytinus hypocistis (L.) L., Syst. ed. 12, 2 : 602 (1767); Boiss., Fl. 4 : 1071 (1879). *Asarum hypocistis* L., Sp. Pl. 442 (1753). [Plate 53]

Perennial glabrous herb, 3–15 cm. Stems many or solitary, erect, cylindrical-clavate. Leaves scale-like, oblong-ovate or elliptical, obtuse, yellow-orange. to scarlet, denticulate or almost entire, sparsely ciliate or not. Bracteoles 2, puberulent. Flowers unisexual (plant monoecious), 5–10 in each short, spike-like head, nearly sessile, the pistillate lateral, the staminate central. Perianth yellowish, with 4 erect, ovate-oblong to elliptical and puberulent lobes. Anthers 8, sessile on the central column. Style 1, bearing a capitate and grooved stigma. Berry many-seeded. Fl. April–May.

Subsp. **orientalis** Wettst., Ber. Deutsch. Bot. Ges. 35 : 97 (1910). Scale leaves usually scarlet. Flowers about 1.8 cm., much longer than bracteoles.

Hab. : Maquis. Parasitic on roots of *Cistus*. Mt. Carmel. Rare.

Area of species : Mediterranean.

Balanophorales

24. CYNOMORIACEAE

Parasitic leafless polygamous or dioecious herbs, with minute flowers, densely crowded on a terminal, club-shaped spadix. Perianth simple of (1–)3–4(–5–8) free or more or less connate segments, sometimes (in pistillate flowers) perianth 0. Stamen 1. Ovary inferior or half-inferior, 1-celled, containing 1 pendulous ovule; style 1. Fruit nut-like or somewhat drupe-like, indehiscent. Seeds solitary in fruit cavity; endosperm oleaginous; embryo lateral, minute.

One genus and 1 species in the Mediterranean, the Sahara, and S. W. Asia.

1. CYNOMORIUM L.

Fleshy leafless herbs, destitute of chlorophyll, parasitic on roots. Spike bracteolate at base, bearing small, staminate, pistillate and hermaphrodite flowers. Staminate flowers; perianth segments mostly 1–5; stamen 1, anthers 2-celled; ovary rudimentary, semi-cylindrical with a groove into which the filament fits. Pistillate flowers : perianth adnate to ovary, lobes mostly 3–6 or perianth 0; ovary half-inferior or superior, sessile or somewhat stipitate, 1-celled, ending in a grooved style and obtuse stigma; ovule solitary, suspended by a short funicle from the top of the cell, suborthotropous. Hermaphrodite flowers few, fertile, with the stamen at the top of the ovary opposite the groove of the style. Fruit nut-like, with the persistent perianth and style at summit; pericarp thin, somewhat leathery, united with the thick testa. Seeds nearly globular; embryo lateral.

One species. Distrib. as for the family.

1. **Cynomorium coccineum** L., Sp. Pl. 970 (1753); Boiss., Fl. 4 : 1072 (1879). [Plate 54]

Fleshy perennial, up to 30 cm., with branching rootstock. Stems unbranched, cylindrical, reddish, with few deciduous scales, ending in a long, thick, club-shaped spike. Cymes confluent, covering the surface of the spike. Bracts peltate, at first imbricated, then remote, deciduous. Staminate flowers usually sessile on common receptacle. Pistillate and hermaphrodite flowers usually in cymes. Fl. April–July.

Hab. : Parasitic on plants of salt marshes and maritime sands. Lower Jordan Valley, Arava Valley, deserts of Edom.

Area : Mediterranean, Irano-Turanian and Saharo-Arabian.

Polygonales

25. POLYGONACEAE

Herbs or woody plants, sometimes climbers, rarely trees. Stems usually jointed, with swollen nodes. Leaves mostly alternate, rarely opposite or whorled, simple, entire or variously lobed or toothed; stipules commonly connate into a sheath (ochrea). Inflorescences – panicles, spikes or heads, composed of cymose clusters or sometimes of single flowers. Bracts 1–2, connate into a sheath (ochreole). Pedicels usually articulate. Flowers hermaphrodite or sometimes unisexual (then plants monoecious or dioecious), actinomorphic. Perianth 2-whorled, sepaloid or coloured, usually persistent, of 3–6 free or connate tepals, the inner whorl sometimes accrescent or modified in fruit. Stamens 6–9, rarely more or less, opposite tepals, with 2-celled anthers. Ovary superior, 1-celled and 1-ovuled; styles 2–4, free or connate at base. Fruit a flat or angular or winged achene, often enclosed within the membranous or leathery perianth. Endosperm copious, mealy; embryo eccentric or lateral, curved or straight.

About 40 genera and over 800 species primarily in north-temperate regions, some in subtropical and tropical regions.

1. Leafless, spartoid shrubs. Stamens 10–18. Achenes tetragonous, with several rows of branching bristles.   **5. Calligonum**
 – Leafy plants (sometimes leaves only at base of plants). Stamens 9 or less. Achenes lenticular or triquetrous or trigonous, without bristles   2
2. Leaves rosulate at base, very large (over 20 cm. long and broad), reniform to cordate or orbicular, palmately nerved. Stamens 9. Achenes 3-winged.   **2. Rheum**
 – Leaves not as above. Stamens 8(–9) or less   3
3. Fruit spiny · perianth with urn-shaped to triquetrous tube and spiny segments. Ochreae truncate, lacerate. Annuals with thick roots.   **1. Emon**
 – Fruit not spiny. Perianth not as above   4
4. Perianth segments usually 5, enclosing fruit but not accrescent. Ochreae often lacerate, fringed or ciliate.   **1. Polygonum**
 – Perianth segments 4 or 6, the inner ones accrescent and sometimes coloured in fruit   5
5. Herbs. Perianth segments 6, the inner 3 enlarged into membranous or leathery valves. Styles 3, bearing fimbriate stigmas. Achene trigonous.   **3. Rumex**
 – Shrubs, sometimes spinescent. Leaves leathery or fleshy, very small. Perianth segments
 – 4, inner 2 growing into scarious valves. Styles 2–3, bearing capitate stigmas. Achene sublenticular.   **6. Atraphaxis**

1. POLYGONUM L.

Annual or perennial herbs, rarely shrubs. Stems erect to prostrate, sometimes climbing. Leaves alternate, simple; stipules membranous, forming clasping ochreae, often lacerate, fringed or ciliate. Inflorescences spike-like, glomerate or paniculate. Flowers usually small, hermaphrodite, either solitary or in cymose axillary clusters. Bracts partly membranous and united into a sheathing ochreole. Pedicels jointed. Perianth simple, divided into 5 (rarely more or less) subequal segments, usually enclosing fruit. Stamens 8 (rarely fewer), with 2-celled anthers. Ovary superior, 1-ovuled; style 2–3-parted, with capitate stigmas. Achene triquetrous, trigonous or lenticular. Embryo lateral.

Over 200 species, many of them weeds or aquatic plants, mainly in the northern hemisphere; many cosmopolitan.

Literature : C. F. Meissner, *Polygonum*, in : DC., *Prodr.* 14 : 83–143 (1856).

1. Inflorescence mostly leafy, rather loose with scattered flowers, never spike-like. Leaves usually not exceeding 6 cm. in length; ochreae 2-fid, later torn. Filaments, at least the inner ones, dilated at base. Achene triquetrous or trigonous   2
 – Inflorescence mostly leafless, forming a terminal spike. Leaves usually longer than 6 cm.; ochreae short-fringed or ciliate. Filaments narrow. Achene mostly lenticular   6
2. Perennial, littoral plants. Leaves 1.5–3.5 × 0.5–0.8 cm., leathery or fleshy, with revolute margins; ochreae silvery-hyaline, many-nerved, shorter or longer than the short internodes. Perianth about 4 mm. Achene 3–4 mm., smooth, somewhat longer than perianth.
                                                                                 3. P. maritimum
 – Plants not as above   3
3. Annual, glabrous, erect, slender herbs. Leaves yellowish-green, lateral nerves not prominent; ochreae much shorter than elongated internodes, 6–8-nerved. Inflorescence interrupted, leafless towards apex. Pedicels as long as perianth. Flowers 2–3.5 mm.,

fragrant. Achene 2–3 mm., minutely punctate.   **4. P. patulum**
- Plants differing from above in one or more characters   4
4. Erect, desert or sand-dune perennials. Leaves narrowly elliptical to linear. Bracts and pedicels shorter than flowers. Upper part of perianth lobes spreading in fruit. Achene exserted.   **2. P. palaestinum**
- Low, mostly prostrate, roadside perennials or annuals. Achene more or less included in perianth   5
5. Perennials. Leaves oblong-lanceolate to lanceolate-linear; ochreae up to 1.8 cm., many-nerved. Bracts and pedicels mostly as long as flowers. Flowers 1–3 in each ochreole.
                                                                   1. P. equisetiforme
- Annuals. Leaves usually elliptical or ovate-lanceolate; ochreae up to 1.2 cm., 6–10-nerved, silvery. Bracts leaf-like, longer than flowers. Flowers usually 2–5 in each ochreole.   **5. P. arenastrum**
6(1). Plants woolly or pubescent   7
- Plants glabrous or almost so   8
7. White cobwebby-woolly plants. Leaves broadly lanceolate; ochreae shortly curly-ciliate. Spikes 2–7.5 cm.   **9. P. lanigerum**
- Appressed-pubescent green plants. Leaves lanceolate, acuminate; ochreae pilose-pubescent, setose-ciliate at apex. Spikes 4–11 cm.   **8. P. acuminatum**
8(6). Leaves usually linear-lanceolate; ochreae up to 3.5 cm., bristly ciliate with cilia almost as long as tube. Leaves and peduncles not glandular-viscid. Spikes 1.5–8 cm., on glabrous peduncles. Perianth about 2.5–4 mm., pink or whitish.   **7. P. salicifolium**
- Leaves generally oblong-lanceolate; ochreae short-fringed or ciliate. Leaves or peduncles or both glandular or viscid   9
9. Leaves (4–)7–14 cm. long, 1–3 cm. broad, often with a dark spot; ochreae up to 2 cm., shorter than internodes. Flowers about 2 mm., very short-pedicelled, white or pink.
                                                                   6. P. lapathifolium
- Leaves (7–)10–30 cm. long, 1.5–7 cm. broad, long-acuminate; ochreae up to 4.5 cm., mostly longer than internodes. Pedicels about as long as 3–4 mm. long flowers.
                                                                   10. P. senegalense

Sect. POLYGONUM. Sect. *Avicularia* Meissn., l.c. 85. Annuals or perennials. Inflorescences axillary, few-flowered. Stamens 5–8. Fruit triquetrous or almost so.

1. Polygonum equisetiforme Sibth. et Sm., Fl. Gr. Prodr. 1 : 266 (1809) et Fl. Gr. 4 : t. 364 (1823); Boiss., Fl. 4 : 1036 (1879). [Plate 55]

Herbaceous or suffruticose glabrous perennial, 50–100 cm. Rhizome thickened, indurated. Stems numerous and richly branched, prostrate to decumbent, rarely ascending or erect. Leaves up to 5 × 0.5–1.5 cm., deciduous, more or less subsessile, jointed at base, the lower oblong-lanceolate or elliptical, the upper narrowly oblanceolate or lanceolate-linear, all acute and obsoletely serrate, sometimes slightly crispulate at margin; ochreae up to 1.8 cm., much shorter than internodes, membranous, white-hyaline toward apex, leathery and brownish at base, 2-fid, torn with age, many-nerved. Inflorescences terminal, long, leafy, interrupted, loose. Bracts almost as long as or longer than flowers; ochreoles 1–3-flowered, silvery-hyaline, torn. Pedicels as long as or longer than perianth. Perianth about (2–)3 mm., white, appressed to and enclosing the achene. Style short, 3-cleft. Achene about 2–2.5 mm., ovoid, triquetrous with somewhat concave faces, brown, shining, minutely punctate. Fl. February–December.

Hab.: Waste places, roadsides. Throughout, notably Acco Plain, Sharon Plain, Philistean Plain, Upper and Lower Galilee, Mt. Carmel, Esdraelon Plain, Samaria, Shefela, Judean Mts., Judean Desert, W., N. and C. Negev, Hula Plain, Upper and Lower Jordan Valley, Dead Sea area, Arava Valley, Gilead, Ammon, Moav, Edom.

Area: Mediterranean and W. Irano-Turanian.

This species varies considerably in length of internodes, thickness of stem, size and shape of leaf, etc. Several varieties have been distinguished in Palestine which also differ ecologically. Their identification presents difficulties, since they may be partly included among the many forms described from other Mediterranean countries.

Eaten by goats.

2. Polygonum palaestinum Zoh.* *P. equisetiforme* Sibth. et Sm. var. *arenarium* Eig et Feinbr., Palest. Journ. Bot. Jerusalem ser. 2 : 99 (1940). [Plate 56]

Tall, erect, glabrous perennial, up to 150 cm. Stems few or numerous, more or less herbaceous, stiff, striate, with elongated internodes. Leaves 1–6 × 0.1–0.8 cm., soon deciduous, subsessile or short-petioled, narrowly elliptical-oblanceolate to linear, jointed at base, entire or nearly so, with more or less prominent nerves on both sides; ochreae up to 2.5 cm., membranous, lacerate above, many-nerved, white-hyaline at apex, brown at base. Inflorescences little branched or nearly simple, elongated, interrupted, made up of many axillary, 2–3-flowered clusters. Bracts small, shorter than flowers, absent above; ochreoles brownish-green at base, hyaline and torn at apex. Pedicels somewhat shorter than flowers. Flowers (2–)3–4 mm. Tepals broadly obovate, white above, greenish-pink beneath, upper part spreading in fruit, forming a wing-like margin around the achene. Style very short, 3-partite. Achene 2–3 mm., exserted from perianth, triquetrous with slightly concave faces, ovoid, tapering, brown, shining. 2n = 40. Fl. March–October.

Var. **palaestinum.** Stems erect, up to 150 cm. Flowers 3–4 × 3.5 mm. Achene about 3 mm. Tepals often apiculate.

Hab.: Sand dunes of coastal plain. Acco Plain, Sharon Plain, Philistean Plain.

Var. **negevense** Zoh. et Waisel.* Stems erect, 40–80 cm., slender. Leaves narrower and smaller than in preceding variety. Flowers and achene smaller, about 2 mm. Tepals obtuse, truncate or notched, never apiculate.

Hab.: Wadis in deserts. Negev, Arava Valley.

Area of species: E. Mediterranean and W. Saharo-Arabian.

3. Polygonum maritimum L., Sp. Pl. 361 (1753); Boiss., Fl. 4 : 1037 (1879). [Plate 57]

Glabrous, greyish-glaucous perennial, 15–50 cm. Rhizome woody, branching. Stems numerous, usually procumbent, with short internodes and few branches, densely leafy. Leaves 1.5–3.5 × 0.5–0.8 cm., subsessile, leathery-fleshy, elliptical to lanceolate or oblanceolate, roundish or more or less acute, with revolute margin and prominent nerves; ochreae shorter or longer than internodes, up to 1.5 cm., membranous or

---

* See Appendix at end of this volume.

silvery-hyaline, dark brown at base, many-nerved, 2-fid, later lacerate. Inflorescences leafy, loose, denser towards apex. Bracts leaf-like, longer than flowers. Flowers 1–4 in axil, short-pedicelled. Perianth about 4 mm., pink or white, greenish at base. Styles 3, very short. Achene 3–4 mm., mostly longer than perianth, broadly ovoid, sharply triquetrous, concave between angles, acute, smooth, dark brown, shining. Fl. May–August.

Hab.: Sea shore, sandy and sandy-calcareous soil. Coastal Galilee, Acco Plain, Sharon Plain, Philistean Plain.

Area: Mediterranean and W. Euro-Siberian; also in N. and S. America.

Used for healing burns.

4. Polygonum patulum M.B., Fl. Taur.-Cauc. 1 : 304 (1808). *P. bellardi* auct. non All., Fl. Ped. 2 : 207; 3 : t. 90 f. 2 (1785); Boiss., Fl. 4 : 1034 (1879). [Plate 58]

Glabrous annual, 40–75 cm. Stems erect, more or less paniculately branched, slender, stiff, angular, striate. Leaves (1–)1.5–4.5(–5) × 0.2–1(–1.5) cm., often soon deciduous, short-petioled or subsessile, jointed at base, narrowly elliptical to linear-lanceolate, acute, with prominent midrib beneath, yellowish-green; ochreae and ochreoles up to 1.5 cm., much shorter than the elongated internodes, cleft or torn, sometimes 2-partite, 6–8-nerved, hyaline-membranous, basal part brownish. Flowers loosely scattered along stem, 1–3 at each node, somewhat denser towards raceme-like, bracteate apex. Pedicels about as long as perianth. Perianth 2–3.5 mm., fragrant, pink-green with reddish margins; segments united near base. Stamens usually 8. Style short, 3-partite. Achene 2–3 mm., included in perianth, ovoid, triquetrous, acuminate, minutely punctate, shining, dark brown. Fl. May.

Hab.: Cultivated fields. Sharon Plain, Philistean Plain, Upper and Lower Galilee, Mt. Carmel, Esdraelon Plain, Judean Mts., Judean Desert, N. Negev, Hula Plain, Upper Jordan Valley, Gilead, Ammon.

Area: Mediterranean and Irano-Turanian, extending into the Euro-Siberian region; also in some tropical regions.

5. Polygonum arenastrum Bor., Fl. Centr. Fr. ed. 3, 2 : 559 (1857); Styles, Watsonia 5 : 204 (1962). *P. aviculare* auct. non L., Sp. Pl. 362 (1753); Boiss., Fl. 4 : 1036 (1879). [Plate 59]

Glabrous annual, up to 70 cm. Stems many, prostrate or decumbent, rarely erect, branching, leafy throughout. Leaves 1–5 × 0.1–1.4 cm., short-petioled or subsessile, jointed at base, elliptical or ovate-lanceolate to linear-lanceolate, acute or obtuse, entire or nearly so, pinnately and prominently nerved, sometimes soon deciduous; ochreae up to 1.2 cm., shorter than internodes, membranous, 6–10-nerved, lower part brown, upper silvery-hyaline, 2-fid, later lacerate. Inflorescences leafy. Bracts leaf-like, longer than flowers. Pedicels shorter than perianth. Flowers 1.5–3.5 mm., 2–5 in each axillary cluster. Perianth greenish-white or pink, partly greenish outside; segments oblong-elliptical, united up to half their length. Styles (2–)3, very short. Achene 1.5–3 mm., enclosed in persistent perianth, triquetrous, minutely punctate-striate, brown, somewhat glossy. Fl. March–October.

Hab.: Roadsides and other waste places. Acco Plain, Sharon Plain, Philistean

Plain, Upper and Lower Galilee, Esdraelon Plain, Samaria, Shefela, Judean Mts., Judean Desert, Negev, Dan Valley, Hula Plain, Upper and Lower Jordan Valley, Dead Sea area, Gilead, Edom.

Area : Mainly temperate regions of the northern hemisphere.

A very variable species; many forms common in this country still await an experimental study.

Contains tannins; used in medicine (herba polygoni).

Sect. PERSICARIA Meissn. in DC., Prodr. 14 : 101 (1856). Annuals or perennials. Inflorescences spike-like. Perianth often coloured. Stamens 1 8. Styles 2–3, often connate below. Achenes lenticular, rarely trigonous. Ochreae entire or ciliate but not lacerate.

6. Polygonum lapathifolium L., Sp. Pl. 360 (1753); Boiss., Fl. 4 : 1030 (1879). [Plate 60]

Perennial herb, subglabrous, 50–150 cm. Stems erect, rarely procumbent, with swollen nodes, usually reddish, rooting from nodes. Leaves up to 14 × 1–3 cm., short-petioled, ovate, oblong-lanceolate to linear-lanceolate, more or less tapering at base, acuminate at apex, greenish, often with a dark spot, appressed-puberulent on reddish midrib and near margin, pinnately nerved with prominent nerves, lower surface with golden-yellow glands; ochreae up to 2 cm., shorter than internodes, membranous, entire or very short-fringed ·at apex, more or less glabrous. Peduncles and pedicels often glandular. Inflorescences spike-like, 2–7 cm., axillary or terminal, cylindrical, slender, sometimes more or less interrupted. Ochreoles shorter than flowers, membranous, rhombic-ovate and tapering at apex, obsoletely ciliolate. Flowers about 2 mm., very short-pedicelled. Tepals white or pink. Style with 2 long, recurved-spreading lobes. Achene about 2 mm., lenticular, suborbicular, slightly concave on both sides, with somewhat thickened rim, brown, shining. Fl. April–November.

Hab. : Swamps and river banks. Acco Plain, Sharon Plain, Philistean Plain, Upper and Lower Galilee, Mt. Carmel, Esdraelon Plain, Samaria, N. Negev, Dan Valley, Hula Plain, Upper Jordan Valley.

Area : Euro-Siberian and Mediterranean; also in America, S. Africa and elsewhere.

Very polymorphic.

7. Polygonum salicifolium Brouss. ex Willd., Enum. Pl. Hort. Berol. 1 : 428 (1809) non Del., Fl. Eg. 12 (1813). *P. scabrum* Poir. in Lam., Encycl. 6 : 148 (1804) non Moench, Meth. 629 (1794). *P. serrulatum* Lag., Gen. et Sp. Nov. 14 (1816); Boiss., Fl. 4 : 1028 (1879). [Plate 61]

Glabrous perennial herb, 60–120 cm. Stems erect, rarely decumbent, branching, rooting from nodes. Leaves 6–15 × 0.5–2 cm., subsessile, narrowly lanceolate to almost linear, tapering or rounded to subcordate at base, acuminate at apex, minutely scabrous-serrulate and ciliate at margin; ochreae up to 2.8(–3.5) cm. (including cilia), membranous, bristly-ciliate at apex, appressed-setulose on nerves or glabrous. Peduncles slender, glabrous. Inflorescences 1.5–8 cm., spike-like, many-flowered, rather loose, linear. Ochreoles usually ciliate at apex. Pedicels about as long as perianth, jointed above. Flowers 2–6 in each axil. Perianth about 2.5–4 mm., pink or whitish, not glandular. Stamens 6 (5–8). Style 3-partite. Achene 2–3 mm., included, usually tri-gonous, ovoid, acuminate, brown-black, shining. Fl. March–November.

Var. **salicifolium** s.s. *P. salicifolium* Brouss. ex Willd. var. *serrulatum* (Lag.) Maire et Weiller in Maire, Fl. Afr. N. 7 : 249 (1961). Ochreae appressed pilose-setulose. Ochreo-les ciliate.

Hab. : Swamps and river banks. Acco Plain, Sharon Plain, Philistean Plain, Upper Galilee, Samaria, Dan Valley, Hula Plain, Upper and Lower Jordan Valley, Gilead, Moav.

Var. **glabrescens** (Feinbr.) Zoh. (comb. nov.). *P. scabrum* Poir. var. *glabrescens* Feinbr., Palest. Journ. Bot. Jerusalem ser. 2 : 97 (1940). *P. salicifolium* Brouss. ex Willd. var. *salicifolium* (Del. ex Meissn.) Maire et Weiller in Maire, Fl. Afr. N. 7 : 249 (1961). Ochreae glabrous or nearly so. Ochreoles not ciliate.

Hab. : As above. Sharon Plain, Philistean Plain, Dan Valley, Hula Plain.

Area of species : Pluriregional. Temperate and tropical regions, especially of the Old World.

8. Polygonum acuminatum Kunth in H.B.K., Nov. Gen. et Sp. 2 : 178 (1817); Meissn. in DC., Prodr. 14 : 114 (1856); Baker and Wright, in Fl. Trop. Afr. 6, 1 : 112 (1909). [Plate 62]

Perennial, shrubby, appressed pilose-pubescent, up to 1 m. Stems erect or ascending, or creeping on water surface, branched, rather stout, often rooting from nodes. Leaves 10–25 × 2–4 cm., short-petioled or subsessile, lanceolate, acuminate; ochreae 1.5–4.5 cm., densely pilose-pubescent, setose-ciliate at apex. Inflorescences 4–11 × 0.5–1 cm., spike-like, cylindrical. Ochreoles shorter than flowers, obovate, obtuse, ciliate. Pedicels hardly exceeding ochreoles in length, somewhat shorter than flowers. Perianth 3–4 mm., pinkish; segments 4–5, obovate, not glandular. Stamens usually 6, included or exserted. Style rather long, 2-partite. Achene about 2 mm. in diam., more or less lenticular, dark brown, shining. Fl. May–September.

Hab. : Swamps, near pools, river banks. Sharon Plain, Philistean Plain, Dan Valley, Hula Plain.

Area : Tropical regions of the Old World.

9. Polygonum lanigerum R. Br., Prodr. Fl. Nov. Holl. 419 (1810); Boiss., Fl. 4 : 1030 (1879). [Plate 63]

White cobwebby-woolly perennial, shrubby, 80–160 cm. Stems erect, branching above, stout, rooting from lower nodes. Leaves up to 25 × 2–7 cm., short-petioled, broadly lanceolate, tapering at base, acuminate at apex, densely woolly beneath and thinly grey-woolly above; ochreae 3.5 cm., truncate, with thin, short, somewhat curly cilia. Peduncles lanate, eglandular. Inflorescences 2–7.5 cm., spike-like, more or less dense. Ochreoles subimbricated, orbicular-obovate, short-denticulate or ciliate. Pedicels short. Perianth 3–4 mm., pinkish. Stamens 6 (–7–8). Style 2 (–3)-partite. Achene lenti-cular, suborbicular, biconvex, blackish, shining. Fl. September–October.

Hab. : Swamps. Sharon Plain, Hula Plain.

Area : Tropical regions of the Old World.

Sect. AMBLYGONON Meissn., Monogr. Polyg. 43 et 53 (1826). Annuals, rarely perennials with fibrous roots. Inflorescences racemose or spike-like. Perianth coloured. Stamens mostly 7, rarely 5–6 or 8. Styles 2, connate, rarely almost free. Achene included, lenticular.

10. Polygonum senegalense Meissn., Monogr. Polyg. 54 (1826) et in DC., Prodr. 14 : 123 (1856); Boiss., Fl. 4 : 1031 (1879). [Plate 64]

Perennial, shrubby, glabrous or nearly so, 80–150 cm. Stems more or less erect, branching, stout, rooting from basal nodes. Leaves (7–)10–30 × 1.5–7 cm., oblong-lanceolate, tapering at base, long-acuminate at apex, scabrous-hispid at margin, mostly appressed-puberulent on midrib, viscid and glandular-punctate on both sides, pinnately nerved; petioles hispid-puberulent; ochreae mostly longer than internodes, up to 4.5 cm., membranous, many-nerved, truncate, short-ciliate at apex, almost glabrous. Peduncles glandular. Inflorescences 2–8 × 1 cm., spike-like, cylindrical, dense. Bracts rudimentary or 0; ochreoles membranous, broadly obovate, slightly ciliolate, sometimes obsoletely fringed at apex. Pedicels becoming as long as perianth and somewhat exceeding ochreole. Flowers 2–4 in each cluster. Perianth 3–4 mm., whitish-pink, with 4–5 ovate-elliptical segments. Stamens (5–)7(–8). Style parted almost to base. Achene 2–3 mm., lenticular, suborbicular, with flattened faces, sometimes with a shallow groove, brown-black, shining. Fl. May–September.

Hab.: Swamps. Sharon Plain, Philistean Plain, Hula Plain.

Area: Trop. Africa and Asia (extending towards the S. Mediterranean territories).

2. RHEUM L.

Robust herbs with thick roots. Stems rather short, scapose. Leaves large, often all radical, entire or repand-dentate or lobed, palmately nerved with 3 to many nerves at base; ochreae leathery-membranous. Inflorescences paniculate. Flowers usually hermaphrodite, pedicellate. Pedicels jointed. Perianth herbaceous, parted into 6 lobes, the outer 3 somewhat larger. Stamens 9, rarely less. Ovary superior; styles 3. Achene triquetrous, 3-winged. Embryo straight.

Twenty five to 30 species in temperate regions of Asia and E. Europe.

1. Rheum palaestinum Feinbr., Palest. Journ. Bot. Jerusalem ser. 3 : 117 (1944). [Plate 65]

Perennial herb. Roots vertical, thick. Stems up to 40 cm., leafless and scapose, bearing a thick, much branched, paniculate inflorescence, with glabrous, grooved branches. Leaves radical, up to 25(–60) × 30(–70) cm., petiolate, reniform to cordate or orbicular, auriculate at base, repand and minutely papillose-ciliate at margins, fleshy-leathery, pulverulent at upper surface, more or less palmately nerved, lower nerves prominent and minutely papillose. Pedicels as long as perianth, filiform, jointed above middle or near apex, accrescent but shorter than fruit, recurved. Flowers about 2 mm. Fruit 1–1.4 cm.; wings about 4 mm. broad, with a continuous vein near margin. Seeds up to 8 mm., triquetrous. Fl. March–April.

Hab. : Steppes and deserts. C. Negev, Ammon, Moav, Edom.

Area : W. Irano-Turanian (endemic in Palestine).

3. Rumex L.

Annual, biennial or perennial herbs. Leaves alternate; ochreae tubular. Inflorescences racemose or paniculate. Flowers unisexual (plants monoecious or dioecious) or herm-aphrodite, on jointed pedicels, arranged in whorls subtended by ochreoles. Perianth segments 6, greenish; outer 3 usually spreading or reflexed in fruit; inner accrescent into valves and later enclosing achene. Stamens 6, in 2 whorls. Ovary trigonous, 1-ovuled; styles 3, with fringed stigmas. Valves membranous or leathery, dentate or entire, with or without swollen tubercle on midrib. Achene enclosed, trigonous. Embryo lateral.

About 150 species, mainly in temperate regions of the northern hemisphere.

Literature : K. H. Rechinger f., Vorarbeiten zu einer Monographie der Gattung *Rumex*. I. *Beih. Bot. Centralbl.* 49, 2 : 1–132 (1932); II. *Repert. Sp. Nov.* 31 : 225–283 (1933); VII. *Candollea* 12 : 9–152 (1949). G. Samuelsson, *Rumex pictus* Forsk. und einige ver-wandte Arten. *Ber. Schweiz. Bot. Ges.* 42 : 770–779 (1933). G. Samuelsson, Bemerkungen über einige *Rumex*-Sippen aus der *Vesicarius*-Gruppe. *Bot. Notis.* 1939 : 505–527 (1939).

1. Valves (inner 3 tepals) 3-lobed, resembling a butterfly or aeroplane. Outer tepals deflexed and partly united. **7. R. rothschildianus**
- Valves entire or dentate but not lobed 2
2. Valves membranous, orbicular, ovate or cordate, wing-like, (0.5–)0.6–2 cm. broad 3
- Valves leathery, dentate or rarely entire, ovate-lanceolate to ovate, oblong, triangular or deltoid, usually not as broad as in above 8
3. Perennial herbs with creeping stolons or with roots ending in tubers. All leaves hastate-sagittate. Valves about 1 cm. or less in diam. 4
- Annuals. Leaves of different shape or hastate-sagittate but then valves longer or broader than 1 cm. 5
4. Valves about 1 cm. **1. R. tingitanus**
- Valves 5–7 mm. **2. R. tuberosus**
5 (3). Leaves entire 6
- All or some leaves pinnatifid to pinnatipartite. Plants of sandy soils 7
6. Valves distinctly unequal, flat, more or less denticulate with vein around margin. Common. **4. R. cyprius**
- Valves subequal, longitudinally folded, entire, and without marginal vein. Rare. **3. R. vesicarius**
7 (5). Pedicels about as long as valves. Flowers 1.5–2.5 mm. Valves 5–9 mm., not folded, with radiating veins branching at margin; wart visible. **5. R. pictus**
- Pedicels shorter than valves. Flowers 2–3.5 mm. Valves up to 1.7(–2.2) cm., folded longitudinally along middle, reticulately veined, with main vein and basal wart hidden by folds. **6. R. occultans**
8 (2). Valves entire or obsoletely denticulate 9
- Valves conspicuously dentate (except in *R. pulcher* ssp. *anodontus*) 10
9. Panicles with rather long divaricate branches; whorls remote, subtended by leaves. Valves 2–3 × 1.5–2.5 mm., oblong-ovate, entire. **9. R. conglomeratus**
- Panicles short-branched, compact towards apex; whorls approximate, only the lower ones subtended by leaves. Valves 4–6 × 3–5 mm., ovate-cordate, mostly obsoletely

denticulate.   **8. R. crispus**

10(8). Fruiting pedicels flat and broad, jointed near base. Racemes simple, spike-like;
whorls with 2–4 flowers. Annuals of sandy soils.   **14. R. bucephalophorus**

— Fruiting pedicels terete. Racemes mostly branched; whorls many-flowered. Perennials
or annuals of heavy soils   **11**

11. Leaves all tapering at base. Fruiting pedicels longer than valves. Valves narrowly
triangular at apex, with 2 setiform teeth on each side and a relatively large yellowish
wart at base.   **13. R. maritimus**

— Leaves, at least the lower ones, rounded, truncate or subcordate at base. Fruiting
pedicels usually shorter than valves. Valves not as above   **12**

12. Flowering stems almost simple; racemes leafless, except at base. Valves triangular
ovate to deltoid, 6 × 4–6 mm., including 3–5, up to 2.5 mm. long teeth on each side,
at least some of them hooked at tip; only 1 of the warts fully developed. Perennial
maquis plants.   **11. R. cassius**

— Plants differing from above at least is one character   **13**

13. Warts well developed on all 3 valves of fruit (sometimes unequal in size). Annuals.
                                                                 12. R. dentatus

— Only 1 of the valves with well developed wart. Perennials or biennials. **10. R. pulcher**

Subgen. ACETOSA (Mill.) Rech. f. Plants dioecious or polygamous, rarely monoecious.
Fruiting perianth longer than achene, with or without tubercles. Leaves often hastate
or sagittate.

1. Rumex tingitanus L., Syst. ed. 10, 2 : 991 (1759); Sam., Bot. Notis. 1939 : 506–509
(1939); Maire, Fl. Afr. N. 7 : 312 (1961). *R. roseus* L., Sp. Pl. 337 (1753) non L., Syst.
ed. 10, 2 : 990 (1759) (=*R. cyprius* Murb.). [Plate 66]

Perennial, glabrescent, 30–50 cm. Rhizome with long stolons. Stems ascending,
branching from base. Leaves petiolate, sinuate-crenate to lobate, more or less hastate;
lower ones ovate or ovate-oblong, obtuse, often auriculate; upper leaves lanceolate-
triangular to narrowly lanceolate. Pedicels 3–6 in each ochreole, jointed at middle.
Flowers 2–3 mm. Valves about 1 cm. broad, suborbicular, warted. Achene 3–4 mm.,
trigonous, lanceolate-ovoid, reddish-brown, smooth. Fl. January–February.

Hab. : Steppes. Judean Desert, Lower Jordan Valley, Moav.

Area : W. Irano-Turanian (Mauritanian Steppes), extending into some W. Mediterranean countries (Spain, Portugal, Italy, France).

2. Rumex tuberosus L., Sp. Pl. ed. 2, 481 (1762); Boiss., Fl. 4 : 1017 (1879); Rech. f.,
Candollea 12 : 29 (1949). [Plate 67]

Dioecious perennial herb, glabrous or somewhat puberulent below, up to 80 cm.
Rhizome thick, bearing spindle-shaped root fibres, ending in ovoid-oblong, thick tubers.
Stems erect, almost simple, paniculate above, sparingly leafy. Leaves green, glabrous,
rarely scabrous, the basal ones up to 11 × 4 cm., long-petioled, ovate or oblong, cordate-
hastate at base, with diverging, entire or almost entire, more or less acute auricles,
obtuse or acute at apex; cauline leaves short-petioled to sessile, oblong-lanceolate to
linear, hastate-sagittate and subclasping, acute; ochreae lacerate. Panicles usually short,
dense, with erect branches, leafless. Pedicels filiform, longer than flowers and fruits,

jointed at middle. Outer tepals reflexed in fruit. Valves up to 7 mm., ovate-orbicular, more or less cordate at base, entire, netted-veined, each usually with an ovate-oblong wart at base. Achene minute, dark brown. Fl. April.

Hab.: Maquis and among rocks. Upper Galilee.

Area: Mediterranean and Irano-Turanian.

3. Rumex vesicarius L., Sp. Pl. 336 (1753); Boiss., Fl. 4 : 1017 (1879); Rech. f., Candollea 12 : 35 (1949). [Plate 68]

Annual, green-glaucescent, glabrous, 10–40 cm. Stems decumbent or ascending, branching from base, rather thick. Leaves up to 7 × 4 cm., petiolate, ovate to deltoid or oblong-triangular, cuneate or truncate, subcordate or subhastate at base, more or less obtuse, entire. Flowers hermaphrodite or unisexual, racemose or paniculate. Pedicels single in axils, often bearing 2 flowers, jointed below middle, elongated and reflexed in fruit. Valves 1–2 cm., longer than pedicels, membranous, subequal, suborbicular, narrowly sinuate-cordate at base and apex, longitudinally folded and hence more or less ovate in outline, concealing the second flower, entire, purplish and netted-veined, without marginal vein, 2 of the 3 valves bearing a basal wart. Achene 3 mm. long or longer, trigonous, ovoid, acuminate, brownish; those of the first flowers somewhat longer than those of the second ones. Fl. February–April.

Hab.: Steppes and batha. Judean Mts., Negev, Lower Jordan Valley. Very rare.

Area: Saharo-Arabian, extending into the Mediterranean and Sudanian regions.

Leaves eaten green and cooked; also used in medicine as a laxative, tonic, etc.

4. Rumex cyprius Murb., Lunds Univ. Arsskr. N. F. Afd. 2, 2, No. 14 : 20 (1907) emend. Sam., Bot. Notis. 1939 : 509 (1939); Rech. f., Candollea 12 : 34 (1949). *R. cyprius* Murb. ssp. *disciformis* Sam., Bot. Notis. 1939 : 512 (1939). *R. roseus* L., Syst. ed. 10, 2 : 990 (1759); Boiss., Fl. 4 : 1018 (1879) non *R. roseus* L., Sp. Pl. 337 (1753) [=*R. tingitanus* L., Syst. ed. 10, 2 : 991 (1759)]. [Plate 69]

Annual, glabrescent, 10–35 (–40) cm. Stems ascending to erect, branching from base, rather thick below. Leaves up to 8.5 × 3 cm., petiolate, somewhat fleshy, ovate-triangular or oblong-deltoid, the upper lanceolate, cuneate-subhastate at base, acute-acuminate at apex, entire; ochreae membranous-hyaline. Pedicels 1–2 (–3) in each axil, longer than perianth, filiform, jointed below middle, elongated and reflexed in fruit. Flowers 2–3 mm., 1–2 on each pedicel. Valves distinctly unequal, membranous but leathery at centre, flat, the larger one up to 2 cm. long and broad, suborbicular, deeply cordate at base, all denticulate-spinescent at margins or obsoletely so, netted-veined with a vein around margin. Achene about 3 mm., trigonous, lanceolate-ovoid, acuminate, brownish-white, smooth. Fl. January–May.

Hab.: Batha and steppes. Lower Galilee, Esdraelon Plain, Samaria, Judean Mts., Judean Desert, W., N., C. and S. Negev, Hula Plain, Upper and Lower Jordan Valley, Dead Sea area, Arava Valley, Gilead, Ammon, Moav, Edom.

Area: Saharo-Arabian and Irano-Turanian, extending into E. Mediterranean territories.

Used as pot herb.

5. Rumex pictus Forssk., Fl. Aeg.-Arab. 77 (1775); Rech. f., Candollea 12 : 33 (1949). *R. lacerus* Balbis, Mém. Acad. Sci. Turin 7 : 19 (1804–1806); Boiss., Fl. 4 : 1017 (1879). [Plate 70]

Annual, glabrous, 10–25 cm. Stems decumbent, ascending, rarely erect, usually much branched at base. Leaves up to 5 cm., petiolate, fleshy, ovate or oblong-triangular in outline, almost all pinnatifid to pinnatipartite into 1–4 oblong-triangular to linear lobes on each side, terminal lobe ovate or oblong-triangular; lower leaves long-petioled, uppermost ones and bracts lanceolate to linear, wavy-margined to almost entire. Flowers unisexual and hermaphrodite, in remote whorls of 1–5, only the lower ones subtended by leaves. Pedicels filiform, longer than flowers, jointed below middle. Perianth 1.5–2.5 mm. Valves 5–9 × 6–9 mm., about as long as pedicels, more or less equal, membranous, not folded, reniform-ovate to oblong-cordate, purplish-red along veins, veins radiating, branching at margin, marginal vein 0; centre of each valve occupied by a yellowish, ovate-oblong wart. Achene about 2.5 mm., trigonous, ovoid-lanceolate, yellowish-brown, smooth, shining. Fl. January–May.

Hab.: Sandy soils. Acco Plain, Sharon Plain, Philistean Plain, W., N. and C. Negev, Lower Jordan Valley, Ammon, Edom.

Area: Mainly E. Mediterranean, extending into adjacent Saharo-Arabian territories.

6. Rumex occultans Sam., Ber. Schweiz. Bot. Ges. 42 : 776 (1933); Rech. f., Candollea 12 : 34 (1949). *R. lacerus* Balbis var. *macrocarpus* Boiss., Fl. 4 : 1017 (1879). [Plate 71]

Annual, glabrous, 10–40 cm. Stems decumbent to almost erect, branching at base. Leaves up to 4.5 (–7) cm., petiolate, fleshy, ovate-hastate to triangular-lanceolate in outline, sometimes subauriculate at base, acute at apex, pinnatifid to pinnatipartite into linear or oblong, obtuse lobes, the terminal one longer, lanceolate, sometimes leaves 2-pinnatifid; uppermost leaves subsessile, lanceolate-subhastate. Flowers unisexual or hermaphrodite, whorled, the lower whorls subtended by leaves. Pedicels longer than perianth but shorter than fruit, filiform, jointed below middle, later reflexed. Valves up to 1.7 (–2.2) × 1.7 cm., subequal, membranous, folded longitudinally along middle and then appearing oblong, ovate or orbicular, purplish-red, netted-veined; main vein and basal ovate-lanceolate wart hidden by folds of valve. Achene up to 5 mm., lanceolate-trigonous, yellowish. Fl. January–May.

Hab.: Sands and sandy clay. Sharon Plain, Philistean Plain.

Area: E. Mediterranean (endemic in Palestine and Lebanon).

7. Rumex rothschildianus Aarons. ex Evenari in Opphr. et Evenari, Bull. Soc. Bot. Genève ser. 2, 31 : 212 (1941); Rech. f., Candollea 12 : 32 (1949). [Plate 72]

Dioecious, glabrous annual, 10–45 (–75) cm. Stems erect, simple. Leaves up to 9 × 3.5 (–4) cm., the radical ones petiolate, oblong-obovate, short-hastate at base, rounded and sometimes mucronulate at apex; cauline leaves sessile, ovate-oblong, sagittate-amplexicaul at base, abruptly short-acuminate at apex; the uppermost narrower, triangular-lanceolate, acuminate. Inflorescences terminal, dense, elongated, cylindrical. Pedicels 3–4 in each axil, jointed below middle. Staminate flowers 3–4 mm. in diam.; pistillate flowers about 2 mm. Valves up to 7 mm. (excluding the reflexed outer tepals), membranous-coriaceous, 3-lobed and aeroplane-like, with more or less

4-angled, denticulate lateral lobes and a smaller, triangular, acute apical lobe; each valve usually with an ovate-orbicular, flat, small wart at base. Achene 2–3 mm., trigonous, spindle-shaped, acute, black-brown, shining. Fl. March–April.

Hab.: Moist sandy soils. Sharon Plain, Philistean Plain.

Area: E. Mediterranean (endemic in Palestine).

Subgen. RUMEX. Flowers mostly hermaphrodite. Fruiting perianth longer than achene, mostly with tubercle (wart). Leaves never hastate or sagittate.

8. Rumex crispus L., Sp. Pl. 335 (1753); Boiss., Fl. 4 : 1009 (1879); Rech. f., Candollea 12 : 80 (1949). [Plate 73]

Perennial, glabrous, 40–120 cm. Stems erect, simple or branched from base, brownish. Leaves usually crisp-undulate, sometimes minutely and obsoletely papillose on veins of lower surface; the lower leaves up to 20 × 8 cm., long-petioled, oblong-lanceolate, rounded or cuneate or truncate at base, tapering towards the somewhat acute apex; upper leaves short-petioled, linear-lanceolate, tapering-cuneate at base. Panicles short-branched, somewhat interrupted at base, compact towards apex; branches almost erect, subtended by linear, crisp leaves; whorls many-flowered, approximate. Pedicels reflexed, filiform, jointed below middle, elongating and exceeding fruiting perianth. Flowers hermaphrodite, sometimes unisexual and plants then dioecious. Valves leathery, (3–)4–6 × 3–5 mm., more or less ovate, subcordate, entire or obsoletely denticulate, netted-veined; warts on 1 or all valves, about 2.5 mm., ovate, brown. Achene 2–3 mm., trigonous, ovoid, acuminate, brown.

Var. **unicallosus** Peterm., Fl. Lips. 266 (1838); Rech. f., Candollea 12 : 84 (1949). Only 1 valve with a prominent wart, others without wart. Fl. April–September.

Hab.: Damp soil and grassy places. Philistean Plain, Judean Mts., Hula Plain.

Area of species: Euro-Siberian, Mediterranean and Irano-Turanian; also in other temperate regions.

Wild salad plant and pot herb with high vitamin C content. Roots and leaves (radix et folia rumicis crispi) used in medicine.

9. Rumex conglomeratus Murr., Prodr. Stirp. Gott. 52 (1770); Boiss., Fl. 4 : 1010 (1879); Rech. f., Candollea 12 : 96 (1949). [Plate 74]

Perennial or biennial, glabrous, 60–120 cm., often with vertical rhizome. Stems erect, often paniculately branching, more or less flexuous, striate-angular. Branches divaricate to ascending. Lower leaves up to 18 × 15 cm., long-petioled, broadly oblong-lanceolate, rounded or truncate or subcordate at base, obtuse at apex, entire or somewhat repand; upper ones smaller, short-petioled, ovate-lanceolate, rounded or subcordate at base, more or less acute or short-acuminate at apex, sometimes minutely crisp. Panicles large, many-whorled; whorls many-flowered, more or less remote, subtended by leaves except for the uppermost ones. Pedicels jointed below middle or near base, reflexed, elongating, equalling or somewhat exceeding fruiting perianth in length. Flowers 1.5–2 mm., usually hermaphrodite. Valves 2–3 × 1.5–2.5 mm., oblong-ovate, entire, each with an ovate-oblong wart, 1–2 mm. long. Achene up to 2 mm., trigonous to ovoid, acute, brown, shining. Fl. April–September.

Hab.: Swamps and damp places. Sharon Plain, Hula Plain, Gilead.

Area: Euro-Siberian, Mediterranean and Irano-Turanian; also in S. Africa and N. America.

Leaves used as pot herb.

10. Rumex pulcher L., Sp. Pl. 336 (1753); Boiss., Fl. 4:1012 (1879); Rech. f., Candollea 12:102 (1949). [Plate 75]

Perennial or annual, 15 100(120) cm., glabrous, sometimes petioles, nodes, ochreae and main nerves rough-puberulent. Stems erect, much branched. Branches more or less ascending, striate. Leaves up to 12(–18) cm., petiolate, more or less fleshy, ovate-oblong, sometimes lyrate, subcordate at base, obtuse to somewhat acute at apex, crispulate; the uppermost leaves and bracts short-petioled or subsessile, lanceolate to linear, acute. Panicles composed of flexuous, elongated, spreading branches; whorls many-flowered, usually remote, mostly subtended by leaves. Bracts leaf-like, the upper rather small. Pedicels jointed below middle or near base, somewhat shorter than valves, reflexed and becoming thickened. Flowers about 2 mm. Valves 4–7 × 3–5 mm., leathery, ovate to oblong, sometimes more or less triangular, obtuse or somewhat acute, reticulately veined, with 2–8 subspinescent, up to 1.5 mm. long teeth on either side of base; warts 2–3 × 1–2 mm., distinctly unequal, suborbicular-ovate to oblong-lanceolate, often wrinkled or somewhat verrucose. Achene 1.5–3 mm., trigonous, broadly ovoid, acute, brown. Fl. March–June.

Subsp. **anodontus** (Hausskn.) Rech. f., Beih. Bot. Centralbl. 49, 2:34 (1932). *R. pulcher* L. var. *palaestinus* Feinbr., Palest. Journ. Bot. Jerusalem ser. 2:98 (1940). Valves 4–7 × 3–5 mm., oblong-triangular, somewhat acute, with (2–)3–5(–6) short teeth on each side (rarely entire). Basal leaves frequently lyrate.

Hab.: Roadsides and damp places. Acco Plain, Sharon Plain, Philistean Plain, Upper and Lower Galilee, Mt. Carmel, Esdraelon Plain, Samaria, Judean Mts., Hula Plain, Upper Jordan Valley, Gilead, Ammon.

Subsp. **divaricatus** (L.) Murb., Lunds Univ. Arsskr. 27:45 (1891). *R. divaricatus* L., Sp. Pl. ed. 2, 478 (1762). Valves 5–6 × 5 mm., ovate-triangular to almost orbicular, broadly and shortly acuminate at apex, with 3–8, 0.5–1.5 mm. long teeth on each side. Basal leaves rarely lyrate.

Hab.: Roadsides and damp places. Upper Galilee.

Area of species: Euro-Siberian, Mediterranean, Irano-Turanian; also in S. Africa. Leaves used as pot herb.

11. Rumex cassius Boiss., Fl. 4:1013 (1879). *R. pulcher* L. ssp. *cassius* (Boiss.) Rech. f., Beih. Bot. Centralbl. 49, 2:38 (1932). [Plate 76]

Perennial, glabrous, 35–100 cm., with short, thick rootstock. Stems erect, almost simple. Lower leaves up to 15 × 5(–6) cm., very long-petioled, oblong or ovate, rounded, truncate or subcordate at base, obtuse at apex; upper leaves with shorter petioles, oblong-lanceolate, somewhat acute; all obsoletely crisp. Racemes long, almost simple, leafy at base; whorls many-flowered, more or less remote. Pedicels rather thick, jointed near base, thickening at length, shorter than valves, reflexed. Valves (5–)6 × 4–6 mm. (including teeth), leathery, triangular-ovate to deltoid, with 3–5 subulate, hooked-spinescent, 1–2.5 mm. long teeth on each side, each valve with an ovate or oblong, slightly verrucose wart; warts unequal, up to 2.5 mm. Achene about 2 mm., trigonous, brown. Fl. April–June.

Hab.: Maquis. Upper Galilee, Mt. Carmel.

Area: E. Mediterranean (endemic in Palestine and Syria).

12. Rumex dentatus L., Mant. Alt. 226 (1771); Boiss., Fl. 4 : 1013 (1879); Rech. f., Candollea 12 : 116 (1949). [Plate 77]

Annual, glabrous, 15–90 cm. or more. Stems erect, simple or somewhat branching. Branches arcuate-divaricate to erect. Basal leaves up to 8 (–12) × 3 cm., ovate-oblong or oblong-lanceolate, sometimes fiddle-shaped, truncate to subcordate at base, more or less obtuse at apex, somewhat wavy; petiole as long as or shorter than blade; upper leaves smaller, short-petioled, oblong to elliptical, lanceolate, somewhat acute, minutely crisp or almost entire. Panicles branched or almost simple, many-whorled; whorls many-flowered, remote, upper ones sometimes confluent, almost all subtended by lincar-lanceolate leaves. Pedicels reflexed, jointed near base, almost equalling fruiting perianth. Valves 3–6 × 2–7 mm. (including teeth), ovate-lanceolate, more or less acute, reticulate, mostly with subulate teeth, sometimes subentire, each with an ovate to oblong-lanceolate, 1.5–3 mm. long wart. Achene about 2 mm., trigonous, broadly ovoid, dark brown. Fl. March–September.

Subsp. **mesopotamicus** Rech. f., Beih. Bot. Centralbl. 49, 2 : 15 (1932) et Candollea 12 : 121 (1949). *R. dentatus* L. var. *pleiodon* Boiss., Fl. 4 : 1013 (1879) p.p. Valves 4–6 × 4–7 (–8) mm. (including teeth), with warts 2–3 × 1–2 mm.; teeth 3–5, equal, strong, divaricate, longer than diameter of valves.

Hab.: Roadsides and moist places. Sharon Plain, Philistean Plain, Upper Jordan Valley, Beit Shean Valley, Lower Jordan Valley, Dead Sea area, Ammon, Moav.

Subsp. **callosissimus** (Meissn.) Rech. f., Beih. Bot. Centralbl. 49, 2 : 13 (1932) et Candollea 12 : 122 (1949). *R. callosissimus* Meissn. in DC., Prodr. 14 : 57 (1856). Valves 3–4 × 2–4 mm., with prominent warts 1.5–2.5 × 1–2 mm.; teeth 1–3, unequal, short, weak, sometimes absent.

Hab.: Roadsides and moist places. Sharon Plain, Philistean Plain, Judean Mts., Hula Plain, Moav.

Area of species: Mediterranean, Euro-Siberian, Irano-Turanian, Sino-Japanese; also in some tropical regions.

13. Rumex maritimus L., Sp. Pl. 335 (1753); Meissn. in DC., Prodr. 14 : 59 (1856); Rech. f., Candollea 12 : 128 (1949). [Plate 78]

Annual or biennial, glabrous, up to 70 cm., turning yellowish-green with age. Stems erect, simple below, paniculate above, sulcate-striate. Leaves up to 13 × 3 cm., petiolate, oblong-lanceolate to lanceolate-linear, more or less tapering at base, acute at apex, minutely crispulate. Whorls many-flowered, the lowest more or less remote, the upper crowded, all subtended by narrow leaves. Pedicels mostly longer than valves, filiform, jointed near base, elongated in fruit, reflexed. Outer tepals usually spreading, sometimes subrecurved in fruit. Valves 2.5–4 mm., narrowly ovate-triangular, somewhat acute, with 2 narrow, 1.5–2.5 mm. long teeth on each side, reticulate-veined, each valve bearing a prominent, yellowish, lanceolate, about 1.5 mm. long wart. Achene 1.3–1.5 mm., trigonous, ovoid, acuminate, yellowish-brown. Fl. April.

    Hab.: At edge of water. Hula Plain. Rare.

    Area: Euro-Siberian; also in N. and S. America.

Subgen. PLATYPODIUM (Willk.) Rech. f. Flowers hermaphrodite. Fruiting perianth leathery, with thick, hooked teeth. Fruiting pedicels flattened or club-shaped, recurved. Leaves entire, not hastate-sagittate.

14. Rumex bucephalophorus L., Sp. Pl. 336 (1753); Boiss., Fl. 4 : 1014 (1879); Rech. f., Candollea 12 : 140 (1949) [Plate 79]

    Annual, 10–35 cm. Stems single or few, ascending to erect. Leaves 1–4 × 0.5–2 cm., the basal ones long-petioled, ovate-spatulate to oblong-lanceolate, tapering at both ends; upper leaves short-petioled, lanceolate to linear, acute; ochreae up to 1 cm., membranous-hyaline. Inflorescences terminal, simple, racemose or spicate, many-whorled, elongated in fruit; whorls (1–)2–3(–4)-flowered, the lower remote, the upper dense. Bracts membranous-hyaline, longer than flowers. Pedicels at first shorter, later longer than flowers, dilated, jointed near base, reflexed. Flowers about 2 mm., hermaphrodite, sometimes dimorphic. Valves 2.5–5 × 2.5–6 mm., leathery, deltoid, obtuse, with 2–4 narrowly triangular spiny teeth on each side and a minute wart at base. Fl. February–May.

Var. **bucephalophorus.** Valves neither papillose nor glandular.

    Hab.: Sandy places. Coastal Galilee, Acco Plain, Sharon Plain, Philistean Plain, Samaria, Judean Mts., N. Negev, Upper Jordan Valley, Gilead, Ammon, Moav.

Var. **papillosus** Feinbr., Palest. Journ. Bot. Jerusalem ser. 2 : 98 (1940). Pedicels and valves papillose or glandular-papillose.

    Hab.: Sandy places. Sharon Plain, Philistean Plain, N. Negev, Gilead (endemic in Palestine).

    Area of species: Mediterranean, extending into W. Europe.

    Leaves used as pot herb.

4. EMEX Campd.

Annual, polygamous-monoecious herbs. Leaves alternate, entire or nearly so, generally ovate; ochreae membranous, truncate and cleft at apex. Flowers clustered at nodes or arranged in axillary, leafless, interrupted racemes; flowers of lower whorls pistillate, those or upper ones hermaphrodite or staminate; a few flowers at base of stem pulled down by contractile roots and ripening under soil surface. Staminate flowers pedicellate; perianth herbaceous, 3–6-partite with equal, spreading segments. Pistillate flowers more or less sessille; perianth herbaceous, urceolate, 6-lobed. Stamens 4–6. Ovary trigonous, 1-ovuled; styles 3, short, with fringed stigmas. Fruiting perianth much accrescent, indurated, with a trigonous tube pitted between ribs, outer 3 segments spinescent, spreading or recurved at apex, inner ones erect, connivent. Achene free, triquetrous, with membranous-leathery pericarp. Embryo lateral or peripheral; cotyledons narrowly linear to club-shaped.

One or 2 species in the Mediterranean countries, S. Africa and Australia.

1. Emex spinosa (L.) Campd., Monogr. Rum. 58 (1819); Boiss., Fl. 4 : 1005 (1879). *Rumex spinosus* L., Sp. Pl. 337 (1753). [Plate 80]

Glabrous annual, 5–50 cm., many- or few-stemmed. Stems decumbent to erect, sometimes reddish. Leaves 1–12 × 1–7 cm., petiolate, ovate to oblong, entire, truncate to cordate at base, more or less rounded and apiculate at apex. Staminate flowers 1.5–2 mm., on filiform pedicels together with some perfect ones, clustered at upper part of interrupted racemes. Pistillate flowers 1.5 mm., sessile or nearly so, clustered at nodes and lower parts of racemes; perianth urn-shaped, greenish-red, much growing in fruit, with outer segments spinescent, more or less recurved at apex and inner ones ovate-triangular, 3-ribbed. Achene about 4 mm., enclosed in perianth, triquetrous. Fruits of two kinds, aerial and subterranean, the latter crowded near root neck, larger and less spinescent than the former. Fl. December–May.

Var. **spinosa.** Leaves many, up to 12 × 6.5 cm.; petiole generally as long as or shorter than blade. Fruit 6–8 mm., more or less spinescent. Cotyledons more or less wide, club-shaped.

Hab.: Fields and roadsides. Coastal Galilee, Acco Plain, Sharon Plain, Philistean Plain, Upper Galilee, Judean Mts., Judean Desert, Upper Jordan Valley, Moav, Edom.

Var. **minor** Zoh. et Waisel.* Leaves 1–5 × 1–2.5 cm., yellowish-green; petiole generally longer than blade. Fruit small, 4–6 mm., more spiny than in var. *spinosa.* Cotyledons narrow, linear.

Hab.: Steppes and deserts, particularly on sandy soils. Sharon Plain, Judean Desert, Negev, Lower Jordan Valley, Dead Sea area, Arava Valley, Edom.

Area of species : Mediterranean, extending into adjacent Saharo-Arabian territories.

The relations between var. *minor* and var. *pusilla* Beg. et Vaccari, Sec. Contr. Fl. Libia 21 (1913) need further clarification.

Leaves and roots used for culinary purposes.

5. CALLIGONUM L.

Shrubs, much branched, with rigid branches, the young ones weak, fasciculate. Leaves minute, soon deciduous; ochreae short, membranous. Flowers small, hermaphrodite, single or few, clustered on jointed, axillary pedicels, whitish. Perianth of 5 subequal, more or less spreading lobes. Stamens 10–18, connate at base. Ovary tetragonous; styles 4, with capitate stigmas. Achene woody, much longer than the subpersistent perianth, tetragonous, 4-winged or echinate and bearing 8–16 rows of branching

* See Appendix at end of this volume.

intertwined bristles. Seeds oblong, 4-grooved; embryo straight, in the centre of a fleshy endosperm.

About 70 species in Asian deserts, a few in N. Africa.

1. Calligonum comosum L'Hér., Trans. Linn. Soc. Lond. 1 : 180 (1791); Boiss., Fl. 4 : 1000 (1879). [Plate 81]

Glabrous shrub, up to 1.5 m. or more. Stems and branches ascending to erect, articulate, tortuous, becoming lignified and white; young branches fasciculate, green. Leaves soon deciduous, up to 2.5 × 1 mm., subulate; ochreae membranous, short, 2-lobed. Pedicels 1–3 in an axil, usually longer than perianth, jointed at about middle. Flowers about 3–5 mm., arranged in clusters along young branches; perianth rotate, greenish-white. Achene up to 1.3 × 0.6 cm. (without bristles), oblong-ovoid, bearing 10–16 rows of soft, long and branching bristles, somewhat dilated at base into 4 short wings. Fl. February–April.

Hab.: Sands and sand dunes. W., C. and S. Negev, Dead Sea area, Arava Valley, deserts of Edom.

Area: Saharo-Arabian and W. Irano-Turanian.

This species is a codominant in the *Anabasis-Calligonum* and *Haloxylon persicum-Calligonum* communities in the N.E. Negev and the Arava Valley, respectively.

An excellent desert sand-binder.

6. ATRAPHAXIS L.

Shrubs, much branching, often spinescent. Leaves alternate, subsessile, solitary or fasciculate on the shorter twigs, mostly leathery or fleshy; ochreae membranous, 2-fid at apex. Pedicels filiform. Flowers hermaphrodite, grouped in small axillary clusters on short twigs. Perianth persistent, petaloid, 4–5-partite, 2-whorled, outer segments smaller, reflexed in fruit, inner segments larger, more or less erect, growing into scarious wing-like valves appressed to achene. Stamens 6–8, with filaments dilated at base, the 2–3 inner ones longer than outer. Ovary 2–3-carpelled, 1-ovuled, compressed or trigonous; styles 2–3, short, free or somewhat connate at base, bearing capitate stigmas. Achene lenticular or trigonous. Embryo curved; cotyledons linear.

About 20 species mainly in S.W. and C. Asia.

1. Atraphaxis spinosa L., Sp. Pl. 333 (1753); Boiss., Fl. 4 : 1020 (1879). [Plate 82]

Shrub, 30–80 cm. Stems and branches more or less ascending, woody at base, mostly spiny at apex, bark greyish-white to brownish. Young branches long, thin, leafy. Leaves up to 1 cm., subsessile, elliptical-rhombic to ovate or suborbicular, generally subcordate at base, more or less obtuse at apex, crisp, netted-nerved, green or glaucous; ochreae small, 2-fid. Pedicels filiform, jointed at middle or below it. Flowers clustered in axillary groups of 1–6, on short twigs of the current year. Perianth 4-partite, pink with white margins, 2 outer segments ovate-orbicular, reflexed in fruit, the 2 inner segments cordate-orbicular to reniform, reaching 0.8–1 cm. in fruit. Stamens 6. Styles 2–3. Achene 3–4 mm., sublenticular, broadly ovate to orbicular, light brown, smooth and shining. Fl. May–June.

Hab.: Stony places, in mountainous steppes. C. and S. Negev, Edom.

Area: Irano-Turanian.

Description fits var. **sinaica** (Jaub. et Sp.) Boiss., l.c. 1021, distinguished by its smaller orbicular leaflets. This is the only variety occurring in Palestine.

Centrospermae

26. PHYTOLACCACEAE

Herbs, shrubs or trees – or rarely woody vines, with alternate, simple, entire leaves; stipules minute or 0. Inflorescences terminal or axillary, racemose or cymose, mostly bracteate. Flowers small, hermaphrodite or unisexual (and then plants monoecious), mostly actinomorphic. Calyx of 4–5 free or somewhat connate sepals. Petals 0. Stamens 3 to many or of same number as sepals, often inserted on a hypogynous disk; filaments free or connate at base; staminodes sometimes petaloid. Gynoecium apocarpous or syncarpous, made of 2 to many carpels; ovary superior, rarely inferior; style 1 or 0; stigmas as many as carpels, usually linear to filiform. Fruit a berry, drupe, schizocarp, utricle or achene. Seeds often arillate with copious endosperm surrounded by the embryo.

Seventeen genera and 125 species mainly in Trop. America.

1. PHYTOLACCA L.

Herbs, shrubs or trees, sometimes climbers, glabrous or younger parts pubescent. Leaves petiolate or subsessile, acuminate or acute, exstipulate. Inflorescence a raceme, panicle or spike. Pedicels bracteate at base, often with 1–3 bracteoles. Flowers small, mostly hermaphrodite. Perianth sepaloid or coloured, persistent and mostly reflexed in fruit; tepals 4–5, almost equal, glabrous or nearly so. Stamens 5–30; filaments mostly free; anthers 2-celled, dorsifixed. Carpels 5–16, free or united at base or completely connate; ovule 1 in each carpel, basal, campylotropous; styles as many as carpels, erect or recurved, stigmatous on inner surface. Fruit a depressed-globular, juicy or fleshy berry of free or connate carpels. Seeds compressed, almost reniform, with crustaceous coat; endosperm mealy; embryo annular; cotyledons semiterete.

About 35 species mostly in the tropics and subtropics of the western hemisphere; some in Africa and Asia.

1. Phytolacca americana L., Sp. Pl. 441 (1753). *P. decandra* L., Sp. Pl. ed. 2, 631 (1762); Boiss., Fl. 4 : 895 (1879). [Plate 83]

Perennial, erect, branching herb, somewhat succulent, glabrous, with strong odour, 80–150 cm. Stems stout, striate, usually purplish. Leaves 10–25 cm., ovate to oblong to broadly lanceolate, cuneate at base, acute or acuminate at apex, entire; petiole 2–3 cm. Inflorescences 5–20 cm., racemed, long-peduncled, many-flowered. Flowers hermaphrodite, pale green, on divergent, 0.4–1.2 cm. long pedicels, provided with a larger subulate-lanceolate bract at base and 2 smaller bracteoles above. Perianth 4–6

mm. broad, white; tepals ovate-orbicular. Stamens 10. Ovary 10-carpelled, 10-celled.
Fruit a depressed-globular, 10-ribbed berry, 0.7–1.2 cm. across, dark purple. Fl.
July–September.

Hab.: Waste places. Philistean Plain, Mt. Carmel, Samaria. Very rare and most
probably adventive.

Area: Presumably N. American in origin; widely distributed in the Mediterranean,
Euro-Siberian and other regions.

Grown in gardens and also used as a dye plant (berries); young leaves have high vitamin
C content and are used as vegetable; also used in medicine as a purgative and stimulant
(radix et baccae phytolaccae).

27. NYCTAGINACEAE

Herbs, climbing shrubs or trees. Leaves opposite, sometimes alternate or whorled,
petiolate, simple, exstipulate. Inflorescences various, usually cymose. Flowers hermaphrodite, sometimes unisexual, usually subtended by free or united bracts. Perianth
petaloid, usually 5-lobed, rotate, campanulate, tubular or funnel-shaped, lower part
persistent in fruit. Stamens 1–5 or many, free or united at base. Ovary superior, 1-
ovuled; style often long. Fruit an anthocarp with enclosed achene, ribbed, winged or
glandular. Embryo straight or curved.

Thirty genera and about 300 species, mostly in tropical regions of both hemispheres,
especially in America.

1. Petaloid part of the perianth campanulate or rotate. Anthocarps without projecting
   warts.   **2. Boerhavia**
 – Petaloid part of the perianth funnel-shaped or tubular, rarely campanulate. Anthocarps
   with projecting, glandular warts.   **1. Commicarpus**

1. COMMICARPUS Standl.

Shrublets or herbs. Stems branching, the young ones herbaceous. Leaves opposite,
petiolate, simple, entire to incised, palmately and pinnately veined. Panicles composed
of whorls or umbels. Bracts small. Flowers hermaphrodite, without involucre. Perianth
more or less funnel-shaped with 5-lobed petaloid limb and persistent tube. Stamens 2–5,
free; anthers 2-celled. Ovary 1-celled, 1-ovuled, with style as long as or longer than
stamens. Anthocarps more or less cylindrical to clavate, obsoletely ribbed, with stalked
or sessile conspicuous warts at apex.

About 8 species, mostly in tropical regions.

1. Flowers 10 or less in each umbel, 1–1.5 cm. Stamens 3, long-exserted. Fruit 8–9 mm.
                                                                            1. C. africanus
 – Flowers 5 or less in a whorl, 3–4 mm. Stamens 2, slightly exserted. Fruit about 5 mm.
                                                                          2. C. verticillatus

1. Commicarpus africanus (Lour.) Dandy in Andrews, Fl. Pl. Angl.-Eg. Sudan 1 : 152
(1950). *Boerhavia africana* Lour., Fl. Cochinch. 1 : 116 (1790). *B. plumbaginea* Cav.,
Ic. 2 : 7, t. 112 (1793); Boiss., Fl. 4 : 1044 (1879). [Plate 84]

Perennial, glabrous to pubescent herb. Stems herbaceous, erect or scrambling, branching, more or less woody below. Leaves 3–7 × 1–4 cm., petiolate, broadly ovate to ovate-lanceolate, rounded, cuneate, truncate or subcordate at base, obtuse to acute at apex, entire to deeply sinuate. Panicles usually terminal, leafy, loose, compound of 1–2 whorls or umbels borne on axillary rather long peduncles. Bracts small, soon deciduous, puberulent. Pedicels as long as or shorter than flowers. Flowers up to 10 in each umbel, 1–1.5 cm. Perianth funnel-shaped, white or lilac above, tubular and greenish below, short-pubescent; limb 4–5 mm. broad. Stamens 3, long-exserted. Fruit up to 9 mm., somewhat ribbed, with stalked or sessile tubercles at apex. Fl. December–June.

Var. **africanus**. *Boerhavia plumbaginea* Cav. var. *glabrata* Boiss., l.c. Green plants, glabrous to sparingly pubescent or appressed-hirtellous. Leaves entire or somewhat sinuate.

Hab.: Shady places. Philistean Plain, Upper and Lower Galilee, Esdraelon Plain, Judean Desert, N. and C. Negev, Upper Jordan Valley, Beit Shean Valley, Lower Jordan Valley, Dead Sea area, Arava Valley, Moav, Edom.

Var. **viscosus** (Ehrenb. ex Aschers. et Schweinf.) Baum (comb. nov.). *Boerhavia plumbaginea* Cav. var. *viscosa* Ehrenb. ex Aschers. et Schweinf. in Schweinf., Beitr. Fl. Aethiop. 166 (1867); Boiss., l.c. Densely crisp-puberulent. Stems viscid, more woody than in the preceding variety. Leaves small, sinuate to repand.

Hab.: Shady places. S. Negev, Arava Valley.

Area of species: Probably primarily Sudanian and extending into other tropical regions as well as into adjacent Mediterranean territories.

2. Commicarpus verticillatus (Poir.) Standl., Contrib. U.S. Nat. Herb. 18 : 101 (1916). *Boerhavia verticillata* Poir. in Lam., Encycl. 5 : 56 (1804); Boiss., Fl. 4 : 1044 (1879). [Plate 85]

Perennial, glabrous or glabrescent climbing herb, up to 1 m. or more. Stems much branched, forked, pale-green. Leaves up to 5.5 × 4 (–5) cm., petiolate, broadly ovate to ovate-lanceolate, more or less truncate at base, obtuse to acute at apex, obscurely sinuate-repand at margin. Panicles terminal and axillary, loose, composed of remote, pedunculate or sessile 3–5-flowered whorls. Bracts small and narrow. Pedicels longer or shorter than flowers. Flowers 3–4 mm. Perianth campanulate to funnel-shaped and pink above, tubular and greenish below, puberulent; limb 2–3 mm. broad. Stamens 2, slightly exserted. Fruit about 5–7 mm., obscurely ribbed, glandular, with reddish brown tubercles in ring-like rows. Fl. March–May.

Hab.: Shady places in deserts. Judean Desert, Dead Sea area, Arava Valley, S. Negev, Edom.

Area: Sudanian, extending into adjacent regions.

2. Boerhavia L.

Herbaceous or suffrutescent plants. Stems more or less divaricately branching. Leaves opposite or whorled, usually petiolate, simple, more or less pinnately veined. Panicles composed of umbels, heads or whorls. Bracts small, sometimes deciduous. Flowers generally hermaphrodite, small, without involucre. Petaloid part of perianth rotate or campanulate, 5-lobed, lower part very short-tubular. Stamens 1–6, mostly 5, unequal; anthers 2-celled. Ovary stipitate, 1-celled, 1-ovuled; style as long as stamens. Fruit an anthocarp with enclosed achene, often 5-ribbed and glandular but without warts.

About 20 species, mainly in tropical regions.

1. Boerhavia repens L., Sp. Pl. 3 (1753); Boiss., Fl. 4 : 1045 (1879). [Plate 86]

Perennial, glabrescent or puberulent, up to 60 cm. Stems prostrate or ascending, much branching, woody below, Branches herbaceous, slender. Leaves up to 4 × 3 cm., petiolate, somewhat fleshy, broadly ovate to oblong or lanceolate or lanceolate-linear, usually rounded, sometimes truncate or subcordate at base, obtuse or nearly acute at apex, slightly sinuate-repand. Panicles leafy, composed of many umbels; umbels head-like, (3–)4–10-flowered, borne on axillary, short, slender peduncles. Bracts small, lanceolate. Flowers up to 2 mm., sessile or nearly so. Perianth pinkish, almost campanulate, with short lobes and persistent shorter tube, elongating in fruit. Stamens 1–3. Fruit 3–4 mm., club-shaped to ellipsoidal-oblong, markedly 5-ribbed, viscid-glandular. Fl. March–November.

Hab. : Shade of rocks. Lower Jordan Valley, Dead Sea area, Moav.

Area : Tropical regions of Africa and Asia, with extensions into Saharo-Arabian territories.

28. MOLLUGINACEAE

Shrubs and herbs. Leaves opposite or in whorls of 3–4 or alternate; stipules when present scarious, soon deciduous or 0. Inflorescences dichasial, glomerate or scorpioid. Flowers hermaphrodite or unisexual, actinomorphic, with simple or double perianth. Sepals 4–5, free or connate at base. Petals (where present) mostly derived from outer stamens. Stamens 5–10, rarely more, free or connate at base. Ovary superior; carpels 2–5, syncarpous, rarely apocarpous; ovules campylotropous or anatropous. Fruit a capsule, schizocarp or nutlet.

About 14 genera and 95 species mostly in tropical or subtropical regions of both hemispheres.

1. Carpels free. Fruit a schizocarp of 5 free achenes.   **2. Gisekia**
– Carpels united. Fruit a capsule.   **1. Glinus**

1. Glinus L.

Annual, tomentose (rarely glabrous) herbs. Leaves alternate or spuriously whorled, exstipulate. Flowers hermaphrodite, axillary, usually crowded at nodes, more or less pedicellate. Sepals 5, equal, imbricated. Petals 0 or indefinite, entire or 2–3-fid. Stamens 8–20, distinct or in groups. Ovary superior, 3–5-celled, many-ovuled; styles 3–5. Capsule incompletely 3–5-celled, membranous, 2–5-grooved, opening by 5 valves. Seeds reniform, with an entire caruncle; embryo curved, cylindrical.

About 8 species in tropical and subtropical regions.

1. Glinus lotoides L., Sp. Pl. 463 (1753); Boiss., Fl. 1 : 755 (1867). [Plates 87, 88]

Annual, pubescent-tomentose, stellately hairy, up to 30 cm. Stems prostrate or ascending, dichasially branched. Leaves spuriously whorled, petiolate, unequal in size and shape, oblong-spatulate to obovate or orbicular, obtuse, densely pubescent to glabrescent. Flowers 3–8 mm., axillary, few or many at each node, on short, unequal pedicels. Sepals oblong-ovate, more or less obtuse, membranous at margin, hairy on outer faces. Petals 0 or indefinite, white, linear, entire or 2–3-fid. Stamens 8–15 (–20). Capsule almost as long as calyx, ovoid-pentagonous, many-seeded. Seeds finely tuberculate, brown. Fl. May–October.

Var. **lotoides**. [Plate 87] Leaves 0.6–3.5 × 0.6–2.5 cm. Flowers and fruit 5–8 mm. Stamens 12–15 (–18). Stigmas 5. Petals usually present.

Hab. : Soils inundated in winter and dried up in early summer. Acco Plain, Sharon Plain, Philistean Plain, Mt. Carmel, Esdraelon Plain, Hula Plain, Upper Jordan Valley, Beit Shean Valley.

Var. **dictamnoides** (Burm. f.) Maire, Fl. Afr. N. 8 : 276 (1962). *G. dictamnoides* Burm. f., Fl. Ind. 113 (1768); L., Mant. Alt. 243 (1771); Boiss., Fl. 1 : 756 (1867). [Plate 88] Leaves much smaller than in preceding variety, 0.3–1.5 (–2) × 0.2–1 cm. Flowers and fruit mostly up to 5 mm. Stamens 8–10. Stigmas 3. Petals usually 0.

Hab. : As above. Sharon Plain, Philistean Plain, Upper Galilee, Judean Mts., Hula Plain, Upper Jordan Valley, Gilead.

Area of species : Mainly Mediterranean and Irano-Turanian; also found in tropical parts of Asia and Africa.

Transitions between the two varieties have been observed.

2. GISEKIA L.

Herbs, mostly annual. Leaves opposite or spuriously whorled, exstipulate. Flowers small, unisexual, sessile or pedicellate, in axillary cymes. Perianth sepaloid, green or slightly coloured, 5-partite; lobes equal. Stamens 5–15, free; filaments subulate, dilated at base. Carpels 3–5, free, 1-celled, sessile on a small torus; ovules solitary; styles short, recurved. Fruit a schizocarp of 5 free, 1-seeded, compressed achenes. Seeds crustaceous; embryo annular, enclosing endosperm.

Five species in Trop. Africa, Arabia and India.

1. Gisekia pharnacioides L., Mant. Alt. 554 et 562 (1771); Boiss., Fl. 4 : 896 (1879).

Glaucous, glabrous annual, branching from base. Stems procumbent, much branching, slender, flexuous. Leaves small, short-petioled, more or less fleshy, lanceolate, obtuse, entire. Cymes 10–20-flowered, subsessile in axils of bracts, shorter than leaves. Pedicels as long as or longer than flowers. Flowers minute. Perianth lobes spreading, later reflexed, oblong-ovate, acute, concave, membranous at margin. Stamens 5. Carpels 5. Achenes somewhat compressed, ovoid-globular, papillose on surface. Fl. April.

Hab. : Transjordan (Mouterde, oral communication).

Area : Tropical.

29. AIZOACEAE

Herbs or rarely low shrubs. Leaves opposite, sometimes alternate, rarely whorled, scale-like or succulent, exstipulate, rarely stipulate. Flowers hermaphrodite, solitary or crowded, actinomorphic. Perianth simple, sometimes double with sepals and petals (petaloid staminodes). Calyx tubular or campanulate, often fleshy; sepals 3–5, united or free. Petals (when present) usually numerous. Stamens 3–5 to many, inserted on calyx tube, and alternating with sepals. Ovary usually superior, sometimes inferior, 3–5-carpelled, 2- or more-celled, with 1 style and stigmas as many as cells, each cell 1- to many-ovuled; ovules campylotropous or anatropous. Fruit a dry or berry-like, many-seeded capsule. Embryo curved or annular, surrounding the endosperm.

Eleven (or, according to recent classification, over 100) genera and 2,500 species. Chiefly in S. Africa but also in some other warm regions, both of the western and eastern hemispheres.

Literature: F. Pax u. K. Hoffmann, Aizoaceae, in: Engl. u. Prantl, *Nat. Pflznfam.* ed. 2, 16c: 179–233 (1934).

1. Petals numerous, linear. **3. Mesembryanthemum**
– Petals 5 or 0 **2**
2. Stamens more than 5. Stigmas 3–5. Capsule 3–5-celled. **1. Aizoon**
– Stamens (in ours) 5. Stigmas 1–2. Capsule 2-celled, opening by an operculum.
2. Trianthema

Subfam. AIZOOIDEAE. Ovary superior. Petals 0.

1. AIZOON L.

Annual or perennial herbs or low shrubs. Leaves alternate or opposite, sessile or petiolate, usually fleshy, simple, exstipulate. Flowers hermaphrodite, axillary, solitary or in cymes. Perianth sepaloid, 5-cleft, with more or less turbinate or hemispherical tube. Stamens many, in groups, alternating with perianth segments; anthers 2-celled. Ovary superior, 3–5-celled, each cell 2- to many-ovuled; stigmas 3–5. Capsule 3–5-celled, dehiscent loculicidally or by a star-shaped slit at apex.

About 10 species mainly in Africa and Australia.

1. Leaves opposite, mostly sessile, oblong-lanceolate. Flowers about 1.5 cm., with perianth segments longer than tube. Stems ascending to erect. **1. A. hispanicum**
– Leaves alternate, petiolate, more or less spatulate. Flowers about 4–6 mm., with segments shorter than tube. Stems procumbent. **2. A. canariense**

1. Aizoon hispanicum L., Sp. Pl. 488 (1753); Boiss., Fl. 2: 765 (1872). [Plate 89]

Annual, papillose, rather fleshy, 5–20 cm. Stems ascending to erect, dichasially branched. Leaves up to 4(–5) × 1 cm., more or less opposite, mostly sessile, oblong-lanceolate to almost linear, obtuse. Flowers 1.2–1.7 cm., short-pedicelled, green. Perianth somewhat fleshy, 5-cleft into lanceolate to ovate about 1.5 cm. long segments, green outside, whitish within. Capsule dehiscing when moistened, broadly ovoid-subconical, flattened at apex. Seeds about 1 mm., reniform, brown. Fl. February–May.

Hab.: Deserts, sometimes on saline ground. Judean Desert, W., N. and C. Negev, Lower Jordan Valley, Dead Sea area, desert E. of Gilead, Ammon.

Area: Saharo-Arabian, extending into dry areas of adjacent regions.

Common especially in the *Zygophyllum dumosum* and *Chenolea arabica* communities.

2. Aizoon canariense L., Sp. Pl. 488 (1753); Boiss., Fl. 2 : 765 (1872). [Plate 90]

Annual or perennating low herb, papillose-pubescent or villose. Stems procumbent, alternately branched, flexuous, woody towards base, 5–15 cm. Leaves usually about 1–3 × 0.8–1.5 cm., alternate, petiolate, fleshy, spatulate to oblong-obovate, cuneate and tapering at base, obtuse at apex, somewhat villose. Flowers 4–8 mm., sessile. Perianth segments 2–5 mm., broadly ovate to short-triangular, greenish outside, yellowish within. Capsule somewhat flattened. Seeds 1 mm., reniform, black. Fl. January–April.

Hab.: Wadis, dry stream beds and oases in hot deserts. E. Judean Desert, C. and S. Negev, Dead Sea area, Arava Valley, deserts of Moav, Edom.

Area: Mainly Sudanian, extending into adjacent territories of the Saharo-Arabian region; also in S. Africa and in the Macaronesian region.

Leaves eaten as salad.

2. TRIANTHEMA L.

Herbs or shrublets, prostrate, glabrous-papillose or sparingly villose. Leaves opposite, petiolate, often fleshy, unequal, entire, exstipulate. Flowers axillary, solitary or clustered, sessile or pedicellate. Calyx campanulate, 5-lobed, each lobe with a dorsal, subapical cusp. Petals 0. Stamens 5–10 or more, alternating with the calyx lobes. Ovary superior, 1–2-celled; stigmas 1–2. Capsule 2-celled, dehiscent by an operculum, membranous or coriaceous. Seeds 1 to many in each cell, orbicular-reniform, compressed, frequently wrinkled; embryo annular.

About 13 species in tropical and subtropical regions.

1. Trianthema pentandra L., Mant. 70 (1767); Boiss., Fl. 2 : 766 (1872). [Plate 91]

Perennial herbs, more or less setulose-puberulent, covered with pustules. Stems prostrate or ascending, branched, 15–30 cm. Leaves 1–4 × 0.3–0.8(–1.5) cm., oblong-obovate to elliptical, obtuse; base of petiole expanded at both sides into membranous wings. Bracteoles membranous. Flowers about 3 mm., clustered, subsessile. Calyx 5-cleft into ovate to oblong lobes, membranous at margin and hooded at apex. Petals 0. Stamens 5. Stigmas (1–)2. Capsule about as long as or somewhat longer as or somewhat longer than calyx, 2-celled, coriaceous. Fl. March–July(–October).

Hab.: Waste places. Philistean Plain, Judean Desert, Upper Jordan Valley, Beit Shean Valley, Lower Jordan Valley, Dead Sea area. Very rare.

Area: Sudanian; also in other Paleotropical regions.

This subfamily also comprises *Tetragonia tetragonoides* (Pall.) 0. Ktze., Rev. Gen. 1 : 264 (1891) (=*T. expansa* Murr.). It is a fleshy procumbent annual with hastate leaves, cultivated as the New Zealand spinach; it also occurs as a garden escapee.

Subfam. APTENIODEAE. Ovary half-inferior. Petals more or less connate at base.

3. MESEMBRYANTHEMUM L.

Annual or perennial, usually fleshy or papillose herbs or shrubs. Stems more or less dichasially branched. Leaves opposite, rarely alternate, flat or terete, exstipulate. Flowers mostly axillary, solitary or in cymes, white, yellow, pink or red. Calyx with 4–5 often unequal segments. Petals (petaloid staminodes) many. Stamens many, often villose. Ovary half-inferior, (4–)5-celled, many ovuled; stigmas 4–5, free or connate at base. Capsule 5-celled, subglobular-pentagonous, depressed at apex, ligneous or fleshy-leathery, opening by a star-shaped slit. Seeds many, minute; embryo curved.

The Linnaean genus *Mesembryanthemum* has recently been divided by various authors into as many as 50 to 100 genera. Reportedly *Mesembryanthemum* s.s. comprises as many as 350 species which are mainly confined to S. Africa.

1. Leaves flat, obovate to spatulate, the radical petiolate, all covered with pellucid, glistening vesicles.   **1. M. crystallinum**
– Leaves thick, terete or linear, sessile, papillose   2
2. Flowers sessile or nearly so. Petals (petaloid staminodes) shorter than calyx. Leaves less than 5 mm. broad, the upper alternate, the lower opposite.   **2. M. nodiflorum**
– Flowers short-pedicelled. Petals longer than calyx. Leaves 0.8–1.5 cm. thick, all opposite.
                                                                               3. M. forsskalii

1. Mesembryanthemum crystallinum L., Sp. Pl. 480 (1753); Boiss., Fl. 2 : 764 (1872). *Cryophytum crystallinum* (L.) N.E. Br. in Phillips, Gen. S. Afr. Fl. Pl. 245 (1926). [Plate 92]

Annual or biennial fleshy herb, densely covered with pellucid, glistening vesicles. Stems procumbent, branching, 20–80 cm. Leaves usually 3–7 × 2–5 cm., sometimes larger, radical ones opposite, petiolate, cauline alternate, narrowed into a short petiole or a clasping base, all leaves spatulate to obovate, obtuse, undulate. Flowers 1 cm. or more in diam., axillary, almost sessile, whitish to pale pink. Calyx with broadly ovate lobes. Petals many, longer than calyx. Capsule about 1 cm. Fl. February–June.

Hab.: Rocks, walls and sandy slopes facing the sea. Acco Plain, Sharon Plain, Philistean Plain.

Area: Mediterranean and W. Euro-Siberian (Atlantic). Origin S. Africa.

Sometimes leaves and flowers used as salad; ash used as washing soda; also grown as an ornamental plant.

2. Mesembryanthemum nodiflorum L., Sp. Pl. 481 (1753); Boiss., Fl. 2 : 764 (1872). *Cryophytum nodiflorum* (L.) L. Bol., S. Afr. Gard. 17 : 327 (1927). [Plate 93]

Annual, minutely-papillose, fleshy low herb. Stems generally ascending, 10–15 cm., moderately branching. Leaves up to 4 × 0.4 cm., the lower opposite, the upper alternate, sessile, linear to semiterete, sparingly ciliate at margins. Flowers axillary, almost sessile. Calyx up to 1 cm., with unequal, linear lobes. Petals almost filiform, white-cream, shorter than calyx. Stigmas 5, sessile. Capsule small, 0.7–1 cm. Fl. March–June.

Hab.: Dry salines in deserts and Mediterranean sea-shore. Sharon Plain, Philistean Plain, Esdraelon Plain, Judean Desert, W. and N. Negev, Lower Jordan Valley, Dead Sea area, Arava Valley, deserts of Ammon, Moav, Edom.

Area: Mediterranean, W. Euro-Siberian (Atlantic) and Saharo-Arabian. Origin S. Africa.

3. Mesembryanthemum forsskalii Hochst. ex Boiss., Fl. 2 : 765 (1872). *M. forskahlei* Hochst. in Schimp., Pl. Arab. Exs. ed. 2 (1832). *Opophytum forskahlii* (Hochst. ex Boiss.) N.E. Br., Gard. Chron. ser. 3, 84 : 253 (1928). [Plate 94]

Annual, papillose, fleshy herb, 10–25 cm. Stems short, erect to ascending, simple or branching from base, often spreading. Leaves up to 5 × 0.8–1.5 cm., thick, opposite, sessile, very fleshy, terete-conical, the upper ones decurrent. Flowers axillary. Pedicels shorter than calyx. Calyx with subequal conical lobes. Petals white-cream, longer than calyx. Capsule 1.2–1.5 cm. Fl. March–May.

Hab.: Hot deserts, mostly on saline soil. Lower Jordan Valley, Dead Sea area, Arava Valley, Edom.

Area: Sudanian, extending into the Saharo-Arabian borderland.

Believed by Löw, Flora der Juden 4 : 28 (1934) to be the אֲהָלִים of the Bible (Num. XXIV : 6). *Samh* in Arabic. Reportedly bread is made from the seeds; the latter are released by immersing the fruits in water so that their valves open hygrochastically (Crowfoot and Baldensperger, 1932).

30. PORTULACACEAE

Annual or perennial herbs or shrubs. Leaves alternate or opposite, simple, sometimes fleshy, mostly with scarious or setaceous stipules. Flowers hermaphrodite, actinomorphic, often showy. Calyx of 2 or more sepals, free or connate at base, often caducous. Corolla of 4–6 (rarely 2–3) petals, free or connate at base. Stamens as many as petals or 2–4 times as many (rarely fewer); filaments mostly free; anthers 2-celled, introrse, dehiscing longitudinally. Ovary superior (or half-inferior), 1-celled, of 2–3 carpels; placentation central or basal; ovules 1 to many, campylotropous; styles and stigmas 2–5 (–8). Fruit a capsule, dehiscing by a lid or by 2–3 apical valves, rarely fruit a nut. Seeds with copious endosperm surrounded by embryo.

Nineteen genera and over 500 species mostly in tropical and Pacific S. America.

1. PORTULACA L.

Diffusely branching annual or perennial herbs. Leaves alternate or almost opposite, mostly fleshy, flat or terete, with scarious or setaceous stipules, sometimes minute. Flowers mostly terminal, solitary or in clusters, sessile or pedicellate, yellow or purplish. Sepals 2, connate at base, partly united with ovary, caducous. Petals 4–6, free or inserted at top of calyx. Stamens 7 to many, inserted on petals. Ovary 1-celled; placentation central; ovules many in vertical rows; style filiform, 2–8-fid. Capsule membranous, opening by a lid. Seeds reniform, shining, often tuberculate.

About 100 species mostly in tropical regions of both hemispheres.

1. Portulaca oleracea L., Sp. Pl. 445 (1753); Boiss., Fl. 1 : 757 (1867). [Plate 95]

Annual, glabrous, prostrate, much branching, 10–50 cm. Branches fleshy, brittle, very leafy. Leaves 1–2 cm., opposite, sessile, fleshy, obovate-oblong, entire, dark green. Flowers sessile in dense forks. Sepals unequal, up to 2 mm. broad, obtuse, keeled beneath apex. Petals about as long as sepals, ovate-oblong, retuse, open in the morning only. Stamens 8–15. Capsule 0.6–1 cm., many-seeded, pyriform-rhomboidal, opening by a lid. Seeds 0.5–1 mm. across, black, tuberculate. Fl. mainly February–September.

Hab.: In irrigated and tilled fields and orchards. Acco Plain, Sharon Plain, Philistean Plain, Upper Galilee, Esdraelon Plain, Negev, Dan Valley, Hula Plain, Upper and Lower Jordan Valley, Arava Valley, Transjordan.

Area : Warm-temperate regions of both hemispheres.

The cultivated variety of this species (var. *sativa*) has not been recorded from Palestine, but a form with larger leaves and larger flowers occurs among the wild growing populations.

Formerly used in medicine; eaten as vegetable and salad.

31. CARYOPHYLLACEAE

Annual or perennial herbs, rarely half-shrubs. Leaves opposite (very rarely alternate), decussate, simple, entire, with or without stipules. Inflorescences mostly of mono-, di- or pleiochasial cymes, rarely flowers solitary and terminal. Flowers hermaphrodite (rarely unisexual and then plants dioecious), actinomorphic. Perianth usually composed of calyx and corolla, rarely simple and sepaloid. Calyx mostly of 5 (4) sepals, connate (by scarious strips of tissue – commissures) or free, sometimes subtended by bracts. Corolla of 5 (4) entire, lobed or dentate, sessile or clawed petals; apex of claw with or without scales. Stamens of same number as petals or twice as many, rarely fewer. Pistil 1; carpels usually 2–5; ovary superior, 1-celled with central placentation or 3–5-celled with axile placentation in lower and central in upper part (in 1- or few-ovuled ovaries placentation basal); styles and stigmas 2–5 (rarely 1). Fruit a capsule dehiscing by valves or teeth equalling the styles in number or twice as many or fruit an indehiscent achene or utricle (or berry). Seeds with copious endosperm, usually surrounded by the embryo.

About 80 genera and about 2,000 species, mainly in temperate regions of the northern hemisphere; some genera also in southern hemisphere and in mountains of the tropics.

Literature : F. Pax u. K. Hoffmann, Caryophyllaceae, in : Engl. u. Prantl, *Nat. Pflznfam.* ed. 2, 16c : 275–367 (1934). H. Rohweder, Weitere Beiträge zur Systematik und Phylogenie der Caryophyllaceen unter besonderer Berücksichtigung der karyologischen Verhältnisse. *Beih. Bot. Centralbl.* 59, 2 : 1–58 (1938). E. Janchen, Naturgemässe Anordnung der mitteleuropäischen Gattungen der Silenoideae. *Oesterr. Bot Zeitschr.* 102 : 381–386 (1955).

1. Sepals or tepals connate at least to about half their length   2
– Sepals or tepals free or connate at their lower part only   13
2. Woody desert shrubs with terete, glabrous and succulent leaves and whitish stems. Flowers in head-like clusters. Calyx about 7 mm.   **26. Gymnocarpos**
– Plants not as above   3
3. Styles 3–5   4
– Styles 2   5
4. Styles 3. Teeth of calyx usually much shorter than tube. Lower part of ovary (or capsule) septate, upper 1-celled. Petals with scales (coronal scales) at apex of claw.   **2. Silene**
– Styles 5. Lobes of calyx much longer or somewhat shorter than tube. Ovary not septate. Petals without coronal scales.   **1. Agrostemma**
5(3). Calyx with many scale-like bracts at base or calyx hidden within involucre of bracts   6
– Calyx not as above   7
6. Flowers mostly solitary, rarely 2–3 in a cluster but then each flower subtended by a group of bracts. Calyx herbaceous.   **9. Dianthus**
– Flowers 2–5 in a cluster, each cluster subtended by a group of bracts. Calyx scarious.
                                                                                              8. Petrorhagia
7(5). Fruiting calyx with 5 green, protruding, wing-like ribs.   **7. Vaccaria**
– Fruiting calyx not winged   8
8. Calyx at least 10 times as long as broad.   **10. Velezia**
– Calyx much shorter   9
9. Petals deeply 3- or 5-cleft.   **5. Ankyropetalum**
– Petals entire or 2-lobed   10
10. Calyx with 5 green, 3-nerved ribs.   **8. Petrorhagia**
– Calyx with or without simple ribs or nerves   11
11. Calyx purple, about 1 cm. long. Petals pink. Annuals.   **6. Saponaria**
– Calyx membranous between nerves, not purple, 2–6 mm. long. Perennials or annuals   12
12. Calyx tubular, pentagonous. Petal claw distinctly winged, with constricted apex. Capsule narrowly ovoid to cylindrical.   **4. Bolanthus**
– Calyx more or less campanulate. Petal claw mostly cuneate, not constricted at apex. Capsule globular or almost so.   **3. Gypsophila**
13(1). Leaves stipulate (stipules mostly scarious, scale-like)   14
– Leaves exstipulate   26
14. Leaves all alternate or the lower ones only sometimes opposite   15
– Leaves all or at least the upper ones opposite or whorled   16
15. Fruit a 1-seeded nutlet. Prostrate herbs with 1–2 mm. long flowers.   **25. Corrigiola**
– Fruit a many-seeded capsule. Flowers larger.   **20. Telephium**
16(14). Flowers crowded in short cymes borne on a common broadened, leaf-like peduncle or accompanied by an involucre made of sterile plumose branches   17
– Flowers not as above   18
17. Common peduncle of dichasium leaf-like.   **31. Pteranthus**
– Common peduncle not leaf-like; involucre plumose.   **30. Cometes**
18(16). Fruit a few- or many-seeded capsule, dehiscing by teeth or valves   19
– Fruit a 1-seeded, indehiscent nutlet or utricle   24
19. Stipules adnate to leaves up to middle. Calyx lobes with a subulate appendage on either side.   **24. Loeflingia**
– Stipules and sepals not as above   20

20. Tomentose plants with brittle stem and with leaves ending in a prickly point. Style thickened at apex. **22. Polycarpaea**
— Plants not tomentose. Other characters not as in above   21
21. Leaves (spuriously) whorled, narrowly linear. All or part of seeds distinctly winged or only narrowly margined but then stems glandular pubescent and styles or valves of capsule 5. **18. Spergula**
— Leaves opposite, rarely whorled but then stems glabrous and seeds wingless   22
22. Stems glandular or pubescent, at least in upper part. All or part of seeds winged or all seeds wingless. Styles 3 or 5. **19. Spergularia**
— Stems glabrous, not glandular. Seeds not winged. Style 1, 3-fid at apex   23
23. Bracts and stipules entirely scarious. Annuals. **21. Polycarpon**
— Bracts and stipules with a green stripe in middle. Perennials. **23. Robbairea**
24(18). Flowers in heads hidden between broad silvery bracts, rarely bracts shorter than flowers but then sepals hooded and provided with an awn near apex. Mostly prostrate annuals or perennials. **27. Paronychia**
— Flowers not as above   25
25. Leaves flat, oblong, elliptical or ovate. Fruiting calyces not crowded in spiny heads. **28. Herniaria**
— Leaves terete, filiform, the upper ones spiny, concrescent with flowers. Fruiting heads spiny. Petals 0. **29. Sclerocephalus**
26(13). Teeth or valves of capsule as many as styles (not exceeding 5)   27
— Teeth or valves of capsule twice (or more than twice) as many as styles, exceeding 5 in number   29
27. Styles and valves of capsule usually 3. Petals 5, entire or slightly notched, white or pink. **12. Minuartia**
— Styles and valves 2 or 4–5. Petals 0 or 4   28
28. Styles 2. Valves 2. Pedicels short. **13. Bufonia**
— Styles 4–5. Valves 4–5. Pedicels very long. **17. Sagina**
29(26). Petals entire, slightly notched or minutely dentate   30
— Petals (if present) distinctly 2-lobed or -parted   31
30. Flowers umbellate. Pedicels unequal in length, deflexed after flowering. Teeth of open capsule reflexed. **15. Holosteum**
— Flowers not umbellate. Pedicels erect. Teeth of open capsule erect. **11. Arenaria**
31(29). Styles 3. Leaves ovate. **14. Stellaria**
— Styles 5(4) or 3 but then leaves narrowly linear-lanceolate. **16. Cerastium**

Subfam. SILENOIDEAE. Perianth double. Calyx gamosepalous. Leaves exstipulate. Styles free. Torus often elongated between calyx and corolla.

Trib. LYCHNIDEAE. Styles 3 or 5. Fruit a many-seeded capsule, dehiscing by teeth or valves, rarely fruit baccate. Calyx with commissural nerves. Corolla imbricated in bud.

1. AGROSTEMMA L.

Annual herbs. Stems erect, villose. Leaves opposite, entire, exstipulate. Flowers conspicuous, hermaphrodite, long-pedicelled, solitary or in few-flowered dichasia. Calyx with 10-ribbed tube and 5 leaf-like, narrow, elongated lobes. Petals 5, often shorter than calyx, purple or white, without coronal scales. Stamens 10 (5 adnate to petals). Ovary 1-celled, with central placentation and many ovules; styles 5, alternating with calyx lobes. Capsule dehiscing by 5 teeth. Seeds reniform, black.

Two or 3 species in temperate Eurasia.

1. Calyx lobes much longer than tube and corolla.   **1. A. githago**
 – Calyx lobes shorter than tube and corolla.   **2. A. gracile**

1. Agrostemma githago L., Sp. Pl. 435 (1753). *Githago segetum* Desf., Fl. Atl. 1 : 363 (1798); Boiss., Fl. 1 : 661 (1867). [Plate 96]

Annual, appressed-hairy, canescent, 30–60 cm. Stems erect, simple or somewhat branched. Leaves 5–12 × 0.3–1 cm., sessile, connate at base, linear-lanceolate, usually acute. Flowers generally terminal, solitary, long-pedicelled. Calyx 4–5 cm., with cylindrical-ovoid tube and spreading, long-linear, acute lobes, distinctly longer than tube, caducous in fruit. Petals shorter than calyx, purple, obovate, somewhat retuse, clawed. Capsule about 1.5 cm., ovoid-oblong, exceeding calyx tube. Seeds 3–3.5 mm. in diam., black, muricate. Fl. April–June.

Hab.: Fields. Sharon Plain, Philistean Plain, Judean Mts., Judean Desert, Upper and Lower Jordan Valley. Rare and sporadic.

Area: Mediterranean, Euro-Siberian and Irano-Turanian.

Elsewhere (not in Palestine) a noxious weed. Seeds contain a poisonous glycoside; used in medicine as diuretic, vermifuge, etc.

2. Agrostemma gracile Boiss., Diagn. ser. 2, 1 : 80 (1853). *Githago gracilis* (Boiss.) Boiss., Fl. 1 : 661 (1867). [Plate 97]

Annual, appressed-hairy, canescent, 30–50 cm. Stems erect, simple or branched. Leaves up to 7 cm., sessile, narrowly linear, acute to acuminate. Flowers 3–4 cm., terminal, solitary, on very long pedicels. Calyx about 3 cm., with cylindrical tube and narrowly linear-subulate lobes shorter than tube. Petals longer than calyx, pink, obovate or obcordate, cuneate-clawed. Capsule and seeds unknown. Fl. Spring.

Hab.: Fields. Philistean Plain. Very rare.

Area: E. Mediterranean (Palestine, Turkey, Greece).

2. Silene L.

Annual or perennial herbs or dwarf-shrubs. Leaves opposite, entire, exstipulate. Flowers hermaphrodite, rarely unisexual, in cymose inflorescences or solitary. Calyx gamosepalous, usually tubular, mostly 10- but sometimes also 20-, 30- or 60-nerved; teeth 5. Petals 5, rarely 0, white, red, pink, purple or green, rarely yellow, long-clawed; limb entire to lobed, usually 2-fid; coronal scales usually present between limb and claw, sometimes rudimentary or 0. Stamens 10, unequal. Ovary mostly 3 (–5)-celled at base, 1-celled in upper part, many-ovuled; styles usually 3, alternating with sepals. Capsule often borne on a carpophore, rarely sessile, oblong to spherical, dehiscing by 6 (–10) teeth. Seeds more or less reniform.

About 400 species, mainly in the northern hemisphere, particularly in S.W. Asia.

Literature : P. Rohrbach, *Monographie der Gattung Silene*. Leipzig (1868). F. N. Williams, A revision of the genus *Silene*. *Journ. Linn. Soc. Lond. Bot.* 32 : 1–196 (1896). P. K. Chowdhuri, Studies in the genus *Silene*. *Not. Roy. Bot. Gard. Edinb.* 22 : 221–278 (1957).

1. Calyx 20–60-nerved 2
– Calyx mostly 10-nerved or nerves obsolete 4
2. Calyx 20-nerved. 33. S. coniflora
– Calyx 30- or 60-nerved 3
3. Calyx 60-nerved. 35. S. macrodonta
  Calyx 30-nerved. 34. S. conoidea
4 (1). Succulent littoral perennials with large flowers (3 cm. or more in diam.).
  7. S. succulenta
– Plants not as above 5
5. Petal limb with a basal tooth on each side. 12. S. aegyptiaca
– Petal limb not toothed at base 6
6. Petal limb cut into 4–6 filiform lobes. 5. S. physalodes
– Petal limb 2- rarely 3-lobed or -fid, or entire 7
7. Petals 3 cm. Calyx – or only the fruiting one – up to 1.5–3 cm. Perennial, viscid, glabrous plants, with 1-flowered inflorescences. 3. S. swertiifolia
– Petals and calyx shorter, rarely up to 2.5 cm. but then plants not glabrous 8
8. Annuals, glabrous and viscid, at least in upper part. Leaves (at least upper ones) linear-plicate or with revolute margin. Littoral or desert plants 9
– Plants with a different set of characters 10
9. Capsule as long as carpophore or a little shorter. 10. S. linearis
– Capsule twice or more than twice as long as carpophore. 11. S. modesta
10 (8). Calyx glabrous 11
— Calyx pubescent (papillose, villose, hirsute, etc.) 16
11. Calyx inflated, 1 cm. broad or more. Glaucescent but not viscid perennial herbs. 4. S. vulgaris
— Plants not as above 12
12. Petal limb purplish-veined. Carpophore longer than capsule. 9. S. reinwardtii
— Petal limb not veined. Carpophore shorter than capsule 13
13. Stems pubescent or scabrous. Calyx membranous. 14. S. rubella
— Stems glabrous and smooth, at least in upper part, often viscid. Calyx not membranous or slightly so but then striate with red veins 14
14. Flowers deflexed. Inflorescences divaricately branched. Petals white (rarely purple). Perennials. 2. S. longipetala
— Flowers erect. Inflorescences not divaricately branched. Petals pink. Annuals 15
15. Plants viscid at least in upper part. Calyx twice as long as broad or longer. Carpophore half as long as the oblong-ellipsoidal capsule. 17. S. muscipula
— Plants not viscid. Calyx and carpophore shorter than in above. Capsule ovoid. 16. S. behen
16 (10). Teeth of fruiting calyx 4–5 mm., acuminate-subulate, about half or more than half as long as calyx tube; calyx nerves thick, prominent. Carpophore 1–2 mm. Petals almost included. 30. S. tridentata
— Plants with a different set of characters 17
17. Seeds deeply grooved on back, with 2 undulate wings 18
— Seeds without undulate wings 19
18. Corolla long-exserted. 31. S. colorata

— Corolla included in calyx or very slightly exserted or corolla 0.   **32. S. apetala**

19(17). Leaves more or less fleshy. Pedicels spreading or deflexed in fruit. Dwarf or procumbent, glandular-hairy, coastal or sandy desert annuals   20
— Plants not as above   21

20. Flowers (6–)7–9 mm. Leaves about 1 cm. Coastal plants.   **15. S. sedoides**
— Flowers 1.5–2.5 cm. Leaves longer than in above. Sandy desert plants.   **22. S. villosa**

21(19). Perennials up to 80 cm. with woody stock. Inflorescences loose, paniculate-dichasial.
Calyx up to 2 cm. Corolla white. Carpophore as long as or shorter than capsule.
                                                                    1. S. italica
— Annuals or perennials different from above   22

22. Petal limb entire, denticulate or slightly 2-lobed with lobes up to one quarter the length
of limb or less   23
— Petal limb deeply 2-fid or 2-partite, with lobes one third as long as limb or more   26

23. Inflorescence a monochasial raceme. Calyx green, with patulous, glandular and glandless hairs.   **28. S. gallica**
— Inflorescences composed of equal or unequal dichasia. Calyx white or coloured or green
and then hairs neither patulous nor glandular   24

24. Calyx green, herbaceous, papillose-hairy. Carpophore very short.   **18. S. crassipes**
— Calyx red, brown, pink or white, glandular-pubescent or appressed-puberulent   25

25. Capsule 2–3 times as long as carpophore. Calyx membranous, pink or white, with
minute, appressed, not glandular hairs.   **14. S. rubella**
— Capsule almost as long as carpophore. Calyx red, glandular-hairy.   **13. S. fuscata**

26(22). Calyx and often also stems glandular-hairy. Hairs mostly patulous   27
— Calyx and often also stems papillose or villose but never glandular-hairy   31

27. Petals purple or pink; limb obovate to oblong-obovate; lobes oblong to obovate.
Calyx red or red-nerved   28
— Petals white to pink or yellowish-green; limb oblong to linear; lobes linear or almost
so. Calyx neither red nor red-nerved   29

28. Petal limb red to purple; lobes of limb obovate. Stems glandular-villose.
                                                                  24. S. palaestina
— Petal limb usually pink; lobes oblong. Stems glandular-pubescent.   **23. S. damascena**

29(27). Rare mountain perennials, with woody stock, caespitose stems and spatulate basal
leaves. Petals yellowish-green.   **6. S. grisea**
— Desert annuals, with white to pink flowers   30

30. Pedicels as long as calyx, spreading or reflexed in fruit. Lower leaves narrowly linear-
lanceolate.   **25. S. arabica**
— Pedicels shorter than calyx, erect in fruit. Lower leaves obovate, cauline lanceolate.
                                                           25. S. arabica var. **nabathaea**

31(26). Dwarf desert annuals. Calyx 4–6 mm. Coronal scales 2–4-toothed.   **8. S. hussonii**
— Plants not as above   32

32. Petals white to cream-coloured   33
— Petals pink to purple, rarely white in coastal annuals   35

33. Biennial or perennial canescent plants. Leaves ovate to oblong. Inflorescences dichasially branched with alar flowers between branches. Calyx papillose-setose along the
prominent nerves.   **21. S. trinervis**
— Annual, not canescent plants. Inflorescences more or less monochasial. Calyx not as
in above   34

34. Leaves linear-setaceous, folded lengthwise. Carpophore almost as long as capsule.
Calyx 1.2–1.5 cm.   **26. S. vivianii**

— Leaves spatulate to oblong-linear, not folded. Carpophore much shorter than capsule. Calyx 0.8–1 cm.   **27. S. nocturna**

35 (32). Flowers intensely purple.   **29. S. oxyodonta**

— Flowers pink   36

36. Petals 2-fid into ovate to orbicular lobes. Inflorescence a true dichasium. Calyx with triangular papillae.   **19. S. papillosa**

— Petals 2-partite into oblong-linear lobes. Inflorescence a simple or a double monochasium. Calyx with undilated papillae.   **20. S. telavivensis**

Sect. SIPHONOMORPHA Otth in DC., Prodr. 1 : 377 (1824) [incl. Sect. *Paniculatae* (Boiss.) Chowdhuri, Not. Roy. Bot. Gard. Edinb. 22 : 233 (1957)]. Perennials with woody stock, rarely biennials. Lower leaves lanceolate to spatulate. Inflorescence paniculate, with opposite branches, bearing (1–)3–7-flowered dichasia. Flowers usually large. Filaments glabrous. Seeds with plane faces and usually shallowly grooved back.

1. Silene italica (L.) Pers., Syn. 1:498 (1805); Boiss., Fl. 1:631 (1867); Rohrb., Monogr. Sil. 218 (1868). *Cucubalus italicus* L., Syst. ed. 10, 2 : 1030 (1759). [Plate 98]

Perennial, tomentose-canescent below, viscid above, up to 80 cm. or more. Stems many, branching from base, some short and sterile, others long with paniculately branched and repeatedly dichasial inflorescences. Lower leaves up to 8 cm., petiolate, spatulate to oblong, upper ones 2–3 cm., sessile, linear-lanceolate, all ciliate. Bracts linear. Pedicels of alar flowers as long as or longer than calyx, the others shorter. Flowers 1.5–2.5 cm. Calyx obconical-cylindrical, umbilicate, club-shaped in fruit, glandular-pubescent, violet, with red or green nerves; teeth triangular-lanceolate, white at margin, ciliate. Petals white, deeply 2-fid into oblong-spatulate lobes; coronal scales much reduced or 0. Capsule about 1 cm., as long as or longer than carpophore, oblong-ovoid. Seeds slightly wrinkled-tuberculate, grooved on back. Fl. April–June.

Hab. : Maquis. Upper Galilee.

Area : Mediterranean, slightly penetrating into adjacent Euro-Siberian and Irano-Turanian territories.

Sect. LASIOSTEMONES (Boiss.) Schischk., in Fl. URSS 6 : 631 (1936). Perennials with woody stock. Inflorescences paniculate. Flowers deflexed at maturity. Filaments hairy.

2. Silene longipetala Vent., Descr. Pl. Jard. Cels t. 83 (1802); Boiss., Fl. 1 : 636 (1867); Rohrb., Monogr. Sil. 211 (1868). [Plate 99]

Perennial, 40–60 cm. Stems erect, solitary or few, divaricately branching, with thickened nodes, pubescent below, glabrous and viscid above. Leaves 3–7 cm., the lower oblong-obovate or oblanceolate, the upper almost linear-lanceolate, tapering to a petiole, acute. Inflorescences divaricately and paniculately branched; branches long, decussate, ending in 2–7-flowered dichasia. Pedicels mostly about as long as calyx. Flowers about 1.3–1.7 cm., deflexed. Calyx 0.8–1 cm., obconical, top-shaped in fruit, glabrous, green-nerved; teeth short, ovate, obtuse, white and ciliate at margin. Petals white or rarely purple (var. *purpurascens* Boiss., l.c.); claw ciliate; limb 2-partite into long-linear lobes; coronal scales oblong. Capsule 6–8 mm., twice as long as carpophore, ovoid-subglobular. Seeds finely wrinkled, furrowed on back. Fl. April–June.

Hab. : Batha, fallow fields and among crops. Upper and Lower Galilee, Mt. Carmel, Esdraelon Plain, Samaria, Shefela, Judean Mts., Judean Desert, W. and N. Negev, Lower Jordan Valley, Gilead, Ammon, Moav, Edom.

Area : E. Mediterranean, slightly extending into W. Irano-Turanian territories.

Sect. SCLEROCALYCINAE (Boiss.) Schischk., in Fl. URSS 6 : 636 (1936) [incl. Sect. *Chloranthae* (Rohrb.) Schischk., l.c. 616 and Sect. *Tatarica* Chowdhuri, Not. Roy. Bot. Gard. Edinb. 22 : 236 (1957)]. Perennials with woody stock, more rarely biennials. Inflorescences usually narrow, spike-like, with opposite branches, bearing 1- to 3 (–5)-flowered dichasia. Calyx often glabrous.

3. Silene swertiifolia Boiss., Diagn. ser. 1, 1 : 32 (1843) et Fl. 1 : 640 (1867). *S. chloraefolia* Sm. var. *swertiaefolia* (Boiss.) Rohrb., Monogr. Sil. 177 (1868). [Plate 100]

Perennial, glaucescent, glabrous, viscid above, 40–50 cm. Stems erect, more or less thickened at nodes. Leaves thick, glaucescent, sparingly ciliate, the lower ones oblong to spatulate, tapering to a petiole, the upper ones ovate-lanceolate to linear, clasping. Peduncles long, 1-flowered. Bracts narrowly linear. Pedicels varying in length, shorter or longer than calyx. Calyx 1.5–3 cm., cylindrical to obconical, club-shaped in fruit, subumbilicate, green to violet, obsoletely nerved; teeth short, triangular, ciliolate. Petals about 3 cm., white to pinkish, 2-fid into obovate to orbicular lobes; claw more or less exserted; coronal scales minute. Capsule oblong, 1.5–2 times as long as carpophore. Seeds about 2 mm. in diam., wrinkled, greenish yellow. Fl. May–August.

Hab. : Batha and stony ground. Samaria, Judean Mts., Upper Jordan Valley, Ammon.

Area : E. Mediterranean and W. Irano-Turanian.

Sect. INFLATAE (Boiss.) Chowdhuri, Not. Roy. Bot. Gard. Edinb. 22 : 241 (1957). Perennials. Flowers solitary or in few-flowered dichasia. Calyx inflated, 10- or 20-nerved, with conspicuous reticulate venation.

4. Silene vulgaris (Moench) Garcke, Fl. Nord u. Mitt. Deutschl. ed. 9, 64 (1869). *Behen vulgaris* Moench, Meth. 709 (1794). *Cucubalus behen* L., Sp. Pl. 414 (1753) non *S. behen* L., Sp. Pl. 418 (1753). *C. venosus* Gilib., Fl. Lithuan. 2 : 165 (1781) *nom. illegit. C. inflatus* Salisb., Prodr. Stirp. 302. (1796) *nom. illegit. S. inflata* (Salisb.) Sm., Fl. Brit. 467 (1800) *nom. illegit.*; Boiss., Fl. 1 : 628 (1867). [Plate 101]

Perennial, glabrous or sometimes pubescent, glaucescent, 25–70 cm. Stems simple or branching from base, not viscid. Leaves obovate or elliptical-oblong to narrowly linear-lanceolate, acuminate. Inflorescences cymose, dichasial. Bracts and bracteoles frequently scarious, lanceolate-triangular. Pedicels half to 4 times as long as calyx. Flowers hermaphrodite and unisexual, about 2 cm. Calyx 1 cm. broad or more (in fruit), bladdery, ovoid, umbilicate, glabrous, green to somewhat violet, about 20-nerved, with broadly triangular, acute teeth. Petals usually much longer than calyx, white to pink, sometimes red (var. *rubriflora* Boiss., l.c. 629); limb 2-partite; coronal scales mostly gibbous, sometimes absent. Fruiting calyx about 1 cm. Capsule 2–4 times as long as carpophore, ovoid-globular, included. Seeds orbicular-reniform, tuberculate. Fl. March–July.

Hab.: Fallow fields, batha and grassy patches. Acco Plain, Sharon Plain, Philistean Plain, Upper and Lower Galilee, Mt. Carmel, Esdraelon Plain, Samaria, Shefela, Judean Mts., Judean Desert, N. Negev, Dan Valley, Gilead, Ammon, Edom.

Area: Mediterranean, Euro-Siberian and Irano-Turanian.

This variable species is fairly homogeneous in Palestine. A variety differing from the type by finely tuberculate seeds, lack of coronal scales and heart-shaped cauline leaves, has been recorded in Palestine (Sharon Plain, Philistean Plain, Mt. Carmel, Samaria, Judean Mts.) as *S. venosa* (Gilib.) Aschers. or *S. venosa* var. *commutata* (Guss.) Dinsmore in Post, Fl. Syr. Pal. Sin. ed. 2, 1 : 184 (1932).

5. Silene physalodes Boiss., Diagn. ser. 1, 8:83 (1849) et Fl. 1:630 (1867); Rohrb., Monogr. Sil. 88 (1868). [Plate 102]

Perennial, minutely hirsute-pubescent, 10–20 cm. Stems thick, hollow, erect, leafy, paniculate above. Leaves large, up to 12 cm., ovate, the lower long-petioled, sub-cordate at base, the upper cuneate at base, acuminate. Inflorescences dichasial. Bracts leaf-like, subcordate-acuminate to oblong-linear. Pedicels short, elongating and deflexed in fruit. Calyx about 2 cm., cylindrical, very inflated, umbilicate, green, velvety-puberulent, netted-veined, becoming campanulate, folded and membranous in fruit; teeth broadly triangular, ciliate. Petals small, with somewhat dilated claw and white, linear limb, cut into 4–6 filiform lobes; coronal scales cut into many filiform lobes. Capsule included, shorter than calyx but hardly longer than carpophore, ovoid. Seeds large, about 2–3 mm., tuberculate, with convex back. Fl. June.

Hab.: Among shrubs. Upper Galilee, Hula Plain. Very rare.

Area: E. Mediterranean (endemic in Palestine?).

Sect. BRACHYPODAE (Boiss.) Chowdhuri, Not. Roy. Bot. Gard. Edinb. 22 : 241 (1957). Velvety-canescent perennials. Radical leaves lanceolate to spatulate. Stems few-flowered. Capsule much longer than carpophore.

6. Silene grisea Boiss., Diagn. ser. 1, 8 : 88 (1849) et Fl. 1 : 646 (1867); Rohrb., Monogr. Sil. 145 (1868). [Plate 103]

Perennial, velvety-pubescent, grey, 20–40 cm. Stems many, tufted, woody at base, geniculate, viscid above. Leaves crowded at base, spatulate, attenuate, mucronulate, upper ones oblanceolate to linear. Inflorescences 3–7-flowered, raceme-like, rarely 1-flowered. Bracts linear-lanceolate, acute. Pedicels shorter than calyx, deflexed in fruit. Calyx about 1.2 cm., cylindrical to club-shaped, glandular-hairy, green-nerved; teeth small, triangular, acute, white and ciliate at margin. Petals 1.2–1.5 cm., yellowish-green, 2-partite into linear lobes; coronal scales small, oblong. Capsule oblong, somewhat longer than calyx, 3–4 times as long as carpophore, deflexed. Seeds minutely wrinkled, furrowed on back. Fl. March–May.

Hab.: Rocks. Upper Galilee, Judean Mts., Hula Plain. Very rare.

Area: E. Mediterranean.

Sect. SUFFRUTICOSAE (Rohrb.) Schischk., in Fl. URSS 6: 648 (1936) [incl. Sect. *Macranthae* Rohrb., Monogr. Sil. 74 (1868)]. Perennials with more or less woody stock and ovate, lanceolate or linear leaves. Flowers solitary or in few-flowered dichasia, usually large. Carpophore rather long.

7. Silene succulenta Forssk., Fl. Aeg.-Arab. LXVI et 89 (1775); Boiss., Fl. 1: 648 (1867); Rohrb., Monogr. Sil. 134 (1868). [Plate 104]

Viscid, papillose-pubescent perennial, up to 25 cm. Stems diffuse, ascending, indurated at base, leafy, richly branching below, often forming cushion-like shrublets. Leaves fleshy, ovate to oblong-spatulate to lanceolate, obtuse, hairy, ciliate. Flowers solitary or forming short, leafy, dichasial racemes. Bracts lanceolate. Pedicels shorter than bracts. Calyx about 2 cm. or more, oblong-cylindrical, somewhat umbilicate, club-shaped in fruit; teeth one third or one fourth as long as calyx, lanceolate, acuminate or subulate, ciliate, slightly recurved. Petals white or pink (var. *eliezeri* Eig, Bull. Inst. Agr. Nat. Hist. 6 : 4, 1927); claw exserted; limb up to 1.5 cm., 2-partite to middle into oblong-spatulate lobes; coronal scales denticulate. Capsule oblong, somewhat longer than carpophore. Seeds rather smooth, with flat back. Fl. March–July.

Hab.: Littoral sands. Coastal Galilee, Acco Plain, Sharon Plain, Philistean Plain.

Area: E. Mediterranean (Egypt-Lebanon, Crete, Sardinia, Corsica).

Roots contain considerable amounts of saponine which can be used in detergents.

Sect. RIGIDULAE (Boiss.) Schischk., in Fl. URSS 6 : 681 (1936). Annuals. Stems branching, rigid, often filiform, viscid above. Inflorescences very loose, composed of divaricately branched equal dichasia. Fruiting calyx not inflated or constricted at throat. Seeds broadly reniform, with flat or slightly convex faces.

8. Silene hussonii Boiss., Diagn. ser. 1, 8 : 76 (1849) et Fl. 1 : 604 (1867); Rohrb., Monogr. Sil. 158 (1868). [Plate 105]

Annual, hirtellous-viscidulous, 4–7 cm. Stems slender, erect, simple or dichotomously branching from base. Leaves up to 2 × 0.4 cm., the lower ones oblong-spatulate or oblanceolate, more or less acute, the upper oblong-lanceolate to linear. Inflorescences few-flowered, dichasially branched. Pedicels usually longer than calyx. Calyx 4–6 mm., more or less top-shaped, club-shaped in fruit, umbilicate, puberulent, green-nerved; teeth one third as long as calyx, triangular-lanceolate, membranous-margined, ciliate. Petals about 8 mm., whitish; limb 2-fid or 2-partite into linear lobes; coronal scales 2–4-toothed. Capsule not exceeding calyx, as long as or shorter than carpophore, ovoid. Seeds tuberculate-wrinkled. Fl. March–April.

Hab.: Deserts. N. Negev (Hart, Fauna and Fl. Sin. Petr. Wad. Arab. 163, 1891). Very rare.

Area: E. Saharo-Arabian.

9. Silene reinwardtii Roth, Catalecta 3 : 42 (1806). *S. picta* Pers., Syn. 1 : 498 (1805) non Desf., Tabl. 159 (1804); Rohrb., Monogr. Sil. 159 (1868). *S. juncea* Sibth. et Sm., Fl. Gr. Prodr. 1 : 295 (1809) non Roth, Catalecta 1 : 54 (1797); Boiss., Fl. 1 : 605 (1867). [Plate 106]

Annual or perennial, 30–60 cm. Stems scabrous-puberulent below, subglabrous and viscid above, with elongated, slender, almost leafless branches. Radical leaves large, obovate; upper leaves linear-lanceolate to narrowly linear. Inflorescences paniculate, dichasial. Bracts white-margined, ciliate. Pedicels shorter than calyx, longer than bracts. Flowers 1.5–2 cm. Calyx obconical to club-shaped, not umbilicate, glabrous, purple to red-nerved; teeth short, acute. Petals white or pink, purple-veined; limb 2-partite; coronal scales truncate, denticulate. Capsule about 9 mm., shorter than pubescent carpophore, ovoid. Seeds with rows of small tubercles. Fl. March–August.

Hab.: Rocky places in devastated maquis. Upper Galilee, Hula Plain, Upper Jordan Valley, Ammon.

Area: E. Mediterranean (Palestine-Syria).

10. Silene linearis Decne., Ann. Sci. Nat. Bot. ser. 2, 3 : 276 (1834); Boiss., Fl. 1 : 602 (1867); Rohrb., Monogr. Sil. 162 (1868). [Plate 107]

Glaucescent annual, puberulent in lower part, glabrous and viscid in upper, up to 50 cm. Stems slender, branching at base, almost leafless above. Leaves lanceolate to linear-subulate, acute, more or less plicate, ciliate. Inflorescences dichasial. Pedicels about as long as or longer than calyx, the upper shorter. Flowers about 1.4 cm., white-yellowish. Calyx about 1 cm., narrowly cylindrical to club-shaped, hardly umbilicate, more or less puberulent or glabrous and green-nerved; teeth short, triangular, acute, ciliate and white at margin. Petals with oblong limb, 2-partite into linear lobes; coronal scales usually denticulate. Capsule ovoid-oblong, as long as or a little shorter than carpophore. Seeds finely wrinkled. Fl. March–May.

Hab.: Deserts and steppes, often on rocky ground, especially in hot ravines. Judean Desert, C. and S. Negev, Lower Jordan Valley, Dead Sea area, Arava Valley, deserts of Moav and Edom.

Area: E. Sudanian, extending into Saharo-Arabian territories.

11. Silene modesta Boiss. et Bl. in Boiss., Diagn. ser. 2, 6 : 33 (1859). *S. chaetodonta* Boiss. var. *modesta* (Boiss. et Bl.) Boiss., Fl. 1 : 606 (1867); Post, Fl. Syr. Pal. Sin. ed. 2, 1 : 182 (1932). [Plate 108]

Annual, puberulent-scabrous below, glabrous-viscid above, 40–60 cm. Stems erect, slender, almost leafless above. Leaves narrowly linear, acute, somewhat ciliate at base, canaliculate, sometimes with revolute margin. Inflorescences repeatedly and often irregularly dichasial, corymbose. Bracts narrow. Pedicels longer than calyx. Calyx about 1.5 cm., cylindrical to club-shaped, somewhat umbilicate, glabrous or hairy, green, purple-nerved; calyx teeth 2–3 mm., acute, scabrous, white and ciliate at margin. Petals white to pink; limb 2-partite to less than middle; coronal scales linear, acute. Capsule about 1.2 cm., narrowly ovoid to oblong, somewhat longer than calyx, about 2–3 times as long as carpophore. Seeds reniform, finely reticulate, with flattened or somewhat concave faces and slightly grooved back. Fl. April–June.

Hab.: Batha and grassland on coastal sandy soils. Coastal Galilee, Acco Plain, Sharon Plain, Philistean Plain.

Area: E. Mediterranean (Palestine, Lebanon).

Sect. ATOCION Otth in DC., Prodr. 1 : 383 (1824). Stems erect, usually viscid above, not rigid. Inflorescence a dichasial or monochasial cyme (except in *S. sedoides*). Fruiting calyx not constricted at throat. Seeds not winged.

12. Silene aegyptiaca (L.) L. f., Suppl. 241 (1781); Rohrb., Monogr. Sil. 156 (1868). *Cucubalus aegyptiacus* L., Sp. Pl. 415 (1753). *S. atocion* Murr., Syst. Veg. ed. 13, 421 (1774); Boiss., Fl. 1 : 600 (1867). [Plate 109]

Annual, 10–40 cm. Stems ascending to erect, crisp glandular-pubescent, viscid. Leaves varying in size, ovate, the lower petiolate, the upper sessile, all mucronulate, subglabrous. Inflorescences dichasial. Bracts narrow, leaf-like; bracteoles scarious, ovate. Flowers pink-violet, sometimes white, the alar flowers with long, the rest with very short pedicels. Calyx 1.4–2 cm., cylindrical, club-shaped in fruit, viscid-puberulent, often red, obsoletely nerved; teeth short, ovate, obtuse, ciliate and white-margined. Petals 2-lobed, 2-dentate at base of limb; coronal scales ovate, obtuse. Capsule 6–8 mm., oblong, much shorter than carpophore. Seeds wrinkled-tuberculate, with flattened or concave faces umbilicate at centre. Fl. January–April.

Hab.: Fields, vineyards and orchards. Acco Plain, Sharon Plain, Philistean Plain, Upper Galilee, Mt. Carmel, Shefela, Judean Mts., Judean Desert, Esdraelon Plain, Hula Plain, Upper Jordan Valley, Gilead, Ammon.

Area: E. Mediterranean.

One of the most common *Silene* species lending a pink colour to wide hill areas in winter and early spring.

Varies in dimensions, branching [e.g. var. *umbrosa* Nab., It. Turc.-Pers. 45 (1923), referring to tall, much branched plants with larger leaves and colourless calyx], in number of flowers per cyme, length of calyx, colour of corolla [e.g. var. *alba* Aarons. et Evenari ex Opphr. et Evenari, Bull. Soc. Bot. Genève ser. 2, 31 : 222 (1941)] and in pubescence.

13. Silene fuscata Link ex Brot., Fl. Lusit. 2:187 (1804); Boiss., Fl. 1: 600 (1867); Rohrb., Monogr. Sil. 153 (1868). [Plate 110]

Annual, crisp-pubescent and viscid, 20–50 cm. Stems erect, few or many, simple or dichotomously branching above. Leaves undulate, the lower ones petiolate, obovate to obovate-oblong, obtuse, the upper sessile, lanceolate, ciliate. Pedicels varying in length, those of alar flowers long, the terminal ones short, all glandular. Flowers about 1.5 cm., in axillary and terminal dichasial cymes. Calyx 1.2–1.6 cm., mostly longer than pedicel, oblong-cylindrical, club-shaped in fruit, umbilicate, glandular-hairy, brownish-red; teeth short, ovate-triangular, ciliate at margin, deflexed in fruit. Petals pink or purple, oblong-obovate, almost entire; coronal scales oblong-linear, notched or crenulate at apex. Capsule 0.7–1 cm., about as long as the glabrous carpophore, ovoid. Seeds reniform, with concave minutely wrinkled-tuberculate faces and canaliculate or ungrooved back. Fl. February–April.

Hab.: Damp and heavy, mostly cultivated ground. Acco Plain, Sharon Plain,

Philistean Plain, Upper Galilee, Esdraelon Plain, Judean Mts., Dan Valley, Hula Plain, Upper Jordan Valley.

Area: Mediterranean.

14. Silene rubella L., Sp. Pl. 419 (1753); Boiss., Fl. 1 : 598 (1867); Rohrb., Monogr. Sil. 155 (1868). [Plate 111]

Annual, glaucescent, minutely puberulent, not glandular, up to 60 cm. Stems simple or branching at base. Lower leaves 4–8 cm., obovate-spatulate to obovate-oblong, tapering at base, upper ones lanceolate or linear, all obtuse, somewhat undulate. Inflorescences branched into unequal dichasia. Pedicels of alar flowers slender, long, those of the rest short. Calyx 0.8–1.2 cm., membranous, oblong to obconical, top- to club-shaped in fruit, not umbilicate, puberulent to glabrous, pinkish-white; teeth short, obtuse, ciliate, recurved in fruit. Petals reddish-pink; limb obovate, retuse; coronal scales linear, obtuse. Capsule 7–9 mm., 2–3 times as long as carpophore, ovoid-cylindrical. Seeds tuberculate-wrinkled. Fl. March–May.

Hab.: Fallow and irrigated fields. Philistean Plain, Upper Galilee, Mt. Carmel, Esdraelon Plain, Judean Mts., Judean Desert, Hula Plain, Upper Jordan Valley, Beit Shean Valley, Lower Jordan Valley.

Area: Mediterranean.

15. Silene sedoides Poir., Voy. Barb. 2 : 164 (1789); Boiss., Fl. 1 : 598 (1867); Rohrb., Monogr. Sil. 164 (1868). [Plate 112]

Annual, glandular-pubescent, 5–10 cm. Stems almost simple or dichotomously branched from base. Branches short or elongated. Leaves about 1 cm., fleshy, oblong-spatulate to oblanceolate-linear, obtuse, glandular-pubescent. Flowers axillary, mostly in monochasial cymes. Pedicels about as long as calyces or longer, spreading or deflexed after flowering. Calyx about (6–)7–8 mm., oblong, club-shaped in fruit, more or less umbilicate, glandular-hairy; teeth short, ovate, obtuse or acute. Petals small, reddish; limb obovate, from almost entire to 2-fid; coronal scales dentate. Capsule about as long as calyx, 4 times as long as carpophore, oblong, membranous. Seeds less than 1 mm., with wrinkled, striate faces and obtusely grooved back. Fl. March–April.

Hab.: Sands and cliffs of coastal spray zone. Coastal Galilee, Acco Plain, Sharon Plain.

Area: Mediterranean.

Sect. BEHENANTHA Otth in DC., Prodr. 1 : 367 (1824). Annuals. Stems usually glabrous and viscid above. Fruiting calyx constricted at throat.

16. Silene behen L., Sp. Pl. 418 (1753); Boiss., Fl. 1 : 583 (1867); Rohrb., Monogr. Sil. 169 (1868). [Plate 113]

Annual, glabrous, glaucescent, 30–40(–60) cm. Stems erect, simple or dichotomously branched. Lower leaves spatulate to oblanceolate, long-tapering, acute, mucronulate, ;paringly dentate or short-ciliate at base; upper ones oblong. Inflorescences loosely and sometimes irregularly dichasial. Bracts membranous at margin, almost lanceolate.

Pedicels equalling or mostly shorter than calyx, thickened in fruit, those of alar flowers long. Flowers about 1.6 cm. Calyx 1.5–1.8 cm., oblong-ovoid, inflated in fruit, umbilicate, somewhat membranous, glabrous, reddish-white with red nerves; teeth short, ovate, white-margined. Petals small; limb pink, 2-partite; coronal scales short, simple or somewhat emarginate. Capsule about 0.8–1 cm., more than twice as long as carpophore, ovoid, tapering at apex. Seeds rugose-tuberculate, faces concave, back 4–5-furrowed. Fl. (January–) March–April.

Hab.: Fallow fields and batha. Acco Plain, Sharon Plain, Philistean Plain, Upper and Lower Galilee, Mt. Carmel, Esdraelon Plain, Judean Mts., W. Negev, Gilead, Ammon, Moav.

Area: C. and E. Mediterranean.

17. Silene muscipula L., Sp. Pl. 420 (1753); Boiss., Fl. 1 : 583 (1867); Rohrb., Monogr. Sil. 170 (1868). [Plate 114]

Annual, glabrous, viscid above, 25–40 cm. Stems 1 or few, erect, dichotomously and strictly branched. Leaves oblanceolate-spatulate to linear-lanceolate. Inflorescences in terminal, many-flowered dichasia. Bracts awl-shaped, white-margined at base, mostly as long as or longer than calyx. Pedicels all short. Flowers 1.3–1.6 cm. Calyx more or less cylindrical, club-shaped in fruit, umbilicate, glabrous, with green or reddish anastomosing nerves; teeth short, triangular-ovate, acute, white-margined. Petals pink; limb small, cuneate, 2-fid or notched; coronal scales rather long, 2-fid. Capsule 0.8–1.3 cm., about twice as long as carpophore, exceeding calyx, oblong-ellipsoidal. Seeds finely tuberculate, faces plane, back obtusely grooved. Fl. March–July.

Hab.: Fallow and crop fields. Sharon Plain, Philistean Plain, Upper and Lower Galilee, Mt. Carmel, Judean Mts., N. Negev, Gilead.

Area: Mediterranean, extending to S. W. Europe.

Sect. LASIOCALYCINAE (Boiss.) Chowdhuri, Not. Roy. Bot. Gard. Edinb. 22 : 246 (1957). Annuals. Nerves of calyx papillose-scabrid, hirsute or with bulbous hairs. Fruiting calyx constricted at throat.

18. Silene crassipes Fenzl, Pugill. 8 (1842); Boiss., Fl. 1 : 586 (1867); Rohrb., Monogr. Sil. 172 (1868). [Plate 115]

Annual, scabrous-pubescent, 15–30 cm. Stems erect, simple or branching from base, branches rigid. Lower leaves oblong-spatulate, the uppermost lanceolate-linear, acute. Inflorescences dichasial. Flowers 1.2–1.6 cm., on short pedicels, becoming very thick in fruit. Calyx 1–1.5 cm., cylindrical, oblong to club-shaped in fruit, umbilicate, scabridulous with conical papillae especially along nerves; teeth short, ovate, obtuse. Petals with pink to purple, linear-cuneate, entire or notched limb; coronal scales oblong-linear, acute. Capsule 0.7–1.1 cm., 2–3 times as long as carpophore, ovoid. Seeds wrinkled, with grooved back. Fl. February–April.

Hab.: Fallow fields and among crops. Philistean Plain, Upper and Lower Galilee, Mt. Carmel, Samaria, Judean Mts., Hula Plain, Upper and Lower Jordan Valley, Gilead, Ammon, Edom.

Area: E. Mediterranean (Palestine–Syria).

19. Silene papillosa Boiss., Diagn ser. 1, 1 : 39 (1843) et Fl. 1 : 587 (1867); Rohrb., Monogr. Sil. 172 (1868). [Plate 116]

Annual, up to 40 cm., puberulent below, subglabrous above. Lower leaves 2–3 × 0.6–1 cm., oblong-lanceolate, tapering; upper ones linear or lanceolate, acute, pubescent, setose-ciliate. Inflorescences dichasial. Bracts longer than flowers. Flowers short-pedicelled. Calyx 1.2–1.5 cm., cylindrical, club-shaped in fruit, umbilicate, green- to purple-nerved, nerves more or less prominent, papillose-setose with broad-based papillae, curved upwards; teeth lanceolate-subulate, white and ciliate at margin. Petals pink; limb obovate-cuneate, 2-lobed. Capsule about 1 cm., almost twice as long as carpophore, ovoid to globular. Seeds about 1 mm., tuberculate, grooved on back. Fl. March–April.

Hab.: Fallow fields on light soils. Sharon Plain, Philistean Plain. Rare.

Area: E. Mediterranean (Palestine-Turkey).

20. Silene telavivensis Zoh. et Plitm. * [Plate 117]

Annual, more or less pubescent, up to 35 cm. Stems many, erect from branched base, slender. Leaves crowded below, oblong-spatulate to oblanceolate-linear, generally tapering into a petiole; petioles somewhat dilated and shortly connate at base, ciliate at margin. Inflorescences monochasial, spike-like. Bracts much shorter than calyx, lanceolate-linear, white-margined. Pedicels very short. Calyx 8–9 mm., narrowly cylindrical, slightly club-shaped in fruit, umbilicate, with prominent, ciliate-hirsute, papillose nerves; teeth short, linear-lanceolate, slightly recurved in fruit. Petals white or pink; claw more or less exserted; limb 2-partite into linear lobes; coronal scales rudimentary. Capsule included, longer than carpophore, cylindrical. Seeds 0.5–1 mm., with rows of tubercles. Fl. April–May.

Hab.: Light soils. Philistean Plain. Rare.

Area: E. Mediterranean, slightly extending towards north and east.

Sect. DICHOTOMAE (Rohrb.) Chowdhuri, Not. Roy. Bot. Gard. Edinb. 22 : 247 (1957). Annuals, biennials or perennials. Stems erect, pubescent. Inflorescences dichasially branched into monochasial cymes. Fruiting calyx constricted, with prominent simple nerves. Seeds reniform, with plane or slightly concave tuberculate faces and bluntly tuberculate back.

21. Silene trinervis Banks et Sol. in Russ., Nat. Hist. Aleppo ed. 2, 2 : 252 (1794). *S. racemosa* Otth in DC., Prodr. 1 : 384 (1824); Boiss., Fl. 1 : 589 (1867). *S. dichotoma* Ehrh. var. *racemosa* (Otth) Rohrb., Monogr. Sil. 95 (1868). *S. sibthorpiana* Reichb., Fl. Germ. Exc. 815 (1832). [Plate 118]

Biennial or perennial, canescent, with crisp, appressed, non-glandular hairs, 30–70 cm. Stems many, dichotomously branching. Radical leaves rosulate, oblong-spatulate, tapering into a petiole; stem leaves mostly sessile, ovate to oblong, acute. Inflorescences branching dichasially into monochasial cymes, with alar flowers between them. Bracteo-

---

* See Appendix at end of this volume.

les small, rather membranous, oblong-lanceolate to ovate, ciliate. Pedicels very short, spreading or deflexed in fruit. Flowers 1.3–1.8 cm. Calyx 0.9–1.2 cm., cylindrical, ovoid-oblong in fruit, papillose-setose along the prominent nerves; teeth small, ovate-lanceolate, ciliate. Petals white-cream; limb obovate, 2-partite almost to base with obtuse coronal scales. Capsule 7–8 mm., oblong-ovoid, much longer than carpophore. Seeds up to 2 mm., tuberculate, with plane faces and grooved back. Fl. March–May.

Hab.: Fallow fields and batha. Sharon Plain, Upper and Lower Galilee, Mt. Carmel, Esdraelon Plain, Mt. Gilboa, Judean Mts., Judean Desert, N. Negev, Hula Plain, Upper Jordan Valley, Ammon, Moav.

Area: E. Mediterranean, extending into adjacent Irano-Turanian territories.

Sect. SCORPIOIDEAE (Rohrb.) Chowdhuri, Not. Roy. Bot. Gard. Edinb. 22:247 (1957). Annuals. Inflorescences monochasial. Fruiting calyx not constricted at throat; nerves anastomosing or not. Seeds not winged.

22. Silene villosa Forssk., Fl. Aeg.-Arab. 88 non 210 (1775); Boiss., Fl. 1:592 (1867); Rohrb., Monogr. Sil. 110 (1868). [Plate 119]

Annual, patulous-hirsute, viscid, 7–20 cm. Stems procumbent to ascending, much branched. Leaves somewhat fleshy, spatulate-oblong to oblong-lanceolate, obtuse. Flowers arranged in more or less one-sided short or elongated cymes. Bracts lanceolate-linear. Pedicels shorter than or as long as or longer than calyx. Calyx 1–2.5 cm., cylindrical, club-shaped and reflexed in fruit, umbilicate; teeth small, white-margined. Petals white, rarely pink, with claw exceeding calyx and with 2-fid or 2-partite limb; coronal scales oblong to ovate-triangular. Capsule ovoid-oblong, about as long as or a little longer than almost glabrous carpophore. Seeds with minute rows of wrinkles. Fl. March–May.

Var. **villosa.** Flowers pinkish. Calyx 1.5–2.5 cm. Carpophore about as long as capsule.
Hab.: Sandy soils. Sharon Plain, S. Negev, Dead Sea area.

Var. **ismaelitica** Schweinf. in Aschers. et Schweinf., Ill. Fl. Eg. Suppl. 748 (1889); Muschl., Man. Fl. Eg. 335 (1912). Flowers white. Calyx 1–1.5 cm. Carpophore somewhat shorter than capsule.
Hab.: As above. W. Negev.
Area of species: Saharo-Arabian.

23. Silene damascena Boiss. et Gaill. in Boiss., Diagn. ser. 2, 6:34 (1859) et Fl. 1:594 (1867). *S. palaestina* Boiss. var. *damascena* (Boiss. et Gaill.) Rohrb., Monogr. Sil. 103 (1868). [Plate 120]

Annual, patulous-pubescent and viscid-glandular, 20–40 cm. Stems erect, branching from base. Leaves oblong-spatulate below, linear-lanceolate above, more or less acute. Inflorescences monochasial. Bracts unequal. Lower pedicels longer, upper shorter than calyx. Flowers about 2 cm., pink, rarely white. Calyx cylindrical, club-shaped in fruit, truncate, umbilicate, reddish and red-nerved; teeth oblong, obtuse. Petals 2-partite into linear-oblong lobes. Capsule 0.8–1.2 cm., as long to twice as long as carpophore. Seeds hardly 1 mm. in diam., finely wrinkled, deeply grooved on back. Fl. February–May.

Hab.: Fields and batha. Philistean Plain, Upper and Lower Galilee, Esdraelon Plain, Judean Mts., Judean Desert, N. Negev, Upper and Lower Jordan Valley, Dead Sea area, Gilead, Ammon, Moav.

Area: E. Mediterranean.

24. Silene palaestina Boiss., Diagn. ser. 1, 8:80 (1849) et Fl. 1:595 (1867); Rohrb., Monogr. Sil. 103 (1868). [Plate 121]

Annual, glandular villose, viscid, 30–50 cm. Stems erect, often simple. Leaves oblong, the upper linear to lanceolate, acute. Inflorescences monochasial, elongated, strictly branching, 15–25 cm. Bracts unequal. Pedicels mostly shorter than calyx. Flowers 1.2–2 cm., red to purple. Calyx about 1–1.5 cm., cylindrical to club-shaped, truncate, red-nerved, patulous glandular-hairy; teeth triangular-oblong. Petals rather large; limb obovate, 2-partite almost to base into obovate lobes; coronal scales truncate, denticulate. Capsule 0.7–1 cm., 2 to 3 times as long as carpophore. Seeds about 0.5 mm. in diam., almost smooth, shining. Fl. March–May.

Hab.: Fallow fields, light soils. Coastal Galilee, Acco Plain, Sharon Plain, Philistean Plain, Samaria, Judean Mts. (Rechinger f., Ark. Bot. ser. 2, 2:332, 1952), Lower Jordan Valley (Bornmueller fide Post, Fl. Syr. Pal. Sin. ed. 2, 1:177, 1932).

Area: E. Mediterranean (Egypt, Palestine, Lebanon).

25. Silene arabica Boiss., Fl. 1:593 (1867). *S. affinis* Boiss., Diagn. ser. 2, 1:72 (1853) non Godr., Mém. Acad. Montp. 1:417 (1853); Rohrb., Monogr. Sil. 104 (1868). [Plates 122, 123]

Annual, with two kinds or patulous hairs: papillose-villose and glandular ones; both from very short to very long. Stems 10–40 cm., simple or branching from leafy base. Leaves sessile, obovate-oblong to lanceolate and linear, acute, glandular-pubescent. Inflorescences loose, monochasial. Bracts narrowly lanceolate-linear, unequal to almost equal. Pedicels almost as long as calyx, the lower longer, some spreading to deflexed in fruit. Flowers about 1.2–1.8 cm. Calyx about 1 cm., cylindrical, oblong to club-shaped in fruit, truncate-umbilicate, glandular-pubescent, whitish-green, and mostly green-nerved; teeth oblong-triangular with membranous and ciliate margin. Petals white to pink, deeply 2-partite into linear lobes; claw exserted; coronal scales small, oblong. Capsule 6–9 mm., longer to shorter than carpophore, ovoid. Seeds about 0.6 mm., with concave, finely striate to almost smooth faces and grooved back. Fl. April.

Hab.: Mainly in sandy deserts and on weathered Nubian sandstone. Negev, Lower Jordan Valley, Dead Sea area, Gilead, Moav, Edom.

1. Leaves linear-lanceolate, at least part of them folded lengthwise. Plants scabrous, very shortly glandular-papillose.   var. **viscida**
– Plants not as above   2
2. Plants covered with very long (up to 2–3 mm.) patulous hairs. Lower pedicels much longer than calyx.   var. **moabitica**
– Hairy cover shorter than in above   3
3. Lower leaves obovate. Capsule about twice as long as carpophore.   var. **nabathaea**

– Lower leaves linear to oblong. Capsule somewhat shorter than to as long as carpophore.
                                                                            var. **arabica**

Var. **arabica**. Stems short, 10–20 cm.; all parts rather sparingly covered with medium-sized villose and glandular hairs. Lower leaves crowded, linear to oblong, the upper short, oblong-lanceolate. Petals white. Capsule somewhat shorter than to as long as hairy carpophore. [Plate 122a]
   Moav, Edom, Lower Jordan Valley.

Var. **nabathaea** (Gomb. et A. Camus) Zoh. (stat. nov.). *S. nabathaea* Gomb. et A. Camus, Bull. Soc. Bot. Fr. 93 : 125 (1947). Stems about 20 cm.; all parts fairly densely covered with patulous and glandular hairs. Lower leaves obovate, cauline lanceolate. Pedicels shorter than calyx. Flowers about 1.5 cm. Capsule about twice as long as hairy carpophore.
   E. Negev, Edom (Petra). Very close to the typical form.

Var. **viscida** (Boiss.) Zoh. (comb. nov.). *S. setacea* Viv. var. *viscida* Boiss., l.c. 594. Plants scabrous with very short and sparse papillae and glandular hairs. Leaves linear-lanceolate, partly folded lengthwise, partly plane. In its outer appearance it approaches *S. vivianii* which has a simple, appressed indumentum. [Plate 122]
   Negev. Drawn from a specimen of Hail, Arabia.

Var. **moabitica** Zoh. * *S. moabitica* Eig (in herb.). Stems mostly branching from base; all parts densely covered with very long patulous hairs (up to 2–3 mm.) and longer or shorter glandular hairs. Lower leaves oblong-lanceolate, cauline broadly linear, not folded lengthwise. Lower pedicels much longer than calyx. Petals white to pink. Capsule as long as or shorter than hairy carpophore. [Plate 123]
   Gilead, Moav, Edom. Very common.
   Area of species : E. Saharo-Arabian (Palestine, Sinai, Arabia and probably also Iraq and Egypt).
   The above forms of *S. arabica* call for a more thorough taxonomic study.

26. Silene vivianii Steud., Nom. ed. 2, 2 : 588 (1841). *S. setacea* Viv., Fl. Lib. Spec. 23, t. 12, f. 2 (1824) non Otth in DC., Prodr. 1 : 372 (1824); Boiss., Fl. 1 : 594 (1867); Rohrb., Monogr. Sil. 105 (1868). [Plate 124]
   Annual, minutely appressed-puberulent to pubescent, not viscid, 10–30 cm. Stems ascending to erect, simple or branched especially at base. Leaves narrowly linear to setaceous, plicate, densely or sparingly pubescent. Cymes simple, monochasial. Bracts short, subequal. Pedicels mostly shorter than bracts. Calyx 1.2–1.5 cm., tubular, later club-shaped, umbilicate at base, pubescent or puberulent, more or less green-nerved; teeth about 3 mm., oblong-lanceolate, more or less acute, white-margined, ciliate. Petals white; claw somewhat exserted; limb 2-partite into widely diverging linear lobes, dark-veined beneath; coronal scales small, obtuse. Capsule ovoid, about as long as hairy carpophore. Seeds 0.5–1 mm., reniform, rugulose, with ear-shaped depression and obtusely grooved back. Fl. March–April.

* See Appendix at end of this volume.

Hab.: Steppes. Philistean Plain, Judean Desert, W. and C. Negev, Lower Jordan Valley, Dead Sea area, Ammon, Moav, Edom.

Area: Saharo-Arabian.

27. Silene nocturna L., Sp. Pl. 416 (1753); Boiss., Fl. 1 : 595 (1867); Rohrb., Monogr. Sil. 100 (1868). [Plate 125]

Puberulent annual, viscid above, 20–40 cm. Stems ascending to erect, sometimes prostrate [var. *prostrata* Post, Fl. Syr. Pal. Sin. 139 (1883–1896)], simple to much branching. Leaves spatulate to oblong-linear, upper linear-lanceolate, acute, pubescent, ciliate at base. Inflorescence a raceme-like, many- or few-flowered, more or less one-sided monochasium. Bracts slightly unequal, longer than pedicels, ciliate. Flowers 1–1.5 cm., the lower long-pedicelled, the upper subsessile. Calyx 0.8–1 × 0.2–0.4 cm., tubular, oblong in fruit, not umbilicate, appressed-pubescent, green, with anastomosing nerves; teeth short, triangular, ciliate, spreading to erect in fruit. Petals whitish, greenish below, exceeding the calyx; limb 2-fid to 2-partite, with 2-partite coronal scales (var. *genuina* Rohrb., l.c.), rarely petals somewhat shorter than calyx and then without coronal scales [var. *brachypetala* (Rob. et Cast.) Rohrb., l.c. 101]. Capsule about as long as calyx, oblong, borne on a very short carpophore. Seeds wrinkled-tuberculate, with concave faces and shallow groove on back. Fl. February–May.

Hab.: Fields and batha. Acco Plain, Sharon Plain, Philistean Plain, Upper and Lower Galilee, Mt. Carmel, Esdraelon Plain, Mt. Gilboa, Judean Mts., Judean Desert, N. Negev, Dan Valley, Hula Plain, Upper and Lower Jordan Valley, Gilead, Ammon, Moav.

Area: Mainly Mediterranean.

Sect. SILENE. Annuals. Inflorescences monochasial. Fruiting calyx constricted at throat. Carpophore pubescent. Seeds reniform, not winged.

28. Silene gallica L., Sp. Pl. 417 (1753); Boiss., Fl. 1 : 590 (1867); Rohrb., Monogr. Sil. 96 (1868). [Plate 126]

Annual, patulous-hirsute, viscid above, 25–50 cm. Stems ascending to erect, few or many. Lower leaves petiolate, obovate-spatulate, middle and upper ones sessile, lanceolate, all pubescent. Bracts subulate or linear-lanceolate. Flowers 1–1.2 cm., in monochasial raceme-like inflorescences, pedicellate below, subsessile above. Calyx 0.7–1 cm., cylindrical, ovoid in fruit, not umbilicate, glandular-hispid with spreading or sometimes appressed hairs, green- or red-nerved; teeth short, linear-lanceolate, ciliolate, subrecurved. Petals pink (rarely white), sometimes with red stripes or with a red blotch at base and whitish at margin; limb obovate, entire to emarginate; coronal scales 2-partite. Capsule 0.7–1 cm., more or less ovoid; carpophore almost 0. Seeds reniform, blackish, tuberculate; faces deeply concave, back plane. Fl. February–April.

Hab.: Fields, mainly on light soils. Coastal Galilee, Acco Plain, Sharon Plain, Philistean Plain, Hula Plain.

Area: Mediterranean and W. Euro-Siberian; adventive in a few other regions.

Very polymorphic. The species has been subdivided into several forms and varieties which have not yet been adequately studied.

29. Silene oxyodonta Barb., Herbor. Levant 121 (1882); Boiss., Fl. Suppl. 92 (1888); Williams, Journ. Linn. Soc. Lond. Bot. 32 : 66 (1896). [Plate 127]

Annual. Stems erect with ascending branches, crisp-puberulent to villose, up to 30 cm. Leaves oblong-spatulate to elliptical-lanceolate or oblong-linear, acute, subglabrous. Inflorescences dichasial or monochasial. Bracts mostly longer than flowers, equal, leaf-like, dilated at base. Flowers pedicellate. Calyx about 1.5 cm., oblong, truncate at base, with broad, prominent, almost crested, green, crisp-hirsute nerves; teeth 5 mm. or more, lanceolate, mucronate, rather rigid and recurved in fruit. Petals purple; claw subexserted; limb oblong, deeply 2-fid into oblong-linear, obtuse lobes; coronal scales oblong. Capsule ovoid, 2 to 3 times as long as carpophore. Seeds about 1 mm., finely tuberculate. Fl. March–April.

Hab.: Steppes and grassy patches. Esdraelon Plain, Samaria, Judean Mts., Judean Desert, Lower Jordan Valley, Ammon, Moav, Edom.

Area: E. Mediterranean (endemic in Palestine).

30. Silene tridentata Desf., Fl. Atl. 1 : 349 (1798); Rohrb., Monogr. Sil. 99 (1868). *S. calycina* Salzm. ex Rohrb., l.c. [Plate 128]

Annual, appressed-pubescent, not viscid, 10–30 cm. Stems simple or branched. Leaves spatulate-lanceolate to linear-oblanceolate, acute, ciliate at base. Flowers 1.3–1.8 cm., subsessile. Calyx 1.2–1.5 cm., oblong-ovoid, rounded at base, ovoid-globular in fruit, hispid along nerves with articulate hairs; teeth 4–5 mm., acuminate-subulate, elongated in fruit. Petals pink, almost included in calyx; claws connate in upper part; limb 2-lobed; coronal scales minute. Capsule about 1 cm., subsessile, ovoid to almost globular; carpophore 1–2 mm. Seeds reniform-orbicular, auriculate, grooved on back. Fl. March–April.

Hab.: Steppes. Judean Desert, Lower Jordan Valley, Edom.

Area: W. Irano-Turanian, extending towards W. Mediterranean territories.

Sect. DIPTEROSPERMAE (Rohrb.) Chowdhuri, Not. Roy. Bot. Gard. Edinb. 22 : 248 (1957). Annuals. Inflorescences mostly monochasial. Fruiting calyx scarcely or not constricted at throat. Seeds reniform, with plane faces, and a deeply and acutely grooved back forming 2 undulate wings.

31. Silene colorata Poir., Voy. Barb. 2 : 163 (1789); Rohrb., Monogr. Sil. 114 (1868). *S. bipartita* Desf., Fl. Atl. 1 : 352, t. 100 (1798); Boiss., Fl. 1 : 597 (1867). [Plate 129]

Annual, puberulent or hispid to appressed-pubescent, not glandular, 10–30 cm. Stems decumbent to erect, simple or more or less branching. Leaves 3–10 × 0.15–1.5 cm., the lower ones obovate-spatulate or oblong-lanceolate to oblong-linear, the rest oblong-linear to narrowly linear-lanceolate, acute or obtuse, all more or less ciliate. Inflorescences monochasial, usually few-flowered. Bracts unequal. Pedicels shorter than calyx, shorter or longer or the lower as long as or longer than bracts. Calyx 1–2 cm., cylindrical, ovoid to club-shaped in fruit, umbilicate, hairy throughout or along nerves only, rarely scabrous and subglabrous, red-, purple- or green-nerved; teeth 1.5–4.5 mm., ovate to lanceolate, obtuse or acute, ciliate. Petals varying in length, with claw usually long-exserted from calyx; limb pink-lilac to whitish, 2-partite into oblong-linear to narrowly obovate-spatulate lobes; coronal scales obovate to oblong-lanceolate, acute or obtuse, rarely retuse. Capsule about as long as to twice as long as carpophore, ovoid. Seeds 1 mm. or more in diam., ear-shaped with undulate wings at margin, almost smooth to somewhat tuberculate on faces and deeply grooved on back. Fl. February–April.

Subsp. **colorata.** Lower leaves obovate, obtuse, the rest oblong-linear, ciliate, broader than 2 mm. Calyx with appressed long hairs, red- or green-nerved; teeth 2–4 mm., oblong, obtuse. Petal limb pink; coronal scales oblong-lanceolate, more or less acute. Capsule as long as to twice as long as carpophore.

Hab.: Batha and fields. Acco Plain, Sharon Plain, Philistean Plain, Upper and Lower Galilee, Mt. Carmel, Samaria, Judean Mts., Judean Desert, N. and C. Negev, Dan Valley, Hula Plain, Upper and Lower Jordan Valley, Gilead, Ammon, Moav, Edom.

Area: Mediterranean.

Subsp. **oliveriana** (Otth) Rohrb., Monogr. Sil. 116 (1868). *S. oliveriana* Otth in DC., Prodr. 1 : 373 (1824); Boiss., Fl. 1 : 597 (1867). Lower leaves oblong-linear, others narrower, acute. Bracts linear. Calyx crisp-hairy, generally green-nerved; teeth 2 mm., lanceolate, acute. Petal limb whitish; claw barely exserted from calyx; coronal scales obovate, obtuse or rarely retuse. Capsule as long as or somewhat shorter than carpophore.

Hab.: Steppes and deserts. Sharon Plain, Philistean Plain, Judean Desert, Negev, Lower Jordan Valley, Gilead, Ammon, Moav, Edom and probably elsewhere.

Area: Mainly Saharo-Arabian, slightly extending into arid parts of the Mediterranean region.

Apart from the above taxa, some other varieties are recorded by Post (Fl. Syr. Pal. Sin. ed. 2, 1 : 178, 1932), Rohrbach (l.c.) and by Oppenheimer and Evenari (Bull. Soc. Bot. Genève ser. 2, 31 : 224, 1941). All these varieties and forms are, so far, very poorly known and deserve a thorough taxonomic study.

This is one of the most common species of *Silene* in the area.

32. Silene apetala Willd., Sp. Pl. 2 : 703 (1799); Boiss., Fl. 1 : 596 (1867); Rohrb., Monogr. Sil. 118 (1868). [Plate 130]

Annual, appressed-pubescent, 10–30 cm. Stems simple or dichotomously branched. Leaves up to 5 (–7) mm. broad, lanceolate to linear, the lower often more or less spatulate, acute, ciliolate. Inflorescences elongated, almost one-sided or irregularly cymose. Bracts ovate-lanceolate. Pedicels varying in length, usually longer than calyx. Flowers 0.7–1 cm. Calyx oblong-obovoid to campanulate, not umbilicate, green or purple or red-nerved; teeth short, lanceolate, acute, ciliate. Petals white or pinkish, very short, not or rarely exserted from calyx, 2-partite and with truncate or retuse coronal scales (var. *grandiflora* Boiss., l.c. 597); sometimes petals 0. Capsule subsessile, 3 times as long as carpophore or more, subglobular. Seeds flat, deeply grooved on back, ear-shaped with undulate margins. Fl. February–April.

Hab.: Fallow fields and waste places.

Var. **apetala.** Calyx appressed short-puberulent.

Sharon Plain, Philistean Plain, Judean Mts., Judean Desert, Negev, Lower Jordan Valley, Dead Sea area, Moav, Edom.

Var. **alexandrina** Aschers. in Aschers. et Schweinf., Ill. Fl. Eg. 46 (1887). Calyx white-hispid along nerves. Pedicels mostly up to twice as long as calyx.

N. Negev.

Area of species: Mediterranean and Irano-Turanian; adventive in Trop. Africa.

Sect. CONOIMORPHA Otth in DC., Prodr. 1: 371 (1824). Annuals. Inflorescences of lax dichasia. Calyx 15–20- or 30- or 60-nerved. Seeds reniform; faces plane or slightly concave; back with shallow groove.

33. Silene coniflora Nees ex Otth in DC., Prodr. 1: 371 (1824); Boiss., Fl. 1: 578 (1867); Rohrb., Monogr. Sil. 89 (1868). [Plate 131]

Annual, viscid-pubescent, 5–20 cm. Stems simple or branching from base. Leaves less than 5 mm. broad, sessile, grass-like, linear-lanceolate, acuminate, puberulent. Inflorescences usually dichasial. Bracts narrowly linear, lanceolate-acuminate. Pedicels varying in length. Flowers 1–1.5 cm., pinkish or white. Calyx 0.8–1.3 cm., 20-nerved, obconical to cylindrical, ovoid-oblong in fruit, more or less umbilicate, glandular-pubescent, especially along nerves; teeth narrowly lanceolate, acute, up to one third as long as tube. Petals obovate-cuneate, entire; coronal scales oblong. Capsule about 1 cm., almost as long as calyx, oblong-conical, almost without carpophore. Seeds tuberculate wrinkled. Fl. March–April.

Hab.: Fields. Judean Mts., Judean Desert, N. Negev, Ammon, Moav, Edom.

Area: W. Irano-Turanian, extending towards S. Mediterranean.

34. Silene conoidea L., Sp. Pl. 418 (1753); Boiss., Fl. 1: 580 (1867); Rohrb., Monogr. Sil. 92 (1868). [Plate 132]

Annual, viscid-pubescent, 10–50 cm. Stems simple or branching from base. Leaves 2–6 cm., sessile, oblong-lanceolate to lanceolate-linear, acute, puberulent. Inflorescences loosely dichasial, few-flowered. Bracts lanceolate. Pedicels mostly shorter than calyx, elongated in fruit. Flowers 1.7–2.4 cm., pink to pale red. Calyx 1.5–2 cm., 30-nerved, conical-cylindrical, umbilicate, in fruit globular-inflated with attenuate apex, pubescent, green, with lanceolate teeth half as long as tube. Petals entire, emarginate or 2-fid. Capsule sessile (without carpophore), almost equalling calyx. Seeds tuberculate-wrinkled. Fl. March–May.

Hab.: Fields. Philistean Plain, Judean Mts., Judean Desert, Negev, Dead Sea area, Gilead, Ammon, Moav, Edom.

Area: Mainly W. Irano-Turanian and E. Mediterranean.

35. Silene macrodonta Boiss., Diagn. ser. 1, 1: 37 (1843) et Fl. 1: 580 (1867); Rohrb., Monogr. Sil. 92 (1868). [Plate 133]

Annual, velvety-canescent, viscid above, 15–30 cm. Stems erect, simple or branched. Basal leaves oblong-oblanceolate, the upper ones linear, all leaves acute, somewhat connate at base. Flowers solitary or few in loose, raceme-like, dichasial branches. Bracts linear-lanceolate. Pedicels generally shorter than calyx. Calyx 1.2–1.5 cm., 60-nerved, cylindrical, ovoid in fruit, subtruncate at base, glandular-pubescent; teeth almost as long as tube, lanceolate-subulate, rigid, later spreading. Petals pink; limb cuneate-obcordate; coronal scales truncate to retuse. Capsule short-stipitate or subsessile, shorter than calyx, oblong-conical to globular. Seeds wrinkled and acutely tuberculate. Fl. April–June.

Hab.: Fields. Moav (Post), Judean Mts. (Dinsmore); both in Post, ed. 2, 1 : 172 (1932).

Area : E. Mediterranean (Palestine-Turkey).

Not observed by the author. Drawn from an Anatolian specimen.

Trib. DIANTHEAE. Capsule many- or 1–2-seeded, dehiscent. Calyx without commissural nerves. Petals contorted or imbricated in bud.

3. GYPSOPHILA L.

Annual or perennial herbs or shrubs. Leaves opposite, exstipulate. Inflorescences cymose, dichasially branched. Bracts mostly scarious, rarely herbaceous. Flowers small, hermaphrodite, sometimes unisexual. Calyx top-shaped or hemispherical or campanulate, rarely cylindrical, with 5-nerved tube membranous between nerves, and with 5 teeth. Petals 5, white or pink, mostly cuneate, tapering into a wingless claw, without coronal scales. Stamens 10, free. Ovary 1-celled, many-ovuled, with 2 (–3) styles. Capsule usually globular or ovoid, dehiscing to base by 4 (–6) valves. Seeds reniform-auriculate.

Over 90 species centred mainly in the Mediterranean and Irano-Turanian regions, but also in other north-temperate regions of the Old World.

Literature : F. N. Williams, Revision of the specific forms of the genus *Gypsophila*. Journ. Bot. Lond. 27 : 321–329 (1889). G. Stroh, Die Gattung *Gypsophila* Linn. Vorläufiger Katalog. *Beih. Bot. Centralbl.* 59, 2 : 455–477 (1939). Y. I. Barkoudah, A revision of *Gypsophila, Bolanthus, Ankyropetalum* and *Phryna. Wentia* 9 : 1–203 (1962).

1. Stems patulous-villose. Calyx hairy.   **3. G. pilosa**
– Stems and calyx glabrous   **2**
2. Perennials (30–70 cm.). Leaves 1-nerved. Bracts herbaceous. Capsule as long as or shorter than calyx.   **1. G. arabica**
– Annuals (15–35 cm.). Leaves indistinctly 3–5-nerved. Bracts scarious. Capsule longer than calyx.   **2. G. viscosa**

Sect. ROKEJEKA (Forssk.) Graebn. in Aschers. et Graebn., Syn. 5, 2 : 235 (1921). Perennials. Inflorescence a lax, diffuse panicle. Calyx more or less campanulate. Capsule globular.

1. Gypsophila arabica Barkoudah, Wentia 9 : 139 (1962). *G. rokejeka* auct. non Del., Fl. Eg. 87 (1813); Boiss., Fl. 1 : 543 (1867). [Plate 134]

Glabrous, glaucescent perennial herb, 30–70 (–80) cm. Stems slender, much branched, dichotomously paniculate from the woody base, leafy only below. Leaves 1–5 × 0.1–

0.3 (–0.6) cm., rather caducous, the lowest larger, oblanceolate or lanceolate to linear, obtuse, the upper smaller, linear-lanceolate to narrowly linear, acute, often canaliculate, all 1-nerved, papillose or not. Inflorescences much branched, many-flowered, flowering branches almost filiform. Bracts small, leaf-like, narrowly linear. Pedicels 4–8 times as long as calyx, filiform. Calyx 2–3 mm., campanulate; lobes oblong, white-margined, obtuse or emarginate. Petals 3–6 mm., longer than calyx, white, pink- to violet-striped, oblong-elliptical. Ovary 4-lobed, 4-ovuled. Capsule almost bare, as long as or shorter than calyx, 1 (–2)-seeded. Seeds 1–1.5 mm., wrinkled and acutely tuberculate. Fl. May–September (–November).

Hab.: Steppes, rarely batha. Sharon Plain, Philistean Plain, Judean Desert, W., N., C. and S. Negev, Upper Jordan Valley, Beit Shean Valley, Lower Jordan Valley, Dead Sea area, Arava Valley, Gilead, Ammon, Moav, Edom.

Area: W. Irano-Turanian, extending into adjacent Saharo-Arabian territories.

Root used in medicine.

G. arabica closely resembles *G. rokejeka* Del., Fl. Eg. 87 (1813) [*G. capillaris* (Forssk.) Christens., Dansk Bot. Ark. 4, 3 : 19 (1922). *Rokejeka capillaris* Forssk., Fl. Aeg.-Arab. 90 (1775)]. The latter differs from *G. arabica* in the following characters: leaves are generally oblanceolate, and partly broader than 3 mm.; internodes are rather long; calyx lobes obtuse; ovary 2-lobed, 12-ovuled; capsule included in the calyx, 4–6-seeded; seeds with flat tubercles.

G. rokejeka has not been observed by us in Palestine but it probably occurs here.

Sect. DICHOGLOTTIS (Fisch. et Mey.) Boiss., Fl. 1 : 536 (1867). Annuals. Inflorescence a dichasial panicle. Calyx short-campanulate, divided to middle or further. Petals truncate, emarginate or 2-fid. Capsule subglobular.

2. Gypsophila viscosa Murr., Commentat. Soc. Sci. Gott. ser. 2, 6:9, t. 3 (1785); Boiss., Fl. 1 : 551 (1867); Barkoudah, Wentia 9 : 137 (1962). [Plate 135]

Annual, glabrous, viscid in upper part, 15–35 (–45) cm. Stems thin, flexible, dichotomously branched, corymbose, paniculate. Leaves 1.5–6 × 0.2–1 cm., crowded at base of stems, sessile and slightly connate at base, oblong-lanceolate to oblong-linear, acute or obtuse, indistinctly 3–5-nerved. Inflorescences divaricately branched, many-flowered. Bracts minute, triangular-acuminate, scarious. Pedicels 3–5 times as long as calyx, filiform. Calyx 1.5–3 mm., short-campanulate, with ovate, obtuse, white-margined teeth. Petals longer than calyx, 2.5–4 mm., pink, obovate to cuneate, truncate to retuse. Capsule longer than calyx. Seeds about 1 mm., tuberculate on back. Fl. March–June.

Hab.: Steppes and sandy fields. Philistean Plain, N. and C. Negev, Edom.

Area: Irano-Turanian.

Sect. HAGENIA (Moench) Boiss., Fl. 1 : 537 (1867). Annuals. Inflorescences paniculate, dichasial. Calyx more or less tubular or campanulate. Capsule ovoid or globular.

3. Gypsophila pilosa Huds., Phil. Trans. Roy. Soc. Lond. 56:252 (1767); Barkoudah, Wentia 9·: 151 (1962). *G. porrigens* (L.) Boiss., Fl. 1 : 557 (1867). [Plate 136]

Annual, patulous-villose, viscid, 25–50 cm. Stems dichotomously branched, fairly thick below. Leaves 1–7 × 0.5–1 (–1.5) cm., the lower larger, sessile, somewhat connate at base, oblong-lanceolate to linear, acute, 3-nerved. Inflorescence – a divaricately branching panicle. Bracts short, narrowly linear-lanceolate, hairy beneath. Pedicels 3–6 times as long as calyx, filiform. Calyx 4–5 mm., oblong-campanulate, later subspherical, hispid, deeply parted into oblong, white-margined lobes. Petals 6–8 mm., white, oblong-cuneate, retuse. Capsule about as long as calyx, globular, deflexed. Seeds 1.5 mm., tuberculate. Fl. March–May.

Hab.: Steppes. Judean Mts., Judean Desert, W., N. and C. Negev, Lower Jordan Valley.

Area: W. Irano-Turanian, slightly extending into adjacent regions.

4. BOLANTHUS (Ser.) Reichb.

Perennial hairy herbs. Leaves opposite, small, exstipulate. Bracts leaf-like. Pedicels as long as or shorter than calyx. Flowers small, hermaphrodite. Calyx tubular, pentagonous, with 5-ribbed tube membranous between ribs and with 5 small teeth. Petals cuneate, with spreading limb and long, finely winged claw constricted at top. Stamens 10. Ovary 1-celled, many-ovuled; styles 2, stigmatose all along the inner surface. Capsule oblong-ovoid or cylindrical to urceolate, dehiscent by 4 teeth. Seeds comma-shaped.

About 8 species, mainly E. Mediterranean.

1. Bolanthus filicaulis (Boiss.) Barkoudah, Wentia 9 : 167 (1962). *Saponaria filicaulis* Boiss., Diagn. ser. 1, 7 : 71 (1847). *Gypsophila filicaulis* (Boiss.) Bornm., Beih. Bot. Centralbl. 31, 2 : 191 (1914). *G. hirsuta* (Labill.) Spreng. var. *filicaulis* Boiss., Fl. 1 : 557 (1867). [Plate 137]

Perennial herb, velvety or short-pubescent, canescent, tufted, 6–20 cm. Stems many, flexible, erect or decumbent, elongated. Leaves 0.3–1.5 (–2) × 0.1–0.2 cm., sessile or almost so, pubescent, oblong-ovate to linear-oblanceolate, more or less acute. Bracts leaf-like. Pedicels generally shorter than calyx. Calyx 4–5 mm., almost tubular, with triangular to lanceolate teeth, patulous-puberulent. Petals 6–7 mm., long-clawed, with oblong-obovate white limb. Capsule cylindrical. Seeds finely wrinkled. Fl. April–May.

Hab.: Batha. Ammon, Moav, Edom.

Area: W. Irano-Turanian (Palestine-Syria).

5. ANKYROPETALUM Fenzl

Perennial rigid plants with very thin branches. Leaves opposite, soon deciduous, linear, exstipulate. Inflorescences dichasial, loose, paniculate. Bracts membranous. Calyx cylindrical, 5-nerved, membranous between nerves; teeth small. Petals 5, inserted on the upper margin of a cup-like disk, with long, naked claw and deeply 3–5-parted limb. Stamens 10, exserted. Ovary 1-celled, with 8–10 long-funicled ovules; styles 2, long. Capsule 1-celled, ovoid, 4-valved, irregularly dehiscent. Seeds 1–3, reniform-globular, wrinkled-granulate.

Some 4 species in the western part of the Irano-Turanian region.

1. Ankyropetalum gypsophiloides Fenzl, Bot. Zeit. 1 : 393 (1843); Boiss., Fl. 1 : 533 (1867); Barkoudah, Wentia 9 : 173 (1962). [Plate 138]

Grey-green perennial herb, 30–80 cm. Stems ascending or decumbent, branched, glabrous or puberulent and sometimes viscid-glandular above. Leaves 1–4 × 0.1–0.2 cm., soon deciduous, linear, glandular-pubescent. Pedicels filiform, much longer than calyx. Calyx 3–6 mm., tubular, glandular-puberulent; teeth very small, ovate-rounded, white-margined. Petals longer than calyx, purplish to pink; limb 5-lobed, the middle lobe deltoid, often limb 3-lobed with middle lobe anchor-shaped; claw linear-cuneate, exserted. Capsule almost included in calyx, ellipsoidal. Seeds with somewhat muricate wartlets. Fl. June–July.

Var. **gypsophiloides.** Basal part of stems glabrous, upper part glandular-hairy. Petal limb with anchor-shaped middle lobe 2-toothed below.

Hab.: Mediterranean border batha. Upper and Lower Galilee, Dan Valley, Hula Plain, Upper Jordan Valley. Rather rare.

Var. **coelesyriacum** (Boiss.) Barkoudah, l.c. 174. *A. coelesyriacum* Boiss., Diagn. ser. 1, 8 : 59 (1849) et Fl. 1 : 534 (1867). Basal part of stem puberulent, upper part glandular-hairy. Petal limb with deltoid middle lobe not 2-toothed below.

Hab.: As above. Upper and Lower Galilee, Negev, Dan Valley, Hula Plain, Upper Jordan Valley.

Area of species : W. Irano-Turanian (Palestine, Syria, Iraq and Iran).

6. SAPONARIA L.

Annual or perennial herbs. Leaves opposite, exstipulate. Flowers in loose or dense dichasia, sometimes solitary. Calyx cylindrical, 15–25-nerved, 5-toothed. Petals 5, long-clawed; coronal scales present or 0. Stamens 10. Ovary sessile or borne on a carpophore, 1-celled, many-ovuled, with central placentation; styles 2. Capsule oblong to obovoid, dehiscent by 4 teeth or valves. Seeds reniform or almost globular; embryo peripheral.

About 30 species in temperate Eurasia, chiefly in the Mediterranean region.

Literature : G. Simmler, Monographie der Gattung *Saponaria. Denkschr. Akad. Wiss. Wien Math.-Nat. Kl.* 75 : 433–509 (1910).

1. Saponaria mesogitana Boiss., Diagn. ser. 1, 1 : 16 (1843) et Fl. 1 : 528 (1867). [Plate 139]

Pubescent-viscid annual, 10–20 cm. Stems erect, dichotomously branching. Lower leaves oblong-spatulate, obtuse, upper and floral ones smaller, oblong-lanceolate to linear-lanceolate. Inflorescences corymbose, dense. Pedicels shorter than calyx, erect to patulous. Calyx about 1 cm., purple, cylindrical, almost oblong in fruit; teeth ovate to oblong-lanceolate, narrowly white-margined. Petals pink, with obovate limb half as long as calyx, tapering into a long claw. Capsule as long as calyx, oblong. Seeds finely tuberculate. Fl. Spring.

Hab.: Mountains. Upper Galilee (Samuelsson, in litteris). Rare. Drawn from an Anatolian specimen.

Area : E. Mediterranean (Palestine-Turkey).

7. Vaccaria Moench

Annual, dichotomously branching herbs. Leaves opposite, exstipulate. Flowers herm-aphrodite, in loose, corymbose or paniculate dichasia. Calyx herbaceous, many-nerved, prominently 5-ribbed, accrescent and becoming 5-winged in fruit; teeth 5. Petals 5, generally purple, clawed, without coronal scales. Stamens 10. Ovary borne on a short carpophore, 1-celled, slightly 2-celled at base, many-ovuled. Styles 2, filiform. Capsule oblong to globular, dehiscent by 4 teeth. Seeds reniform or subspherical.

Some 4 species in the Mediterranean region.

1. Vaccaria pyramidata Medik., Phil. Bot. 1 : 96 (1789). *V. parviflora* Moench, Meth. 63 (1794). *V. segetalis* (Neck.) Garcke ex Aschers., Fl. Prov. Brandenb. 1 : 84 (1864). *Saponaria vaccaria* L., Sp. Pl. 409 (1753); Boiss., Fl. 1 : 525 (1867). [Plate 140]

Annual, glabrous, glaucous, 15–60 cm. Stems erect, branched in upper part. Leaves 5–10 × 0.5–4 (–5) cm., radical oblong or lanceolate, tapering at base, upper ovate, clasping or connate at base. Inflorescences paniculate, of loose, spreading dichasial branches. Bracts narrowly lanceolate, acute, green to almost membranous, 1-nerved. Pedicels mostly as long as or longer than calyx. Calyx 1–1.7 cm., oblong to spindle-shaped, ovoid in fruit, glabrous, whitish-green between the 5 prominent ribs; teeth short, ovate-triangular or lanceolate, scarious at margin and often at apex. Petals 1.5–2.2 cm., pink; limb short, obovate, entire or irregularly dentate or notched; claw long, exserted or not. Styles 2. Capsule 0.7–1 cm., ovoid or subglobular, with very prominent ribs extending into wings. Seeds about 2 mm. in diam., subglobular, black, finely tuberculate. Fl. February–May.

Var. **pyramidata.** Calyx 1.4–1.7 cm., with short-triangular teeth, scarious at margin and apex. Claw somewhat exserted.

Hab.: Grain fields. Acco Plain, Sharon Plain, Philistean Plain, Upper and Lower Galilee, Esdraelon Plain, Samaria, Judean Mts., N. Negev, Lower Jordan Valley, Gilead, Ammon.

Var. **oxyodonta** (Boiss.) Zoh. (comb. nov.). *V. oxyodonta* Boiss., Diagn. ser. 2, 1 : 68 (1853). *Saponaria oxyodonta* Boiss., Fl. 1 : 525 (1867). Calyx 1–1.4 cm., with teeth lanceolate, longer than above, not scarious at apex. Limb small; claw included. To-gether with typical form but very rare and intergrading with it.

Hab.: As above. Shefela, Judean Desert, Negev, Hula Plain, Upper and Lower Jordan Valley, Ammon, Moav, Edom.

Area of species: Mainly Mediterranean, extending into the Irano-Turanian and Euro-Siberian regions; also introduced elsewhere (N. America, Australia, New Zea-land).

One of the most characteristic weeds of winter crops.

Sometimes grown as an ornamental plant. Roots contain saponine. Reputed as a medi-cinal herb with depurative and other healing qualities.

8. PETRORHAGIA (Ser.) Link

Annual or perennial slender herbs. Stems rigid. Leaves narrow. Inflorescences paniculate to capitate. Calyx with or without bracts at base (epicalyx), 5-toothed, 5–15-nerved, with scarious commissures (intervals between nerves or groups of nerves). Petals 5, with entire to 2-fid limb and without coronal scales. Stamens 10. Ovary 4-celled at base, 1-celled at apex; styles 2. Capsule many-seeded, oblong or ovoid, dehiscent, with 4 teeth. Seeds shield-shaped; embryo eccentric, straight.

About 35 species in north-temperate regions of the eastern hemisphere and mainly in the Mediterranean region.

Literature : P. W. Ball and V. H. Heywood, A revision of the genus *Petrorhagia. Bull. Brit. Mus. Bot.* 3 : 121–172, pls. 13–15 (1964).

1. Inflorescences capitate; heads surrounded by an involucre of scarious bracts.
3. P. velutina
- Inflorescences paniculate. Bracts absent **2**
2. Petals exserted. Calyx 7–8 mm., interspace between nerves hairy. Leaves up to 4 cm. Plants higher than 10 cm. **1. P. cretica**
- Petals more or less included. Calyx 6–7 mm., interspace between nerves glabrous. Leaves mostly shorter than 1 cm. Plants up to 10 cm. high. **2. P. arabica**

Sect. PSEUDOTUNICA (Fenzl) P. W. Ball et Heywood, Bull. Brit. Mus. Bot. 3 : 132 (1964). Annuals or perennials. Inflorescences paniculate or fasciculate. Leaves 3-nerved. Bracts absent. Petals not clawed. Seeds black, smooth.

1. Petrorhagia cretica (L.) P. W. Ball et Heywood, Bull. Brit. Mus. Bot. 3 : 142 (1964). *Saponaria cretica* L., Sp. Pl. ed. 2 : 584 (1762). *Tunica pachygona* Fisch. et Mey., Ind. Sem. Hort. Petrop. 4 : 50 (1837); Boiss., Fl. 1 : 522 (1867). [Plate 141]

Glandular-pubescent, viscid annual, 10–25 cm. Stems erect, dichotomously branched in upper part. Leaves up to 4 × 0.1–0.4 cm., the radical ones oblong-linear or linear, the cauline narrowly linear, tapering, all sparingly pubescent, more or less scabrous at margin and especially beneath, 3-nerved. Inflorescences divaricately paniculate. Pedicels rigid and rather thick, almost as long as or longer than calyx. Calyx 7–8 mm., obconical, 15-nerved, with nerves arranged in 5 groups of 3, membranous and pubescent in interspaces between nerve groups, often with spreading glandular hairs; teeth short, ovate-triangular, acute-mucronate, scarious-margined. Petals 0.9–1.1 cm., later exceeding calyx, linear-spatulate, entire, white, sometimes reddish. Capsule about as long as calyx. Fl. April–May.

Hab.: Steppes and batha. Philistean Plain, Upper and Lower Galilee, Samaria, Judean Mts., Judean Desert, Lower Jordan Valley, Gilead, Ammon, Moav, Edom.

Area : E. Mediterranean, extending into adjacent areas of the Irano-Turanian region.

2. Petrorhagia arabica (Boiss.) P. W. Ball et Heywood, Bull. Brit. Mus. Bot. 3 : 143 (1964). *Tunica arabica* Boiss., Diagn. ser. 1, 8 : 62 (1849) et Fl. 1 : 523 (1867). [Plate 141a]

Glandular-pubescent annual, 5–10 cm. Stems dichotomously branched, branches not divaricate. Leaves 0.6–1 × 0.1–0.4 cm., the radical ones oblong, the cauline narrowly linear, all nearly glabrous, scabrous-margined, obsoletely 3-nerved. Inflorescences divaricately paniculate. Pedicels rather thick, about as long as or longer than calyx. Calyx 6–7 mm., cylindrical-obconical, 15-nerved, glandular-pubescent along nerves, not hairy in the membranous parts between the nerve groups; teeth triangular, acute-mucronulate. Petals included in or slightly exserted from calyx, white, linear. Fl. April–May.

Hab. : Deserts. Lower Jordan Valley, Moav, Edom.

Area : Saharo-Arabian (endemic in Palestine).

The taxonomic value of the above taxon needs further clarification. The data by Post (1932) and others on this species from Upper Galilee are doubtful.

Sect. KOHLRAUSCHIA (Kunth) P. W. Ball et Heywood, Bull. Brit. Mus. Bot. 3 : 160 (1964). Annuals. Leaves 3-nerved. Inflorescence head-like, with broad, brown, scarious bracts. Petals abruptly clawed. Seeds blackish-brown, reticulate to papillose.

3. Petrorhagia velutina (Guss.) P. W. Ball et Heywood, Bull. Brit. Mus. Bot. 3 : 166 (1964). *Kohlrauschia velutina* (Guss.) Reichb., Ic. Fl. Germ. 6 : 43 (1844). *Dianthus velutinus* Guss., Ind. Sem. Hort. R. Boccadifalco 1825 : 2 (1825) et Pl. Rar. 166, Ic. 7, t. 32 (1826). *Tunica velutina* (Guss.) Fisch. et Mey., Ind. Sem. Hort. Petrop. 6 : 66 (1839); Boiss., Fl. 1 : 516 (1867). [Plate 142]

Velvety to subglabrous annual, 15–50 cm. Stems erect, almost simple. Leaves 1–5(–7) × 0.1–0.4(–0.7) cm., the lower oblanceolate-spatulate or oblong-linear, more or less scabrous at margin, the upper linear, connate at the somewhat scarious base, usually acute or mucronulate. Flower heads 1.2–2 cm., terminal, 2–3-flowered. Bracts 0.7–1.5 cm., ovate to obovate, hyaline, mucronulate. Pedicels short. Flowers 1.2–1.3 cm., dark pink. Calyx tubular, with very short teeth. Petals exserted; limb small, notched or 2-fid. Capsule about half as long as calyx, oblong. Seeds boat-shaped, tuberculate. Fl. February–May.

Hab. : Batha and fields. Coastal Galilee, Acco Plain, Sharon Plain, Philistean Plain, Upper Galilee, Mt. Carmel, Judean Mts., Gilead.

Area : Mediterranean.

9. DIANTHUS L.

Perennial herbs or dwarf-shrubs, rarely annuals. Stems mostly thickened at nodes. Leaves opposite, grass-like. Flowers conspicuous, hermaphrodite or rarely unisexual, solitary or in cymose, terminal inflorescences. Calyx subtended by an epicalyx of 1 to many pairs of mucronate or awned scales (bracts); tube cylindrical to conical; teeth 5, usually short. Petals 5, pink or red, sometimes white, rarely yellow, long-clawed; limb entire, dentate or fringed, hairy or glabrous at base, without coronal scales. Stamens 10.

Ovary 1-celled; styles 2. Capsule many-seeded, cylindrical to ovoid, dehiscing by 4 teeth. Seeds compressed, discoid, concave on one side; embryo straight, eccentric.

About 300 species, in Asia, Europe, Africa and America; especially rich in species are the Mediterranean and Irano-Turanian regions.

Literature: F. N. Williams, A monograph of the genus *Dianthus* L. *Journ. Linn. Soc. Lond. Bot.* 29 : 346–478 (1893). H. Rohweder, Beiträge zur Systematik u. Phylogenie d. Genus *Dianthus,* unter besonderer Berücksichtigung der karyologischen Verhältnisse. *Bot. Jahrb.* 66 : 249–368 (1934). F. Lemperg, Studies in the perennial species of the genus *Dianthus* L. I. (Sektion *Barbulatum*). *Meddel. Göteborgs Bot. Trädg.* 11 : 71–134 (1936). F. Weissmann–Kollmann, Taxonomic study in *Dianthus,* etc. *Israel Journ. Bot.* (in press).

1. Calyx tube tuberculate   2
– Calyx tube not tuberculate   5
2. Annuals. Petals glabrous, rarely petals sparingly hairy but then calyx not striate-nerved   3
– Perennials. Petals hairy at base. Nerves of calyx in close groups with nerveless interspaces between groups   4
3. Calyx tube nerved. Epicalyx shorter than calyx.   **4. D. tripunctatus**
– Calyx tube nerveless. Epicalyx about as long as to longer than calyx.   **5. D. cyri**
4(2). Calyx (1–)1.5(–2) cm. long. Petal limb obovate-cuneate and dentate at margin, rarely oblong and emarginate, pink, usually with deep crimson spots and stripes.   **2. D. strictus**
– Calyx 0.6–9(–1.2) cm. long. Petal limb linear, almost entire or retuse, pale pink or white.   **3. D. polycladus**
5(1). Petal limb almost entire or slightly crenate or dentate, glabrous at base.   **6. D. judaicus**
– Petal limb deeply incised-dentate or fimbriate   6
6. Epicalyx scales 4–6, spreading, tapering into a recurved, lanceolate, spiny tip. Cauline leaves 3–7 mm. broad, often spreading or recurved.   **7. D. libanotis**
– Epicalyx of 10–14 appressed, imbricated scales   7
7. Leaves up to 7 cm. Petal limb obovate, hairy at base, deeply incised-dentate.   **1. D. pendulus**
– Leaves shorter than in above. Petal limb narrowly oblong, glabrous, fimbriate up to three quarters of its length.   **8. D. sinaicus**

Sect. DENTATI Boiss., Fl. 1 : 480 (1867). Perennials. Calyx striate-nerved. Petal limb pink or purple, mostly barbate at base, very deeply incised-dentate (up to about one quarter of its length).

1. Dianthus pendulus Boiss. et Bl. in Boiss., Diagn. ser. 2, 6 : 28 (1859) et Fl. 1 : 499 (1867). [Plate 143]

Glabrous perennial, 20–60 cm. Stems many, erect, leafy, woody below. Leaves up to 7 × 0.1–0.4 cm., connate and forming short sheaths at their white-margined ciliolate bases, linear, patulous or recurved. Inflorescences short-branched. Flowers solitary or 2–3 in a cluster, short-pedicelled. Scales of epicalyx 10–14, 3–8 mm., imbricated, oblong-lanceolate to ovate, acute to mucronate, scarious-margined, the uppermost about one third as long as calyx. Calyx 2.5–3 cm.; tube striate, much longer than lanceolate, acute teeth. Petals 4.5–5 cm., pinkish, long-clawed; limb obovate in outline, deeply incised-dentate, barbate at base. Capsule obovoid. Fl. April–September.

Hab.: Hanging on cliffs. Upper and Lower Galilee, Mt. Carmel, Upper Jordan Valley, Beit Shean Valley, Gilead.

Area: E. Mediterranean with slight eastward extensions.

Showy; worthy of introduction into ornamental gardening.

Plants of the Lower Galilee, Upper Jordan Valley and Beit Shean Valley differ from the type by somewhat broader, shorter and abruptly mucronate scales of epicalyx which are sometimes coloured and broadly membranous-margined.

Sect. VERRUCULOSI Boiss., Fl. 1 : 479 (1867). Annuals and perennials. Calyx conical, striate and minutely tuberculate-verrucose. Scales of epicalyx 4–6. Petal limb barbate or glabrous.

2. Dianthus strictus Banks et Sol. in Russ., Nat. Hist. Aleppo, ed. 2, 2 : 252 (1794); Eig, Journ. Bot. Lond. 75 : 187 (1937) non *D. strictus* Sibth. et Sm., Fl. Gr. Prodr. 1 : 288 (1809). *D. multipunctatus* Ser. in DC., Prodr. 1 : 362 (1824); Boiss., Fl. 1 : 482 (1867). [Plate 144]

Perennial herb, glabrous or sometimes short-hairy, 20–60 cm. Stems many, ascending to erect, simple or branching. Leaves 2–4 mm. broad, linear, ciliate or scabrous at margin and narrowly scarious-margined towards base. Pedicels longer or shorter than calyx, sometimes almost 0. Flowers usually terminal, mostly solitary. Epicalyx much shorter than calyx, composed of 4–6 ovate, obtuse, broadly scarious scales, ending with a short awn. Calyx (1–)1.5(–2) cm., with almost nerveless to wholly nerved tube and short, lanceolate, striate teeth. Petals 2–3 cm.; limb pink, often marked with crimson spots and stripes, obovate-cuneate, rarely oblong, dentate or emarginate, barbate at base. Capsule oblong. Seeds broadly elliptical to rounded, black, tuberculate. Fl. April–October.

Var. **strictus.** Glabrous. Calyx almost without nerves except for the ones grouped below sinuses (between calyx teeth). Petals obovate-cuneate. Pedicels conspicuous.

Hab.: Batha and stony ground. Coastal Galilee, Acco Plain, Sharon Plain, Philistean Plain, Upper and Lower Galilee, Mt. Carmel, Esdraelon Plain, Samaria, Judean Mts., Judean Desert, W. and C. Negev, Upper and Lower Jordan Valley, Gilead, Ammon, Moav.

Var. **velutinus** (Boiss.) Eig, l.c. 191. *D. multipunctatus* Ser. var. *velutinus* Boiss., Diagn. ser. 1, 8 : 65 (1849). *D. deserti* Post, Pl. Post. 2 : 6 (1891). *D. palaestinus* Freyn, Bull. Herb. Boiss. 5 : 583 (1897). Stems and leaves short-pubescent or velvety-pruinose. Leaves somewhat scabrous at margin. Flowers solitary or 2–3 in a cluster. Scales of epicalyx cuspidate. Calyx and petals as in var. *strictus.*

Hab.: As above. Judean Mts., Judean Desert, N. Negev, Lower Jordan Valley, Ammon, Edom.

There are also forms with calyx tube wholly striate-nerved.

Area of species: E. Mediterranean, extending slightly towards adjacent areas of the Irano-Turanian region.

Very polymorphic. Var. *axilliflorus* (Fenzl) Eig, l.c. with subsessile, solitary and clustered flowers seems to be merely a late-summer form of the typical taxon.

3. Dianthus polycladus Boiss., Diagn. ser. 1, 8:65 (1849) et Fl. 1: 483 (1867). [Plate 145]

Perennial herb, glabrous or somewhat puberulent below, 20–80 cm. Stems many, slender, erect or ascending, diffusely and dichotomously branched in upper part. Leaves 1–3 mm. broad, linear, scabrous-ciliate and narrowly scarious-margined towards base, those of lateral branches very short to scale-like. Pedicels generally shorter than calyx. Scales of epicalyx 4–6, much shorter than calyx, ovate, broadly membranous margined, abruptly short-mucronate. Calyx 0.6–1.2 cm.; tube nerved; teeth lanceolate. Petals 1–1.7 cm.; limb pale pink or white, oblong-linear, almost entire or retuse, hairy at base. Capsule oblong. Fl. May–September.

Hab.: Batha. Upper Galilee.

Area: E. Mediterranean.

Very rare and not adequately delimited from *D. strictus*. Probably only a subspecies of the latter.

4. Dianthus tripunctatus Sibth. et Sm., Fl. Gr. Prodr. 1 : 286 (1809); Boiss., Fl. 1 : 482 (1867). [Plate 146]

Glabrous annual, 15–45 cm. Stems erect, branching from base and all along. Leaves up to 8 × 0.2–0.4 cm., somewhat rigid, white-margined and ciliate at base, the upper narrower, rather appressed. Flowers solitary, long-pedicelled. Epicalyx somewhat shorter than calyx, composed of 4 ovate, broadly membranous-margined scales, tapering into awns. Calyx 1.6–2 cm., striate-nerved to base, with lanceolate, acute teeth much shorter than tube, narrowly membranous-margined. Petals 1.2–2 cm., long-clawed; limb obovate-cuneate, dentate, 3-lined at base. Capsule cylindrical-conical, somewhat tetragonous. Seeds compressed, elliptical, with elevated and undulate margin. Fl. March–July.

Hab.: Batha and fallow fields. Acco Plain, Sharon Plain, Philistean Plain, Upper and Lower Galilee, Mt. Carmel, Esdraelon Plain, Samaria, Judean Mts., Hula Plain, Upper Jordan Valley.

Area: E. Mediterranean, with westward extension.

5. Dianthus cyri Fisch. et Mey., Ind. Sem. Hort. Petrop. 4 : 34 (1837); Boiss., Fl. 1 : 482 (1867). [Plate 147]

Glabrous annual, 30–75 cm. Stems erect, usually branching from base. Branches rigid, strict. Leaves up to 10 × 0.2–0.6 cm., linear, scabrous, narrowly scarious-margined, the upper narrower. Inflorescences many-flowered, branched, sometimes corymbose. Pedicels shorter to longer than calyx. Flowers terminal. Epicalyx composed of 4 ovate, white-margined scales tapering into awns, about of same length or longer than calyx. Calyx 1.2–1.5 cm.; tube nerveless, densely and finely tuberculate; teeth shorter than tube, lanceolate, striate, scarious at apex. Petals about 1.5–1.8 cm., cuneate-oblong; limb pink, 4–6-dentate, glabrous or slightly hairy at base. Capsule shorter than calyx. Seeds black, finely tuberculate. Fl. April–July.

Hab.: Batha and heavy alluvial soils. Upper and Lower Galilee, Hula Plain, Upper Jordan Valley, Gilead.

Area: W. Irano-Turanian, slightly extending into adjacent Mediterranean territories.

Among local representatives of this species there are plants with paniculate-corymbose inflorescences corresponding to var. *corymbosus* Opphr., Bull. Soc. Bot. Genève, ser. 2, 22 : 298 (1931). In others the epicalyx is shorter than the calyx, and in still other plants the petals are slightly hairy at base. The taxonomic value of these deviations from the type has not yet been assessed.

Sect. LEIOPETALI Boiss., Fl. 1 . 479 (1867) Perennials. Calyx striate-nerved. Petal limb white or pale pink, glabrous at base, entire or dentate but never fimbriate.

6. Dianthus judaicus Boiss., Diagn. ser. 1, 8 : 66 (1849) et Fl. 1 : 485 (1867). [Plate 148]

Perennial, glaucous herb, velvety-scabrous, sometimes glabrous, 20–35 cm. Stems many, tufted, erect, simple or somewhat branched. Leaves up to 4 × 0.2–0.3 cm., linear, usually plicate and recurved, ciliolate-scabrous at margin, membranous-margined towards base. Flowers terminal, solitary, on pedicels shorter than calyx. Scales of epicalyx oblong-ovate, scarious-margined, ending with a slightly spreading awn much shorter than scale. Calyx about 3 cm., with striate tube and short, lanceolate, white-margined teeth. Petals 4–4.5 cm., whitish or cream-coloured; limb oblong-obovate, almost entire. Capsule narrowly ellipsoidal, half as long as calyx. Seeds flat, smooth, brown. Fl. March–April.

Hab.: Steppes. Judean Desert, N. and C. Negev, Ammon, Moav, Edom.

Area: W. Irano-Turanian.

D. pallens Sibth. et Sm. var. *oxylepis* Boiss., Fl. 1 : 485 (1867), as collected by Dinsmore from Moav and Ammon Desert and recorded by Rechinger f. (Ark. Bot. ser. 2, 2 : 329, 1952), seems to be more or less identical with *D. judaicus*.

Sect. PLUMARIA (Opiz) Aschers. et Graebn., Syn. 5, 2 : 408 (1921). Perennials. Petals with glabrous or barbate limb; margin deeply dissected or fimbriate.

7. Dianthus libanotis Labill., Ic. Pl. Syr. Dec. 1 : 14, t. 4 (1791); Boiss., Fl. 1 : 492 (1867).

Perennial, woody at base, 20–30 cm. Stems many, angular, simple or corymbosely branching above, crisp-hirtellous. Leaves 2.5–4 × 0.3–0.7 cm., linear-lanceolate, glabrous or hirtellous, with prominent middle nerve and obscure lateral nerves; margin ciliolate, often spreading to recurved. Flowers solitary, long-pedicelled or 1–2 at ends of dichotomous branches. Bracts of epicalyx 4–6, spreading, leathery, ovate at base, ending in a long-lanceolate, patulous to recurved, sharp-pointed cusp. Calyx 4 cm., twice as long as bracts; teeth lanceolate, acute. Petals with pink, obovate-cuneate limb, reddish-spotted at base and fringed to near middle. Fl. April.

Hab.: Moav (Hafayer Djerra a Modjib, Aaronsohn, No. 2280).

Area: E. Mediterranean and Irano-Turanian.

The specimen quoted is in flower buds. It deviates from the description of *D. libanotis* by its indumentum and may perhaps be considered a special variety.

8. Dianthus sinaicus Boiss., Diagn. ser. 1, 1 : 23 (1843) et Fl. 1 : 497 (1867). [Plate 149]

Glaucous, glabrous perennial, 25–40 cm. Stems many, more or less woody at base, ascending to erect, dichotomously branched in upper part. Leaves linear-subulate, almost acute, more or less scabrous at margin, connate at the somewhat dilated base and forming a short, whitish sheath; the radical ones up to 3 (–4.5) cm., withering before flowering time, stem leaves 0.5–1.5 (–2) cm., acute to mucronulate. Flowers usually solitary and terminal, short-pedicelled. Scales of epicalyx 10 or more, one third to one fourth as long as calyx or shorter, appressed, imbricated, oblong to ovate, scarious-margined, tapering into a short mucro. Calyx 2.5–3 cm., narrowly cylindrical, with lanceolate, acute teeth. Petals 3–3.5 cm.; limb pink, narrowly oblong, glabrous, somewhat broader than the long claw, deeply fimbriate-laciniate. Capsule narrowly oblong. Fl. August–September.

Hab. : Rocks. W. and N. Negev.

Area : W. Irano-Turanian (Palestine, Sinai).

Our specimens differ from the original description in having longer leaves (radical ones up to 4.5 cm., stem leaves up to 1.5 cm.) and pink flowers.

10. VELEZIA L.

Annual rigid herbs, dichotomously branched. Leaves opposite, awl-shaped, exstipulate. Bracts narrow. Flowers small. Calyx tubular, usually 15-nerved, persistent; teeth 5, acute. Petals 5, long-clawed, with small, emarginate or 2-fid or dentate limb; coronal scales 0 or present as short, crisp hairs. Stamens 5–10. Ovary 1-celled, many-ovuled; styles 2. Capsule long-cylindrical, 4-toothed, borne on a very short carpophore. Seeds few, with a straight embryo.

Four species, mainly in the Mediterranean and Irano-Turanian regions.

1. Velezia rigida L., Sp. Pl. 332 (1753); Boiss., Fl. 1 : 478 (1867). [Plate 150]

Annual, glandular-puberulent to glabrescent, 10–40 cm. Stems erect or ascending, branching from base. Leaves 1–3 × 0.1–0.3 cm., 3–5-nerved, basal oblong-lanceolate, cauline linear, ciliolate, narrowly scarious-margined towards base. Bracts leaf-like, shorter than flowers. Pedicels much shorter than calyx, somewhat thickened. Flowers solitary or in pairs at nodes, rarely subsessile and densely clustered. Calyx 1–1.3 × 0.1 cm., 15-nerved, tube more or less angulate, glabrous or glandular-puberulent; teeth much shorter than tube, narrowly lanceolate. Petals 1.3–1.5 cm.; limb narrow, 2-fid, naked or with few hairs at base. Stamens 5 or 10, unequal. Styles 2, filiform. Capsule about as long as calyx. Seeds oblong, compressed-peltate, black. Fl. March–June.

Var. **rigida.** Flowers pedicellate, scattered along branches. Petals sparingly barbate.

Hab. : Batha. Sharon Plain, Philistean Plain, Upper and Lower Galilee, Mt. Carmel, Mt. Gilboa, Judean Mts., Upper and Lower Jordan Valley, Gilead, Ammon, Edom.

Var. **fasciculata** (Boiss.) Post, Fl. Syr. Pal. Sin. 122 (1883–1896). *V. fasciculata* Boiss., Diagn. ser. 1, 8 : 92 (1849). Flowers subsessile, densely clustered at the apex of the short branches. Petals shorter than in type, beardless.

Hab. : As above. Lower Galilee (Nazareth, Bornmueller, fide Post 1932).

Area of species : E. Mediterranean, slightly extending into adjacent Irano-Turanian and Euro-Siberian territories.

Subfam. ALSINOIDEAE. Perianth mostly compound. Calyx mostly cleft to base. Stipules 0.

11. ARENARIA L.

Annual or perennial herbs or dwarf-shrubs. Stems slender. Leaves opposite, small, exstipulate. Flowers many in terminal or axillary dichasial cymes or sometimes solitary. Sepals 5, free. Petals 5, rarely 0, white or pink, entire or emarginate. Stamens 10 (5), inserted on a nectariferous disk. Ovary 2–5-carpelled, 1-celled, many-ovuled; styles 2–3 (rarely 4–5). Capsule dehiscing to base by 6 entire or 3-fid valves. Seeds usually reniform, tuberculate.

Over 160 species, in many regions, chiefly in the northern hemisphere.

Literature : J. McNeill, Taxonomic studies in the Alsinoideae : I. Generic and infra-generic groups. *Not. Roy. Bot. Gard. Edinb.* 24 : 79–155 (1962). J. McNeill, Taxonomic studies in the Alsinoideae : II. A revision of the species in the Orient. *Not. Roy. Bot. Gard. Edinb.* 24 : 241–404 (1963).

1. Perennials. Leaves petiolate. Petals longer than sepals.   **1. A. deflexa**
– Annuals. Leaves sessile or almost so. Petals shorter than sepals   **2**
2. Calyx glabrous. Leaves oblong. Pedicels 2–3 cm. long, rigid.   **3. A. tremula**
– Calyx hairy. Leaves ovate (rarely oblong). Pedicels much shorter than in above.
    2. A. leptoclados

1. Arenaria deflexa Decne., Ann. Sci. Nat. Bot. ser. 2, 3 : 277 (1834); McNeill, Not. Roy. Bot. Gard. Edinb. 24 : 273 (1963). *A. graveolens* auct. non Schreb., Nov. Act. Nat. Cur. 3 : 478 (1767); Boiss., Fl. 1 : 700 (1867). [Plate 151]

Perennial herb, glandular-pubescent, up to 25 cm. Stems slender, brittle, tufted, decumbent to ascending, much branched. Leaves 4–9 × 2–5 mm., long- to short-petioled, ovate, elliptical or oblong, acute-apiculate, pubescent to puberulent. Inflo-rescences often dichotomously branched. Bracts narrow. Pedicels much longer than calyx. Sepals about 3 mm., lanceolate, acuminate, scarious-margined, more or less pubescent. Petals somewhat longer than sepals, white, oblong-linear, entire or retuse. Capsule about as long as calyx. Seeds reniform, more or less tuberculate. Fl. May–June.

Hab. : Mainly on shady rocks. Upper Galilee, Judean Mts., Samaria, Gilead.

Area : E. Mediterranean.

Post (Fl. Syr. Pal. Sin. 15, 1883–1896), records from Gilead (Hab. : woods) a variety with 2 mm. long flowers (*A. graveolens* Schreb. var. *minuta* Post).

2. Arenaria leptoclados (Reichb.) Guss., Fl. Sic. Syn. 2 : 824 (1845). *A. leptoclados* (Reichb.) Boiss., Fl. 1 : 701 (1867). *A. serpyllifolia* sensu Guss., l.c. 1 : 495 (1842) non

L., Sp. Pl. 423 (1753). *A. serpyllifolia* L. var. *leptoclados* Reichb., Ic. Fl. Germ. 5 : 32 (1841). [Plate 152]

Annual, puberulent, sometimes viscid-glandular, 4–20 cm. Stems rigid, slender, decumbent or ascending, usually branched. Leaves 3–8 × 2–4 mm., sessile or almost so, usually ovate, short-acuminate, ciliate. Cymes dichasial. Bracts leaf-like but smaller. Pedicels longer than calyx, spreading to erect, filiform. Flowers 3–6 mm. in diam. Sepals ovate to lanceolate, acute, hirsute, scarious-margined, 3-nerved. Petals shorter than sepals, white, oblong-cuneate. Capsule as long as or somewhat longer than calyx, flask-shaped or oblong. Seeds about 0.5 mm., black, finely tuberculate. Fl. March–May.

Hab. : Fields and batha. Sharon Plain, Philistean Plain, Upper and Lower Galilee, Esdraelon Plain, Samaria, Judean Mts., Judean Desert, Upper Jordan Valley, Gilead, Ammon, Edom.

Area : Mediterranean, Irano-Turanian and Euro-Siberian.

3. Arenaria tremula Boiss., Diagn. ser. 1, 8 : 101 (1849) et Fl. 1 : 702 (1867); McNeill, Not. Roy. Bot. Gard. Edinb. 24 : 288 (1963). [Plate 153]

Annual, scabrous-puberulent, pruinose below, 12–25 cm. Stems usually many, rigid, thin, dichotomously branching above. Leaves 0.5–2 cm., oblong, acute. Pedicels much longer than calyx, rigid, filiform. Sepals about 3 mm., ovate-lanceolate, acuminate-acute, 3-nerved, broadly white-margined, glabrous. Petals shorter than sepals, white, oblong. Capsule about 4 mm., ovoid-conical, longer than sepals. Seeds globular-reniform, brown, with rows of tubercles. Fl. April–May.

Hab. : Batha and maquis. Upper Galilee.

Area : E. Mediterranean (Palestine-Turkey).

12. MINUARTIA L.

Annual or perennial herbs or sometimes dwarf-shrubs. Leaves opposite, linear or subulate, rarely oblong or lanceolate, exstipulate. Flowers small, in dichasial cymes or solitary, terminal. Sepals 5 (4), free. Petals 5 (4 or 0), white or pinkish, entire or retuse. Stamens 10 or rarely less, inserted on the perigynous disk with simple or 2-fid lobes. Ovary 1-celled, usually many-ovuled; styles 3, sometimes more. Capsule dehiscing by 3 (–5) valves or teeth. Seeds many, almost reniform.

About 130 species, mainly in the extratropical regions of the northern hemisphere.

Literature : J. Mattfeld, Geographisch-genetische Untersuchungen über die Gattung *Minuartia* (L.) Hiern. *Repert. Sp. Nov. Beih.* 15 : 1–228 (1922). J. McNeill, Taxonomic studies in the Alsinoideae : I. Generic and infra-generic groups. *Not. Roy. Bot. Gard. Edinb.* 24 : 79–155 (1962). J. McNeill, Taxonomic studies in the Alsinoideae : II. A revision of the species in the Orient. *Not. Roy. Bot. Gard. Edinb.* 24 : 241–404 (1963).

1. Pedicels twice as long as calyx or longer   2
 – Pedicels 0 or shorter to slightly longer than calyx   4
2. Sepals 3-nerved. Corolla white, shorter than or as long as calyx. Leaves not crowded.
     2. M. hybrida
 – Sepals mostly 1-nerved, more or less keeled, with membranous margins. Corolla usually pink, longer than calyx. Lower leaves crowded in whorl-like groups   3

3. Sepals 2–3 mm., obtuse. Bracts short, ovate-oblong, acute-mucronulate. Glandular-
   pubescent, fairly common plants.   **6. M. picta**
 – Sepals 3–5 mm., acuminate-cuspidate. Bracts lanceolate-ovate, long acuminate-mucronate. Glabrous plants. Rare.   **7. M. formosa**
4(1). Pedicels as long as or longer than calyx.   **1. M. mediterranea**
 – Pedicels 0 or one third to half as long as calyx   5
5. Calyx not longer than 5 mm. Flowers sessile or subsessile.   **5. M. decipiens**
 – Calyx 6 mm. or longer   6
6. Calyx conical. Pedicels spreading or deflexed in fruit.   **4. M. globulosa**
 – Calyx campanulate. Pedicels erect in fruit.   **3. M. meyeri**

Sect. SABULINA (Reichb.) Graebn. in Aschers. et Graebn., Syn. 5, 1 : 700 (1910).
Annuals. Calyx not indurated at base; sepals 3-nerved. Leaves 3-nerved, at least at
base. Petals shorter to somewhat longer than calyx. Stamens 3–10.

1. Minuartia mediterranea (Ledeb.) K. Maly, Glasnik Muz. Bosn. Herceg. 20 : 563
(1908). *Arenaria mediterranea* Ledeb. in Link, Enum. Hort. Berol. Alt. 1 : 431 (1821).
Alsine mediterranea (Ledeb.) J. Maly, Enum. Pl. Austr. 296 (1848); non *A. (Arenaria)*
mucronata Sibth. et Sm., Fl. Gr. t. 293 (1819). *A. tenuifolia* (L.) Crantz var. *mucronata*
(Sibth. et Sm.) Boiss., Fl. 1 : 686 (1867). *A. tenuifolia* (L.) Crantz var. *maritima* Boiss.
et Heldr. in Boiss., Diagn. ser. 1, 8 : 95 (1849). [Plate 154]

   Annual, subglabrous or somewhat hairy-viscid, 5–15 cm. Stems thin, simple or
branching from base. Leaves 0.5–1.7 × 0.1 cm., linear-filiform. Inflorescences dichasially branched, congested. Lowermost flowers long-pedicelled, the others with pedicels
about as long as calyx. Calyx 3–4 mm., pilose-glandular; sepals ovate-lanceolate, acu-
minate-mucronulate, 3-nerved, white-margined. Petals much shorter than sepals or
0. Stamens mostly 5. Capsule about as long as sepals, conical-cylindrical. Seeds finely
wrinkled. Fl. February–April.
   Hab. : Batha and fallow fields, mostly on rocky ground. Coastal Galilee, Sharon
Plain, Philistean Plain, Judean Mts.
   Area : Mediterranean, slightly extending into adjacent regions.

2. Minuartia hybrida (Vill.) Schischk., in Fl. URSS 6 : 488 (1936). *Arenaria hybrida*
Vill., Prosp. Pl. Dauph. 48 (1779). *A. tenuifolia* L., Sp. Pl. 424 (1753). *M. tenuifolia*
(L.) Hiern., Journ. Bot. Lond. 37 : 321 (1899) non Nees. *Alsine tenuifolia* (L.) Crantz,
Inst. 2 : 407 (1766); Boiss., Fl. 1 : 685 (1867). [Plate 155]

   Annual, glabrous or more or less puberulent-glandular, 5–20 cm. Leaves 0.5–2 cm.,
linear-subulate, acute-acuminate, somewhat broadened at base. Inflorescences dichasial.
Pedicels much longer than calyx. Sepals 2–3.5 mm., lanceolate-acuminate, 3-nerved,
glabrous or more or less pilose-glandular, scarious-margined. Petals shorter to as long as
sepals, white, ovate- or oblong-lanceolate. Stamens 5–10. Capsule as long as or longer
than calyx, ovoid-cylindrical. Seeds reniform, finely tuberculate. Fl. January–May.
   Hab. : Batha, fallow fields, mainly in pioneer communities. Sharon Plain, Philistean
Plain, Upper Galilee, Mt. Carmel, Esdraelon Plain, Samaria, Judean Mts., Judean
Desert, Upper Jordan Valley, Dead Sea area, Gilead, Ammon, Moav, Edom.

Subsp. **hybrida** var. **hybrida**. McNeill, Not. Roy. Bot. Gard. Edinb. 24 : 394 (1963). Petals ovate-lanceolate, cuneate. Plants moderately robust, usually with a single main stem not branching at base.

Upper Galilee, Gilead, Moav, Edom.

Subsp. **hybrida** var. **palaestina** McNeill, Not. Roy. Bot. Gard. Edinb. 24 : 395 (1963). Petals narrowly deltoid, abruptly narrowed at base into a very short claw. Plants usually very robust, vigorously branching from base.

Common everywhere.

Area of species : Mediterranean and W. Irano-Turanian, extending into Euro-Siberian territories.

M. mesogitana (Boiss.) Hand.-Mazzetti ssp. *mesogitana* is recorded from Galilee by McNeill, l.c. 387. We have seen from the above locality only specimens somewhat intermediate between *M. mesogitana* and *M. hybrida,* that are closer to the latter than to the former.

Sect. MINUARTIA Graebn. in Aschers. et Graebn., Syn. 5, 1 : 710 (1918). Annuals or perennials. Calyx strongly indurated; sepals acute, 1–3 (–5)-nerved. Stamens 3–10.

3. Minuartia meyeri (Boiss.) Bornm., Beih. Bot. Centralbl. 27, 2 : 318 (1910). *Alsine meyeri* Boiss., Diagn. ser. 1, 8 : 96 (1849) et Fl. 1 : 682 (1867). [Plate 156]

Annual, patulous glandular-pubescent, 5–10 cm. Stems many, branching from base. Branches filiform. Leaves 0.8–1.5 × 0.2 cm., linear, mostly 7-nerved. Flowers arranged in dichasial cymes. Pedicels shorter than calyx, erect in fruit. Calyx 6 mm. or more, campanulate; sepals unequal, lanceolate-acuminate, white-margined, 3–5-nerved. Petals one third as long as calyx, white, ovate-oblong. Capsule shorter than calyx, oblong. Seeds finely tuberculate. Fl. April.

Hab. : Steppes. Moav, Edom.

Area : Irano-Turanian.

4. Minuartia globulosa (Labill.) Schinz et Thell., Bull. Herb. Boiss. ser. 2, 7 : 403 (1907). *Arenaria globulosa* Labill., Ic. Pl. Syr. Dec. 4 : 6, t. 3, f. 1 (1812). *Alsine globulosa* (Labill.) C. A. Mey., Verz. Pfl. Cauc. 219 (1831). *Alsine smithii* Fenzl, Verbr. Alsin. 57 (1833); Boiss., Fl. 1 : 683 (1867). [Plate 157]

Annual, villose-pubescent, glandular, 3–20 cm. Stems stiff, simple or branching at base. Leaves 0.5–2 (–2.5) × 0.1–0.2 cm., lanceolate-linear, acute, (3–)5–7-nerved. Flowers clustered in axillary, more or less dichasial cymes. Pedicels about half as long as calyx (longer than 2 mm.), spreading or deflexed in fruit. Calyx 6–8 mm., conical, more or less truncate at base; sepals unequal, lanceolate-linear, acute-mucronulate, 3-nerved. Petals one quarter to one third the length of sepals, white, obovate to oblong. Capsule shorter than calyx. Seeds echinate-tuberculate. Fl. April–June.

Hab. : Batha. Upper Galilee, Gilead.

Area : E. Mediterranean.

5. Minuartia decipiens (Fenzl) Bornm., Beih. Bot. Centralbl. 31, 2 : 193 (1914). *Alsine decipiens* Fenzl, Pugill. 12 (1842); Boiss., Fl. 1 : 684 (1867). [Plate 158]

Annual, velvety-pruinose, viscid, 5–25 cm. Stems slender, simple or branching at base. Leaves 1.2–3 × 0.1–0.2 cm., narrowly lanceolate-linear, 3–7-nerved. Flowers clustered in dichasial, axillary or terminal cymes, sessile or subsessile. Calyx up to about 5 mm., conical, truncate at base; sepals lanceolate-acuminate, gibbous at base, more or less white-margined, 3-nerved. Petals about half as long as to somewhat shorter than sepals, lanceolate-linear, white. Capsule shorter than calyx, conical. Seeds finely tuberculate. Fl. February–June.

Hab. : Rock crevices and stony ground. Acco Plain, Sharon Plain, Philistean Plain, Upper and Lower Galilee, Mt. Carmel, Esdraelon Plain, Samaria, Shefela, Judean Mts., Judean Desert, Upper and Lower Jordan Valley, Gilead, Ammon, Moav.

Area : E. Mediterranean (Palestine – Balkan Peninsula).

Sect. SPERGELLA Fenzl in Endl., Gen. 965 (1840). Annuals. Sepals mostly 1-nerved, keeled. Leaves 3-nerved at base. Flowers pink (rarely white).

6. Minuartia picta (Sibth. et Sm.) Bornm., Beih. Bot. Centralbl. 28, 2 : 148 (1911). *Arenaria picta* Sibth. et Sm., Fl. Gr. Prodr. 1 : 304 (1809). *Alsine picta* (Sibth. et Sm.) Fenzl, Verbr. Alsin. 57 (1833); Boiss., Fl. 1 : 687 (1867). [Plate 159]

Annual, glandular-pubescent, 5–15 cm. Stems slender, simple or branching from base. Leaves (1.2–)1.5–2 (–5) cm., crowded and almost whorled in lower part, narrowly linear-setaceous, dilated and ciliate at base. Inflorescences dichotomously branched. Bracts short, ovate-oblong, acute-mucronulate, scarious. Pedicels much longer than calyx, erect to spreading or reflexed, filiform. Sepals 2–3 mm., ovate, obtuse, scarious-margined, 1 (–3)-nerved. Petals one and a half times to twice as long as sepals [var. *brachypetala* (Boiss.) Zoh. (comb. nov.)], pink, rarely white [var. *albiflora* (Eig) Zoh. (comb. nov.)], obovate-oblong, retuse. Capsule somewhat longer than calyx, ovoid. Seeds wrinkled and finely tuberculate, brownish-black. Fl. February–May.

Hab. : Batha, fields and steppes. Samaria, Judean Mts., Judean Desert, N. and C. Negev, Upper and Lower Jordan Valley, Gilead, Ammon, Moav.

Area : Irano-Turanian, extending into Mediterranean and Saharo-Arabian territories.

A showy plant varying in colour and size of petals.

7. Minuartia formosa (Fenzl) Mattf., Bot. Jahrb. 57 Beibl. 126 : 33 (1921). *Alsine formosa* Fenzl, Flora 26 : 403 (1843); Boiss., Fl. 1 : 688 (1867). [Plate 160]

Annual, glandular-pubescent or glabrous (in ours), 7–20 cm. Stems erect, filiform, branched, especially at base. Leaves 1–3 × 0.2–0.5 cm., narrow, linear-setaceous to almost filiform, somewhat dilated at base, lower leaves crowded. Inflorescences many-flowered, dichotomously branched, diffuse. Bracts lanceolate-ovate, long acuminate-mucronate, scarious; pedicels longer than calyx, spreading to erect. Sepals 3–5 mm., ovate-lanceolate, acuminate-cuspidate, scarious-margined, 1 (–2)-nerved. Petals about twice as long as sepals, pink, obovate-cuneate, more or less retuse at apex. Capsule about as long as calyx, ovoid-globular. Seeds minutely tuberculate-wrinkled.

Var. **glabra** Opphr., Bull. Soc. Bot. Genève ser. 2, 22 : 294 (1931). Glabrous, eglandular. Petals obovate-deltoid, somewhat undulate-crenulate at apex. Fl. March–May.

Hab. : Fallow fields, basalt soils. Upper Galilee, Gilead.

Area : E. Mediterranean (Palestine, Syria, Turkey).

13. BUFONIA L.

Annual or perennial herbs or shrubs. Leaves opposite, subulate, exstipulate. Flowers small, grouped in cymose spikes or panicles. Sepals 4, scarious-margined. Petals 4, entire or 2-dentate to 2-fid. Stamens (2–)4–8; filaments inserted on a perigynous disk. Ovary 1-celled, 2–10-ovuled; styles 2. Capsule 2-seeded, 2-valved, compressed-lenticular. Seeds often 2, horseshoe-shaped.

About 10 species in the Mediterranean and Irano-Turanian regions.

1. Bufonia virgata Boiss., Fl. 1 : 665 (1867). [Plate 161]

Annual, glabrous or almost so, 10–50 cm. Stems erect, simple or moderately branched in upper part, rigid. Leaves 0.5–2 cm., sessile, appressed to stem, narrowly subulate, dilated and somewhat ciliate at base. Inflorescences cymose, spike-like. Sepals 3–4 mm., oblong-lanceolate, acute-mucronulate, 3-nerved. Petals white, much shorter than calyx. Stamens 4, shorter than petals. Ovary ovoid; styles short; stigma recurved. Capsule almost as long as or shorter than calyx. Seeds brown, minutely tuberculate. Fl. March–May.

Hab. : Batha and steppes. Judean Mts., Ammon.

Area : E. Mediterranean (Palestine, Syria, Turkey).

14. STELLARIA L.

Annual or perennial slender herbs. Leaves opposite, simple, entire, exstipulate. Flowers in dichasial cymes, rarely solitary. Sepals 5 (rarely 4), free, persistent. Petals 5 (rarely 4), usually 2-fid to 2-partite, sometimes 0. Stamens 10, rarely fewer or 0, hypogynous or perigynous, mostly inserted on an annular disk. Ovary 1-celled, many- or few-ovuled; styles (2–)3(–4). Capsule spherical or oblong, usually dehiscing by 6 valves. Seeds many, rounded-reniform, more or less compressed, papillose or echinulate.

A cosmopolitan genus of about 100 species.

Literature : S. Murbeck, Die nordeuropäischen Formen der Gattung *Stellaria*. *Bot. Notis.* 1899 : 193–219 (1899). A. Béguinot, Ricerche intorno al polimorfismo della *"Stellaria media"* (L.) Cyr. I, II. *Nuovo Giorn. Bot. It.* (N.S.), 17 : 299–326, 348–390 (1910).

1. Upper part of stem hairy-glandular all around and sometimes also with 1–2 rows of hairs or stem entirely glabrous.   **1. S. media**
– Stem (including upper part) with 1 (–2) rows of hairs.   **2. S. pallida**

1. Stellaria media (L.) Vill., Hist. Pl. Dauph. 3 : 615 (1789); Boiss., Fl. 1 : 707 (1867). *Alsine media* L., Sp. Pl. 272 (1753). [Plate 162]

Annual, 10–50 cm. Stems weak, decumbent to ascending, more or less branched, cymose at top, glabrous or pubescent or glandular and often also with 1–2 lines of hairs. Leaves 1–6 × 0.6–3 cm., long-petioled to sessile, ovate to oblong-elliptical, acute or short-acuminate, glabrous or ciliate. Pedicels as long as or longer than calyx, usually puberulent. Flowers many, in more or less dichasial cymes. Sepals 2–6.5 mm., ovate-lanceolate or oblong-lanceolate, narrowly white-margined, pubescent to glabrous. Petals longer or as long as or shorter than sepals, white, 2-fid to 2-partite, sometimes 0. Stamens usually 3–5, sometimes 5–10. Styles 3, very short, to 1.5 mm. long. Capsule exceeding calyx or equalling it, ovoid-oblong. Seeds 0.7–1.4 mm. in diam., dark to pale brown, with acute or obtuse tubercles. Fl. January–April.

1. Lower part of stem with 1 row of hairs, upper part hairy-glandular all round. Stamens 10, rarely 8–9. subsp. **postii**
– Stems entirely glabrous. Stamens 2–5. subsp. **eliezeri**

Subsp. **postii** Holmboe, Bergens Mus. Skr. 1, 2 : 70 (1914). *S. media* (L.) Vill. var. *neglecta* (Weihe) Mert. et Koch in Röhling, Deutschl. Fl. ed. 3, 3 : 253 (1831). *S. neglecta* Weihe in Bluff et Fingerh., Comp. Fl. Germ. 1 : 560 (1825). *S. media* (L.) Vill. var. *major* Koch, Syn. Fl. Germ. Helv. 118 (1835); Boiss., Fl. 1 : 707 (1867). Stems up to 60 cm., with 1 row of hairs in lower part and glandular-pubescent all round in upper part. Leaves puberulent above. Inflorescences dense, puberulent. Sepals 4.5–6.5 mm. Petals as long as or slightly shorter than sepals. Stamens usually 5–10. Seeds 1–1.3 mm. in diam., with conical tubercles.

Hab. : Roadsides and fences, shady and moist places, mostly under trees. Upper and Lower Galilee, Mt. Carmel, Hula Plain, Upper Jordan Valley, Moav.

Subsp. **eliezeri** (Eig) Zoh. (stat. nov.) *S. media* (L.)Vill. var. *eliezeri* Eig, Bull. Inst. Agr. Nat. Hist. 6 : 6 (1927). Plant entirely glabrous. Petals half as long as sepals. Stamens 2–5. Seeds 0.7–1.2 mm. in diam.

Hab. : Roadsides, shady and moist places. Philistean Plain.
Area of species : Pluriregional.
Used as a wild salad plant; also used medicinally.

2. Stellaria pallida (Dumort.) Piré, Bull. Soc. Bot. Belg. 2 : 49 (1863). *S. apetala* auct. non Ucria in Roem., Arch. 1, 1 : 68 (1796). *S. media* (L.) Vill. var. *apetala* (Ucria) Ledeb., Fl. Ross. 1 : 378 (1842); Boiss., Fl. 1 : 707 (1867).

Annual, up to 30 cm. Stems very slender, terete, with 1 (–2) rows of hairs along stem, rarely glabrous. Leaves 1–3 × 0.8–2 cm., usually all petiolate, ovate, somewhat acute, glabrous. Inflorescences few- or many-flowered. Pedicels short, filiform. Sepals up to 3 mm., often grey-tomentose. Petals very short or 0. Stamens 1–3, very rarely 5. Seeds 0.7–0.8 mm. in diam., pale yellowish-brown, rarely dark brown, with rounded or conical tubercles. Fl. January–April.

Hab. : Roadsides or shady and damp places, mostly in orchards. Sharon Plain, Philistean Plain, Upper and Lower Galilee, Mt. Carmel, Esdraelon Plain, Samaria, Shefela, Judean Mts., Hula Plain, Upper Jordan Valley, Ammon.
Area : Mediterranean and Euro-Siberian.

15. HOLOSTEUM L.

Small, annual herbs. Lower leaves more or less rosulate, the rest opposite, all exstipulate. Cymes terminal, umbellate. Pedicels unequal, deflexed after flowering. Sepals 5,
free, persistent. Petals 5, entire or denticulate. Stamens 3–5 or 8–10. Hypogynous disk
present. Ovary 1-celled, many-ovuled; styles 3(–4–5). Capsule oblong-cylindrical,
opening by 6 revolute teeth. Seeds plano-convex or concavo-convex, longitudinally
keeled, covered with rows of tubercles.

Some 6 species, mainly in semiarid regions of Eurasia.

1. Stamens 8–10. Petals ovate or elliptical.   **1. H. glutinosum**
– Stamens 3–5. Petals oblong-cuneate.   **2. H. umbellatum**

1. Holosteum glutinosum (M.B.) Fisch. et Mey., Ind. Sem. Hort. Petrop. 6 : 52 (1839).
H. linifolium Fisch. et Mey., Ind. Sem. Hort. Petrop. 3 : 39 (1837); Boiss., Fl. 1 : 710
(1867). *Arenaria glutinosa* M.B., Fl. Taur.-Cauc. 1 : 344 (1808). [Plate 163]

Annual, glandular-pubescent, 8–15 cm. Leaves 1.5 × 0.2 cm., oblong to oblong-
linear. Flowers in umbellate cymes. Bracts very small, broadly ovate, greenish-scarious,
glandular. Pedicels longer than calyx. Sepals 4 mm., ovate-oblong, scarious at margin,
elongated in fruit. Petals longer than calyx, white, ovate-elliptical, entire. Stamens
8–10, almost as long as calyx. Capsule 0.8–1 cm. Seeds finely tuberculate. Fl. March–
April.

Hab.: Steppes and grazed areas. Upper Galilee, Judean Mts., Ammon, Edom.

Area: Mainly Irano-Turanian.

2. Holosteum umbellatum L., Sp. Pl. 88 (1753); Boiss., Fl. 1 : 709 (1867). [Plate 164]

Annual, glaucous, glabrous or rather sparingly puberulent-viscid, 10–20 cm. Stems
erect, simple or branched from base. Leaves 1–2 × 0.2–0.5 cm., oblong or elliptical,
acute. Flowers in umbellate cymes. Bracts small, ovate, mostly scarious. Pedicels unequal, much longer than calyx, glandular-hairy. Sepals about 3–4 mm., ovate, scarious
at margin. Petals longer than calyx, white or pale pink, oblong-cuneate, sometimes
3-dentate at apex. Stamens 3–5, shorter than calyx. Capsule 5–7 mm., twice as long
as calyx, with pedicels first deflexed, later upright. Seeds peltate, reddish-brown, finely
tuberculate. Fl. February–April.

Hab.: Steppes and fallow fields. Samaria, Judean Mts., Judean Desert, N. Negev,
Gilead, Ammon.

Area: Mediterranean and Irano-Turanian, extending far into the Euro-Siberian
region.

16. CERASTIUM L.

Annual or perennial herbs, sometimes shrubby, usually hairy. Leaves opposite, mostly
sessile, entire, exstipulate. Flowers white, mostly in terminal, dichasial cymes. Sepals
5 (4), free. Petals 5 (4), rarely 0, emarginate to 2-fid, rarely entire. Stamens 10 (8), sometimes 5 or less. Hypogynous disk present. Ovary 1-celled; styles 5, sometimes 3–4.
Capsule scarious, usually longer than calyx, cylindrical or oblong, dehiscing by short

teeth the number of which is twice that of styles. Seeds many, globular or reniform, mostly tuberculate; embryo almost annular.

About 100 species, mainly in north-temperate regions.

1. Styles 3. Leaves 5 mm. or less broad. Pedicels, all or part of them, longer than calyx.
                                                                                              1. C. dubium
– Styles (4–)5. Leaves 1 cm. or more broad. Pedicels as long as or shorter than calyx   2
2. Sepals hairy at apex. Capsule 1 cm. or less, slightly curved above; teeth acute, with revolute margin.   **3. C. glomeratum**
– Sepals scarious and glabrous at apex. Capsule 1.5–2 cm., not curved; teeth not acute and not revolute at margin.   **2. C. dichotomum**

Subgen. DICHODON Bartl. Styles 3 (or in some flowers 4 or 5).

1. Cerastium dubium (Bast.) O. Schwarz, Mitt. Thür. Bot. Ges. 1, 1 : 98 (1949). *Stellaria dubia* Bast., Fl. Maine-et-Loire, Suppl. 24 (1812). *C. anomalum* Waldst. et Kit. ex Willd., Sp. Pl. 2 : 812 (1799) et Pl. Rar. Hung. 1 : 21, t. 22 (1800) non Schrank, Briefe Donaum. 75 (1795); Boiss., Fl. 1 : 714 (1867). [Plate 165]

Annual, glandular-pubescent, especially in upper part, up to 20 cm. Stems erect or ascending, sometimes rooting at nodes, simple, rarely branching from base. Leaves up to 2–5 × 0.3 cm., linear, the lower more or less oblanceolate to spatulate, tapering to a petiole, the upper sessile, linear-lanceolate, short-connate, ciliate at margin. Inflorescences dichasial, corymbose. Bracts herbaceous, linear-lanceolate, similar to leaves. Pedicels unequal, the lower much longer than calyx, the upper shorter, all glandular. Sepals 4–5 mm., oblong-lanceolate, white-membranous at margin, 3-nerved. Petals longer than sepals, white, 2-fid or 2-partite. Stamens 10, shorter than sepals. Styles 3. Capsule about one and a half times as long as calyx, oblong-ovoid, with erect or spreading teeth. Seeds finely tuberculate. Fl. February–April.

Hab.: Damp places. Upper Galilee, Hula Plain, Upper Jordan Valley, Transjordan.
Area: Euro-Siberian, Mediterranean and Irano-Turanian.

Subgen. CERASTIUM. Styles 5.

2. Cerastium dichotomum L., Sp. Pl. 438 (1753); Boiss., Fl. 1 : 721 (1867). [Plate 166]

Annual, glandular-pubescent, 10–20 cm. Stems erect or ascending, patulous glandular-hairy, dichotomously branched. Leaves up to 3.5 cm., sessile, oblong-ovate to lanceolate-linear, the lower ones subspatulate. Flowers in terminal dichasia. Pedicels shorter than to as long as calyx. Calyx about 1 cm.; sepals oblong-lanceolate, acute, white-margined, scarious and glabrous at apex. Petals shorter than sepals, white, obovate-oblong, 2-lobed, glabrous. Stamens 10. Styles (4–)5. Capsule 1.5–2 cm., 2–3 times as long as calyx, cylindrical, 10-nerved, opening by more or less obtuse, short teeth. Seeds finely tuberculate. Fl. February–May.

Hab.: Fields. Philistean Plain, Upper and Lower Galilee, Judean Mts., Judean Desert, Hula Plain, Upper Jordan Valley, Gilead, Ammon, Moav.
Area: Mediterranean and Irano-Turanian.

3. Cerastium glomeratum Thuill., Fl. Par. ed. 2, 226 (1799). *C. viscosum* L., Sp. Pl. 51 (1753) p.p.; Boiss., Fl. 1 : 722 (1867). [Plate 167]

Annual, glandular-pubescent herb, 10–35 cm. Stems erect or ascending, simple or branching. Leaves up to about 3 × 1 cm., the lower obovate-spatulate, tapering into a petiole, the upper all sessile, oblong-elliptical, apiculate. Inflorescences dichasial, flowers clustered. Pedicels shorter than calyx, later spreading or recurved. Calyx about 5 mm.; sepals oblong-lanceolate, acute-acuminate, white-margined, with a few long hairs at tip. Petals about as long as sepals, white, 2-fid, sometimes 0, with one or more cilia on each side near base. Stamens 10. Styles 5. Capsule about 1 cm., narrowly cylindrical, slightly curved above, opening by short, acute teeth with revolute margins. Seeds finely tuberculate. Fl. February–April.

Hab. : Fields and waste places. Acco Plain, Sharon Plain, Philistean Plain, Upper Galilee, Esdraelon Plain, Judean Mts., Judean Desert, N. Negev, Hula Plain, Upper Jordan Valley, Ammon.

Area : Mediterranean, Euro-Siberian and Irano-Turanian.

Two forms can be distinguished : one with petals that are shorter or scarcely longer than calyx, the other without petals or sometimes with 2–5 very short ones.

The data by Post (1932) on the occurrence in Palestine of *C. inflatum* Link ex Sweet, Hort. Brit. ed. 2, 57 (1830) are erroneous.

17. SAGINA L.

Dwarf annuals or perennials, often tufted. Leaves opposite, subulate, connate at base, exstipulate. Flowering shoots slender, prostrate or ascending. Flowers in dichasial cymes or solitary and terminal, usually on long pedicels. Sepals 4–5, free. Petals 4–5, entire or notched, sometimes 0. Stamens 4–10. Ovary 1-celled, many-ovuled; styles 4–5, alternating with sepals. Capsule dehiscent to base into 4–5 valves. Seeds minute, reniform.

Over 20 species, chiefly in north-temperate regions.

1. Sagina apetala Ard., Animadv. Bot. Spec. Alt. 2 : 22 (1763) non L., Mant. Alt. 559 (1771); Boiss., Fl. 1 : 663 (1867). [Plate 168]

Subglabrous annual, 3–20 cm. Stems slender, decumbent to erect, branching at base. Leaves 2–8 mm., usually not longer than internodes, narrowly linear-subulate, aristate, glabrous or ciliate at the dilated base. Pedicels longer than flowers, elongated in fruit, filiform, glandular-pilose or glabrous. Flowers terminal and axillary. Sepals 4, about 2 mm., subequal, ovate-oblong, obtuse, white-margined, spreading in fruit. Petals 4, minute, white or 0. Stamens 4, shorter than sepals. Capsule somewhat longer than sepals, ovoid-subglobular. Seeds about 0.4 mm., reniform, papillose. Fl. February–April.

Hab. : Shady and rocky places. Sharon Plain, Philistean Plain, Lower Galilee, Mt. Carmel, Judean Mts., Upper and Lower Jordan Valley, Gilead, Ammon.

Area : Mediterranean, Euro-Siberian, Irano-Turanian and other north-temperate regions.

Subfam. PARONYCHIOIDEAE. Sepals free or slightly connate. Leaves stipulate.

Trib. SPERGULEAE. Sepals free. Styles free. Fruit a 1-celled, many-seeded capsule, dehiscing by teeth or valves.

18. SPERGULA L.

Annual, slender herbs. Leaves spuriously whorled, linear; stipules small, scarious. Cymes terminal, dichasial, loose. Sepals 5, free, herbaceous and often somewhat fleshy, scarious at margin. Petals 5, white or whitish, entire. Stamens (5–)10, inserted on a perigynous ring. Ovary 1-celled, many-ovuled; styles (3–)5, alternating with sepals. Capsule (3–)5-valved or toothed. Seeds more or less lenticular, keeled to winged.

Some 5 (–7) species in north-temperate regions.

1. Leaves grooved beneath. Stems and sepals more or less pubescent. Styles 5. Capsule 5-valved. Seeds wingless or narrowly margined, tuberculate.   **1. S. arvensis**
– Leaves not grooved. Stems and sepals almost glabrous. Styles 3. Capsule 3-valved. Seeds broadly winged, obsoletely tuberculate or almost smooth.   **2. S. fallax**

1. Spergula arvensis L., Sp. Pl. 440 (1753); Boiss., Fl. 1 : 731 (1867). [Plate 169]

Annual, more or less pubescent, 10–30 cm. Stems many, ascending, branching from base, glandular-pubescent. Leaves 1–3 × 0.05–0.1 cm., narrowly linear, grooved beneath; stipules very short but conspicuous, ovate. Panicles cymose, loose. Bracts minute, scarious. Pedicels thin, much longer than calyx, deflexed to pendulous in fruit. Sepals about 3 mm., ovate, narrowly white-margined, glandular-pubescent. Petals half as long as to slightly longer than sepals, whitish, ovate, obtuse. Stamens 10 or less. Styles 5. Capsule about 4 mm., ovoid, generally 5-valved. Seeds 1–2 mm. in diam., lenticular-globular, narrowly margined, black, minutely tuberculate and white-papillose. Fl. January–April.

Hab.: Fallow fields and among crops on sandy and calcareous soils. Sharon Plain, Philistean Plain, Upper and Lower Galilee, Judean Mts., Lower Jordan Valley.

Area: Mainly Euro-Siberian and Mediterranean; also in other temperate and tropical regions.

2. Spergula fallax (Lowe) Krause in Sturm, Deutschl. Fl. ed. 2, 5 : 19 (1901). *Spergularia fallax* Lowe in Hook., Kew Journ. 8 : 289 (1856). *Spergula flaccida* (Roxb.) Aschers., Verh. Bot. Ver. Prov. Brandenb. 30 : 311 (1889). *Arenaria flaccida* Roxb., Hort. Bengal. 34 (1814) non Clairv. (1811). *Spergula pentandra* L. var. *intermedia* Boiss., Diagn. ser. 2, 1 : 93 (1853) et Fl. 1 : 731 (1867). [Plate 170]

Glaucescent annual, glabrous or somewhat puberulent, 10–25 cm. Stems many, ascending. Leaves 0.5–4 × 0.05–0.15 cm., linear, not grooved beneath; stipules minute, ovate-triangular. Panicles branched. Bracts minute, more or less scarious. Pedicels filiform, generally longer than calyx. Sepals 3–4 mm., ovate to oblong, white-margined, more or less glabrous. Petals somewhat shorter than sepals, white, ovate to lanceolate, somewhat acute or obtuse. Stamens 10, rarely 6–8. Styles 3. Capsule 4–5 (–6) mm., ovoid or subglobular, 3-valved. Seeds lenticular, black, with hyaline wing almost as broad as seed, obsoletely tuberculate or almost smooth. Fl. January–April (–June).

Hab.: Steppes. Philistean Plain, Judean Desert, W., N. and C. Negev, Upper and Lower Jordan Valley, Dead Sea area, Arava Valley, Moav, Edom.

Area: Saharo-Arabian, extending into Mediterranean and Sudanian territories.

19. SPERGULARIA (Pers.) J. et C. Presl

Annual or perennial herbs. Stems often decumbent or ascending. Leaves opposite, linear to filiform; stipules scarious. Flowers hermaphrodite, rarely unirexual, in cymose inflorescences. Sepals 5, free, with scarious margins. Petals 5, white or pink, entire. Stamens (2)5–10, inserted on a perigynous ring. Ovary 1-celled, many-ovuled; styles 3(–5). Fruiting pedicels reflexed. Capsule 3(–5)-valved. Seeds subspherical to almost lenticular-reniform, winged or wingless.

About 20 species, almost cosmopolitan.

1. Capsule usually exceeding calyx. Perennials or annuals   2
 – Capsule not exceeding calyx. Annuals   3
2. All seeds winged. Capsule about 8 mm. Stamens usually 10. Perennials.   **1. S. media**
 – Lower seeds of capsule winged, the upper usually wingless. Capsule about 6 mm. Stamens usually 5–8. Annuals or perennials.   **2. S. salina**
3(1). Pedicels as long as or shorter than flowers. Stipules white. Sepals 2–3.5 mm. Rare herbs with inflorescence resembling a one-sided raceme.   **5. S. bocconii**
 – Pedicels longer than flowers. Common   4
4. Flowers and capsules 3–5 mm. Stamens 5–10. Leaves 1–2 mm. broad; stipules white-silvery.   **4. S. rubra**
 – Flowers and capsules about 3 mm. Stamens 2–3. Leaves 0.5–1 mm. broad; stipules dull coloured.   **3. S. diandra**

1. Spergularia media (L.) C. Presl, Fl. Sic. 161 (1826). *Arenaria media* L., Sp. Pl. ed. 2, 606 (1762). *A. marginata* DC., Ic. Gall. Rar. t. 48 (1808). *S. marginata* (DC.) Kitt., Taschenb. Fl. Deutschl. ed. 2, 1003 (1844); Boiss., Fl. 1 : 733 (1867). [Plate 171]

Perennial or biennial, nearly glabrous in lower part, glandular-pubescent towards apex, 10–40 cm. Stems usually many, stout, prostrate to ascending. Leaves 0.6–3 (–5) × 0.1–0.2 cm., somewhat fleshy, linear, obtuse or acute, sometimes short-mucronulate, usually glabrous; stipules conspicuous, lanceolate to triangular, connate at base, white. Inflorescences terminal, dichasial, loose. Bracts leaf-like, much smaller than leaves. Pedicels as long as to much longer than calyx. Sepals about 5 mm., lanceolate to oblong-ovate, scarious-margined, glandular-pubescent to glabrous, 3-nerved. Petals about as long as sepals or shorter, pale pink, ovate. Stamens 10, rarely less, shorter than petals. Capsule about 8 mm., exceeding calyx, ovoid. Seeds 1.5 mm. across, compressed, orbicular-obovate, brownish, smooth, winged. Fl. March–June.

Hab.: Muddy and also somewhat saline soils. Sharon Plain, Philistean Plain, Upper Galilee, Esdraelon Plain, W. and C. Negev, Upper and Lower Jordan Valley, Edom.

Area: Euro-Siberian, Mediterranean and Irano-Turanian. Also introduced into other temperate regions of both hemispheres.

There are transitional forms between this species and *S. salina* J. et C. Presl, e.g., plants with large capsules and partly wingless seeds growing in the Hula Plain and the Upper Jordan Valley.

2. Spergularia salina J. et C. Presl, Fl. Čech. 93 (1819). *S. marina* (L.) Griseb., Spicil. Fl. Rumel. 1 : 213 (1843). *S. media* Boiss., Fl. 1 : 733 (1867). *Arenaria rubra* L. var. *marina* L. Sp. Pl. 423 (1753) p.p. [Plate 172]

Annual or perennial herb, glabrescent to glandular-pubescent, 8–25 cm. Stems decumbent to ascending, simple or branched. Leaves 0.6–3 (–5) × 0.1 cm., somewhat fleshy, linear, more or less acute, glabrous to glandular-pubescent; stipules short, triangular-ovate, acuminate, connate at base, white. Inflorescences terminal, more or less dichasial, rather loose. Bracts similar to stipules. Pedicels as long as calyx or somewhat longer. Sepals 3–5 mm., lanceolate to ovate, obtuse, scarious-margined, glabrous to glandular-pubescent. Petals slightly shorter than to as long as sepals, pink or whitish, ovate to obovate, obtuse. Stamens (2–)5–8 (–10). Capsule 5–6 mm., slightly exceeding calyx, ovoid. Seeds 0.7–0.8 mm., more or less compressed, ovoid, brownish, smooth to finely tuberculate, mostly wingless, some with erose or fringed wings. Fl. March–June.

Hab.: Muddy and saline soils. Coastal Galilee, Acco Plain, Sharon Plain, Philistean Plain, W. Negev, Upper Jordan Valley, Beit Shean Valley, Lower Jordan Valley, Ammon, Moav.

Area: Euro-Siberian, Mediterranean and Irano-Turanian. Also in some other temperate and tropical regions.

3. Spergularia diandra (Guss.) Heldr. et Sart. in Heldr., Herb. Graec. Norm. n. 492 (1855). *Arenaria diandra* Guss., Fl. Sic. Prodr. 1 : 515 (1827). [Plate 173]

Annual slender herb, glandular-pubescent above, puberulent or glabrescent below, 7–20 cm. Stems many, decumbent to ascending. Leaves 1–4 × 0.05–0.1 cm., somewhat fleshy, filiform or narrowly linear, mucronulate; stipules short, triangular, caducous. Inflorescences usually leafless or with small, short leaves, dichasially branched, many-flowered. Bracts similar to stipules. Pedicels longer than flowers. Sepals 2–3 mm., more or less oblong, obtuse, scarious-margined, glandular-puberulent. Petals as long as or slightly shorter than sepals, pink, ovate. Stamens 2–3. Capsule 2–3 mm., about as long or longer than calyx. Seeds minute, triangular-obovoid, brown, scabrous, minutely tuberculate, wingless. Fl. February–May.

Hab.: Batha and steppes. Sharon Plain, Philistean Plain, Esdraelon Plain, Mt. Gilboa, Judean Mts., Judean Desert, W., N., C. and S. Negev, Upper and Lower Jordan Valley, Dead Sea area, Arava Valley, Moav, Edom.

Area: Mediterranean, Irano-Turanian and Saharo-Arabian.

4. Spergularia rubra (L.) J. et C. Presl, Fl. Čech. 94 (1819); Boiss., Fl. 1 : 732 (1867). *Arenaria rubra* L. et var. *campestris* L., Sp. Pl. 423 (1753). *S. campestris* (L.) Aschers., Fl. Prov. Brandenb. 94 (1864). [Plate 174]

Annual, glandular-pubescent above, glabrescent below, 10–25 cm. Stems usually many, decumbent to ascending. Leaves 1–4 (–5) × 0.1–0.2 cm., narrowly linear, mucronate, glabrous or glandular-puberulent, not fleshy; stipules short but conspicuous, ovate-acuminate, more or less lacerate at tip, white-silvery, scarious. Inflorescences terminal, more or less leafy, dichasially branched. Bracts leaf-like. Pedicels longer than flowers. Sepals 3–5 mm., oblong or ovate-lanceolate, obtuse, broadly scarious-margined, glandular-pubescent. Petals 5, as long as or somewhat shorter than sepals, pink-red, ovate to obovate. Stamens 5–10. Capsule about as long as calyx. Seeds 0.5 mm., triangular-obovoid, thick-margined but wingless, brownish, finely tuberculate. Fl. February–June.

Hab. : Sandy or gravelly damp ground. Coastal Galilee, Acco Plain, Sharon Plain, Philistean Plain, Mt. Carmel, Esdraelon Plain, Mt. Gilboa, Judean Mts., W. Negev, Upper and Lower Jordan Valley, Moav.

Area : Mediterranean and Euro-Siberian; also in some other parts of Asia, N. America, and elsewhere.

Used in medicine (herba arenariae rubrae).

5. Spergularia bocconii (Sol. ex Scheele) Aschers. et Graebn., Syn. 5, 1 : 849 (1919). *Arenaria bocconi* Sol., Pl. Cors. (1825) *nom nud. Alsine bocconi* Scheele, Flora 26 : 431 (1843). [Plate 175]

Annual, subglabrous below, glandular-pubescent above, 8–25 cm. Leaves 1–3 × 0.1 cm.; floral leaves shorter, narrowly linear, mucronate; stipules triangular, white. Inflorescences many-flowered, resembling one-sided racemes, elongating in fruit. Pedicels usually as long as or shorter than sepals. Sepals 2–3.5 mm., glandular-pubescent. Petals shorter than sepals, pinkish. Capsule not exceeding calyx. Seeds about 0.4–0.5 mm., greyish-brown, tuberculate, wingless. Fl. May.

Hab. : Fallow fields. Judean Mts. Very rare (casual?).

Area : Mediterranean and Euro-Siberian.

20. Telephium L.

Annual or perennial herbs, somewhat fleshy, often glaucous. Leaves alternate; stipules small, scarious. Flowers small, in terminal, capitate or corymbose cymes. Sepals 5, free, obtuse, keeled, with scarious margin, persistent. Petals 5, white, entire. Stamens 5. Ovary 1-celled with 3–4 incomplete septa, many-ovuled, ovoid-trigonous; styles 3, recurved. Capsule incompletely 3–4-celled, 3–4-valved, included in calyx. Seeds many, globular or reniform; embryo almost annular, surrounding endosperm.

Six species, in the Mediterranean and the adjacent tropical regions.

1. Telephium sphaerospermum Boiss., Diagn. ser. 1, 10 : 12 (1849) et Fl. 1 : 754 (1867). [Plate 176]

Glaucous, glabrous annual or biennial, 8–25 cm. Stems tufted, prostrate, leafy. Leaves 0.3–1 × 0.1–0.2 cm., oblong-elliptical, attenuate, somewhat acute to almost obtuse, the lower ones crowded; stipules minute, scarious. Cymes corymbose; floral branches short, dichasial. Bracts minute, scarious, similar to stipules. Pedicels short. Flowers 3–4 mm. Sepals oblong-linear, keeled, white-margined. Petals almost as long as sepals, white, oblong. Capsule about as long as sepals, ovoid-trigonous. Seeds many, 0.5 mm. in diam., subglobular, minutely tuberculate. Fl. March–June.

Hab. : Hot deserts. Negev, Lower Jordan Valley, Dead Sea area, Moav.

Area : E. Sudanian and Saharo-Arabian.

Trib. POLYCARPEAE. Fruit a many-seeded capsule, dehiscent by 3 valves. Styles united in lower part.

21. POLYCARPON L.

Annual or perennial small herbs. Leaves opposite or in whorls of 4, usually oblong or obovate; stipules scarious, silvery. Cymes dichasial, many-flowered. Bracts scarious. Flowers small. Sepals 5, free, keeled, hooded, scarious-margined. Petals 5, small, oblong to obovate, entire or emarginate. Stamens 3–5, inserted on a perigynous disk. Ovary 1-celled, many-ovuled; style short, 3-fid. Capsule 3-valved. Seeds many or few, with convex back.

About 15 species, mainly in the warm and temperate regions.

1. Sepals ovate, acute-mucronate, twice as long as petals. Leaves ovate, obovate to elliptical, sometimes fleshy.   **1. P. tetraphyllum**
– Sepals oblong-ovate, obtuse, about as long as petals. Leaves oblong-spatulate to oblong-
   linear, always fleshy.   **2. P. succulentum**

1. Polycarpon tetraphyllum (L.) L., Syst. ed. 10, 2, 881, 1360 (1759); Boiss., Fl. 1 : 735 (1867). *Mollugo tetraphylla* L., Sp. Pl. 89 (1753). [Plate 177]

Glabrous annual, 3–15 cm. Stems prostrate to ascending, usually branched below. Leaves 5–15 × 0.2–0.6 cm., in opposite pairs or in false whorls of 4 or 6, obovate to elliptical or ovate, generally tapering into a petiole; stipules and bracts scarious through-out. Inflorescences simply or repeatedly forked, many-flowered, loose or dense. Pedicels as long as or longer than calyx. Sepals about 2 mm., ovate-lanceolate, acute-mucro-nate, white-margined and with a scabrous keel. Petals half as long as sepals, white, oblong, more or less emarginate. Stamens 3 or 5. Ovary longer than style. Capsule somewhat shorter than calyx, ovoid. Seeds minute, triangular-reniform, finely tuber-culate. Fl. January–April.

Var. **tetraphyllum**. *P. tetraphyllum* (L.) L. var. *verticillatum* Fenzl in Ledeb., Fl. Ross. 2 : 165 (1843). Leaves oblong-spatulate to elliptical, often not fleshy. Cymes rather loose. Stamens often 3.

Hab.: Roadsides, gardens and waste places. Coastal Galilee, Acco Plain, Sharon Plain, Philistean Plain, Upper Galilee, Mt. Carmel, Esdraelon Plain, Mt. Gilboa, Samaria, Judean Mts., Judean Desert, Upper and Lower Jordan Valley, Gilead. Common.

Var. **alsinifolium** (Mill.) Arcang., Comp. Fl. It. 112 (1882). *Herniaria alsines-folia* Mill., Gard. Dict. ed. 8 no. 3 (1768). *Polycarpon alsinefolium* (Biv.) DC., Prodr. 3 : 376 (1826); Boiss., Fl. 1 : 736 (1867). Leaves obovate to oblong, somewhat fleshy. Cymes rather dense. Stipules, bracts and flowers somewhat longer than in type. Stamens 5.

Hab.: Coastal sands. Coastal Galilee, Acco Plain, Sharon Plain, Philistean Plain. Confined to the coastal belt only.

Area of species: Mediterranean and Euro-Siberian; adventive also in other regions.

2. Polycarpon succulentum (Del.) J. Gay in Duchartre, Rev. Bot. 2 : 372 (1846) non Boiss., Fl. 1 : 736 (1867). *Alsine succulenta* Del., Fl. Eg. 211 (1813). *Polycarpon arabicum* Boiss., l.c. [Plate 178]

Glabrous annual, up to 10 cm. Stems many, prostrate, branched. Leaves 0.3–1.2 × 0.1–0.2 cm., 2 or more at each node, fleshy, oblong-spatulate to oblong-linear, tapering into a petiole; stipules and bracts scarious at margin, green on back. Cymes many, loose or dense. Pedicels as long as or longer than calyx. Sepals 2 mm., oblong-ovate, obtuse, white-margined. Petals about as long as calyx, white, oblong-lanceolate to oblong-obovate. Style longer than ovary. Capsule shorter than calyx. Seeds reniform-roundish, indistinctly tuberculate, shining. Fl. January– April.

Hab.: Desert and also littoral sandy soils. Acco Plain, Sharon Plain, Philistean Plain, W., N. and C. Negev, Edom.

Area : E. Saharo-Arabian.

22. POLYCARPAEA Lam.

Annual or perennial herbs or shrubs. Leaves opposite or spuriously whorled, with scarious stipules. Flowers many, in paniculate or corymbose cymes. Sepals 5 (–8), free, not keeled, scarious-margined. Petals 5, shorter than calyx, lanceolate, entire or emarginate. Stamens 5, free or connate at base, sometimes with 3–5 staminodes opposite petals. Perigynous disk present. Ovary 1-celled, few- or many-ovuled; style with capitate or 3-lobed stigma. Capsule 3-valved. Seeds pear-shaped to almost reniform.

About 30 species, mainly in tropical regions.

1. Polycarpaea repens (Forssk.) Aschers. et Schweinf., Oesterr. Bot. Zeitschr. 39 : 126 (1889). *Corrigiola repens* Forssk., Fl. Aeg.-Arab. 207 (1775). *P. fragilis* Del., Fl. Aeg. Ill. 241, t. 24, f. 1 (1813); Boiss., Fl. 1 : 737 (1867). [Plate 179]

Perennial, appressed crisp-tomentose herb, 7–25 cm. Stems prostrate to ascending, branched, brittle, woody at base. Leaves 0.3–1.5 × 0.1–0.3 cm., opposite or whorled, somewhat rigid, lanceolate-linear, acute-mucronate, with revolute margins; stipules scarious, silvery, lanceolate or ovate-lanceolate. Flowers many, in dense cymes, short-pedicelled. Bracts scarious with a prominent median nerve, ovate, acute, ciliolate. Sepals 5, 2–3 mm. long, ovate, mucronate, broadly scarious-margined, with prominent green median nerve. Petals 5, shorter than sepals. Stamens 5, shorter than petals. Stigma capitate. Capsule 3-valved, about as long as calyx. Seeds smooth, grooved on back. Fl. March–April.

Hab.: Hot deserts. S. Negev, Arava Valley, Edom.

Area : N. Sudanian, extending into Saharo-Arabian borderland.

23. ROBBAIREA Boiss.

Perennial, glabrous, often prostrate herbs. Leaves opposite or spuriously whorled, stipulate. Inflorescences loose. Calyx 5-parted; sepals narrowly scarious-margined. Petals 5, clawed, with ovate or obcordate limb. Stamens 5, connate below into a perigynous ring. Ovary 1-celled, many ovuled; style 3-fid. Capsule 3-valved. Seeds cuneate, incurved, convex and grooved on back. Resembling *Polycarpaea* but differing from it mainly by the clawed petals and absence of staminodes.

Two species in the Saharo-Arabian region and adjacent tropical deserts.

1. Robbairea delileana Milne-Redhead, Kew Bull. 1948:452 (1949). *R. prostrata* (Forssk.) Boiss., Fl. 1 : 735 (1867) p.p. [Plate 180]

Glabrous perennial, up to 30 cm. Stems many, woody at base, often prostrate, thin, dichotomously branched. Leaves 0.5–1 (–1.5) × 0.1 cm., narrowly oblong to linear-oblanceolate, the lower ones spatulate; stipules short, triangular, membranous. Racemes loose, many-flowered. Bracts small, ovate-triangular, broadly membranous at margin. Pedicels usually as long as or longer than calyx. Calyx 2–3 mm.; sepals ovate, the inner 3 sometimes somewhat longer than the outer 2. Petals about as long as calyx, whitish-pink. Stamens 5. Capsule more or less included. Seeds smooth, shining. Fl. January–April.

Hab. : Deserts, especially in hot depressions. Judean Desert, Negev, Lower Jordan Valley, Dead Sea area, Arava Valley, Gilead, Moav, Edom.

Area : Saharo-Arabian and Sudanian.

24. LOEFLINGIA L.

Small annuals with rigid, glandular-pubescent, dichotomously branched stems. Leaves opposite or spuriously clustered, setaceous, subulate or filiform, with stipules adnate to blade. Flowers axillary, solitary or clustered, sessile. Calyx 5-parted; sepals acuminate, carinate, rigid, scarious-margined with a lateral, subulate and ciliate appendage on either side. Petals 3–5, minute, sometimes 0. Stamens 3–5. Ovary 1-celled, many-ovuled; style short, with a capitate or 2–3-lobed stigma. Capsule 3-valved. Seeds obovoid; embryo somewhat curved.

Some 8 species in the Saharo-Arabian region, C. Asia and N. America.

1. Loeflingia hispanica L., Sp. Pl. 35 (1753); Boiss., Fl. 1 : 738 (1867). [Plate 181]

Low annual, glandular-pubescent, green or pinkish, up to 10 cm. Stems decumbent or ascending, dichotomously branched, somewhat inflated at nodes. Leaves 0.3–1 × 0.05–0.2 cm., opposite or clustered along the stem, sessile, setaceous-subulate, mucronate; stipules setaceous-filiform, connate with leaves to middle, ciliate at margin. Flowers small, clustered in the axils of the upper part but solitary in lower, sessile. Bracts leafy, acuminate, mucronate. Calyx 2–4 mm., sepals somewhat unequal, rigid, lanceolate, mucronate, all with lateral, setaceous awns. Petals 5, much shorter than calyx, white, obovate, entire, retuse. Capsule shorter than calyx, more or less ovoid. Seeds many, minute, somewhat compressed, yellowish. Fl. February–April.

Hab. : Steppes and sandy deserts. Coastal and C. Negev, Lower Jordan Valley, Dead Sea area, Arava Valley, Edom (desert).

Area : Saharo-Arabian; also occurring in dry and sandy habitats of some Mediterranean countries and W. Europe.

Trib. PARONYCHIEAE. Sepals free or united at base. Fruit 1-seeded, indehiscent.

25. CORRIGIOLA L.

Annual or perennial glabrous herbs with decumbent stems. Leaves entire, all alternate or the lower ones sometimes opposite; stipules small, scarious. Flowers small, in axillary or terminal, dense, symose inflorescences, minutely bracteate. Calyx persistent, parted into 5 white-margined lobes. Petals 5, free, as long as or longer than calyx. Stamens 5, inserted on the margin of the short perigynous disk. Ovary ovoid, 1-celled, with 1-ovule inserted at the apex of a long funicle; style 3-parted, very short. Fruit an indehiscent, 1-seeded, trigonous nutlet, enclosed in persistent calyx. Seed with membranous coat; embryo annular, surrounding the copious endosperm.

About 10 species in Europe, Africa, W. Asia, Australia and S. America.

1. Corrigiola litoralis L., Sp. Pl. 271 (1753); Boiss., Fl. 1 : 749 (1867). [Plate 182]

Glaucous, glabrous herb. Stems many, more or less decumbent, up to 40 cm. Leaves (0.5–)1–3.5 × 0.1–0.5 cm., slightly succulent, linear or linear-oblanceolate to linear-oblong or oblanceolate-spatulate, tapering at base into a petiole or subsessile; stipules short, scarious, semisagittate, ovate, acuminate, more or less denticulate. Bracts stipule-like. Flowers 1–2 mm., pedicellate, crowded in small, head-like, axillary and terminal cymes. Calyx lobes ovate, obtuse. Petals almost as long as calyx, white, sometimes red-tipped. Stamens shorter than petals; anthers violet. Nutlet about 1–1.5 mm., more or less trigonous. Fl. January–June.

Subsp. **litoralis**. *C. litoralis* L. ssp. *eulittoralis* Briq., Prodr. Fl. Corse 1 : 480 (1910). Annual. Stems longer than 15 cm. Leaves mostly 1.5–3.5 × 0.1–0.5 cm.

Hab.: Mainly fields and roadsides. Sharon Plain, Philistean Plain.

Subsp. **telephiifolia** (Pourr.) Briq., Prodr. Fl. Corse 1 : 481 (1910). *C. telephiifolia* Pourr., Mém. Acad. Toulouse 3 : 316 (1788). Perennial. Stems short. Roots thick. Leaves mostly 5 × 2 mm., only a few longer.

Hab.: Heavy soils inundated in winter. Sharon Plain, Philistean Plain.

Area of specics: Mediterranean, extending into the Euro-Siberian and Sudanian regions.

26. GYMNOCARPOS Forssk.

Low shrubs. Stems woody, much branched. Leaves opposite or the lower whorled, fleshy, terete-linear; stipules small, scarious. Flowers clustered in short-peduncled, axillary or terminal, head-like cymes, reddish. Calyx with urn-shaped base and 5 somewhat hooded and aristate lobes, later indurated. Petals 5, short, setaceous. Stamens 5, alternating with petals. Ovary 1-celled, 1-ovuled; ovule pendulous from apex of funicle; style long, with 3 stigmas. Utricle membranous, indehiscent, at length rupturing at base, included within calyx. Seed oblong-reniform, with horseshoe-shaped embryo surrounding the endosperm.

One species in the Saharo-Arabian region.

1. Gymnocarpos decandrum Forssk., Fl. Aeg.-Arab. 65 (1775). *G. fruticosum* (Vahl) Pers., Syn. 1 : 262 (1805); Boiss., Fl. 1 : 748 (1867). [Plate 183]

Half-shrub, almost glabrous, 10–35 cm. Stems with greyish-white bark. Branches flexuous. Leaves 0.4–1.4 × 0.1–0.2 cm., subsessile, more or less terete-linear, mucro-nate-aristate; stipules short, ovate-triangular, ciliolate. Inflorescences crisp-pubescent. Bracts stipule-like, much shorter than calyx. Calyx about 7 mm., pruinose to hairy at base, reddish or green; lobes linear, narrowly white-margined, glabrescent or papillose, glandular-hairy. Petals about as long as sepals. Ovary pubescent above. Utricle included in calyx tube. Fl. January–April.

Hab.: Stony slopes and hammadas. Judean Desert, N., C. and S. Negev, Lower Jordan Valley, Dead Sea area, Arava Valley, Moav, Edom.

Area: Saharo-Arabian.

A form with a more hairy calyx and glandular calyx lobes has been observed in the surroundings of the Dead Sea.

27. PARONYCHIA Mill.

Annual or perennial herbs or dwarf-shrubs. Leaves mostly opposite; stipules and bracts conspicuous, scarious, often silvery. Flowers small, usually clustered in head-like inflo-rescences. Calyx lobes (4–)5, almost free, more or less hooded, sometimes aristate at apex. Petals 5, sometimes less, inserted at base of calyx. Ovary 1-celled, containing a basal, anatropous ovule; style short, 2-cleft at apex, sometimes the 2 stigmas sessile. Utricle included in calyx, 1-seeded, membranous. Seeds oblong, lenticular to globular.

About 45 species in temperate and dry regions of both hemispheres.

1. Calyx lobes aristate or long-mucronate   2
 – Calyx lobes neither aristate nor long-mucronate   5
2. Bracts minute, much smaller than flowers.   **6. P. echinulata**
 – Bracts large, much broader and longer than flowers   3
3. Heads 1 cm. or more in diam. Calyx with straight hairs or hairless.   **3. P. argentea**
 – Heads about 5 mm. in diam. Calyx with hooked hairs at base   4
4. Leaves glabrous. Stems hairy on one side only. Annuals.   **4. P. arabica**
 – Leaves hairy. Stems hairy all around. Perennials.   **5. P. desertorum**
5(1). Littoral, cushion-like plants. Leaves thick, leathery or more or less succulent. Heads
      about 5 mm. in diam. Bracts oblong-ovate to oblong-lanceolate.   **1. P. palaestina**
 – Plants not littoral, nor cushion-like. Leaves thin, herbaceous. Heads mostly 1 cm.
   (rarely 5 mm.) in diam. Bracts broadly ovate to orbicular.   **2. P. sinaica**

Sect. ANOPLONYCHIA Fenzl in Endl., Gen. 958 (1840). Calyx lobes herbaceous, plane-concave, not hooded nor aristate at apex. Flowers densely capitate.

1. Paronychia palaestina Eig, Agr. Rec. Inst. Agr. Nat. Hist. 2 : 205 (1929). [Plate 184]

Perennial, appressed velvety-pubescent herb, forming dense low cushions, 10–20 cm. in diam. Stems caespitose, prostrate to ascending, covered with densely imbricated leaves and stipules. Leaves 3–8 × 1–2 mm., somewhat fleshy, lanceolate or elliptical, ap-pressed-hairy; stipules ovate to lanceolate. Heads about 5 mm. in diam., few-flowered, more or less terminal. Bracts longer than flowers, oblong-ovate or oblong-lanceolate. Calyx lobes 2–3 mm., unequal, linear-oblong to ovate-lanceolate, appressed-puberulent, not or only narrowly white-margined, not hooded and awnless. Seeds 1 mm., lenticular, brown, shining. Fl. March–May.

Hab. : Calcareous sandstone and sandy loams facing the sea-shore. Coastal Galilee, Acco Plain, Sharon Plain, Philistean Plain.

Area : E. Mediterranean (endemic in Palestine).

2. Paronychia sinaica Fresen., Mus. Senckenb. 1 : 180 (1834); Boiss., Fl. 1 : 744 (1867). [Plate 185]

Perennial, velvety-pubescent, 5–12 cm. Stems more or less ascending or decumbent, branching from woody base. Leaves 0.4–1 × 0.1–0.2 cm., lanceolate to linear-oblong, acute; stipules lanceolate. Heads 0.5–1 cm. in diam. Bracts 3–7 mm., broadly ovate to orbicular, short-acuminate to muticous. Calyx hidden between bracts, 2–3 mm., with unequal, oblong to linear, pubescent, muticous or more or less acute and indistinctly keeled lobes. Seeds 1.5–2 mm., light brown, somewhat punctate. Fl. March–May.

Var. **sinaica.** Bracts about 5 mm., white-silvery.

Hab. : Steppes and deserts. Judean Mts., Judean Desert, W. and C. Negev, Lower Jordan Valley, Arava Valley, Moav, Edom.

Var. **flavescens** Boiss., Fl. 1 : 745 (1867). Bracts 4–7 mm., yellowish.

Hab. : As above. Samaria, Judean Mts., Judean Desert, N. and C. Negev, Lower Jordan Valley, Edom.

Area of species : Irano-Turanian, extending into the Mediterranean and E. Saharo-Arabian border zones.

Polymorphic. A form with smaller heads and 3 mm. long bracts has been collected in the N. Negev (var. **negevensis** Zoh.*).

The records in Post (1932) on *P. capitata* (L.) Lam., Fl. Fr. 3 : 229 (1778) from Palestine are most probably erroneous.

Sect. PARONYCHIA. Calyx lobes partly scarious, hooded below apex, mucronate or aristate. Flowers capitate or glomerate.

3. Paronychia argentea Lam., Fl. Fr. 3 : 230 (1778); Boiss., Fl. 1 : 745 (1867). [Plate 186]

Perennial or sometimes annual herb, velvety-puberulent to glabrescent. Stems prostrate to ascending, up to 50 cm. Leaves 0.5–2 × 0.1–0.5 cm., oblong-elliptical to linear-lanceolate, sometimes obovate or ovate, acute-mucronulate, scabrous-margined, glabrescent; stipules silvery-hyaline, lanceolate or ovate, acute. Heads 1 cm. or more across, terminal and lateral. Bracts longer than flowers, ovate. Calyx lobes about 2 mm., oblong, scarious-margined, more or less puberulent with straight hairs, rarely glabrous, hooded, ending in a mucro or awn shorter than lobe. Seeds 1–1.5 mm., orbicular, brown, shining, slightly punctate-tuberculate. Fl. January–April.

* See Appendix at end of this volume.

Hab.: Batha and fallow fields. Coastal Galilee, Acco Plain, Philistean Plain, Upper and Lower Galilee, Mt. Carmel, Esdraelon Plain, Shefela, Judean Mts., Judean Desert, N. and C. Negev, Hula Plain, Upper and Lower Jordan Valley, Dead Sea area, Gilead, Ammon, Edom.

Area: Mediterranean.

Plants growing in inland parts of the country are usually perennial, with leaves 0.5–1 × 0.1–0.3 cm., while the coastal form is either annual or perennial, with a denser head and larger (0.6–2 × 0.2–0.5 cm.), somewhat fleshy leaves. Transitions are found in Upper Galilee, Judean Mts. and Lower Jordan Valley.

4. Paronychia arabica (L.) DC. in Lam., Encycl. 5 : 24 (1804); Boiss., Fl. 1 : 746 (1867). *Illecebrum arabicum* L., Mant. 51 (1767). [Plate 187]

Annual, more or less puberulent, 5–30 cm. Stems prostrate, elongated, pilose on one side only. Leaves 0.4–1.1 × 0.1–0.2 cm., oblong-oblanceolate, tapering at base, mucronate, glabrous, usually ciliolate; stipules and bracts silvery-hyaline, ovate to oblong-lanceolate, acuminate, ciliolate. Flowers in axillary small clusters, often hidden by bracts. Calyx about 1.5 mm.; lobes oblong, broadly scarious-margined, with hooked hairs at base and a subapical awn of same length as lobes [var. *longiseta* (Bertol.) Aschers. et Schweinf., Oesterr. Bot. Zeitschr. 39 : 301 (1889)] or up to half as long as sepals (var. *breviseta* Aschers. et Schweinf., l.c. 298). Seeds globular or lenticular, brown, shining. Fl. February–May.

Hab.: Sandy deserts and littoral sandy soils. Philistean Plain, Coastal, W. and C. Negev, Edom.

Area: Saharo-Arabian, slightly extending into Mediterranean and Sudanian territories.

5. Paronychia desertorum Boiss., Diagn. ser. 1, 3 : 11 (1843) et Fl. 1 : 746 (1867). *P. lenticulata* (Forssk.) Aschers. et Schweinf., Oesterr. Bot. Zeitschr. 39 : 300 (1889) *nom. illegit. Herniaria lenticulata* Forssk., Fl. Aeg.-Arab. 52 (1775) non L., Sp. Pl. 218 (1753). [Plate 188]

Perennial, rarely annual (by flowering in first year), patulous velvety-pubescent, 4–15(–20) cm. Stems many, prostrate, with densely crowded-imbricated leaves and stipules, pilose all around. Leaves 2–7 ×1–1.5 mm., oblong-lanceolate, generally short-mucronate, pubescent or puberulent, ciliate; stipules about as long as leaves, ovate-lanceolate, acute to acuminate. Heads few-flowered, often forming spikes. Bracts slightly longer than flowers, ovate, acute. Calyx lobes about 2 mm., oblong-obovate, hooded and awned near apex, more or less puberulent with hooked hairs at base, green part somewhat narrower than membranous margin. Seeds 0.5 mm., globular to lenticular, brown, shiny. Fl. March–May.

Hab.: Deserts. W. Negev, Arava Valley, Edom.

Area: E. Saharo-Arabian extending into E. Sudanian territories.

6. Paronychia echinulata Chater, Repert. Sp. Nov. 69 : 52 (1964). *P. echinata* auct. non (Desf.) Lam., Fl. Fr. 3 : 232 (1778); Boiss., Fl. 1 : 747 (1867). [Plate 189]

Annual, hairy, 6–15 cm. Stems erect or ascending, rigid, simple or branched at base.

Leaves ovate to oblong-lanceolate, reddish, with pale, membranous, denticulate-scabrous margin, shortly aristate. Flowers in clusters 3–8 mm. in diam. Bracts much shorter than flowers, lanceolate, acuminate. Calyx 2–2.5 mm.; lobes equal or almost equal, spatulate, hooded, with crisp or hooked hairs at base and with long, patulous subapical awn. Fl. March.

Hab.: Sandy soils. Sharon Plain. Rare.

Area: Mediterranean.

28. HERNIARIA L.

Annual or perennial herbs or dwarf-shrubs. Stems prostrate. Leaves opposite, the upper sometimes alternate, sessile or subsessile, with small scarious stipules. Flowers small, hermaphrodite or unisexual, in dense, axillary clusters. Calyx lobes (4–)5. Petals 5, short, subulate-filiform. Stamens 2–5. Ovary 1-celled, with 1 basal ovule, borne on a long funicle; style 2-partite to 2-fid. Fruit a subspherical utricle, exserting from or included in calyx. Seeds globular to lenticular, blackish, shining.

About 20 species in temperate Europe, N. and E. Africa and W. Asia.

Literature: F. N. Williams, A systematic revision of the genus *Herniaria. Bull. Herb. Boiss.* 4: 556–570 (1896). F. Hermann, Übersicht über die *Herniaria*-Arten d. Berliner Herbars. *Repert. Sp. Nov.* 42: 203–224 (1937).

1. Glabrous or glabrescent rare herbs.	**1. H. glabra**
– Hirsute or tomentose annuals or perennials	2
2. Calyx 5-merous. Annuals.	**2. H. hirsuta**
– Calyx 4-merous. Perennials.	**3. H. hemistemon**

Subgen. HERNIARIA. Calyx lobes usually 5, equal or almost so. Flowers pentamerous.

1. Herniaria glabra L., Sp. Pl. 218 (1753); Boiss., Fl. 1: 740 (1867). [Plate 190]

Annual, rarely perennial, glabrous or glabrescent herb, 4–15 cm. Stems prostrate, caespitose. Leaves 0.3–1.2 × 0.1–0.3 cm., alternate, more or less petiolate, oblong-oblanceolate, tapering at base, rarely ciliate; stipules ovate, greenish-white, ciliolate. Flowers sessile or almost so, 5–12 in alternate clusters opposite leaves. Calyx about 1.5 mm., mostly glabrous, with short tube and 5 equal lobes; lobes oblong-ovate, obsoletely white-margined, 3-nerved, obtuse or rarely terminating in a short bristle. Stamens 5. Style short or 0, with 2 rather divergent stigmas. Utricle exceeding calyx. Fl. May–July.

Hab.: Batha and fallow fields. Coastal Galilee, Acco Plain, Upper Galilee. Rare.

Area: Mediterranean, Irano-Turanian and Euro-Siberian.

Used medicinally as diuretic, astringent, etc. (herba herniariae).

2. Herniaria hirsuta L., Sp. Pl. 218 (1753). *H. cinerea* DC. in Lam. et DC., Fl. Fr. ed. 3, 5: 375 (1815). [Plate 191]

Hirsute-hispid annual, sometimes perennating, 4–15 cm. Stems caespitose, prostrate. Leaves 0.5–1 cm., the lower opposite, lanceolate-elliptical to oblong-oblanceolate or oblong-linear, tapering at base, patulous-hairy at margin; stipules and bracts scarious, ciliolate. Flowers sessile in axillary head-like clusters. Calyx lobes 1–2 mm., oblong, covered with bristly hairs shorter or longer than width of sepal, hooked or straight, the terminal bristles usually longer. Stamens 2(–3) or 5(4). Stigmas subsessile. Fl. March–April.

Hab.: Roadsides, fields and steppes. Acco Plain, Sharon Plain, Philistean Plain, Upper and Lower Galilee, Shefela, Judean Mts., W., N. and C. Negev, Upper and Lower Jordan Valley, Dead Sea area, Gilead, Moav, Edom.

Area: Mediterranean, Irano-Turanian, Euro-Siberian; also in other regions.

In the arid parts of Palestine a form with 2(–3) stamens is predominant. This form may correspond to *H. diandra* Bge., Del. Sem. Hort. Bot. Dorp. 7 (1843), which can hardly be separated from *H. hirsuta*.

Used in medicine in same way as *H. glabra*.

Subgen. HETEROCHITON (Graebn. et Mattf.) Hermann. Perennials. Calyx lobes 4, unequal. Flowers tetramerous.

3. Herniaria hemistemon J. Gay in Duchartre, Rev. Bot. 2:371 (1847); Boiss., Fl. 1:742 (1867). [Plate 192]

Perennial, pubescent-tomentose herb, 3–8 cm. Stems many, prostrate, caespitose, abundantly branched, brittle. Leaves 3–7 × 1–3 mm., short-petioled, oblong-elliptical; stipules ovate-triangular, brown, ciliate at margin and often also on back. Flowers sessile or nearly so, in axillary clusters. Calyx about 1.5–2 mm., with short, hispid tube and 4 hispid-ciliate lobes, the outer 2 somewhat fleshy, deltoid-spatulate to almost orbicular, much larger than the somewhat membranous ovate-oblong inner pair. Stamens 2, opposite the interior calyx lobes, rarely stamens 4. Styles 2. Utricle membranous, included in calyx tube. Fl. January–April.

Hab.: Steppes and deserts. Judean Desert, W. and N. Negev, Lower Jordan Valley, Dead Sea area, Arava Valley, Moav, Edom.

Area: Saharo-Arabian.

One of the most characteristic components of the *Chenolea arabica* community on gypsaceous desert ground, especially in the Judean Desert.

29. SCLEROCEPHALUS Boiss.

Annual herbs. Stems rigid, branched. Leaves opposite, fleshy, terete-linear; stipules scarious. Heads 4–7-flowered, spherical, becoming detached from plant when ripe together with the short, jointed peduncle. Floral leaves united with flowers, indurated, spiny in fruit. Calyx cup-shaped at base, adnate to bracts; lobes 5, hooded and sub-spinescent. Petals 0. Stamens 2–5, short, inserted at the discoid margin of calyx tube. Ovary adnate at base to calyx tube, free above, 1-celled, 1-ovuled; ovule suspended from basal funicle; style 2-fid into 2 somewhat recurved stigmas. Utricle membranous, adnate to calyx tube, irregularly lacerate at apex. Seeds somewhat compressed, ovoid, with annular embryo.

One species, Saharo-Arabian.

1. Sclerocephalus arabicus Boiss., Diagn. ser. 1, 3 : 12 (1843) et Fl. 1 : 748 (1867). *Paronychia sclerocephala* Decne., Ann. Sci. Nat. Bot. ser. 2, 5 : 262 (1836). [Plate 193]

Annual, glabrous, 3–15 cm. Stems indurated, more or less ascending. Leaves 0.5–2 × 0.1 cm., terete-linear, short-mucronate; stipules white-membranous, ovate-lanceolate, acute to acuminate. Heads 0.5–1 cm., comprising 3–7 flowers concrescent with floral leaves and bracts. Floral leaves somewhat longer than calyx; bracts shorter than or as long as calyx, stipule-like. Calyx about 4 mm., crisp-woolly; lobes narrow, straight, hooded, with subapical spine. Stamens shorter than calyx lobes. Fruiting heads about 1 cm., echinate. Seeds 2–3 × 1–2 mm. Fl. February–May.

Hab. : Deserts. Judean Desert, Lower Jordan Valley, Dead Sea area, Arava Valley.
Area : Saharo-Arabian.

Trib. PTERANTHEAE. Dichasia 3-flowered with middle flower hermaphrodite and lateral ones sterile or staminate. Fruit 1-seeded. Style 2–3-fid.

30. COMETES L.

Annual herbs. Stems branching dichotomously from base. Leaves opposite or whorled with scarious stipules. Inflorescences cymose-paniculate; cymes head-like, made of 3 sessile flowers borne on a common peduncle; middle flower hermaphrodite, the lateral ones rudimentary, accompanied by plumosely parted sterile branchlets, elongating and indurating at maturity, enclosing fruit at dispersal. Sepals 5, linear-oblong, mucronate. Petals 5. Stamens 5, united at base. Ovary 1-celled; ovule straight, semianatropous; style long, curved, with 3 short stigmas. Fruit a scarious, 1-seeded, included utricle. Seeds erect with membranous coat; endosperm scanty; embryo straight.

Two species in the Sudanian region.

1. Cometes abyssinica R. Br. in Wall., Pl. As. Rar. t. 18 (1830); Boiss., Fl. 1 : 753 (1867). [Plate 194]

Annual, hirsute-scabrous with branching hairs, 20–30 cm. Stems erect, much branching, corymbose-paniculate, especially above. Leaves 1.2–2.5 cm., opposite and whorled, subsessile, remote, linear-lanceolate, tapering at base, mucronate; stipules minute, membranous, lanceolate-subulate. Flowers in 3-flowered dichasia, sessile, accompanied by an involucre of long, flexuous, sterile, plumosely divided branchlets. Calyx 4–5 mm., hirsute; sepals oblong-linear, hooded and membranous-fimbriate at apex, ending with a patulous, slightly recurved mucro. Petals longer than filaments. Fruiting heads about 1.3 cm. across (incl. involucral branchlets). Fl. April.

Hab. : Sandstone rocks. Arava Valley.
Area : E. Sudanian (Nubo-Sindian).

31. PTERANTHUS Forssk.

Annual herbs, somewhat fleshy. Leaves opposite to whorled; stipules small. Cymes composed of 3-flowered dichasia borne on a common dilated peduncle, becoming leaf- or wing-like in fruit and falling with the fruiting cyme. Middle flower of each dichasium fertile, symmetrical, lateral ones sterile or fertile, asymmetrical, spinescent in fruit and accompanied on either side by sterile branches with recurved spines. Calyx of fertile flower closed, with 4 straight, oblong-linear, hooded and spiny-tipped lobes, the outer lobes broader, more prominently keeled, the inner 2 smaller. Petals 0. Stamens 4, opposite calyx lobes, connate at base. Ovary 1-celled, containing 1 erect basal ovule; Style 2- or 3-fid. Utricle membranous, included. Seeds compressed, with membranous coat; endosperm scanty; embryo lateral.

One species, mainly Saharo-Arabian.

1. Pteranthus dichotomus Forssk., Fl. Aeg.-Arab. LXII et 36 (1775). *P. echinatus* Desf., Fl. Atl. 1 : 144 (1798); Boiss., Fl. 1 : 752 (1867). [Plate 195]

Annual, almost glabrous below, rather puberulent above, 5–25 cm. Stems procumbent to ascending, repeatedly forked. Leaves 0.5–3 (–4) × 0.1 cm., more or less fleshy, linear; stipules lanceolate. Inflorescences dichasially branched, usually forming a corymbose panicle. Flowers sessile in papillose-puberulent 3-flowered cymes. Bracts concave, minute. Calyx 3–4 mm., with 4 oblong-linear lobes, the outer 2 lobes of the middle flower and the outer lobe of each lateral flower deeply keeled, narrowly white-margined, with wing-like appendages. Utricle and seed about 2 mm., oblong-ellipsoidal. Fl. January–April.

Hab.: Deserts, mostly on dry, saline ground and also in ruderal sites. Samaria, Judean Desert, W. and C. Negev, Upper and Lower Jordan Valley, Dead Sea area, Arava Valley, Ammon, Moav, Edom.

Area: Saharo-Arabian, slightly extending into some Irano-Turanian, Mediterranean and Sudanian territories.

Fairly common in typical desert communities, e.g. that of *Zygophyllum dumosum*, the *Suaeda palaestina* community, and others.

32. CHENOPODIACEAE

Annual or perennial herbs or shrubs, rarely trees, often more or less succulent. Leaves usually alternate, sometimes opposite, simple, often fleshy and terete or reduced to scales, exstipulate. Flowers small, hermaphrodite or unisexual, solitary or in clusters or in spicate or paniculate inflorescences, often bracteolate, actinomorphic, greenish. Perianth simple, sepaloid, persistent, of 3–5 tepals or lobes, rarely 0 or replaced by bracteoles. Stamens 2–5, opposite perianth segments, usually free, rarely united at base into a cup-like ring sometimes produced into staminodes between fertile stamens. Ovary usually superior, free, rarely adnate to perianth, 1-celled, with a single basal ovule; styles 2, rarely 3–4, free or united at base. Fruit an utricle or achene, rarely dehiscent capsule, free or adherent to unchanged or accrescent, succulent or membranous perianth; segments of fruiting perianth sometimes winged or appendiculate. Seeds horizontal or vertical; embryo arcuate or circular around endosperm or spiral where endosperm is meagre or 0.

The family is almost cosmopolitan but largely centred in arid regions, mostly on saline soils. It comprises about 100 genera and 1,400 species.

Literature : H. Graf zu Solms-Laubach, Über die in der Oase Biskra und deren nächster Umgebung wachsenden spiroloben Chenopodeen. *Bot. Zeit.* 59 : 159–186 (1901). E. Ulbrich, Chenopodiaceae, in : Engl. u. Prantl, *Nat. Pflznfam.* ed. 2, 16c : 379–584 (1934). A. Eig, A revision of the Chenopodiaceae of Palestine and neighbouring countries, *Palest. Journ. Bot. Jerusalem* ser. 3 : 119–137 (1945).

1. Stems or branches or both jointed, succulent. Leaves succulent, mostly opposite, often reduced to scales   2
– Stems and branches not jointed. Leaves always present, mostly alternate, rarely opposite, variously formed, sometimes scale-like and succulent   7
2. Fruiting perianth with transverse scarious wings   3
– Fruiting perianth wingless   5
3. Seeds vertical. Wings of fruiting perianth 3 or 5.   **22. Anabasis**
– Seeds horizontal. Wings of fruiting perianth 5   4
4. Tall shrubs or trees with very thin, glabrous, spartoid twigs. Staminodes not glandular.
                                                                            16. Haloxylon
– Lower shrubs with hairy or glabrous, not spartoid branches. Staminodes glandular.
                                                                            15. Hammada
5 (2). Leaves small, succulent, scale-like, globular or broadly-ovoid, forming small, decussate cupules. Flowers in cone-like or globular or very short, cylindrical spikes.
                                                                         10. Halocnemum
– Leaves 0 or reduced to triangular ends of joints. Flowers in more or less elongated spikes   6
6. Annuals. Branches all terminating in spikes. Androecium maturing before gynoecium. Stigmas tufted.   **12. Salicornia**
– Perennials. Branches not all terminating in spikes. Gynoecium maturing before androecium. Stigmas 2- or 3-fid.   **11. Arthrocnemum**
7 (1). Plants with short spiny branches. Leaves filiform or linear. Fruiting perianth winged.
                                                                               20. Noaea
– Branches not spiny. Leaves spiny or unarmed   8
8. Leaves spiny, opposite, falcate, not succulent. Fruiting perianth 3-winged.
                                                                         21. Girgensohnia
– Leaves not spiny (rarely leaves succulent with a spiny tip)   9
9. Flowers unisexual. Pistillate flowers without perianth but with 2 flattened, entire, dentate or lobed bracteoles; fruit enclosed within the 2 bractcoles   10
– Flowers usually hermaphrodite (rarely unisexual). All flowers with 3–5-parted (very rarely 2-parted) perianth   11
10. Leaves all opposite. Fruit adnate to the bracteoles; the latter connate almost to top.
                                                                            4. Halimione
— Leaves alternate (rarely partly opposite). Bracteoles connate at base only or up to middle.   **3. Atriplex**
11 (9). Flowers in groups of 2–4, connate at their perianth bases. Fruits connate in pairs or triplets, each adnate to indurated perianth; dispersal unit made of a group of fruits.
                                                                                  1. Beta
— Flowers and fruits not as above   12
12. Fruiting perianth with 3–5 conspicuous membranous wings   13
— Fruiting perianth wingless, sometimes with minute tubercles or spines, very rarely with minute wings but then plants annual, green, with leaves not succulent   17
13. Leaves all opposite   14
— All or most of the leaves alternate   15

14. Staminodes 5, semiorbicular. Leaves cylindrical thickening towards apex. Stems white-shining.   **17. Seidlitzia**
— Staminodes mostly 0. Leaves or/and stems not as above.   **19. Salsola**
15(13). Staminodes 4–5, connate with filaments.   **24. Halogeton**
— Staminodes 0   16
16. Lower part of perianth indurated in fruit, base broad, truncate, 5-angled with 5 pits.
                                                                                                                          18. Aellenia
— Perianth herbaceous or membranous; base not as above.   **19. Salsola**
17(12). Anthers ending in a conspicuous vesicular or inflated appendage. Leaves semiterete-triquetrous, entire, rigid, with cuspidate apex. Flowers scattered, solitary in axils. Low hairy annuals.   **23. Halotis**
— Anthers without inflated connective   18
18. Lobes of perianth hairy or fleecy. Stems and leaves pubescent or tomentose   19
— Lobes of perianth glabrous or covered with mealy scales, rarely lobes velvety or glandular, but then leaves broad, dentate, aromatic   22
19. Flowers in dense spikes, hidden by long fleece or perianth provided with short spines.
                                                                                                                          7. Bassia
— Flowers neither hidden in fleece nor spiny   20
20. Fruiting perianth lobes hemispherical, fleshy, without appendages.   **6. Chenolea**
— Fruiting perianth lobes not hemispherical, but with minute wings or leafy appendages or tubercles   21
21. Grey tall plants (up to 1 m. high). Leaves linear-lanceolate. Appendages of fruiting perianth minute, leaf-like, triangular-lanceolate.   **8. Kochia**
— Green low plants (10–20 cm. high). Leaves linear to oblong-lanceolate. Lobes of fruiting perianth each with small tubercle or minute wing on back.   **5. Panderia**
22(18). Fruiting perianth with rudimentary protuberances. Leaf axils fleecy.
                                                                                                                          14. Traganum
— Fruiting perianth without protuberances. Leaf axils glabrous   23
23. Dwarf annuals. Flowers 3 together, connate with each other and with the subtending fleshy, scale-like bract. Inflorescences composed of oblong, cone-like, alternate spikes. Stamens 1–2 in each flower. Leaves almost globular, clasping.   **9. Halopeplis**
— Not as above   24
24. Leaves terete, semiterete, or sometimes almost globular or linear to subulate, strongly succulent, entire.   **13. Suaeda**
— Leaves flat, ovate, lanceolate, rhombic or elliptical, not succulent, often lobed or dentate.   **2. Chenopodium**

Trib. BETEAE. Fruiting perianth with indurated base. Flowers of each cluster united at their bases. Stamens arising from a fleshy circular disk.

1. Beta L.

Annual or perennial herbs with furrowed stems. Leaves alternate, entire or nearly so. Inflorescences simple or paniculate, with spike-like branches, composed of small, (1–)2–4-flowered clusters, the flowers of each cluster united at their bases. Bracts herbaceous, small to abortive. Flowers hermaphrodite, sessile, greenish. Perianth of 5 lobes thickening in fruit. Stamens 5, inserted on the rim of the glandular, perigynous disk. Ovary adherent at base to perianth, 3-carpelled, more or less trigonous; stigmas 2–3. Utricles enclosed in perianths that are connate by their swollen bases to form a pseudocarp which becomes detached at maturity. Seeds horizontal, lenticular or reniform, glossy; embryo annular or almost so.

About 12 species mainly in the Euro-Siberian and the Mediterranean regions.

1. Beta vulgaris L., Sp. Pl. 222 (1753); Boiss., Fl. 4 : 898 (1879). [Plate 196]

Annual to perennial, glabrous or sometimes hirsute, 20–100 cm. Stems decumbent to erect, branched and leafy, rather stout, furrowed, sometimes coloured. Leaves mostly up to 12 × 6 cm., petiolate, dark green to reddish; radical leaves frequently rosulate, ovate, cuneate to somewhat cordate, obtuse; cauline leaves rhombic-oblong to lanceolate-linear. Flowering clusters (1–)2–4-flowered, arranged in long, slender, more or less interrupted spikes. Bracts longer or shorter than clusters, linear-lanceolate, abortive towards apex. Perianth lobes 2–5 mm., as long as or longer than diameter of fruit, often incurved, more or less keeled, especially in fruit. Fl. March–June.

Subsp. **maritima** (L.) Arcang., Comp. Fl. It. 593 (1882); Thell., Mém. Soc. Sci. Cherbourg 38 : 189 (1912). *B. vulgaris* L. var. *maritima* (L.) Boiss., Fl. 4 : 899 (1879). Perennial, rarely annual, glabrous or loosely pilose, branching from base. Perianth lobes about 2 mm., not or only rarely incurved, rarely cartilaginous at base, short-triangular at apex, slightly keeled or without keel. Clusters 1–2(–3)-flowered, in rather dense spikes, leafy to top or leafless at apex.

Hab.: Fields and roadsides. Acco Plain, Sharon Plain, Philistean Plain, Lower Galilee, Mt. Carmel, Esdraelon Plain, Samaria, Judean Mts., Judean Desert, Negev, Hula Plain, Lower Jordan Valley, Dead Sea area, Gilead, Moav.

Of the above subspecies which includes, according to Aellen (1938) and some others, also ssp. *foliosa* auct.; the following 3 varieties have been distinguished:

Var. **maritima.** *B. maritima* L., Sp. Pl. ed. 2, 322 (1762). *B. vulgaris* L. var. *glabra* (Del.) Aellen, Ber. Schweiz. Bot. Ges. 48 : 476 (1938). Bracts diminishing towards apex of inflorescences. Glabrous plants. Common.

Var. **pilosa** (Del. ex Moq.) Aellen, l.c. 474. Lower leaves and stem base rather hairy. Not common in this area.

Var. **orientalis** Moq. in DC., Prodr. 13, 2 : 56 (1849). *B. vulgaris* L. var. *foliosa* Aellen, l.c. 474. Glabrous plant, with conspicuous bracts all along the inflorescence, mostly overtopping clusters. Not rare.

Subsp. **macrocarpa** (Guss.) Thell., l.c. 190; Aellen, l.c. 478. Annual, glabrous, somewhat fleshy. Perianth lobes up to 5 mm., longer than diameter of fruit, more or less subulate, keeled, hooded, cartilaginous at base. Clusters large, 3–4-flowered, forming rather interrupted spikes.

Hab.: As above. Sharon Plain, Mt. Carmel, Judean Desert, N. Negev, Upper and Lower Jordan Valley.

Area of species: Mediterranean, Euro-Siberian and Irano-Turanian.

Trib. CHENOPODIEAE. Perianth generally herbaceous, with segments often free or united to middle. Stamens 1–5, free or united at base. Flowers without bracts. Leaves flat, variously shaped.

2. CHENOPODIUM L.

Annual or biennial herbs or small shrubs, glabrous or mealy. Stems usually grooved or angular. Leaves generally petiolate, entire or more frequently lobed or dentate. Inflorescences composed of small cymose clusters arranged in spike-like or paniculate racemes. Flowers all hermaphrodite or part of them pistillate only, sessile, bractless, green. Tepals 5 (rarely 2–3), free or united at base or to middle or nearly to apex. Stamens 5 (rarely 2–4), free or connate at base. Stigmas 2(–3–5), simple or rarely 2-lobed. Utricle frequently included but free, usually depressed-globular; pericarp thin, membranous. Seeds lenticular, horizontal, infrequently vertical; embryo peripheral. Over 200 species, cosmopolitan but mainly in temperate regions.

Literature: J. Murr, Versuch einer natürlichen Gliederung der Mittel-europäischen Formen des *Chenopodium album* L. *Ascherson-Festschrift* 216–230 (1904). T. Kowal, Klucz do oznaczania nasion rodzajów *Chenopodium* L. i *Atriplex* L. (A key for the determination of the seeds of the genera *Chenopodium* L. and *Atriplex* L.). *Polsk. Towarz. Bot. (Monogr. Bot.)* 1 : 87–163 (1953). P. Aellen, in : Hegi, *Ill. Fl. Mitteleur.* ed. 2, 3, 2 : 569 et seq. (1960).

1. Leaves with yellowish glands on lower surface. Aromatic plants.   **1. C. ambrosioides**
 – Leaves glandless. Plants not aromatic or with evil odour   2
2. Leaves almost all dentate or lobed   3
 – Leaves predominantly entire or rarely with 1 or 2 lateral lobes   6
3. Flowers (except the terminal ones) 2–3-merous.   **7. C. rubrum**
 – Flowers 5-, rarely 4-merous   4
4. Leaves strongly dentate with many unequal, coarse, acute teeth. Seeds not shining, with acute margins.   **6. C. murale**
 – Leaves usually with few obtuse teeth. Seeds with obtuse margins   5
5. Leaves mostly rhombic, not much longer than broad or as broad as or broader than long.   **5. C. opulifolium**
 – Leaves at least one and a half times as long as broad, often 2.5–5 cm. long.   **4. C. album**
6 (2). Plants smelling strongly of decaying fish. Stems and leaves covered with mealy scales.   **3. C. vulvaria**
 – Plants not as above   7
7. Perianth open in fruit; tepals membranous, glabrous.   **2. C. polyspermum**
 – Perianth closed in fruit; tepals with a mealy cover.   **4. C. album**

Sect. AMBRINA (Sp.) Hook. f. in Benth. et Hook. f., Gen. Pl. 3 : 51 (1880). Perianth (4–)5-parted, somewhat spongy; segments free or united to middle. Stigmas 3–5. All parts of plant with club-shaped glandular hairs. Aromatic plants.

1. Chenopodium ambrosioides L., Sp. Pl. 219 (1753); Boiss., Fl. 4 : 904 (1879). [Plate 197]

Annual or biennial, velvety-puberulent, aromatic, 25–90 cm. Stems more or less erect, simple or branching. Leaves short-petioled, oblong to lanceolate, sinuate-dentate, upper leaves entire or nearly so, all with yellowish glands on lower surface. Flowers in dense clusters, forming elongated spikes arranged in a long, more or less leafy panicle. Perianth about 1.5 mm., enclosing fruit; tepals 4–5, free or united at base, not keeled, velvety-glandular. Stamens 4–5. Seeds mostly horizontal, glossy. Fl. March–October (–December).

Hab.: Roadsides and waste places. Acco Plain, Sharon Plain, Philistean Plain, Mt. Carmel, Esdraelon Plain, Samaria, Judean Mts., N. Negev, Upper and Lower Jordan Valley.

Area: Pluriregional; origin probably Trop. America.

Used medicinally mainly as a vermifuge; in India essential oil is extracted from this plant; leaves used as condiment in soups (S. America).

Sect. CHENOPODIASTRUM Moq. in DC., Prodr. 13, 2 : 61 (1849). Perianth 5-parted; segments flat or keeled on back. Stigmas 2, short. Seeds all horizontal, black.

2. Chenopodium polyspermum L., Sp. Pl. 220 (1753); Boiss., Fl. 4 : 900 (1879). [Plate 198]

Annual, almost glabrous, not mealy, up to 1 m. Stems decumbent to erect, angular. Leaves up to 6 cm., petiolate, the lower ones ovate or oblong-ovate, the upper oblong to lanceolate, cuneate at base, acute or obtuse, entire or obsoletely 2–3-lobed. Cymes of many small glomerules, usually subtended by longer leaves, arranged in elongated, rather loose, divaricately branched inflorescences. Tepals 5, about 1–1.5 mm., united at base only, oblong-ovate, broadly membranous at margin, not keeled, open in fruit. Utricle about 1 mm., more or less lenticular, with easily separable pericarp. Seeds discoid, obscurely punctate, black. Fl. Spring.

Hab.: Waste places and cultivated ground. Judean Mts. Rare.

Area: Euro-Siberian, Mediterranean and Irano-Turanian.

3. Chenopodium vulvaria L., Sp. Pl. 220 (1753); Boiss., Fl. 4 : 901 (1879). [Plate 199]

Annual, evil smelling, with mealy indumentum, 10–50 cm. Stems ascending to erect or prostrate, branched, striate-angular. Leaves mostly 0.5–3 × 0.5–3 cm., petiolate, rhombic- or deltoid-ovate, entire or with 2 lateral lobes near base, mealy especially beneath. Flower clusters all of hermaphrodite flowers or with a few pistillate ones, leafless, arranged in raceme-like, axillary and terminal inflorescences. Tepals 5, about 1 mm., united at base, not keeled, mealy, enclosing fruit. Filaments dilated at base into a very short ring. Style short with filiform stigmas. Pericarp very thin. Seeds 1–1.5 mm. in diam., finely papillose-verrucose, black. Fl. April–October.

Hab.: Waste places, fields, gardens and refuse heaps. Acco Plain, Sharon Plain, Philistean Plain, Upper Galilee, Esdraelon Plain, Samaria, Judean Mts., Coastal and N. Negev, Ammon.

Area: Euro-Siberian, Mediterranean, Irano-Turanian, extending into the Saharo-Arabian region.

Formerly used medicinally; heavy odour due to trimethylamine.

4. Chenopodium album L., Sp. Pl. 219 (1753); Boiss., Fl. 4 : 901 (1879). [Plate 200]

Annual, more or less mealy, whitish, grey or greenish, 30–80 cm. Stems erect, branched, rather rigid, sometimes reddish. Leaves 1–7 × 0.3–4 cm., petiolate, the middle and lower ones rhombic, ovate or often oblong or lanceolate, the upper oblong-lanceolate to linear, cuneate, rarely subhastate, dentate-sinuate, the uppermost and not rarely all leaves almost entire, with mealy under-surface. Flower clusters densely crowded and arranged in paniculately branched, elongated inflorescences, leafy in lower part. Tepals 5, keeled, hooded, scarious-margined, mealy, greenish, enclosing fruit. Seeds about 1.5 mm. in diam., lenticular, obtusely margined, smooth or very finely furrowed, blackish. Fl. May–November.

Hab. : Fields, gardens, and waste places. Acco Plain, Sharon Plain, Philistean Plain, Upper and Lower Galilee, Mt. Carmel, Esdraelon Plain, Samaria, Judean Mts., Judean Desert, Hula Plain, Negev, Upper and Lower Jordan Valley, Dead Sea area, Ammon.

Area : Pluriregional.

Highly polymorphic; divided by various authors into subspecies, according to colour and mealiness, size and shape of leaves, structure of inflorescence, configuration of testa, etc. The local populations comprise at least 6 forms, the taxonomic status of which cannot be established without an experimental study.

A common weed in tilled or irrigated crops. Formerly cultivated as a bread plant because of its highly nutritive seeds; has a high vitamin C content and is used as a salad plant; also used in medicine.

5. Chenopodium opulifolium Schrad. ex Koch et Ziz, Cat. Fl. Palat. 6 (1814) ; Boiss., Fl. 4 : 901 (1879). [Plate 201]

Annual, more or less mealy, green to almost white, 30–80 cm. Stems erect, branching or almost simple. Leaves 0.6–3 × up to 2 cm., often nearly as broad as or broader than long, petiolate, rhombic-ovate, obscurely 3-lobed or more or less triangular, uppermost leaves oblong-lanceolate, all cuneate at base, obtuse to acute, subentire to repand-dentate, mealy mainly on under-surface and especially when young. Cymes of dense glomerules arranged in narrow panicle-like inflorescences. Tepals 5, about 1 mm., keeled, mealy. Stamens 5. Utricle enclosed by appressed tepals. Seeds 1–1.5 mm., horizontal, discoid, black, obtusely keeled, finely dotted-papillose. Fl. April–October.

Hab. : Fields, gardens and waste places. Acco Plain, Sharon Plain, Philistean Plain, Upper and Lower Galilee, Esdraelon Plain, Samaria, Judean Mts., Hula Plain, Upper Jordan Valley, and probably also in Transjordan.

Area : Mediterranean, Irano-Turanian and Euro-Siberian. Also occurring in almost all tropical and temperate regions of Africa.

Used as a pot herb.

6. Chenopodium murale L., Sp. Pl. 219 (1753); Boiss., Fl. 4 : 902 (1879). [Plate 202]

Annual, green, sparingly mealy, 25–70 cm. Stems ascending to erect, generally branching, more or less angular and thickened at base. Leaves 1–7 × 0.5–4 cm., petiolate, rhombic-ovate to rhombic-oblong, cuneate at base, acute to acuminate at apex, irregularly, unequally and acutely serrate-dentate, glabrous or somewhat mealy, mainly on lower surface. Inflorescences axillary and terminal, paniculate, divaricately branched, with dense or loose clusters. Flowers hermaphrodite. Tepals 5, green, bluntly keeled, more or less enclosing fruit. Pericarp membranous, hardly separable from seed. Seeds 1–1.5 mm. in diam., lenticular, acutely keeled at margin, minutely pitted, black. Fl. February–September.

Hab.: Roadsides, waste places, refuse heaps and irrigated fields. Acco Plain, Sharon Plain, Philistean Plain, Upper Galilee, Esdraelon Plain, Judean Mts., Judean Desert, W., N. and S. Negev, Hula Plain, Upper and Lower Jordan Valley, Dead Sea area, Gilead, Ammon, Moav. Very common.

Area: Pluriregional.

The 2 varieties of this species: var. *microphyllum* Boiss., l.c. and var. *humile* Peterm. (cited from Aschers. u. Graebn., Syn. 5, 1 : 35, 1913), should, in our opinion, be included within the variability range of the typical form.

Used as a salad herb.

Sect. CHENOPODIUM. Terminal flower of each cluster with 3–5-perianth segments free almost to base, and with horizontal seeds; the lateral flowers with 3 perianth segments connate to middle or to apex, and with vertical seeds.

7. Chenopodium rubrum L., Sp. Pl. 218 (1753); Boiss., Fl. 4 : 905 (1879). [Plate 203]

Annual, glabrous or nearly glabrous, usually reddish, 15–80 cm. Stems prostrate to erect, simple or branched, angular. Leaves up to 9 × 7 cm., long-petioled, much varying in size and shape, rhombic, ovate or triangular to lanceolate, more or less acute or acuminate, obscurely 3-lobed, sinuate-dentate, the uppermost often entire. Flower clusters crowded in branched, axillary or terminal, generally leafless inflorescences. Terminal flower of each glomerule with 3–5 almost free, strongly incurved tepals and horizontal seeds; other flowers with 2–3 membranous tepals connate almost to middle, 1–2 stamens and vertical seeds. Seeds about 1 mm. in diam., minutely pitted to nearly smooth, brown-black, glossy. Fl. March–May.

Hab.: Waste places, cultivated ground. Upper Jordan Valley. Rare.

Area: Pluriregional.

Trib. ATRIPLICEAE. Flowers mostly unisexual. Staminate flowers with perianth, ebracteolate; pistillate flowers generally naked, bracteolate. Stamens 1–5. Fruit laterally or dorsally compressed, included in the 2, partly connate bracteoles. Embryo circular.

3. ATRIPLEX L.

Perennial or annual herbs or shrubs, mostly with farinose or vesiculose indumentum. Leaves alternate or rarely opposite, sessile or petiolate, flat, green or greyish-green or mealy-white. Flowers unisexual (plants monoecious or dioecious), arranged in terminal or axillary clusters forming spike-like or paniculate inflorescences. Staminate flowers with 5-, rarely 3-parted perianth; stamens 5 or rarely 3, free or connate at base into a ring. Pistillate flowers usually in axillary clusters or solitary or also in spike-like or paniculate inflorescences, often without perianth, but with 2 herbaceous, membranous or leathery bracteoles (valves), connate at base or to above middle, entire, dentate or sinuate; ovary ovoid, globular or flattened; styles 2; stigmas 2, more or less free. Fruit an utricle enclosed in the 2 valves (bracteoles); pericarp membranous, free or slightly adhering to the vertical seed. Seeds erect or pendulous; embryo circular, surrounding the mealy endosperm.

About 120 species, mainly in deserts, salines and waste places, both in temperate and warm regions of the Old and New World. Australia is one of the larger centres of this genus.

Some of the species described below are very important in the vegetal landscape of deserts. Most of them are potential desert pasture plants. Some species of *Atriplex* have recently been introduced to this country from Australia for pasturing purposes. They are not recorded here.

Literature : P. Aellen, Die *Atriplex*-Arten des Orients. *Bot. Jahrb.* 70 : 1–66 (1939).

1. Valves (pistillate bracteoles) all or part ovate or orbicular or cordate, slightly dentate
   or entire, 0.5–1.5 cm. in diam. Annuals   2
 – Valves deltoid, rhombic or semiorbicular or lanceolate, rarely orbicular, generally
   smaller than in above. Annuals or perennials   3
2. Fruiting valves of two forms : small ones about 5 mm., and larger up to 1.5 cm. in diam.
   on the same plant. Desert annuals.   **5. A. dimorphostegia**
 – Fruiting valves uniform, all 1–1.5 cm. in diam. Pistillate flowers of two forms : part with
   a 5-merous perianth, others with 2 membranous bracteoles.   **4. A. nitens**
3(1). Flowers in axillary clusters, never forming leafless, spike-like racemes   4
 – Flowers in axillary and/or terminal clusters, the latter forming continuous or interrupted
   spike-like or paniculate racemes   5
4. Leaves mostly oblong or lanceolate, not over 1 cm. broad, entire or sinuate-dentate.
   Valves entire or slightly toothed.   **10. A. semibaccata**
 – Leaves triangular-deltoid, usually 2 cm. or more broad, deeply and irregularly sinuate
   or toothed-lobed. Valves irregularly toothed-lobed.   **6. A. rosea**
5(3). Annuals. Lower leaves triangular-hastate to deltoid with 2 horizontal basal lobes,
   upper lanceolate-hastate to lanceolate-linear, entire. Pistillate and staminate flowers in
   remote clusters forming long, loose panicles. Valves (in ours) 1–2 mm.   **9. A. hastata**
 – Perennials or annuals not as above   6
6. Leaves entire or remotely sinuate-dentate or obscurely lobed. Shrubs or perennials
   with woody base   7
 – Lower and sometimes also other leaves deeply toothed or laciniate-lobed; teeth mucronate. Annuals   9
7. Valves broadly ovate or semiorbicular, entire or toothed, without tubercles or scales on
   back. All leaves petiolate, mostly entire or slightly lobed. Shrubs, 1–2 m.
                                                                               1. A. halimus
 – Valves deltoid or rhombic, unequally toothed or lobed, with scales or tubercles on back.
   All leaves, or at least the upper ones, sessile   8
8. Leaves mostly oblong to linear-lanceolate (rarely ovate or orbicular but then very
   densely crowded). Valves longer than broad. Dwarf-shrubs.   **2. A. stylosa**
 – Leaves triangular-deltoid, truncate to hastate or cuneate at base, irregularly sinuate-
   dentate (in summer forms leaves small, almost entire, ovate to orbicular). Shrubs or
   perennial herbs with woody base.   **3. A. leucoclada**

9(6). Terminal clusters forming long, spike-like or paniculate inflorescences. Upper leaves
  lanceolate to linear, often entire, rarely short-dentate. Stems wand-like, little branching.
  8. A. lasiantha
- Terminal clusters forming rather short spikes. Leaves hastate-deltoid, most of them
  incised-laciniate. Stems much branching. **7. A. tatarica**

Sect. CORIACEA Aellen, Bot. Jahrb. 70 : 8 (1939). Robust shrubs, 1–3 m. high. Leaves
large, leathery. Inflorescences leafless, mainly terminal and paniculate. Valves orbicular
or semiorbicular or broadly ovate, entire or dentate, not tuberculate or appendiculate
on back.

1. Atriplex halimus L., Sp. Pl. 1052 (1753); Boiss., Fl. 4 : 916 (1879). [Plate 204]
    Shrub with vesicular hairs, about 1–2 m. Stems erect, much branched, woody, terete
or angular, whitish. Leaves 1–6 × 0.5–4 cm., alternate, sometimes opposite below, ovate
to ovate-rhombic to triangular, sometimes cuneate or hastate at base, entire or obsoletely
repand-lobed or dentate, silvery-white, without prominent nerves, the upper narrower,
lanceolate; petiole 0.3–1.2 cm. Flower clusters densely spicate; spikes in terminal, almost
leafless panicles. Staminate flowers inconspicuous, with 5 membranous tepals, generally
at top of cluster. Pistillate flowers at base of cluster, with valves 4–5 mm. long and
broad, not stipitate, scarcely united at base, orbicular or semiorbicular or reniform to
short-cuneate at base, entire or dentate, smooth or reticulate but not tuberculate.
Stigmas filiform, free. Seeds 1–2 mm. in diam., vertical, lenticular, dark brown. Fl.
April–October.

Var. **halimus.** Leaves up to 3 cm. long, short-petioled. Fruiting valves entire or obso-
letely toothed.
    Hab.: Salines, wadi beds and sandy soils. Sharon Plain, Philistean Plain, Judean
Mts., Judean Desert, Negev, Upper and Lower Jordan Valley, Dead Sea area, Arava
Valley, Moav, Edom.

Var. **schweinfurthii** Boiss., Fl. 4 : 916 (1879). Leaves larger, about 4 × up to 3.5 cm.,
long-petioled, almost hastate at base, sinuate to dentate. Fruiting valves toothed.
    Hab.: Wadi beds and sandy soils. Sharon Plain, Philistean Plain, Judean Desert,
S. Negev, Upper and Lower Jordan Valley, Dead Sea area, Arava Valley.
    Transitional forms between the two varieties, as well as other as yet undescribed
forms, occur in both the coastal plain and the desert.
    Area of species : Mediterranean and Saharo-Arabian.

    A. halimus is especially common in inundated saline depressions and around oases of
the Jordan Valley.
    Rather a palatable browse shrub; the leaves are sometimes eaten by hungry shepherds;
the salt content of the leaves increases with the aridity of the habitat, which makes the
plant less palatable. The ash of *A. halimus* is used for manufacture of soap.
    Believed to be the מַלּוּחַ of the Bible (Job. xxx : 4).

Sect. STYLOSA Aellen, Bot. Jahrb, 70 : 16 (1939). Half-shrubs and dwarf-shrubs.
Leaves oblong, deltoid to orbicular, mealy-scurfy. Inflorescences axillary and terminal,

forming interrupted spikes. Valves deltoid, rarely campanulate, mostly tuberculate or appendiculate on back.

2. Atriplex stylosa Viv., Pl. Aeg. 23 (1831); Boiss., Fl. 4 : 917 (1879). *A. palaestinum* Boiss., Diagn. ser. 1, 12 : 96 (1853) et Fl. 4 : 914 (1879). *A. alexandrinum* Boiss., Fl. 4 : 914 (1879). [Plate 205]

Dwarf-shrub, papillose mealy-canescent, 10–30 cm. Stems numerous, erect or ascending, branching, leafy, terete, white. Leaves (0.5–)2–4 × 0.5–1 cm., sessile, oblong to oblong-linear, rarely ovate to orbicular, somewhat tapering at base, obtuse, entire or sinuate-repand, the upper ones linear, acute. Flowers arranged in clusters of two kinds : those mixed of pistillate and staminate flowers form terminal spike-like or branched inflorescences; those consisting of pistillate flowers only are crowded in the leaf axils. Fruiting valves up to 5–7 × 3–5 mm., free up to middle, deltoid or triangular to rhombic-cuneate at base, with 1 tooth on each side, mostly with tubercles, scales or appendages on back. Stigmas free. Seeds about 1.5 mm. in diam., brownish. Fl. March–May (–August).

Eig, Palest. Journ. Bot. Jerusalem ser. 3 : 123 (1945), divides this species into three varieties which should be named as follows :

Var. **stylosa**. *A. parvifolium* Lowe var. *palaestinum* (Boiss.) Eig, l.c. Whitish-mealy, with dense spikes and with less prominent fruiting valves.

Var. **alexandrina** (Boiss.) Zoh. (comb. nov.). *A. parvifolium* Lowe var. *alexandrinum* (Boiss.) Eig, l.c. Plants greenish with more diffuse, prostrate and slender branches.

Var. **conferta** (Eig) Zoh. (comb. nov.). *A. parvifolium* Lowe var. *confertum* Eig, l.c. Plants with dense, nearly round or broadly ovate and minute leaves, 0.5–1 cm. in diam.

Further observations are needed to confirm the constancy of these varieties.

Hab. (of species) : Deserts, particularly on gypsaceous ground. Judean Desert, Negev, Lower Jordan Valley, Dead Sea area, Arava Valley, Ammon, Moav, Edom.

Area of species : Saharo-Arabian.

A. stylosa is very abundant in the eastern part of the Judean Desert. It is the dominant plant in the *Atriplex palaestina* community and is also a codominant in the *Suaeda asphaltica* community on highly gypsaceous ground.

3. Atriplex leucoclada Boiss., Diagn. ser. 1, 12 : 95 (1853) et Fl. 4 : 915 (1879); Aellen, Bot. Jahrb. 70 : 22 (1939). [Plates 206–208]

Perennial herb with woody base or half-shrub, mealy-canescent, 30–100 cm. Stems prostrate to erect. Leaves 0.8–3 × 0.2–2.5 cm., triangular-deltoid, cuneate or truncate or hastate at base, acuminate at apex, sinuate-dentate; lower leaves short-petioled, upper mostly sessile. Flowers in terminal and axillary inflorescences; the terminal of mixed staminate and pistillate flower clusters, spike-like or paniculate; the axillary of pistillate flowers only, solitary or in groups. Staminate perianth membranous. Valves of solitary pistillate flowers 3–6 × 2–6 mm. (in fruit), campanulate, upper part of valve free, 3–5-toothed or -lobed, mostly without tubercles; valves of spicate pistillate flowers with well developed tubercles, deltoid, 3–5-lobed, each of the lateral lobes 2–4-toothed, terminal lobe longer than lateral ones. Stigmas free to base. Pericarp membranous. Seeds 2 mm. across, dark brown. Fl. March–October.

Very polymorphic. The following varieties are not clearly delimited and await a more thorough study. Summer forms of this species differ greatly from winter forms in their smaller, more rounded, less dentate and almost sessile leaves. The main differentiating character between the varieties is the form of the valves.

Var. **leucoclada**. *A. leucoclada* Boiss. ssp. *eu-leucoclada* Aellen, l.c. [Plate 206]. Valves 3-lobed, the terminal lobe longer than the 2-toothed lateral ones. Plate 206 represents a summer form of this variety.

Hab.: Deserts. Judean Desert, Negev, Dead Sea area, Ammon Desert.

Area: E. Saharo-Arabian.

Var. **turcomanica** (Moq.) Zoh. (comb. nov.). *A. leucoclada* Boiss. ssp. *turcomanica* Aellen, l.c. *A. laciniata* L. var. *turcomanica* Moq. in DC., Prodr. 13, 2 : 93 (1849). [Plate 207]. Valves 3-lobed, with lateral lobes 3–5-toothed. Plate 207 represents a spring form of this variety.

Hab.: Deserts. Negev, Dead Sea area, Transjordan.

Area: E. Saharo-Arabian and W. Irano-Turanian.

Var. **inamoena** (Aellen) Zoh. (comb. et st. nov.). *A. inamoena* Aellen, Bot. Jahrb. 70 : 20 (1939). [Plate 208]. Valves usually 5-lobed, the terminal lobe somewhat longer. Plate 208 represents a summer form of this variety.

Hab.: Wadi beds. C. and S. Negev, Arava Valley, Moav Desert.

Area: E. Saharo-Arabian and Irano-Turanian.

Sect. ATRIPLEX. Sect. *Leiotheca* Aellen, Bot. Jahrb. 70 : 25 (1939). Annuals. Fruiting valves orbicular to cordate, membranous, netted-veined, almost entire, free to base, without tubercles or appendages.

4. Atriplex nitens Schkuhr, Bot. Handb. 3 : 541 (1803); Boiss., Fl. 4 : 908 (1879). [Plate 209]

Annual, 0.5–2 m. Stems erect, simple or branching, angular. Leaves large, 5–10 cm., alternate, long-petioled, triangular, truncate or cordate to hastate at base, sinuate-lobed or dentate, rarely entire, scurfy-mealy or glabrous beneath; the upper ones oblong-lanceolate, tapering. Racemes loose, simple or paniculate, leafless. Staminate flowers with 5-lobed membranous perianth. Pistillate flowers of two forms: part with 5-merous perianth and horizontal seed, others with 2 glabrous, membranous, ovate or orbicular, obtuse or more or less acute, reticulately veined, 0.5–1.5 cm. long valves, the larger with vertical seed. Stigmas free to base. Seeds 1–5 mm. in diam., brown or black. Fl. August–September.

Hab.: Waste places. Sharon Plain, Samaria, Judean Mts. Very rare (most probably casual).

Area: Irano-Turanian, Euro-Siberian and Mediterranean.

5. Atriplex dimorphostegia Kar. et Kir., Bull. Soc. Nat. Mosc. 15 : 438 (1842); Boiss., Fl. 4 : 909 (1879). [Plate 210]

Annual, 20–50 cm. Stems prostrate or ascending, white, glabrous, much branching from base. Leaves 1–6 × 1–3 cm., alternate, ovate or deltoid, truncate or short-cuneate at base, acute or obtuse, entire or somewhat irregularly sinuate-dentate, scurfy-mealy beneath, almost glabrous above. Clusters axillary, 5–12-flowered, the lower ones with pistillate flowers only, the upper bearing also staminate ones, terminal clusters spicate. Perianth of staminate flowers membranous, yellowish. Valves of pistillate flowers of two forms: those of the lower clusters up to 1.5 × 1 cm., stipitate, orbicular-cordate, entire or slightly sinuate-lobed, more or less smooth; those of the upper ones smaller, 5 × 3 mm., triangular-ovate with 1–3 teeth on each side, with crested back. Stigmas free to base. Seeds of two forms, smaller 1–? mm. in diam., shining, dark brown and larger, 2–2.5 mm. in diam., opaque, yellowish-brown. Fl. February–May.

Hab.: Sandy and often saline sites in desert depressions. Negev, Arava Valley, Edom.

Area: Irano-Turanian and Saharo-Arabian.

Sect. ROSEA Aellen, Bot. Jahrb. 70:39 (1939). Annuals. Fruiting valves becoming indurated at base, with fan-shaped appendages, terminating with a leaf-like body, or only with tuberculate appendages, rarely without appendages.

6. Atriplex rosea L., Sp. Pl. ed. 2, 1493 (1763); Boiss., Fl. 4:911 (1879). [Plate 211]

Annual, 30–80 cm. Stems leafy, erect or ascending, divaricately much branched, more or less mealy and canescent, later indurated. Leaves up to 6 × 3 cm., triangular-deltoid to ovate-rhombic, unequally sinuate or dentate; upper leaves ovate-oblong. Clusters axillary, forming leafy racemes, sometimes leafless at ends of branches, 2–20-flowered; upper clusters of staminate flowers together with some pistillate ones, the lower ones of pistillate flowers only. Fruiting valves 0.4–1 cm., leathery, whitish, triangular-deltoid or rhombic, somewhat cuneate and indurated at base, irregularly toothed-lobed, reticulately veined, with large, smooth or tuberculate appendages on the back. Stigmas 2, free to base. Pericarp membranous. Seeds 1.5–3 mm. in diam., lenticular, dark brown, smooth, shining. Fl. June–October.

Hab.: Waste places and around refuse heaps. Acco Plain, Sharon Plain, Upper and Lower Galilee, Samaria, Judean Mts., Judean Desert, Negev, Hula Plain, Lower Jordan Valley, Edom.

Area: Mediterranean, extending into Irano-Turanian and Euro-Siberian territories.

7. Atriplex tatarica L., Sp. Pl. 1053 (1753); Boiss., Fl. 4:910 (1879). [Plate 212]

Annual, scurfy-mealy, later glabrescent or glabrous, 30–80 cm. Stems erect or ascending, richly branching, indurated, angular-striate. Leaves up to 6 × 3 cm., long-petioled, triangular-deltoid to lanceolate-hastate, rounded or acuminate at apex, mostly irregularly sinuate-dentate or incised to laciniate (lobes sometimes lobulate), prominently nerved and scurfy-canescent on both sides. Flowers in clusters and solitary: the clusters consist of staminate and pistillate flowers, forming mostly leafless short spikes or panicles; the solitary flowers are pistillate, scattered in the axils of middle and upper leaves. Perianth of staminate flowers with 5 obovate, obtuse tepals. Fruiting valves 5–8 × 2–5 mm., orbicular or oblong-rhombic, dentate or entire, with or without appendages on back. Stigmas 2, connate at base. Seeds 1.5 mm. across, lenticular, black-brown. Fl. January–October.

Hab.: Roadsides and waste places. Judean Mts., Judean Desert, Lower Jordan Valley, Dead Sea area, Arava Valley, Ammon.

Area: Euro-Siberian, Mediterranean and Irano-Turanian.

Var. **desertorum** Eig, Palest. Journ. Bot. Jerusalem ser. 3 : 122 (1945) is referring to lower plants with cuneate-triangular or cuneate-oblong, more or less dentate leaves and short spikes; this form can hardly be separated from the typical form.

8. Atriplex lasiantha Boiss., Diagn. ser. 1, 12 : 95 (1853). *A. tataricum* L. var. *virgatum* Boiss., Fl. 4 : 910 (1879). *A. tataricum* L. var. *hierosolymitanum* Eig, Palest. Journ. Bot. Jerusalem ser. 3 : 122 (1945). [Plate 213]

Annual, up to 1 m. Stems erect, rigid, wand-shaped, whitish, sparsely branched. Leaves up to 3–4 × 0.5–1 cm., the lower and middle ones oblong-deltoid, cuneate, tapering to a petiole, mucronate, irregularly dentate; upper leaves sessile, lanceolate to lanceolate-linear, entire, mealy-canescent. Flowers in clusters and solitary : the clusters consist of staminate and pistillate flowers forming long, interrupted, spike-like or paniculate, leafless inflorescences; the solitary ones are pistillate and scattered in the axils of middle and upper leaves. Staminate flowers with membranous yellowish perianth. Valves of axillary pistillate flowers 3–7 × up to 5 mm., rhombic-deltoid to oblong-deltoid, upper half free with 1 or more lateral teeth, lower part mostly furnished with tubercles; those of the terminal pistillate flowers considerably smaller. Stigmas free to base. Seeds 1 mm. in diam., dark brown. Fl. (May–) June–October.

Hab.: Waste places and roadsides. Sharon Plain, Judean Mts., Negev, Dead Sea area, Gilead, Moav, Edom.

Area: E. Mediterranean and W. Irano-Turanian.

Sect. PATULA Aellen, Bot. Jahrb. 70 : 50 (1939). Sect. *Teutliopsis* Dumort., Florula Belg. 20 (1827) p.p. Annuals. Fruiting valves green, herbaceous, mostly with 1–2, simple or forked tubercles.

9. Atriplex hastata L., Sp. Pl. 1053 (1753); Boiss., Fl. 4 : 909 (1879). [Plate 214]

Annual, 30–100 cm., green or scurfy-canescent, glabrescent or glabrous. Stems branching, often erect or ascending. Leaves mostly alternate, triangular-hastate to deltoid, sometimes cuneate or cordate at base, entire or slightly sinuate, the lower up to 10 × 7 cm., the upper lanceolate-hastate to lanceolate-linear, acute-mucronate. Flowers in axillary or terminal clusters forming long spikes or panicles; pistillate flowers at the base, staminate at the top of each cluster. Valves of pistillate flowers 1–4 × 1–3 mm., triangular-deltoid to elliptical-deltoid or lanceolate, entire or repand-dentate, connate below middle, with smooth or tuberculate surface. Stigmas free to base. Seeds vertical, lenticular, 1–2.5 mm. in diam., dark brown; embryo annular, surrounding endosperm. Fl. June–November.

Var. **microtheca** Schum., Enum. Pl. Saell. 1 : 299 (1801). *A. patulum* L. var. *palaestinum* Eig, Palest. Journ. Bot. Jerusalem ser. 3 : 121 (1945). Climbing or trailing plants with very long and richly branching inflorescences and minute fruits. Lower leaves opposite, upper ones alternate. Fruiting valves 1–2 mm., mostly narrowly lanceolate.

Hab.: Saline meadows and marsh edges. Acco Plain, Sharon Plain, Philistean Plain, Esdraelon Plain, Hula Plain, Upper and Lower Jordan Valley, Dead Sea area.

Area: Mediterranean, Euro-Siberian and Irano-Turanian; also occurring in other regions, but probably introduced there.

Sect. SEMIBACCATA Ulbrich in Engl. et Prantl, Nat. Pflznfam. ed. 2, 16c: 515 (1934). Fruiting valves united to about middle, baccate or strongly indurated and turning brown or black.

10. Atriplex semibaccata R. Br., Prodr. Fl. Nov. Holl. 1 : 406 (1810). [Plate 215]

Annual, green to greyish, with vesicular hairs, 50–80 cm. Stems procumbent or ascending, much branched. Leaves small, about 2–4 × 0.6–1 cm., oblong to lanceolate to ovate-cuneate, entire or sinuate-toothed. Staminate flowers surrounded by a few pistillate ones, in small, globular clusters in axils of upper leaves; other pistillate flowers in very small groups in axils of lower leaves. Fruiting valves more or less deltoid or rhombic, 2–5 × 2–4 mm., connate in their lower half, somewhat thickened and prominently 3-nerved, entire or toothed, grey-green to blackish, with or without tubercles. Seeds 1–2 mm. in diam., brown. Fl. July–December.

Var. semibaccata. Fruiting valves 4–5 mm., green, yellow or brown, without tubercles. Leaves oblanceolate or obovate, entire.

Hab.: Waste places. Sharon Plain, Philistean Plain.

Var. microcarpa (Benth.) Aellen, Bot. Jahrb. 68 : 412 (1938). *A. microcarpa* Benth., Fl. Austral. 5 : 176 (1870). Fruiting valves 2–3 mm., obtuse or with 3 entire or toothed lobes, the middle one larger. Leaves oblong-ovate, cuneate, sinuate-toothed.

Hab.: As above. N. and C. Negev.

Var. melanocarpa Aellen, Bot. Jahrb. 68 : 411 (1938). Differs from the above varieties by the black fruiting valves.

Hab.: As above. N. Negev.

Area of species: Australia. Probably introduced in Palestine as a pasture plant; subspontaneous.

4. HALIMIONE Aellen

Annual or perennial herbs or shrubs, sometimes mealy. Stems angular. Leaves opposite or alternate, sessile or petiolate, ovate-spatulate to oblong-elliptical, entire or nearly so. Flowers unisexual, arranged in monoecious clusters forming loose paniculate inflorescences. Staminate flowers with 4–5-fid perianth and 4–5 stamens. Pistillate flowers without perianth, but with 2 obdeltoid, 3-lobed bracteoles united nearly to top; style short to 0, stigmas 2. Fruit an utricle enclosed in the bracteoles; pericarp thin, membranous, adnate to the bracteoles. Seeds vertical, smooth. Differs from *Atriplex* by the fruiting valves (bracteoles) connate almost to the apex.

Three species, in the Mediterranean, Euro-Siberian, and Irano-Turanian regions.

1. Halimione portulacoides (L.) Aellen, Verh. Naturf. Ges. Basel 49 : 126 (1938). *Atriplex portulacoides* L., Sp. Pl. 1053 (1753); Boiss., Fl. 4 : 913 (1879). *Obione portulacoides* (L.) Moq., Chenop. Monogr. Enum. 75 (1840). [Plate 216]

Shrubby perennial, more or less mealy, glabrescent, 40–100 cm. Stems procumbent, indurated below. Branches ascending, striate to angular. Leaves rather fleshy, 1–6 × 0.3–1.5 cm., opposite, short-petioled, oblong-elliptical to linear-lanceolate, tapering-cuneate at base, slightly acuminate to obtuse or apiculate at apex, grey-green. Flower clusters arranged in loose or dense, terminal and axillary spikes, forming paniculate inflorescences. Bracteoles 3–4 mm., obdeltoid-cuneate, 3-lobed with middle lobe equal to or smaller than lateral ones, green-canescent. Seeds about 1.5 mm. in diam. Fl. June–November.

Hab. : Saline marshes. Acco Plain, Sharon Plain, Beit Shean Valley, Moav.

Area : W. Euro-Siberian, Mediterranean and Irano-Turanian; also in S. Africa and N. America.

Trib. CAMPHOROSMEAE. Leaves sessile, mostly linear, entire, usually covered with long, silky or felty hairs. Flowers hermaphrodite, often intermixed with unisexual ones, the latter often with rudiments of stamens or pistils. Perianth 4–5-merous, often changing in fruit. Bracteoles 0. Embryo annular or horseshoe-shaped.

5. PANDERIA Fisch. et Mey.

Herbaceous silky to villose annuals with aspect of *Kochia*. Leaves alternate, sessile or subsessile, mostly linear, entire. Flowers hermaphrodite or hermaphrodite and pistillate together, solitary or in clusters of 2–4, forming a spike-like inflorescence. Perianth urceolate, with 5 equal lobes connivent in fruit, each producing a triangular tubercle or minute wing on back. Stamens 5, inserted at bottom of perianth; anthers ovoid, exserted; staminodes 0. Ovary ovoid; style bearing 2 long subulate stigmas. Utricle included; pericarp membranous. Seeds vertical; endosperm central; embryo almost annular.

Four species in the Irano-Turanian region.

1. Panderia pilosa Fisch. et Mey., Ind. Sem. Hort. Petrop. 2 : 46 (1835); Boiss., Fl. 4 : 919 (1879). [Plate 217]

Greenish, long-hairy annual, 10–20 cm. Stems decumbent or ascending, simple or branched at base. Leaves 0.5–1.5 × 0.1–0.5 cm., sessile, linear to oblong-lanceolate and elliptical, tapering at base, more or less acute, entire, densely hairy. Bracts (floral-leaves) linear, short but exceeding flowers. Flowers 2–2.5 mm., sessile in clusters of 2–3 (–4), forming short, more or less dense, leafy spikes. Perianth hairy, somewhat accrescent in fruit; lobes one quarter to one third the length of tube, with an appendage or wing arising from back of each lobe. Fl. May–September.

Hab. : Saline deserts and wastes. Moav. Very rare.

Area : W. Irano-Turanian.

6. Chenolea Thunb.

Low, woolly-canescent shrubs. Leaves alternate, entire. Flowers hermaphrodite, rarely polygamous, solitary or clustered, sessile, without bracteoles. Perianth persistent, small, urn-shaped, with 5 short, fleshy, woolly, somewhat keeled lobes not hardening in fruit and not producing appendages. Stamens inserted at bottom of perianth; anthers exserted. Ovary ovoid; style divided into 2–3 filiform lobes. Utricle included in the perianth; pericarp membranous. Seeds horizontal with membranous testa; endosperm little or 0; embryo peripheral.

Some 3 species, in S. Africa and in the Saharo-Arabian region.

1. Chenolea arabica Boiss., Diagn. ser. 1, 12 : 97 (1853) et Fl. 4 : 922 (1879). [Pl. 218]

Woolly-canescent dwarf-shrub, 10–40 cm. Stems woody at base; branches prostrate to ascending. Leaves 0.5–1 × 0.1–0.2 cm., sessile, oblong-linear to oblong, obtuse, puberulent, floral leaves subglabrous. Flower clusters forming dense, leafy, 5–10 cm. long spikes. Flowers hermaphrodite. Perianth about 3 mm., woolly, with 5 hemispherical, obtuse lobes. Style divided into 2 filiform stigmas. Utricle about 3 mm., included in the unchanged but closed perianth. Fl. March–July.

Hab. : Steppes and deserts, especially on gypsaceous ground. Judean Desert, Negev, Lower Jordan Valley, Arava Valley, Ammon, Moav, Edom.

Area : E. Saharo-Arabian.

One of the leading species of the *Chenolea arabica* community; very common in the Judean Desert.

7. Bassia All.

Annual or perennial herbs or shrubs. Leaves alternate, sessile, entire, flat or semiterete, usually linear, oblong to lanceolate. Flowers frequently hermaphrodite or pistillate by abortion, axillary and terminal, without bracteoles. Perianth urceolate, 5-lobed, more or less hairy. Stamens (3–)5; anthers oblong. Style divided into 2–3 filiform stigmas. Fruiting perianth with connivent lobes, mostly furnished with prickles or tubercles. Utricle included, depressed, with membranous pericarp. Seeds usually horizontal, with membranous testa; endosperm copious to 0; embryo annular.

About 10 species mainly in the Saharo-Arabian and the Irano-Turanian regions.

1. Flowers hidden in thick white fleece. Fruiting perianth usually not spiny or with very short, curved protuberances. **2. B. eriophora**
– Flowers not hidden in fleece. Fruiting perianth with long, yellow, spreading, straight spines. **1. B. muricata**

1. Bassia muricata (L.) Aschers. in Schweinf., Beitr. Fl. Aethiop. 1 : 187 (1867). *Salsola muricata* L., Mant. 54 (1767). *Kochia muricata* Schrad., Neues Journ. Bot. Schrad. 3 : III, IV et 86 (1809); Boiss., Fl. 4 : 926 (1879). [Plate 219]

Annual, densely villose, 10–50 cm. Stems usually many, erect or decumbent, branching from base, indurated in lower part. Leaves 0.5–1.5 × 0.1–0.2 cm., linear-lanceolate to oblanceolate, densely hairy. Flowers in clusters subtended by oblong bracts and forming loose, leafy spikes; each cluster consisting of 1 pistillate and 1–2 hermaphrodite flowers. Perianth of 5 tepals connate to middle. Fruiting perianth becoming indurated, tepals furnished with a 3–4 mm. long spine, 2 to 3 times as long as fruit, rarely shorter (var. *brevispina* Bornm., Mitt. Thür. Bot. Ges. 30 : 82, 1913) or scarcely longer, spreading, straight, needle-shaped, yellow. Seeds 1 mm. across, discoid, greenish-grey, opaque, smooth. Fl. February–June.

Hab. : Steppes and deserts, mainly on sandy soil. W. and C. Negev, Lower Jordan Valley, Dead Sea area, Arava Valley, Moav, Edom.

Area : E. Saharo-Arabian and W. Irano-Turanian.

2. Bassia eriophora (Schrad.) Aschers. in Schweinf., Beitr. Fl. Aethiop. 1 : 187 (1867). *Kochia eriophora* Schrad., Neues Journ. Bot. Schrad. 3 : III, IV et 86, t. 3 (1809). *K. latifolia* Fresen., Mus. Senckenb. 1 : 179 (1834); Boiss., Fl. 4 : 927 (1879). *B. eriophora* var. *rosea* Eig, Palest. Journ. Bot. Jerusalem ser. 3 : 125 (1945). [Plate 220]

Annual, fleecy-villose, 8–40 cm. Stems ascending to erect, branching from base. Leaves 0.6–2 × 0.2–0.3 cm., somewhat fleshy, oblong-linear or oblong-lanceolate to elliptical, thinly hairy. Bracts green or sometimes pink, soon deciduous. Flowers hidden in thick, white fleece and forming dense, leafy spikes. Lobes of fruiting perianth sometimes with a short protuberance (about 1.5 mm.), incurved at apex. Seeds about 1 mm. across, discoid, brown. Fl. March–June.

Hab. : Hot deserts. E. Judean Desert, S. Negev, Lower Jordan Valley, Dead Sea area, Arava Valley, deserts of Ammon, Moav and Edom.

Area : Sudanian and Saharo-Arabian, slightly extending into adjacent Irano-Turanian territories.

8. KOCHIA Roth

Herbs or shrubs. Leaves alternate, sessile, linear to lanceolate, more or less hairy. Flowers hermaphrodite and pistillate, solitary, in pairs or in clusters, sessile, ebracteolate. Perianth urceolate with 5 connivent lobes, coriaceous in fruit, with or without wings on back. Stamens 5, exserted, with linear filaments attached to bottom of perianth. Ovary ovoid; style divided into 2–3 filiform stigmas. Utricle depressed, included in the more or less indurated perianth; pericarp membranous. Seeds horizontal, rarely vertical; testa membranous; endosperm central; embryo annular, green.

About 80 species, mainly in temperate zones of the Old World and also in N. America and Australia.

1. Kochia indica Wight, Ic. Pl. Ind. Or. 5, 2 : 5, t. 1791 (1852). *Bassia joppensis* Bornm. et Dinsmore in Bornm., Repert. Sp. Nov. 17 : 274 (1921). [Plate 221]

Bushy annual, with long, whitish, shining hairs, 40–120 cm. Stems erect, much branched. Branches slender, somewhat recurved or spreading to erect. Leaves 0.5–1.5 × 0.1–0.5 cm., lanceolate to linear, more or less acuminate, soft-hairy mainly beneath. Floral branches white, with scattered, 1–3-flowered clusters arranged in loose, leafy spikes. Bracts linear, longer than cluster. Perianth lobes more or less woolly, connivent in fruit, with about 1 mm. broad, more or less triangular, scarious or foliaceous wings. Fl. September–November.

Hab.: Roadsides and waste places. Philistean Plain (rare), N. and C. Negev (very common). Probably casual. In the Negev it appeared for the first time in 1948.

Area: E. Sudanian (Nubo-Sindian) and Irano-Turanian (Egypt, Sinai, Palestine, Afghanistan, Punjab) and probably elsewhere.

A good pasture herb in deserts; it requires, however, a fairly high amount of moisture.

Trib. SALICORNIEAE. Bracts alternate or opposite, persistent or caducous. Flowers usually hermaphrodite, often arranged in 3-flowered dichasia, mostly embedded in a cavity of the stem. Perianth usually 3–4-merous, herbaceous, with mostly connate segments. Stamens 1–2. Embryo circular, semicircular or hook-shaped.

9. HALOPEPLIS Bge. ex Ung.-Sternb.

Shrubs or annual herbs. Branches not jointed. Leaves alternate (except for the lower ones which are almost opposite), fleshy, clasping, subglobular to ovoid. Inflorescences densely spicate. Bracts fleshy, spirally arranged. Flowers all hermaphrodite or some pistillate, clustered in groups of 3, more or less connate with each other and adnate to the bract subtending each group. Perianth small, obconical, of 3 segments, not winged. Stamens 1–2. Ovary pear-shaped, somewhat laterally compressed; stigmas 2, subulate. Utricle included, with membranous pericarp. Seeds ellipsoidal; endosperm central, abundant; embryo hook-shaped.

Some 3 species in the Mediterranean and Irano-Turanian regions.

1. Halopeplis amplexicaulis (Vahl) Ung.-Sternb. ex Ces., Passer. et Gib., Comp. Fl. It. 271 (1869); Boiss., Fl. 4 : 934 (1879). *Salicornia amplexicaulis* Vahl, Symb. Bot. 2 : 1 (1791). [Plate 222]

Glaucous, glabrous annual, 10–30 cm. Stems thin, branching from base, often indurated below. Branches procumbent to ascending, whitish, not jointed. Leaves about 3 mm., alternate, clasping, almost globular or semiglobular, obtuse, with rudimentary blades. Flower clusters in short (0.5–1.5 cm. long), lateral and terminal, alternate, sessile, dense, oblong spikes. Bracts fleshy, ovate-orbicular, acute or acuminate. Flowers connate. Stamen 1. Seeds about 0.5 mm., with cylindrical papillae on back. Fl. May–August.

Hab.: Saline depressions. Acco Plain, Dead Sea area.

Area: Mediterranean, extending into some Saharo-Arabian territories.

10. HALOCNEMUM M.B.

Fleshy low shrubs. Stems branched, with numerous bud-like branches. Leaves opposite, with rudimentary blades connate at base. Bracts early deciduous, opposite, free, concave. Flowers hermaphrodite, usually in clusters of 3, not connate; clusters forming lateral and terminal, short and dense, oblong-globular to oblong spikes. Perianth of 3 unequal segments united in their lower third, brownish, inflexed at tip. Stamen 1, with ovoid anther and somewhat flattened filament. Ovary ovoid; style thick; stigmas

2, subulate. Utricles obovoid-compressed, tapering; pericarp membranous. Seeds vertical; embryo arcuate.

One species, mainly in the Mediterranean and Irano-Turanian regions.

1. Halocnemum strobilaceum (Pall.) M.B., Fl. Taur.-Cauc. 3 : 3 (1819); Boiss., Fl. 4 : 936 (1879). *Salicornia strobilacea* Pall., Reise 1 : 412 (1771). [Plate 223]

Glabrous, fleshy dwarf-shrub, 20–50 cm. Stems ascending to erect, much branched. Branches with short, thick, cylindrical to club-shaped internodes ending with 2, about 1 mm. long, obovate, more or less obtuse, scarious-margined leaves, connate at base, often subtending short, sterile, globular, bud-like branches, with 4 rows of very short, rounded, sessile leaves, broader than long. Bracts of flower clusters reniform to orbicular, shed after flowering. Spikes lateral and terminal, sessile, opposite, cone-like or globular to oblong. Perianth about 1.5 mm. Seeds about 0.5–1 mm., compressed, brown, smooth to minutely tuberculate. Fl. May–September.

Hab. : Salines. Recorded by Post (1933) from Coastal Negev and the Dead Sea area but not seen by us.

Area : Mediterranean, Irano-Turanian and Saharo-Arabian, extending into the Euro-Siberian region.

11. ARTHROCNEMUM Moq.

Succulent, much branching shrubs, with jointed, erect or prostrate stems and branches, sometimes rooting at nodes. Leaves reduced to minute opposite scales, connate below to a cupule on the top of each joint. Bracts opposite, connate, persistent, each subtending a group (cyme) of 3 hermaphrodite flowers, equal in size, free or connate; each cyme immersed in a cavity in the axis of the spike-like, jointed, terminal or lateral inflorescence. Perianth brownish, ovoid or angular, 3–4-dentate or 3–4-fid, somewhat swollen and spongy in fruit. Stamens (1–)2. Ovary ovoid, somewhat compressed; style rather long; stigmas 2(–3); ovule with short funicle. Utricles membranous or indurated, enclosed in perianth. Seeds ovoid to subspherical, smooth or papillose or hairy; endosperm starchy or 0; embryo arcuate to semicircular.

About 12 species on the warmer sea coasts and in interior marshes in the Mediterranean and Irano-Turanian regions, N. America and Australia.

Literature : C. E. Moss, The species of *Arthrocnemum* and *Salicornia* in Southern Africa. *Journ. S. Afr. Bot.* 20 : 1–22 (1954).

1. Flowers in each cyme free or almost free; cavity in the floral axis undivided. Seeds with endosperm. Apex of perianth 3-dentate, obpyramidal.   **3. A. macrostachyum**
– Flowers in each cyme connate; cavity in the floral axis 3-partite. Seeds without endosperm. Apex of perianth plano-convex, 3–4-denticulate   2
2. Erect or almost erect plants. Stigmas usually 2. Middle flower of cyme not reaching the top of joint. Seeds with conical protuberances (short hairs). Glaucous plants.
                                                                                          1. A. fruticosum
– Plants with prostrate main stems and erect branches. Stigmas usually 3. Middle flower of cyme almost reaching the top of joint. Seeds covered with hooked, thin hairs. Greenish to somewhat reddish plants.   **2. A. perenne**

1. Arthrocnemum fruticosum (L.) Moq., Chenop. Monogr. Enum. 111 (1840) et in DC., Prodr. 13, 2 : 151 (1849). *Salicornia fruticosa* (L.) L., Sp. Pl. ed. 2, 5 (1762); Boiss., Fl. 4 : 932 (1879). *S. europaea* L. var. *fruticosa* L., Sp. Pl. 3 (1753). [Plate 224]

Shrubby, glabrous, glaucous perennial, 30–100 cm. Stems caespitose, ascending to erect, woody below, more or less decussately branched, rarely rooting at base; joints cylindrical, 0.6–1.5 cm. Leaves about 2 mm., membranous-margined, forming a cupular sheath. Spikes cylindrical, terminal and lateral. Flowers in 3's in adjacent cells of the floral cavity. Perianth of middle flower obscurely pentagonous or trapezoidal at top. Stigmas 2. Seeds grey, covered with small, thick, conical protuberances. Fl. June–September (–November).

Hab. : Marshes. Acco Plain, Sharon Plain.

Area : Mediterranean; probably also in some other regions.

Dominant shrub in the *Salicornia* community of the Acco Plain.
Eaten by camels.

2. Arthrocnemum perenne (Mill.) Moss, Journ. S. Afr. Bot. 14 : 40 (1948). *Salicornia perennis* Mill., Gard. Dict. ed. 8, no. 2 (1768). *S. radicans* Sm., Engl. Bot. 24 : t. 1691 (1807). [Plate 225]

Glabrous shrub, greenish to reddish, woody at base, 10–20 cm. Main stem prostrate, rooting. Branches ascending to erect. Leaves about 2 mm., forming a cupular sheath. Spikes terminal and lateral. Flowers in 3's, in adjacent cells of the floral cavity, with middle flower projecting above the lateral ones. Perianth of middle flower rhomboidal or trapezoidal at top. Stigmas 3, rarely 2. Seeds subglobular, covered with thin, relatively long, somewhat hooked or curved hairs. Fl. July–November.

Hab. : Coastal marshes. Acco Plain, Sharon Plain.

Area : Mediterranean; penetrating into Euro-Siberian territories.

3. Arthrocnemum macrostachyum (Moric.) Moris et Delponte, Ind. Sem. Hort. Taur. 35, t. 2 (1854). *Salicornia macrostachya* Moric., Fl. Venet. 1 : 2 (1820). *A. glaucum* (Del.) Ung.-Sternb., Atti Congr. Bot. Firenze 283 (1876) quoad planta; Boiss., Fl. 4 : 932 (1879). *S. glauca* Del., Fl. Aeg. Ill. 69 (1813) non Stokes, Bot. Mat. Med. 1 : 8 (1812). *S. mucronata* Lag., Mem. Pl. Barill. 58 (1817). [Plate 226]

Succulent, glabrous, glaucous, much branched shrub, 0.3–1 m. Branches erect to ascending, succulent, cylindrical, the lower ones often rooting at base, jointed; joints cylindrical or club-shaped. Leaves (free part of cupules) very short. Flowering branches erect, ending in thick, cylindrical, obtuse spikes, 3–6 × 0.3–0.4 cm. Flowers in 3's, free, protruding from the undivided cavity to up to one third the length of joint. Perianth 2–4 mm., 3-dentate, obpyramidal. Pericarp membranous. Seeds about 1 mm., black, shining, covered with short papillae, verrucose on back. Fl. May–September.

Hab. : Inundated salines, banks of saline water bodies. Acco Plain, Sharon Plain, N. Negev, Lower Jordan Valley, Dead Sea area, Edom.

Area : Mainly Mediterranean and Saharo-Arabian, extending far into adjacent regions.

Occupies large stretches both in Mediterranean littoral salines and in those of the Dead

Sea foreshore. In both areas it is the dominant in the *Arthrocnemum* and codominant in the *Arthrocnemum-Tamarix* communities.

12. SALICORNIA L.

Fleshy, glabrous annual herbs. Stems jointed, branched. Leaves reduced to scales, opposite and connate in pairs, forming a short sheath. Clusters (cymes) mostly 3-flowered, 2 on each node, embedded in the cavity of the axis and more or less adnate to its wall, subtended by bracts and forming terminal, simple or branched spikes. Bracts persistent, opposite, connate, cup-like. Flowers hermaphrodite, connate, the lateral ones of each cyme usually sterile, rarely pistillate. Perianth fleshy, brownish, 3–4-dentate, or -lobed. Stamens 2, rarely 1. Stigmas 2, subulate. Achenes ovoid or oblong, included in the spongy perianth, with membranous pericarp. Seeds vertical, almost ovoid, covered with short, often curved or hooked hairs; endosperm 0; embryo conduplicate.

About 50 species, in moist saline habitats of the temperate and subtropical regions.

Literature : P. W. Ball and T. G. Tutin, Notes on annual species of *Salicornia* in Britain. *Watsonia* 4 : 193–205 (1959). D. Koenig, Beitraege zur Kenntnis der deutschen Salicornien. *Mitt. Fl.-Soziol. Arbeitsgemeinschaft* N.F. 8 : 5–58 (1960).

1. Salicornia europaea L., Sp. Pl. 3 (1753) excl. β. *S. herbacea* (L.) L., Sp. Pl. ed. 2, 5 (1762); Boiss., Fl. 4 : 933 (1879). [Plate 227]

Annual, glabrous, 15–40 cm. Stems indurated at base, procumbent to erect, more or less decussately and divaricately branched. Branches light green, rarely scarlet. Leaves about 1.5 mm., membranous-margined. Spikes cylindrical, slightly tapering at top. Flowers in groups (cymes) of 3, arranged in the 3-celled floral cavity; middle flower projecting above lateral ones. Perianth of middle flower obovoid-rhomboidal. Stamen mostly 1. Styles and/or stigmas 2. Seeds ovoid, with short hairs, more or less hooked at apex. Fl. August–November.

Hab.: Marshes. Acco Plain, Sharon Plain, C. Negev, Lower Jordan Valley, Dead Sea area, Edom.

Area : Mediterranean and Euro-Siberian (introduced elsewhere).

One of the leading pioneers, occupying drying up saline puddles.
Fleshy stems eaten. Used in medicine (herba salicorniae herbaceae).

Trib. SUAEDEAE. Bracteoles mostly hyaline. Flowers hermaphrodite, rarely also unisexual. Perianth 5-merous, segments sometimes fleshy. Seeds horizontal or vertical, embryo spirally coiled; endosperm 0 or almost so.

13. SUAEDA Forssk. ex Scop.

Annual or perennial herbs, shrubs or trees. Leaves alternate, fleshy, terete, semiterete, sometimes flat to globular or lenticular. Flowers small, usually hermaphrodite, rarely unisexual, axillary, solitary or in clusters, 2–3-bracteolate. Perianth parted or cleft into 5 equal, more or less green, herbaceous or fleshy segments. Stamens 5, inserted on perianth segments. Ovary sessile, free or rarely adnate to perianth; stigmas 2 or more. Fruiting perianth unchanged or becoming fleshy or spongy, enclosing utricle. Utricle free or adnate to perianth, compressed or depressed. Seeds horizontal or vertical; endosperm scanty or 0; embryo flat, spirally coiled.

About 100 species in coastal, interior and desert salines, almost cosmopolitan.

Literature: M. M. Iljin, Systematike roda *Suaeda* Forsk. *Sovietsk. Bot.* 5 : 39–49 (1936).

1. Trees or shrubs 2–4 m. high. Leaves flat on both sides, 1–3 cm. Perianth reddish in fruit. Flowers unisexual and hermaphrodite. **6. S. monoica**
– Perennials or annuals not exceeding 1 m. Leaves and flowers different from above 2
2. Leaves terete-subulate, tapering, with a short caducous bristle at tip. Seeds mostly horizontal. Coastal annuals with slightly thickened, not gibbous nor spongy fruiting perianth. **7. S. splendens**
– Leaves various but without bristle. Shrubs or dwarf-shrubs with woody base or annuals, but then utricle adnate to fruiting perianth which is often spongy-baccate. Seeds vertical or horizontal 3
3. Upper leaves nearly globular or lenticular. Upper part of young branches hairy. **4. S. vermiculata**
– All leaves linear, terete or semiterete 4
4. Flowers inserted on the short petioles of the leaves. **1. S. asphaltica**
– Flowers inserted at axils of sessile leaves 5
5. Utricle adnate to perianth, sometimes forming together with it a gibbous or spongy or berry-like fruit 6
– Utricle free 7
6. Annuals. Leaves terete or semiterete. Fruiting perianth inflated-gibbous or spongy, forming together with utricle a berry-like fruit. **8. S. aegyptiaca**
– Annuals or perennials (very rare). Leaves linear. Fruiting perianth unchanged. **9. S. hortensis**
7 (5). Stigmas usually more than 3. Leaves 2–3 times as long as broad, imbricated, part of them flattish. Bracteoles denticulate. Flowers solitary or 2–3 in a cluster, in dense leafy spikes. **2. S. vera**
– Stigmas 3. Plants not as above 8
8. Leaves usually terete. Flowers mostly in clusters of 3–5. **3. S. fruticosa**
– Leaves semiterete. Flowers solitary, scattered in upper part of branches in the axils of mostly arcuate, not imbricated, usually (0.6–)1–1.2 cm. long, glaucous leaves. **5. S. palaestina**

Sect. SUAEDA. Sect. *Salsina* Moq., Chenop. Monogr. Enum. 121 (1840). Sect. *Eusuaeda* Gren. et Godr., Fl. Fr. 3 : 29 (1855). Perianth with 5 more or less equal lobes, unchanged and not appendiculate in fruit. Utricle mostly free. Seeds horizontal or vertical. Mostly perennials.

1. Suaeda asphaltica (Boiss.) Boiss., Fl. 4 : 938 (1879). *Chenopodina asphaltica* Boiss., Diagn. ser. 1, 12 : 98 (1853). [Plate 228]

Glabrous dwarf-shrub, up to 80 cm. Stems much branched. Branches whitish. Leaves 0.5–1.8 (–3) × about 0.1 cm., short-petioled, terete, green, turning blackish when dry, covered with minute whitish scales. Clusters of 1–3 flowers sessile on the petioles, sometimes forming leafy spikes. Bracteoles minute, scarious; the 2 lateral flowers of each cluster 3-bracteolate, the middle one ebracteolate. Flowers hermaphrodite. Perianth segments 1–2 mm., ovate, obtuse, white-margined, connate to middle. Stigmas 3. Seeds horizontal, beaked, smooth, glossy. Fl. January–May.

Hab. : Deserts, frequently on gypsaceous soils and steep slopes. Judean Desert, C. Negev, Lower Jordan Valley, Dead Sea area, deserts of Moav and Edom.

Area : Saharo-Arabian (almost endemic in Palestine).

One of the most striking shrubs in the Judean Desert; the dominant plant of the *Suaeda asphaltica* community. Before the foliage is shed in summer, the plants appear as black dots in the grey desert landscape.

2. Suaeda vera Forssk. ex J. F. Gmel., Syst. ed. 13, 2 : 503 (1791); Forssk., Fl. Aeg.-Arab. 69 (1775); Boiss., Fl. 4 : 939 (1879); Eig, Palest. Journ. Bot. Jerusalem ser. 3 : 126 (1945). "*S. fruticosa* (L.) Forssk." auct. mult. non *S. fruticosa* Forssk., Fl. Aeg.-Arab. 70 (1775). *Chenopodium fruticosum* L., Sp. Pl. 221 (1753). [Plate 229]

More or less glabrous or somewhat mealy low shrub, 20–50 cm. Stems woody, erect or prostrate, much branched. Leaves 0.5–1.5 (–2) × 0.1–0.15 cm., more or less sessile, dense, almost imbricated, fleshy, semiterete to flat, linear to lanceolate, rounded at apex. Bracts subsessile, somewhat longer than flowers, leaf-like, oblong-linear, smaller than leaves; bracteoles scarious, irregularly denticulate. Flowers hermaphrodite, about 1.5–2 mm., usually solitary, sometimes 2–3 in a cluster, forming dense leafy spikes. Perianth segments ovate, obtuse, sometimes lanceolate and acute, connivent in fruit. Ovary pear-shaped; style dilated in upper part into a disk on which the stigmas are inserted; stigmas usually more than 3, club-shaped to spatulate. Seeds smooth, usually vertical. Fl. February–May.

Var. **vera.** Perianth segments ovate, obtuse. Bracteoles with a well marked midrib.

Hab. : Spray zone of the Mediterranean littoral. Acco Plain, Sharon Plain, Philistean Plain.

Var. **deserti** Zoh. et Baum *. Perianth segments lanceolate, acute. Bracteoles with an inconspicuous midrib.

Hab. : Deserts, mostly saline. C. Negev.

Area of species : Mediterranean and Saharo-Arabian.

Eaten by camels.

3. Suaeda fruticosa Forssk. ex J. F. Gmel., Syst. ed. 13, 2 : 503 (1791); Forssk., Fl. Aeg.-Arab. 70 (1775) non "*S. fruticosa* (L.) Forssk." auct. mult. nec *Chenopodium fruticosum* L., Sp. Pl. 221 (1753). [Plate 230]

Shrubs, more or less glabrous, 40–100 cm. Stems woody, much branched. Young branches scabrous-puberulent. Leaves 0.4–2.5 (–3) × 0.05–0.15 cm., more or less sessile, fleshy, terete, rarely semiterete, straight or arcuate, often deflexed. Clusters axillary, mostly 3–5-flowered, arranged in rather dense, leafy spikes which together form loose, paniculate inflorescences. Bracts leaf-like but smaller than leaves, longer than flowers,

---

* See Appendix at end of this volume.

short-petioled, oblong-linear; bracteoles shorter than flowers, membranous. Flowers 1–2 mm., hermaphrodite. Perianth segments ovate, concave, obtuse and incurved at tip, connate in lower part. Stigmas 3, filiform. Seeds usually vertical, smooth and glossy. Fl. September–May.

Hab.: Hot desert saline marshes. Judean Desert, Lower Jordan Valley, Dead Sea area, Arava Valley, deserts of Moav.

Area: Mainly Sudanian, extending into Saharo-Arabian territories.

On the Dead Sea foreshore this species forms the leading plant of the hydro-halophytic *Suaeda fruticosa* community.

4. Suaeda vermiculata Forssk. ex J. F. Gmel., Syst. ed. 13, 2 : 503 (1791); Forssk., Fl. Aeg.-Arab. 70 (1775); Boiss., Fl. 4 : 940 (1879). [Plate 231]

Half-shrub, glabrous at base, papillose-hirsute in younger parts, 20–50 cm. Stems glaucous, divaricately and very profusely branched. Branches whitish. Leaves 0.3–1 × 0.1–0.4 cm., succulent, the lower obovate-oblong, the upper nearly globular or lenticular, obtuse. Bracts and upper leaves somewhat recurved, becoming black after desiccation. Clusters axillary, sessile, 1- or 2–3-flowered, forming loose, short, spike-like inflorescences. Flowers hermaphrodite, shorter than bracts. Fruiting perianth about 1 mm. in diam., ovoid, segments connivent. Stigmas 3, yellow. Seeds vertical, not beaked. Fl. March–April.

Hab.: Saline soils in deserts. Lower Jordan Valley, Arava Valley, deserts E. of Ammon.

Area: Saharo-Arabian, extending into adjacent Sudanian territories.

5. Suaeda palaestina Eig et Zoh. in Eig, Palest. Journ. Bot. Jerusalem ser. 3 : 126 (1945). [Plate 232]

Glabrous dwarf-shrub, 20–60 cm. Stems divaricately much branched. Leaves 0.6–1.2 × 0.1–0.2 cm., sessile, at some distance from one another in lower part, semi-terete to slightly flattened on upper side only, mostly arcuate to semicircular, glaucous. Flowers hermaphrodite, axillary, solitary, sessile. Bracteoles 2, lanceolate, scarious, one quarter the length of the perianth. Perianth about 2 mm., with 5 (rarely 4) ovate-oblong, narrowly white-margined segments. Stamens shorter than perianth. Stigmas 3. Seeds vertical, beaked, glossy. Fl. January–April.

Hab.: Hot desert salines. Lower Jordan Valley, Dead Sea area.

Area: E. Saharo-Arabian and E. Sudanian (endemic in Palestine, probably also in S. Iraq).

6. Suaeda monoica Forssk. ex J. F. Gmel., Syst. ed. 13, 2 : 503 (1791); Forssk., Fl. Aeg.-Arab. 70 (1775); Boiss., Fl. 4 : 940 (1879). [Plate 233]

Glabrous polygamous shrub or tree, 2–4 m. Stems thick, branching from base or above. Branches erect or spreading, densely leafy. Leaves 1–3 × 0.1–0.4 cm., sessile to short-petioled, fleshy, linear, flattish on both sides. Clusters axillary, sessile or short-peduncled, 1- to few-flowered, forming leafy, loose, spike-like inflorescences. Bracts longer to shorter than clusters, leaf-like, oblong-linear to elliptical; bracteoles small, membranous, usually ciliolate. Flowers unisexual and hermaphrodite. Perianth segments 1–2 mm., ovate, fleshy, reddish in fruit. Stigmas yellow, feathery. Seeds vertical, somewhat beaked, glossy. Fl. April–October.

Hab.: Wet salines. Lower Jordan Valley, Dead Sea area, Arava Valley.

Area: Sudanian, extending into adjacent tropical and Saharo-Arabian regions.

Leading plant of the *Suaeda monoica* community in the Dead Sea and Red Sea surroundings.

Sect. CHENOPODINA Moq. in DC., Prodr. 13, 2 : 159 (1849). Perianth 5-lobed, unchanged or inflated in fruit. Ovary constricted at base. Seeds horizontal or vertical. Mostly annuals.

7. Suaeda splendens (Pourr.) Gren. et Godr., Fl. Fr. 3 : 30 (1855). *Salsola splendens* Pourr., Mém. Acad. Toulouse 3 : 327 (1788). *Suaeda setigera* (DC.) Moq., Ann. Sci. Nat. Bot. ser. 1, 23 : 309 (1831); Boiss., Fl. 4 : 942 (1879). *Chenopodina setigera* (DC.) Moq. in DC., Prodr. 13, 2 : 160 (1849). [Plate 234]

Annual, subglabrous or somewhat pulverulent, 25–60 cm. Stems erect, simple or diffusely branched, slender, indurated at base. Leaves 0.8–2 × 0.1 cm., fleshy, more or less terete-subulate, acute-acuminate, ending in a short, caducous bristle. Clusters axillary, 3–5-flowered, forming rather long, loose spikes. Bracts leaf-like, usually longer than flowers, more or less lanceolate; bracteoles shorter than flowers, membranous. Flowers hermaphrodite. Perianth segments about 1.5 mm., fleshy, ovate, obtuse, appressed to utricle. Seeds horizontal or vertical, lenticular, short-beaked, glossy. Fl. June–October.

Hab.: Saline soils of coastal plain. Acco Plain, Sharon Plain.

Area: Mediterranean.

Sect. SCHANGINIA (C. A. Mey.) Volkens in Engl. et Prantl, Nat. Pflznfam. 3, 1a : 80 (1893). Perianth spongy and gibbous or unchanged in fruit. Utricle adnate to perianth. Seeds vertical. Mostly annuals.

8. Suaeda aegyptiaca (Hasselq.) Zoh., Journ. Linn. Soc. Lond. Bot. 55 : 635 (1957). *Chenopodium aegyptiacum* Hasselq., It. Palaest. 460 (1757). *Schanginia baccata* (Forssk.) Moq., Chenop. Monogr. Enum. 119 (1840); Boiss., Fl. 4 : 944 (1879). *Suaeda baccata* Forssk. ex J. F. Gmel., Syst. ed. 13, 2 : 503 (1791); Forssk., Fl. Aeg.-Arab. 69 (1775). [Plate 235]

Annual herb or low shrub, glabrous, glaucous or somewhat mealy, up to 40 cm. Stems procumbent to erect, branching from base. Leaves 2 × 0.2 cm., fleshy, terete or semiterete, obtuse, incurved. Bracts much longer than flowers; bracteoles minute, scarious. Flowers sessile or short-pedicelled, hermaphrodite, clustered, arranged in long, leafy spikes. Perianth lobes about 2 mm. or more, becoming spongy-baccate, gibbous-inflated in fruit. Seeds 1 mm., black, smooth. Fl. February–August.

Hab.: Saline soils and marshes, often also a weed in desert oases. Judean Desert, ·

C. Negev, Upper Jordan Valley, Beit Shean Valley, Lower Jordan Valley, Dead Sea area, Arava Valley, deserts of Moav and Edom.

Area : E. Saharo-Arabian.

9. Suaeda hortensis Forssk. ex J. F. Gmel., Syst. ed. 13, 2 : 503 (1791); Forssk., Fl. Aeg.-Arab. 71 (1775). *Schanginia hortensis* (Forssk.) Moq., Chenop. Monogr. Enum. 119 (1840); Boiss., Fl. 4 : 945 (1879).

Annual or perennial, glabrous, 50–60 cm. Stems erect, woody, ramose with many hardened, white, glaucous-mealy branches. Leaves 1.5–2 × 1 cm., often in clusters along the stem, thin, almost filiform, linear, somewhat acute, incurved, white-punctate. Inflorescences spike-like, leafy, loose below, dense above. Bracts longer than hermaphrodite flowers. Perianth 3–5-fid, lobes about 2 mm. Fruiting perianth unchanged, not inflated nor gibbous. Utricle ovoid, adnate to perianth. Seeds pyriform, beaked, shining, black, obsoletely punctate. Fl. August–September.

Hab. : Gardens and sandy places. Acco Plain. Very rare (probably casual).

Area : Saharo-Arabian and Sudanian.

Suaeda maris-mortui Post, Fl. Syr. Pal. Sin. 687 (1883–1896).

Annual, glabrous, 30–100 cm. Stems branching paniculately from base. Leaves 0.5–1.2 × 0.1 cm., fleshy, terete, scurfy. Bracts oblong-spatulate to obovate, obtuse, the lower a little longer, the upper as long as flowers. Clusters 2–3-flowered, sessile in axils, forming 10–20 cm. long, leafy spikes. Perianth lobes ovate, obtuse, connivent. Fl. August.

The data from Palestine on the above species by Post (l.c. and 1933) as also on *S. salsa* Pall. by Oppenheimer (1931) and on *S. maritima* (L.) Dumort. and *S. linifolia* Pall. by Oppenheimer and Evenari (1941), have not been confirmed by us.

Trib. SALSOLEAE Moq. Flowers often hermaphrodite, 2-bracteolate. Fruit an utricle, rarely a berry; testa membranous. Fruiting perianth mostly winged or horned. Seed horizontal or vertical; embryo conical-spiral; endosperm 0.

14. TRAGANUM Del.

Low shrubs. Stems much branched, villose-woolly at nodes. Leaves alternate, sessile, fleshy, more or less terete. Flowers small, hermaphrodite, axillary, solitary, 2-bracteolate. Perianth with 5 membranous lobes; lobes indurated and thickened in fruit, furnished with somewhat horn-shaped, hard protuberances. Stamens 5; filaments broad, exserted; anthers linear-sagittate; staminodes 0 or rudimentary. Style 2-partite into 2 subulate stigmas. Utricle included in the woody perianth, more or less globular, somewhat depressed, with membranous, free pericarp. Seeds horizontal.

Some 3 species, mainly in the Saharo-Arabian and S. Mediterranean territories.

1. Traganum nudatum Del., Fl. Aeg. Ill. 60 (1813); Boiss., Fl. 4 : 946 (1879). [Plate 236]

Low shrub, 20–50 cm. Stems woody, divaricately branched, whitish-glossy, glabrous or scabrous. Leaves 0.5–1.1 cm., fleshy, oblong-linear, more or less mucronate, somewhat recurved, fleecy at axils, papillose. Bracteoles generally longer than flowers, fleshy, linear, somewhat keeled, concave above. Flowers about 4 mm., solitary in axils. Perianth lobes erect, oblong-lanceolate. Anthers exserted, elongated. Fruiting perianth 3–5 mm., with rudimentary protuberances on back. Fl. February–April.

Hab.: Deserts. W., C. and S. Negev, Lower Jordan Valley, Dead Sea area, Arava Valley, deserts of Ammon, Moav and Edom.

Area: Saharo-Arabian, extending towards the Sudanian borderland.

15. HAMMADA Iljin

Shrubs or dwarf-shrubs with jointed opposite branches. Leaves opposite, subulate or scale-like, often reduced to a triangular tip at end of joint. Inflorescences spicate or paniculate. Bracteoles 1–2. Flowers small, hermaphrodite, axillary, 5-merous. Perianth segments free, somewhat chaffy, the fruiting ones with horizontal wings. Anthers not or very slightly appendiculate; staminodes 5, semiorbicular, thick, papillose-glandular. Style short; stigmas 2–3, subulate, recurved, papillose on inner side. Utricle included in perianth. Seeds horizontal, with spirally coiled embryo.

Very close to *Haloxylon* but differs from it in the wings being attached to the middle and not to the apex of the fruiting perianth, as well as in the thick papillose-glandular staminodes.

About 11 species in W. Irano-Turanian, Saharo-Arabian and E. Sudanian territories.

Literature: M. M. Iljin, Novi rod *Hammada* Iljin, *Bot. Zhurn.* 33 : 582–583 (1948). M. M. Iljin, Chenopodiaceae novae e Palaestina, in *Novit. Syst. Pl. Vasc.* 71–75 (1964).

1. Leafy, asperulous-velvety, canescent plants. Leaves linear, up to 1 cm. **1. H. negevensis**
  – Glabrous (rarely minutely papillose), leafless plants. Leaves reduced to minute scales
  forming a cup at top of joint 2
2. Branches dark-coloured to black when older; joints, especially of younger twigs, tapering
  at base. Fruiting perianth 4–6(–8) mm. across (incl. wings). Spikes generally 2–3 cm.
  2. H. scoparia
  – Branches becoming yellowish to ivory-coloured or light-coloured when older; joints
  cylindrical all along. Fruiting perianth (0.6–)0.7–1 cm. across (incl. wings). Spikes much
  longer than in above 3
3. Herbaceous perennials, woody only at base; annual branches 40–60 cm., arising from
  base. Inflorescences consisting of a main axis and divaricate lateral branches which
  together form a pyramidal panicle about 30 × 20 cm.; lateral flowering branches up
  to 10 cm. or more. Plants of Transjordan. **5. H. eigii**
  – Woody shrubs or dwarf-shrubs; annual branches much shorter than above, arising
  from upper or middle parts of plant. Inflorescences not as in above; lateral flowering
  branches shorter 4
4. Flowering branches arising as laterals of 1-year-old or older, woody, grey, long stems and
  bearing flowers and fruits all along; the whole inflorescence forming an elongated, much
  branched panicle. Fruiting perianth 0.8–1 cm. (incl. wings); wings distinct, all more
  or less equal, white. **4. H. schmittiana**
  – Flowering branches arising irregularly as laterals of the green, current-year shoots
  which become yellowish and decay after fruit dispersal; flowers and fruits mostly at
  end of branches. Fruiting perianth 7–8 mm. across (incl. wings); wings largely overlapping, unequal, mostly somewhat brownish to livid. **3. H. salicornica**

1. Hammada negevensis Iljin et Zoh., in Novit. Syst. Pl. Vasc. 71 (1964). [Plate 237]

Dwarf-shrub, asperulous-velvety, canescent, 20–40 cm. Stems erect, woody at base, with many erect, woody, stiff, not succulent branches; internodes remote. Leaves 0.8–1 cm., thick, opposite, linear, concave above, convex beneath, the older ones turning black when dry, dropping readily. Inflorescences terminal on the young branches of the current year, simple or somewhat branched, with flowers solitary in the axils of short bracts. Bracteoles 1–2, as long or nearly as long as flowers, orbicular to ovate, concave, ciliate at margin. Perianth lobes 2–3 mm., ovate, apiculate, concave, hairy within. Staminodes papillose at margin. Lobes of fruiting perianth herbaceous and scarious-margined; wings 3–4 mm. broad, ovate, nearly equal. Fruit 7–8 mm. across (incl. wings), depressed. Seeds horizontal. Fl. October–November. Fr. up to February.

Hab.: Stony deserts, often on somewhat gypsaceous soils. C. Negev (Nahal Ovdat, common), Arava Valley.

Area: E. Saharo-Arabian (endemic in Palestine).

2. Hammada scoparia (Pomel) Iljin, Bot. Zhurn. 33 : 583 (1948). *Haloxylon scoparium* Pomel, Nouv. Mat. Fl. Atl. 335 (1875). *Haloxylon articulatum* (Cav.) Bge., Mém. Sav. Etr. Pétersb. 7 : 468 (1851) *nom. illegit.*; Boiss., Fl. 4 : 949 (1879). *Salsola articulata* Cav., Ic. 3 : 43, t. 284 (1794) non Forssk., Fl. Aeg.-Arab. 55 (1775). [Plate 238]

Glabrous shrub, 20–40 cm. Stems erect, woody, intricately branched, with few or many thick basal branches and thinner upper ones; all branches fleshy, grey or grey-brown, later turning black. Leaves reduced to minute triangular scales, connate at top of nodes into a cup minutely villose within and somewhat membranous at margin. Inflorescences mostly consisting of short, spike-like flowering branches. Flowers usually solitary, with 2 bracteoles. Perianth subglobular, composed of 5 more or less free, herbaceous, scarious-margined segments. Stamens 5, alternating with 5 semiorbicular staminodes densely papillose at margin. Ovary globular, tapering into a very short style; stigmas 2–3, papillose on their inner side, revolute. Fruit (including wings) 4–6(–8) mm. in diam.; wings almost equal, spreading, broadly obovate to suborbicular, with slightly erose margin. Fl. October–November.

Hab.: Steppes and deserts, mainly on alluvial loess soil; sometimes in cultivated depressions. W., N. and C. Negev, Lower Jordan Valley, Dead Sea area, desert parts of Gilead, Ammon, Moav and Edom.

Area: Saharo-Arabian and Irano-Turanian, extending westwards to Spain.

A dominant component in a series of steppe communities in N. Negev. Sold in markets and widely used as detergent by Bedouins.

According to Botschantzev, in Novit. Syst. Pl. Vasc. 363 (1964), *H. scoparia* is not identical with *Haloxylon articulatum* of the Spanish authors, the legitimate name of which is *Hammada hispanica* Botsch.

3. Hammada salicornica (Moq.) Iljin, Bot. Zhurn. 33 : 583 (1948). *Haloxylon salicornicum* (Moq.) Bge. ex Boiss., Fl. 4 : 949 (1879). *Caroxylon salicornicum* Moq. in DC., Prodr. 13, 2 : 164 (1849). [Plate 239]

Diffuse half-shrub, 25–50 cm., with annual shoots arising irregularly from old woody low stems. Shoots more or less ascending, rigid, light-coloured to ivory-glaucous or waxy-yellow when dry. Leaves reduced to minute, short-triangular scales, connate into a short cup, membranous at margin, woolly within. Inflorescences diffuse, consisting of short, 3–6 cm. long, scattered spikes, mostly at the ends of green, main or lateral shoots of the current season. Flowers with 2 ovate, concave bracteoles, woolly at base. Stamens 5, alternating with linear-ovate staminodes, which are papillose at tip. Styles short; stigmas 2, strongly papillose on inner side. Fruit (0.6–)7–8 mm. across (incl. wings); wings obovate-orbicular, directed upwards, largely overlapping, unequal, somewhat brownish or livid. Fl. September–November.

Hab.: Sandy ground, debris of sandstone and granite in hotter deserts. C. and S. Negev, Dead Sea area, Arava Valley, Edom.

Area: E. Sudanian (Nubo-Sindian: Egypt, Sinai, Palestine, Arabia, S. Iraq and S. Iran).

A very common shrub in the north-eastern border belt of the Sudanian region and a leading plant of a series of shrubby plant communities within the *Haloxylon salicornicum* class. Known as the "rimth" by Bedouins in the Middle Eastern deserts who use it widely as camel browse and fuel; young plants also consumed by man in time of famine.

Put to various uses in folk medicine (potions in case of snake bites in Sinai, wound dressings, etc.). Used by the Bedouins as a detergent.

Some specimens included in this polymorphic species may belong to *H. elegans* (Bge.) Botsch., in Novit. Syst. Pl. Vasc. 362 (1964). The occurrence of intermediate forms makes it difficult to distinguish the latter from *H. salicornica*.

4. Hammada schmittiana (Pomel) Botsch., in Novit. Syst. Pl. Vasc. 363 (1964). *Haloxylon schmittianum* Pomel, Nouv. Mat. Fl. Atl. 334 (1875). *Haloxylon schweinfurthii* Aschers. in Aschers. et Schweinf., Ill. Fl. Eg. 128 (1887). [Plate 240]

Erect strict dwarf-shrub, 30–70 cm., with the current year's shoots arising from erect branches which become light-coloured or yellowish to ivory-coloured when older. Leaves reduced to minute, triangular, acute, obsoletely ribbed and scarious-margined scales, connate at base to form a short cup woolly within. Inflorescences compact, consisting of numerous laterals (arising from the main shoots of last or previous years), the lower often longer, the upper shorter so as to form an elongated pyramidal panicle. Flowering branches spreading, bearing flowers and fruits all along. Flowers with 2 ovate-orbicular, obtuse, membranous-margined bracteoles with woolly base. Lobes of perianth ovate-oblong, with broad membranous margin in the upper part. Stamens 5. Stigmas 2–3, with papillae on inner side. Fruit (0.8–)1 cm. across (incl. wings); wings spreading, equal, distinct, white. Fl. September–November.

Hab.: Sandy soil. C. and E. Negev. Rare.

Area: W. Saharo-Arabian (Palestine to Morocco).

Readily distinguished from *H. salicornica* by the large, uninterrupted inflorescence, which is particularly striking in fruiting specimens; by the larger, white, and spreading, more or less equal wings of fruit – and by the fact that the flowering branches arise mostly from older shoots. Calls for more intensive studies.

5. Hammada eigii Iljin, in Novit. Syst. Pl. Vasc. 72 (1964). *Haloxylon articulatum* (Cav.) Bge. ssp. *ramosissimum* (Benth. et Hook. f.) Eig, Palest. Journ. Bot. Jerusalem ser. 3 : 128 (1945). *Haloxylon ramosissimum* Benth. et Hook. f., Gen. Pl. 3 : 70 (1880) *nom. illegit. Hammada ramosissima* (Benth. et Hook. f.) Iljin, Bot. Zhurn. 33 : 583 (1948) *comb. illegit. Hammada syriaca* Botsch., in Novit. Syst. Pl. Vasc. 364 (1964). [Plate 241]

Minutely papillose or glabrous perennial, woody only at base, 40–60 cm. Stems herbaceous, all arising from base, erect, terete, light green to glaucous, the older ones yellow to brown-yellow (not becoming black). Leaves reduced to 1.5–2 mm. long, lanceolate to triangular scales, connate at nodes into a cup, villose within. Panicles large, up to 30 × 20 cm., broadly pyramidal, consisting of a main axis and laterals of up to 10 cm. or more. Flowers usually solitary, with 2 bracteoles. Perianth subglobular, composed of 5 segments. Stamens 5, alternating with 5 broadly ovate papillose staminodes. Stigmas usually 3, minutely papillose on inner side. Fruit 0.7–1 cm. across (incl. wings); wings brownish, 2–3.5 mm., broad-ovate, entire. Fl. August–September.

Hab. : Cultivated and abandoned fields in semidesert areas. Gilead, Ammon, Moav. Area : W. Irano-Turanian (Mesopotamian), endemic in Palestine and Syria.

In his description of *H. eigii* in Novit. Syst. Pl. Vasc. 72 (1964), Iljin is mistaken in referring the pilose specimen shown in t. 72, f. 6 to the species described. The stems and branches of the type of *H. eigii* – which is in our possession – have no pilose indumentum, and thus the type specimen does not differ from the Syrian one which was previously described by Iljin, Bot. Zhurn. 33 : 583 (1948) as *Hammada ramosissima.*

16. HALOXYLON Bge.

Shrubs or low trees. Branches cylindrical, jointed, joints fleshy. Leaves opposite, minute or rudimentary, connate. Flowers small, hermaphrodite, axillary and usually solitary, 2-bracteolate. Perianth lobes 5, free or connate below, later developing spreading, scarious wings near tip. Stamens generally 5; anthers muticous; staminodes 5, membranous, thin and glabrous, united with bases of filaments into a lobed cup-like disk. Ovary with short style and 2–5 stigmas. Utricle included in open perianth, globular or top-shaped, depressed, concave above, fleshy. Seeds horizontal, with spirally coiled embryo.

Some 4 species, in the Irano-Turanian region.

1. Haloxylon persicum Bge. in Buhse, Nouv. Mém. Soc. Nat. Mosc. 12 : 189 (1860); Iljin, in Fl. URSS 6 : 311 (1936). [Plate 242]

Shrub or tree, nearly glabrous, generally (1–)2–4 m. or more. Stems thick at base, much branched, young branches green, thin. Leaves 0.5–1.5 mm., more or less scale-like, short-triangular, acute to cuspidate at apex, connate into a cup villose within. Flowers minute, arranged in short, slender, spike-like branchlets. Stamens exserted; staminodes membranous, ovate, exceeding ovary. Stigmas 5, sessile. Fruit (0.6–)0.7–0.9(–1) cm. in diam., including 3–4 mm. long, entire to crenulate wings. Fl. February–April. Fr. November.

Var. **idumaeum** Zoh. * Leaves almost acute, very short-cuspidate (not long-cuspidate as in type). Wings almost entire at margin.

Hab. : Sandy wadis. Arava Valley, Edom.

Var. **maris-mortui** Zoh. * Leaves as above but wings with more or less regularly crenulate margin.

Hab. : Wadis crossing a saline depression. Dead Sea area.

Area of species : Irano-Turanian, extending into adjacent Saharo-Arabian territories (in S. Iraq, N. Arabia and S. Palestine).

17. Seidlitzia Bge. ex Boiss.

Annual glabrous herbs or shrubs. Branches opposite. Leaves opposite, fleshy, cylindrical. Flowers hermaphrodite, axillary, solitary or 2–3 (–5) in a cluster, 2-bracteolate. Perianth segments 5, united at base, elliptical, obtuse or 2-lobed, parchment-like in fruit, with broad, unequal, membranous, transverse wings near middle. Stamens 5, exserted; anthers muticous; staminodes 5, semiorbicular, glandular-ciliate, united with bases of filaments into a ring. Style minute, with 2 stigmas. Utricle depressed. Seeds horizontal; embryo spirally coiled.

Two Irano-Turanian and E. Saharo-Arabian species.

1. Seidlitzia rosmarinus Bge. ex Boiss., Fl. 4 : 951 (1879). *Suaeda rosmarinus* Ehrenb. ex Boiss., l.c. *pro syn. Salsola rosmarinus* (Bge. ex Boiss.) Solms-Laub., Bot. Zeit. 59 : 171 (1901); Eig, Palest. Journ. Bot. Jerusalem ser. 3 : 132 (1945). [Plate 243]

Glabrous low shrub, up to 60 cm. Stems branched, lower internodes longer than upper. Branches whitish, glossy. Leaves 0.5–3 cm., fleshy, thickening towards apex. Clusters 2–3 (–5)-flowered, opposite, in fleecy leaf axils. Perianth lobes obtuse. Fruit 0.8–1.2 cm. broad (incl. wings). Fl. March–May.

Hab. : Saline soils in hot deserts. Negev, Lower Jordan Valley, Dead Sea area, Arava Valley, deserts of Edom.

Area : E. Saharo-Arabian, extending into adjacent territories of the Irano-Turanian and E. Sudanian regions.

18. Aellenia Ulbrich

Annual or perennial herbs, glabrous or with more or less long papillae. Stems erect, divaricately branched. Leaves alternate, sparse, fleshy, ovate to lanceolate, somewhat auriculate at base. Flowers hermaphrodite, arranged along upper parts of branches and forming leafy inflorescences. Bracts solitary, green, ovate, acute; bracteoles 2, keeled, broad at base. Perianth 5-lobed, indurated in fruit; lobes erect, connivent in fruit, acute, with obovate-cuneate to orbicular scarious wings, arising from about middle of lobes. Stamens 5; filaments inserted on the disk, dilated at base; anthers oblong, without appendages. Ovary globular, depressed; style short, with 2 lanceolate, thick and obtuse stigmas. Utricle enclosed within the hard, bony perianth, 5-pitted at base;

---

* See Appendix at end of this volume.

pericarp membranous. Seeds horizontal; embryo green, plano-spiral, without endosperm.

Some 8 species in the Irano-Turanian region.

Literature: P. Aellen, Ergebnisse einer botanisch-zoologischen Sammelreise durch den Iran, *Verh. Naturf. Ges. Basel* 61 : 172–192 (1950).

1. Perennial, glabrous herbs, paniculately branched above. Bracteoles triangular. Perianth lobes triangular, acute. **1. A. lancifolia**
- Annual, often soft-pubescent herbs, branching from base of stem. Bracteoles orbicular. Perianth lobes ovate, obtuse. **2. A. autrani**

1. Aellenia lancifolia (Boiss.) Ulbrich in Engl. et Prantl, Nat. Pflznfam. ed. 2, 16c : 567 (1934). *Salsola lancifolia* Boiss., Fl. 4 : 958 (1879). [Plate 244]

Perennial, glabrous, glaucous-green, 30–60 cm. Stems erect, woody at base, branching mostly in upper part. Branches angulate. Leaves 0.5–4 × 0.2–0.5 cm., succulent, more or less clasping-decurrent, lanceolate to subulate; the floral ones linear-lanceolate, persisting after flowering. Bracteoles succulent, triangular, acute. Flowers in leafy spike-like inflorescences. Flowering perianth with membranous, triangular, acute lobes, (0.5–)0.8–1 cm.; lobes of fruiting perianth indurated, connivent, with orbicular-obovate, 5–7 mm. long wings. Fl. April–July.

Hab.: Stony and gypsaceous ground in steppes and deserts. Judean Desert, N. and C. Negev, Lower Jordan Valley, Dead Sea area, Ammon, Edom.

Area: W. Irano-Turanian (Palestine, Syria).

2. Aellenia autrani (Post) Zoh. (comb. nov.). *Salsola autrani* Post, Fl. Syr. Pal. Sin. 690 (1883–1896); Eig, Palest. Journ. Bot. Jerusalem ser. 3 : 129 (1945). [Plate 245]

Annual, pubescent, rarely glabrous, 20–50 cm. Stems erect; branches spreading from base. Stem leaves usually 8 × 3 mm., sessile, lanceolate, somewhat auriculate at base; the lower 1–2.5 cm., early deciduous, the floral ones (bracts) persistent. Bracteoles 3–6 mm., orbicular, obtuse to acute, green, puberulent. Flowers solitary along branches, forming leafy spikes. Fruiting perianth (0.6–)1–1.2(–1.5) cm. in diam. (incl. wings); lobes ovate, obtuse, connivent, with orbicular-obovate wings. Fl. July–September.

Hab.: Steppes and cultivated ground, especially on loess soil. Judean Desert, N. Negev, Lower Jordan Valley.

Area: W. Irano-Turanian (Mesopotamian).

A. autrani varies widely in indumentum (from glabrous to densely hairy) and size of leaves. *A. hierochuntica* (Bornm.) Aellen, Verh. Naturf. Ges. Basel 61 : 180 (1950) or *A. autrani* (Post) Zoh. var. *hierochuntica* (as recorded by Eig, l.c. under *Salsola*) should be included within the range of this species.

A. autrani has recently become one of the most common weeds on the loess soils of the N. Negev. Striking as a tumble weed in the Beersheva area.

19. Salsola L.

Herbaceous or shrubby plants. Leaves small, alternate or opposite, sessile, often fleshy, narrow, sometimes scale-like. Flowers small, hermaphrodite, axillary, rarely unisexual, sessile and 2-bracteolate. Perianth 5-fid or -parted into ovate, oblong or lanceolate segments membranous or indurated at base, in fruit usually with a transverse, scarious wing on the back. Hypogynous disk present. Stamens (4–)5; filaments mostly exserted; anthers muticous or mucronate; staminodes usually 0, rarely developed. Ovary ovoid to subglobular; style long or short, with 2 (–3) stigmas. Fruit a dry utricle, rarely berry-like, included in perianth. Seeds horizontal, rarely vertical, orbicular; testa membranous; endosperm 0; embryo green, spiral.

About 120 species mainly in Eurasia and Africa but also in America and Australia, especially in steppes and deserts.

1. Leaves and bracteoles spiny-tipped.   **1. S. kali**
– Leaves and bracteoles not spiny   2
2. Glabrous coastal annuals with semiterete, half-clasping, 1–6 cm. long leaves. Segments of perianth pectinate-ciliate at margin. Wings of fruiting perianth small (about 1 mm.), often reduced to transversal keel.   **2. S. soda**
– Plants not as above   3
3. Leaves opposite, scale-like or terete, semiterete or linear. Half-shrubs   4
– Leaves alternate, rarely the lower ones opposite and then leaf apex ending in a bristle. Annuals or half-shrubs   6
4. Leaves long, at least 2 times as long as broad, glabrous.   **9. S. longifolia**
– Leaves almost scale-like, much shorter than in above, hairy   5
5. Leaves with a broad, scarious and glabrous margin. Flowers hermaphrodite, 5-merous.   **7. S. tetragona**
– Leaves with a narrow, scarious and villose margin; hairs long, crisp. Flowers of two kinds: hermaphrodite, 5-merous and staminate, 4-merous.   **8. S. tetrandra**
6(3). Branches white, glabrous, glossy. Leaves fleshy, more or less terete, 0.5–3 cm. long, ending in a bristle. Perennials woody at base, or half-shrubs.   **10. S. schweinfurthii**
– Branches or leaves or both hairy. Leaves not as above   7
7. Bracts minute, about 1–3 mm., imbricated on a fairly dense, spike-like inflorescence. Fruiting perianth (incl. wings) 2–4 mm. across. Tall, fleshy, usually foetid perennials.   **12. S. baryosma**
– Plants not as above. Fruiting perianth mostly larger   8
8. Shrubby perennials, woody at base, 20–60 cm. Stems more or less indurated, villose with yellowish hairs. Lower leaves terete-filiform, densely hairy, usually persisting also when plant in fruit.   **11. S. vermiculata**
– Annuals. Stems and leaves not as above   9
9. Cauline leaves half-clasping, persistent, fleshy, semiterete, obtuse, the lower 1.5–2.5 cm. long. Flowers solitary. Bracts cuspidate. Very rare.   **3. S. crassa**
– Plants not as above   10
10. Adult stems hairy, not mealy. Bracts and bracteoles orbicular. Leaves semiterete. Perianth hairy.   **6. S. volkensii**
— Only younger branches hairy, adult stems mealy   11
11. Bracteoles mostly persistent, 3–3.5 mm. broad. Fruiting perianth (incl. wings) 0.8–1.2 cm. across. Leaves 1–2 cm. long.   **4. S. jordanicola**
— Bracteoles deciduous in fruit, 1.5–2 mm. broad. Fruiting perianth (incl. wings) 4–8 mm. across. Leaves 0.2–1 cm.   **5. S. inermis**

Sect. SALSOLA. Sect. *Kali* (Dumort.) Ulbrich in Engl. et Prantl, Nat. Pflznfam. ed. 2, 16c : 564 (1934). Annuals. Leaves opposite or alternate. Bracteoles rigid, often ending in a spine. Fruiting perianth parchment-like below, membranous above; wings often poorly developed.

1. Salsola kali L., Sp. Pl. 222 (1753); Boiss., Fl. 4 : 954 (1879). [Plate 246]

Annual, succulent, hispid-puberulent to scabrous or glabrescent, 20–60 cm. Stems decumbent or ascending, divaricately branched, especially at base, obtusely angular. Leaves 0.6–3 × (0.1–)0.2–0.3 cm., the lower usually opposite, the upper alternate, fleshy, semiterete, linear-subulate, rarely filiform, somewhat clasping at base, mucronate-spiny. Bracts and bracteoles somewhat longer than perianth, oblong-ovate to triangular, spiny-tipped, keeled. Flowers hermaphrodite, 1–3 in an axil, forming loose and leafy spikes. Perianth segments 3–4 mm., free, membranous, oblong, subacute or obtuse. Fruiting perianth 0.4–1 cm. across (incl. wings); wings obovate to reniform, striate, sometimes very reduced. Stamens (4–)5. Stigmas 2–3, long, filiform. Seeds horizontal, turbinate-subspherical. Fl. July–October.

Hab. : Sandy shores and waste places. Coastal Galilee, Acco Plain, Sharon Plain, Philistean Plain, Esdraelon Plain, Judean Mts. (casual), W. Negev, Dead Sea area.

Very polymorphic. Some of the local forms correspond to the following varieties :

Var. **crassifolia** Reichb., Fl. Germ. Exc. 583 (1832). Glabrescent. Leaves thick and short. Spines long and sturdy.

Var. **tenuifolia** Tausch, Flora 11 : 326 (1828). Lower or all leaves filiform. Flowers in fascicles of 1–3. Spines weak, short.

Area of species : Pluriregional, mainly in the northern hemisphere.

A tumble weed. Young shoots sometimes used as pot herb; ashes used as soap.

2. Salsola soda L., Sp. Pl. 223 (1753); Boiss., Fl. 4 : 953 (1879). [Plate 247]

Glabrous, often reddish annual, 20–50 cm. Stems branched, especially at base. Branches diffuse, opposite or alternate, ascending. Leaves 1–6 cm., opposite below, alternate above, fleshy, semiterete to linear, more or less dilated and half-clasping at base, mucronulate, glabrous at axils. Bracts longer than bracteoles, leaf-like; bracteoles longer than flowers, ovate-lanceolate, keeled. Flowers axillary, solitary, remote. Segments 5, more or less membranous, lanceolate, somewhat pectinate-ciliate at margin. Stamens 5. Stigmas longer than style. Segments of fruiting perianth with a very reduced wing or with a transversal keel on back. Fruit large. Fl. July–September.

Hab. : Littoral marshes. Acco Plain, Sharon Plain.

Area : Pluriregional; mainly north-temperate regions of the eastern hemisphere. Boiled stems and leaves eaten as vegetable (Japan); formerly used for soap production.

Sect. NITRARIA Ulbrich in Engl. et Prantl, Nat. Pflznfam. ed. 2, 16c : 564 (1934). Annuals. Leaves alternate, rarely opposite, terete-linear, sometimes succulent. Fruiting perianth membranous to parchment-like, closed, broadly winged.

3. Salsola crassa M.B., Mém. Soc. Nat. Mosc. 1 : 100 (1806); Boiss., Fl. 4 : 956 (1879); Post, Fl. Syr. Pal. Sin. ed. 2, 2 : 448 (1933).

Annual, pubescent-tomentose, greyish or whitish, 25–50 cm. Stems erect, rigid, much branched. Branches alternate, except the lowermost. Cauline leaves 0.6–2.5 cm., mostly alternate, fleshy, narrowly linear, half-clasping. Bracts leaf-like, short, broadly ovate, more or less cuspidate. Flowers axillary, solitary. Perianth segments ovate, hairy. Anthers with short appendages. Fruiting perianth about 1.5 cm. in diam., lobes broadly triangular; wings 0.8–1 cm. broad, semiorbicular, brownish-purple. Fl. July.

Hab. : Deserts. Very rare. Edom (after Nabelek, It. Turc.-Pers. 4 : 10, 1929).

Area : Irano-Turanian.

4. Salsola jordanicola Eig, Palest. Journ. Bot. Jerusalem ser. 3 : 130 (1945). [Plate 248]

Annual, white-mealy, papillose, pilose to villose, 15–40 cm., lower parts glabrous with age. Stems ascending to erect, rigid, much branched into dense, slender twigs. Leaves 1–2 cm., soon deciduous, alternate, linear, dilated at base, densely papillose or villose. Bracts short, broadly ovate-triangular, often persistent; bracteoles as long as bracts, 3–3.5 mm. broad, orbicular, white-margined, mostly persistent. Flowers mostly solitary, forming spike-like inflorescences. Perianth segments 2–3 × 1–2 mm., ovate, more or less hooded and apiculate. Anthers with short appendages. Stigmas 2, unequal. Fruiting perianth 0.8–1.2 cm. in diam. (incl. wings); wings broadly obovate-suborbicular. Fl. June–October.

Hab. : Steppes and deserts. Judean Desert, Lower Jordan Valley, Dead Sea area.

Area : E. Saharo-Arabian (Palestine-Syria).

5. Salsola inermis Forssk., Fl. Aeg.-Arab. 57 (1775); Boiss., Fl. 4 : 955 (1879). [Plate 249]

Annual, densely papillose-mealy, somewhat pilose with short or long hairs mainly on younger parts of stem, greyish to pale-green, 10–40 cm. Stems prostrate or ascending, much branched from base. Leaves small, 0.2–1 × 0.1 cm., alternate, linear, broadened and half-clasping at base, nearly acute, the lower soon deciduous. Bracts short, lanceolate to ovate-triangular, somewhat white-margined; bracteoles about as long as bracts, 1.5–2 mm. broad, orbicular, caducous after fruit set. Flowers solitary, forming loose or dense spikes. Perianth segments ovate-triangular, mealy. Fruiting perianth 4–8 mm. in diam. (incl. wings); segments connivent; wings obovate to suborbicular. Fl. July–September.

Hab. : Deserts; mostly on disturbed ground and ruderal sites. Judean Desert, W., N. and C. Negev, Lower Jordan Valley, Dead Sea area, Arava Valley, deserts of Ammon, Moav and Edom.

Area : E. Saharo-Arabian, extending into adjacent territories of the Irano-Turanian region.

6. Salsola volkensii Schweinf. et Aschers. in Aschers. et Schweinf., Ill. Fl. Eg. 130 (1887); Muschl., Man. Fl. Eg. 296 (1912); Post, Fl. Syr. Pal. Sin. ed. 2, 2 : 447 (1933). [Plate 250]

White-glaucous annual, more or less villose with articulate hairs, not mealy, 10–40 cm. Stems erect, indurated, much branched. Leaves 3–8 × 1–2 mm., alternate, fleshy, semiterete or linear, dilated at base, hairy. Bracts fleshy, more or less orbicular; bracteoles as long as or longer than bracts, more or less fleshy, orbicular, concave. Flowers solitary, in loose or dense spikes. Perianth segments connivent, oblong-lanceolate to ovate, acute, villose. Fruiting perianth 5–8 mm. in diam. (incl. wings); wings broadly obovate, imbricated. Fl. July–September.

Hab.: Steppes and deserts. W. and C. Negev, Lower Jordan Valley, Dead Sea area, desert of Moav.

Area: E. Saharo-Arabian.

Sect. TETRAGONA Ulbrich in Engl. et Prantl, Nat. Pflznfam. ed. 2, 16c: 565 (1934). Perennial plants with opposite, thick, succulent, scale-like leaves. Stems more or less tetragonous.

7. Salsola tetragona Del., Fl. Eg. 59 (1813). *S. pachoi* Volkens et Aschers. in Aschers. et Schweinf., Ill. Fl. Eg. 130 (1887). [Plate 251]

Dwarf-shrub, 20–40 cm., much branched. Branches opposite or alternate, the younger ones silky, tetragonous. Leaves 2–3 mm. long and broad, opposite, succulent-coriaceous, scale-like, ovate-rhombic, broader and half-clasping at base, mucronate at apex, concave above, carinate beneath. Flowers solitary in axils, sessile, with bracteoles shorter than perianth. Perianth of 5 free segments, silky-villose outside. Stigmas 2, subulate, papillose on inner surface. Fruiting perianth membranous, up to 1 cm. in diam. (incl. wings). Seeds horizontal, subglobular. Fl. October.

Hab.: Salines. Dead Sea area, Arava Valley. Uncommon.

Area: Saharo-Arabian.

8. Salsola tetrandra Forssk., Fl. Aeg.-Arab. 58 (1775). *S. tetragona* Del. var. *tetrandra* (Forssk.) Boiss., Fl. 4: 957 (1879). [Plate 252]

Dwarf-shrub, more or less white and crisp villose-tomentose, 15–50 cm. or more. Stems woody, with many tortuous, opposite and alternate branches. Leaves small, about 2 mm., opposite, scaly and fleshy, ovate-triangular, densely imbricated along the tetragonous-cylindrical flowering branchlets; leaf axils hairy. Bracteoles leaf-like, ovate, concave. Flowers hermaphrodite and unisexual, solitary, in dense, catkin-like spikes. Staminate flowers with 4 more or less free segments, villose outside; stamens 4, alternating with 4 tubercle-like staminodes. Hermaphrodite flowers rare, 5-merous, with 5 segments, developing short dorsal wings at maturity; stamens 5. Seeds horizontal. Fl. March–November.

Hab.: Deserts; dry saline soils. Judean Desert, W. and C. Negev, Lower Jordan Valley, Dead Sea area, Arava Valley, deserts of Ammon, Moav and Edom. More common than *S. tetragona*.

Area: Saharo-Arabian.

One of the few desert xero-halophytic shrubs tolerating both extreme drought and salinity. It is the dominant plant of the *Salsola tetrandra* community characterizing the halophytic zonation complex in the northern Dead Sea foreshore. Elsewhere it is met with under the most extreme desert conditions.

Sect. ARBUSCULA Ulbrich in Engl. et Prantl, Nat. Pflznfam. ed. 2, 16c : 565 (1934). Shrubs with long, opposite or partly opposite leaves. Perianth indurated in fruit.

9. Salsola longifolia Forssk., Fl. Aeg.-Arab. 55 (1775); Boiss., Fl. 4 : 957 (1879). *S. oppositifolia* Desf., Fl. Atl. 1 : 219 (1798). *S. sieberi* C. Presl, Bot. Bemerk. 108 (1844). [Plate 253]

Glabrous or sparingly mealy shrubby perennial, woody at base, 20–50 cm., sometimes higher. Stems ascending to erect, branching into short twigs, the young ones tetragonous, smooth, green. Leaves 0.5–2.5 × 0.2 cm., opposite, rarely some leaves much longer, fleshy, semiterete-linear, broadened at base, obtuse or mucronulate, more or less canaliculate on upper surface. Bracts leaf-like, oblong-linear; bracteoles shorter than bracts, ovate-orbicular, concave. Flowers axillary, opposite along twigs, forming loose or fairly dense spikes. Perianth segments free, whitish, ovate, obtuse. Fruiting perianth 0.8–1.2 cm. or more in diam. (incl. wings), lower part nearly globular, 5-angled at base, wings more or less obovate, sinuate. Seeds almost globular. Fl. May–September.

Hab. : Deserts. C. Negev, Dead Sea area.

Area : Saharo-Arabian, extending into some Mediterranean coastal territories.

Whether *S. oppositifolia* Desf. should be considered a separate species or be included within *S. longifolia* is a question which still needs to be clarified. Botschantzev in Novit. Syst. Pl. Vasc. 374 (1964) separates the two species.

10. Salsola schweinfurthii Solms-Laub., Bot. Zeit. 59 : 173 (1901); Eig, Palest. Journ. Bot. Jerusalem ser. 3 : 132 (1945). [Plate 254]

Shrubby, glabrous perennial, 15–30 cm. Stems ascending, woody, branching, especially at base, white-glossy. Branches alternate or the lowermost almost opposite. Leaves 0.5–3 cm., arising from hairy axils, all except lowermost alternate, fleshy, linear-subterete, more or less arcuate with a caducous bristle at tip. Bracts and bracteoles short, suborbicular, concave above. Flowers 1–2 in each axil, forming loose or dense spikes. Perianth segments ovate-suborbicular, white-margined. Staminodes well developed. Fruiting perianth usually 0.6–1 cm. in diam. (incl. wings); wings suborbicular. Fl. April–September.

Hab. : Deserts, mostly stony ground. Judean Desert, W., N. and C. Negev, Lower Jordan Valley, Dead Sea area.

Area : E. Saharo-Arabian.

Sect. ERICOIDES Ulbrich in Engl. et Prantl, Nat. Pflznfam. ed. 2, 16c : 565 (1934). Shrubs or perennial herbs with small, alternate, rarely opposite leaves. Perianth indurated in fruit; wings often unequal.

11. Salsola vermiculata L., Sp. Pl. 223 (1753); Boiss., Fl. 4 : 962 (1879). [Plate 255]

Shrubby perennial, 20–60(–70) cm., grey-pubescent or yellowish-villose with long denticulate hairs. Stems ascending to erect, paniculate, with many flowering branches. Lower leaves 0.3–1.3 cm., alternate, terete or semiterete, filiform, half-clasping at the somewhat dilated base, obtuse or acute, villose; upper leaves or those of the shorter branches ovate, densely imbricated, scale-like, very short with broader base, obtuse. Bracts ovate, sometimes short-cuspidate, concave, scarious-margined; bracteoles almost as long as bracts, suborbicular, concave above, keeled beneath or slightly so, scarious-margined. Flowers about as long as or longer than bracts, solitary, forming more or less dense spikes. Perianth segments almost free, more or less connivent, ovate-triangular, more or less hairy, broadly scarious-margined. Ovary ovoid. Fruiting perianth 0.7–1.2 cm. in diam. (incl. wings); wings obovate to semiorbicular, imbricated. Seeds horizontal.

Var. **villosa** (Del. ex Roem. et Schult.) Moq., Chenop. Monogr. Enum. 141 (1840) et in DC., Prodr. 13, 2 : 181 (1849). *S. villosa* Del. ex Roem. et Schult., Syst. Veg. ed. 15, 6 : 232 (1820) quoad descr. et nom. excl. specim. sieberianae; Del., Fl. Aeg. Ill. 57, n. 309 (1813) *nom. nud. S. vermiculata* L. ssp. *villosa* (Del.) Eig, Palest. Journ. Bot. Jerusalem ser. 3 : 132 (1945). *S. rigida* Pall. var. *tenuifolia* Boiss., Fl. 4 : 963 (1879). *S. delileana* Botsch., in Novit. Syst. Pl. Vasc. 371 (1964). Villose plants, usually with long, yellowish to greyish, often denticulate hairs. Lower leaves rather long, about 1 cm. Fl. Almost throughout the year, but mainly in spring.

Hab.: Calcareous stony steppes and also somewhat saline soils. Judean Desert, N. Negev, Upper and Lower Jordan Valley, Dead Sea area, Arava Valley, deserts of Ammon, Moav and Edom.

Area : Saharo-Arabian and Irano-Turanian.

A series of ecological races occur among the local populations. A notable desert pasture plant. A dominant of the local, widespread Salsoletum villosae.

12. Salsola baryosma (Roem. et Schult.) Dandy in Andrews, Fl. Pl. Angl.-Eg. Sudan 1 : 111 (1950). *Chenopodium baryosmon* Roem. et Schult., Syst. Veg. ed. 15, 6 : 269 (1820). *S. foetida* Del. ex Spreng., Syst. Veg. ed. 16, 1 : 925 (1824) non Pall. ex Vest. in Roem. et Schult., Syst. Veg. ed. 15, 6 : 238 (1820). *S. foetida* Del., Fl. Aeg. Ill. n. 310 (1813) *nom. nud.*; Boiss., Fl. 4 : 961 (1879). [Plate 256]

Shrub or sometimes annual, mostly evil smelling, whitish or rarely grey-greenish, 25–70 cm. or more. Stems herbaceous or woody, very much branched. Branchlets shortly tomentose-pubescent. Leaves minute, 1–2.5 mm., alternate, fleshy, ovate-orbicular or nearly so, narrowly scarious-margined. Inflorescences many, spike-like, more or less densely flowered, tomentose-pubescent. Bracts minute, 1–3 mm., generally imbricated. Flowers solitary, subtended by 2 bracteoles similar to bracts. Fruiting perianth 2–4 mm. in diam. (incl. wings), glabrous, with white wings expanding from the middle of the short, ovate-triangular connivent lobes. Fl. June–January.

Var. **baryosma.** Whitish-grey foetid plants. Leaves ovate-orbicular. Anthers with short appendages. Ovary ovoid. Flowers 1.5–2 mm. Fruiting perianth 3–4 mm. in diam.

Hab.: Saline, ruderal and segetal sites. C. Negev, Lower Jordan Valley, Dead Sea area, Arava Valley.

Var. **viridis** Zoh.* Greenish plant, more or less regularly branching, not foetid. Flowers and fruiting perianth somewhat smaller than in type. Anthers not appendiculate. Ovary almost globular.

Hab.: As above. Arava Valley.

Area of species: Sudanian, extending into the peripheral territories of the Saharo-Arabian region.

20. NOAEA Moq.

Annual herbs or dwarf-shrubs. Stems rigid, branched, glabrous. Branches divaricate; twigs transformed into thorns. Leaves alternate, sessile. Flowers hermaphrodite, solitary, 2-bracteolate. Perianth segments 5, short-connate at base. Stamens 5, inserted on a fleshy, lobed, papillose disk; anthers linear-sagittate, appendiculate; staminodes 0. Ovary ovoid; style elongated; stigmas 2, lanceolate-recurved. Fruiting perianth winged. Utricle membranous, included in perianth but not adherent to it. Seeds vertical, compressed, orbicular; endosperm 0; embryo spiral, green.

About 7 species in the Irano-Turanian region.

1. Noaea mucronata (Forssk.) Aschers. et Schweinf., Ill. Fl. Eg. 131 (1887). *N. spinosissima* (L.f.) Moq. in DC., Prodr. 13, 2 : 209 (1849); Boiss., Fl. 4 : 965 (1879). *Salsola mucronata* Forssk., Fl. Aeg.-Arab. 56 (1775). [Plate 257]

Glabrous low shrub, 25–50 cm. Stems indurated at base, erect. Branches rigid, flexuous, spreading, spiny at tip. Leaves about 0.5–1 (–1.5) cm., upper ones shorter, somewhat decurrent, all filiform, semiterete, mucronate, more or less papillose-scabrous at margin. Bracts and bracteoles ovate at base, triquetrous-subulate. Flowers axillary, alternate along the branches, often placed at base of twigs. Perianth lobes about 4 mm. Wings of fruiting perianth unequal, about 5–6 mm., membranous, white or purpurescent, obovate to oblong, with more or less crenulate or sinuate margin. Fl. August–October.

Hab.: Steppes, mainly on grey, calcareous soil. Judean Mts., Judean Desert, N., C. and S. Negev, Lower Jordan Valley, deserts of Gilead, Ammon, Moav and Edom.

Area: Irano-Turanian, slightly extending into the Mediterranean region.

One of the most abundant dwarf-shrubs of the Irano-Turanian territory. An important component of the *Artemisia herba-alba* community both in Palestine and outside the country.

21. GIRGENSOHNIA Bge.

Annual or perennial herbs or low shrubs. Stems (in ours) not articulate. Leaves opposite, small, simple, more or less coriaceous, acute to spinescent. Flowers hermaphrodite, axillary, solitary, sometimes grouped in terminal spikes, sessile, 2-bracteolate. Perianth

* See Appendix at end of this volume.

segments 5, connate at base, the outer 3 (or 2, rarely all) with a dorsal wing in fruit. Stamens 5, inserted on a small disk; anthers obtuse or apiculate; staminodes (disk lobes) 5. Style short; stigmas capitate, 2-lobed or 2-fid. Utricle included, membranous. Seeds vertical, compressed; embryo spiral.

Six species, mainly in the Irano-Turanian region.

1. Girgensohnia oppositiflora (Pall.) Fenzl in Ledeb., Fl. Ross. 3 : 835 (1851); Boiss., Fl. 4 : 967 (1879). *Salsola oppositiflora* Pall., Reise 2 : App. 735 (1773). [Plate 258]

Scabrous-puberulent annual, 20–40 cm. Stems ascending to erect, much branching, indurated, somewhat woody at base. Leaves usually 0.5–1 × 0.1–0.3 cm., opposite, recurved, linear-subulate, somewhat connate and dilated at base, mucronate-spinulose at apex, with membranous, ciliate margins; axils frequently bearing short branches or buds. Bracts ovate, short-cuspidate; bracteoles boat-shaped. Flowers opposite in short and rather dense spikes. Perianth segments free almost to base, oblong-ovate, somewhat acute, hairy. Anthers included, short-apiculate. Stigmas capitate to 2-lobed. Fruiting perianth with 3 unequal, triangular to orbicular, entire or denticulate wings. Seeds 1–2 mm., ovoid to oblong-ovoid.

Hab. : Steppes, mainly on saline soil. Edom. Rare (only non-flowering specimens collected; drawn from an Iranian specimen).

Area : Irano-Turanian.

Our material differs from that recorded in literature by its densely puberulent indumentum; it may thus represent a new variety.

22. ANABASIS L.

Perennial, succulent herbs or shrubs. Stems articulate. Leaves opposite, often reduced to scales. Flowers hermaphrodite, solitary or rarely clustered, subtended by 2 pubescent bracteoles. Segments of perianth 5, free to base, 3 or all winged in fruit. Stamens 5, alternating with the 5 staminodal lobes; anthers not appendiculate or only minutely apiculate. Style short, divided to base into 2 short, obtuse stigmas. Utricle membranous, sometimes berry-like, included in the winged perianth. Seeds vertical, compressed, lenticular, without endosperm; testa membranous; embryo spiral.

About 25 species in steppes and deserts, mainly in Irano-Turanian and Saharo-Arabian regions.

Each of the 3 species represented in Palestine belongs to a different section and is confined to a specific habitat.

1. Leaves 0.5–1 cm. or more long, terminating in a caducous bristle. Flowers 3–7 in a cluster. Fruiting perianth 5-winged. **3. A. setifera**
– Leaves absent or very minute, scale-like. Flowers solitary, at least in upper axils. Fruiting perianth 5- or 3-winged 2
2. Fruiting perianth 3-winged. Stems long, herbaceous, sparsely branching. Ovary smooth. Staminodes thin. **2. A. syriaca**
– Fruiting perianth 5-winged. Stems woody, profusely and intricately branching. Ovary papillose. Staminodes thick. **1. A. articulata**

1. **Anabasis articulata** (Forssk.) Moq. in DC., Prodr. 13, 2 : 212 (1849) emend. Aschers. et Schweinf., Ill. Fl. Eg. 128 et 131 (1887); Boiss., Fl. 4 : 970 (1879). *Salsola articulata* Forssk., Fl. Aeg.-Arab. 55, t. 8a (1775) non *S. articulata* Cav., Ic. 3 : 43, t. 284 (1794). [Plate 259]

Dwarf-shrub (chamaephyte), 30–60 cm. Stems woody to half their length or more, erect or tortuous. Branches opposite, with more or less equal internodes, brittle, the older ones with split and peeling bark. Leaves reduced to a short, 2-lobed cupule, villose within. Flowers up to 5 mm., solitary, opposite, the upper arranged in spikes at ends of branches. Perianth segments membranous. Stamens 5, long-exserted; staminodal lobes thick, orbicular-obovate, papillose. Ovary papillose; stigmas thick, papillose. Fruiting perianth membranous; wings 5, nearly equal, 5–7 mm. long, ovate-orbicular, striate, white or pink. Utricle erect, free. Seeds vertical; embryo spirally coiled. Fl. October–November. Fr. December–January.

Hab. : Stony, gravelly and sandy deserts. Judean Desert, Negev, Lower Jordan Valley, Dead Sea area, Arava Valley, deserts of Ammon, Moav and Edom.

Area : Saharo-Arabian, slightly extending into the Irano-Turanian region.

One of the most common desert dwarf-shrubs able to endure extreme drought for several consecutive years. It is the dominant plant in the *Anabasis articulata* community in runnels of plantless hammadas. It is also dominant in the *Anabasis–Retama* community of the sandy soils in C. Negev and a codominant in the Sudanian *Acacia–Anabasis* pseudo-savannas.

Heavily browsed by camels and goats and also collected as fuel; contains potassium and is therefore used as a detergent.

2. **Anabasis syriaca** Iljin, Not. Syst. Leningrad 20 : 138 (1960). *A. haussknechtii* auct. Fl. Palaest. non Bge. ex Boiss., Fl. 4 : 969 (1879). *A. aphylla* auct. nonn. Fl. Palaest. non L., Sp. Pl. 223 (1753). [Plate 260]

Glabrous perennial (hemicryptophyte), white-glaucous to green-greyish, 20–50 cm. Stems woody in lower part, sparsely branching, mainly from base. Branches long, herbaceous, fleshy, with long joints, erect or ascending. Leaves reduced to a short cupule with 2 triangular, acute lobes, the younger ones longer, mucronate; axils slightly villose. Inflorescences much branched, often forming a panicle of many spikes, each with 5–20 joints. Flowers solitary, opposite at each node. Perianth segments oblong-obovate. Stamens more or less exserted; staminodes thin, thicker at apex, somewhat papillose or smooth. Ovary erect, smooth. Fruiting perianth with 3 ascending wings at lower part; wings up to 6 × 5–9 mm., somewhat unequal, semiorbicular-reniform, entire or somewhat crenate. Utricle erect, free, membranous, included. Seeds vertical; embryo spirally coiled. Fl. September. Fr. September–December.

Var. **syriaca.** Plants 20–35 cm. Main branches of inflorescence thick, long or short. Spikes rather short, usually about 10-jointed.

Hab. : Steppes; heavy alluvial loess or loess-like soil, often on cultivated land or otherwise disturbed ground. N. and C. Negev.

Var. **zoharyi** (Iljin) Zoh. (stat. nov.). *A. zoharyi* Iljin, in Novit. Syst. Pl. Vasc. 74 (1964). *A. haussknechtii* var. *longispicata* Eig in herb. Plants 30–50 cm. high. Inflorescences much branched, branches thin and long; spikes long, usually 10–20-jointed.

Hab. : As above. Negev, Moav, Edom.

The above varieties intergrade.

Area of species : W. Irano-Turanian (Palestine, Syria).

An important component of the segetal *Anabasis syriaca–Plantago coronopus* community in cultivated fields of semidesert regions. The exceedingly long vertical roots (up to 5 m. or more) give off aerial shoots where denuded. Not eaten by animals.

3. Anabasis setifera Moq., Chenop. Monogr. Enum. 164 (1840); Boiss., Fl. 4 : 970 (1879). [Plate 261]

Glabrous perennial, 20–60 cm. Stems numerous, erect or ascending, Branches fleshy, 4-angled, elongated. Leaves 0.5–1.5 (–2.5) × 0.2–0.6 cm., thick, opposite, spreading or divaricate, connate at base, fleshy, cylindrical or club-shaped, obtuse, ending in a caducous bristle. Flowers 3–7 in a cluster, mainly in the upper axils; the lower clusters distant, the upper ones approximate. Staminodes rounded-truncate, fimbriate. Ovary papillose. Utricle erect. Fruiting perianth 4–6 mm. in diam., 5-winged; wings 3–5 mm., somewhat unequal, membranous, orbicular or somewhat reniform. Seeds vertical; embryo spirally coiled. Fl. August–November.

Hab. : Hot deserts, moist salines and wadi beds, often on gypsaceous ground. Judean Desert, Negev, Lower Jordan Valley, Dead Sea area, Arava Valley, Moav, Edom.

Area : E. Saharo-Arabian, slightly extending into Irano-Turanian and Sudanian territories.

A fairly common component in various plant communities in the Dead Sea and Arava Valley salines.

23. HALOTIS Bge.

Annual, hispid herbs. Leaves (except the lowest) alternate, sessile, terminating in a conical, cartilaginous, often caducous cusp. Flowers hermaphrodite, solitary in axils, 2-bracteolate. Perianth 4–5-merous, indurated at base, 3-nerved, not producing wing-like appendages in fruit. Stamens 4–5, inserted at margin of disk; anthers each with a membranous, bladdery appendage; staminodes 0. Ovary ovoid; style with 2 short, dilated, torn stigmas. Fruit an utricle, included, compressed, with fleshy pericarp. Seeds vertical, with membranous integument, without endosperm; embryo green, spiral.

Two species in the Irano-Turanian region.

1. Halotis pilosa (Moq.) Iljin, Act. Bot. Inst. Acad. Sci. URSS 1 : 3 (1936). *Halimocnemis pilosa* Moq., Chenop. Monogr. Enum. 152 (1840); Boiss., Fl. 4 : 976 (1879). [Plate 262]

Hispid-pilose annual, 10–20 cm., with long articulate hairs and dense shorter ones. Stems intricately branched, especially from base, forming a spherical, low bush. Leaves 1–3 × 0.1–0.3 cm., semiterete-triquetrous, ending in a sharp-pointed, cartilaginous, caducous cusp; the floral leaves shorter. Bracteoles somewhat shorter than bracts, adnate to flower at base, more or less lanceolate, prickly at apex, hairy. Perianth 6–9 mm., slightly longer than bracteoles; perianth segments 4, slightly unequal, lanceolate, cuspidate, concave; the fruiting ones rounded at base, membranous at margin, coriaceous along middle, glabrous or slightly hairy. Appendages of anthers pink, longer and broader than anthers, stipitate, inflated-clavate. Fruit enclosed within connivent perianth segments forming an indurated cone. Fl. July–August.

Hab. : Salt deserts. Ammon, Moav.

Area : W. Irano-Turanian.

24. HALOGETON C. A. Mey.

Shrubs or annual herbs. Stems branched. Leaves alternate, sessile, fleshy, semicylindrical to subglobular, mucronate, mostly terminating in a stiff bristle. Flowers hermaphrodite and pistillate together, crowded in axils among dense wool, middle flowers bractless, the lateral with 2–3 bracteoles. Perianth segments 5, nearly free, without midrib, the outer 2 or 3 (rarely all) winged or gibbous on back. Stamens 5 (rarely 3 or 2); anthers with short appendages; staminodes 4–5, alternating with stamens. Ovary ovoid; style short; stigmas 2, filiform. Utricle included in perianth; pericarp membranous, adherent to seed. Seeds vertical or horizontal, without endosperm.

Some 5 species in the Irano-Turanian and Saharo-Arabian regions.

1. Halogeton alopecuroides (Del.) Moq., Chenop. Monogr. Enum. 161 (1840); Boiss., Fl. 4 : 985 (1879). *Salsola alopecuroides* Del., Fl. Aeg. Ill. 56, t. 21 (1813). *Agathophora alopecuroides* (Del.) Bge., Mém. Acad. Sci. Pétersb. ser. 7, 4, 11 : 92 (1862). [Plate 263]

Glabrous or papillose-hispid dwarf-shrub, 20–40 cm. Stems woody, much branching from base. Branches spreading to erect, white, with short internodes. Leaves 0.5–1.5 cm., fleshy, mostly from almost globular to prismatic-cylindrical, obtuse, mostly spiny-tipped (spine caducous), glabrous to papillose-hispid, fleecy at axils. Flowers few, arranged in globular clusters in leaf axils along upper part of branches. Bracteoles ovate-triangular. Perianth with hyaline, erose-dentate segments; fruiting perianth about 3 mm., usually with 2–3 wings inserted above the middle of the segments, rarely 5-winged. Stamens 5; filaments connate with the 4–5 papillose staminodes. Fruit ovoid, somewhat gibbous above, with persistent style; pericarp membranous. Seeds vertical. Fl. March–July.

Var. **alopecuroides**. Branches and leaves glabrous; cusp of leaf short.

Hab. : Steppes and saline soil. Judean Desert, C. and S. Negev, Lower Jordan Valley, Arava Valley, deserts of Ammon, Moav and Edom.

Var. **papillosus** Maire, Fl. Afr. N. 8 : 188 (1962). Branches and leaves densely and shortly papillose-hispid; leaf cusp often longer than in preceding variety.

Hab. : Steppes. C. Negev.

Area of species : Saharo-Arabian.

33. AMARANTHACEAE

Herbs or shrubs with green, sometimes reddish stems. Leaves opposite or alternate, usually simple, exstipulate. Flowers small, hermaphrodite or unisexual (plants then monoecious or dioecious), in axillary or terminal cymes, sometimes solitary, subtended by 1 bract and 2(–3) bracteoles. Perianth sepaloid, 3–5-merous, with free or united segments, more or less membranous, often persistent and chaffy. Stamens (1–)3–5, usually opposite segments, free or sometimes connate, inserted on an annular disk. Ovary superior, 2 3-carpelled, 1-celled, usually 1-ovuled; style 1 or 2–4-fid. Fruit usually a capsule, sometimes an utricle, achene or a berry. Seeds crustaceous, often vertical; endosperm mealy; embryo curved, peripheral, annular or horseshoe-shaped.

About 64 genera and 900 species of wide distribution, mainly in Trop. America and Africa.

1. Leaves mostly alternate   2
– Leaves opposite   4
2. Tepals fleecy. Utricle membranous. Densely woolly herbs or shrubs.   **2. Aerva**
– Tepals glabrous. Plants glabrous or pubescent but not woolly   3
3. Fruit membranous. Flowers mostly unisexual. Style 0; stigmas 2–4.   **1. Amaranthus**
– Fruit crustaceous. Flowers hermaphrodite, in clusters of 3, the lateral ones sterile, transformed into a branched crest. Style distinct; stigmas 2.   **3. Digera**
4(1). Flowers in axillary glomerules or in very short spikes. Perianth (in ours) white-membranous, shining.   **5. Alternanthera**
– Flowers in long spikes. Perianth mostly green, later indurated.   **4. Achyranthes**

1. Amaranthus L.

Annual, rarely perennating, monoecious or dioecious herbs. Leaves alternate, mostly petiolate, simple, usually entire, green or reddish. Flowers mostly unisexual, green or reddish, in axillary glomerules or arranged in axillary or terminal, spike-like, simple or paniculate, leafy or naked inflorescences. Bracteoles 2–3, persistent, scarious, green or purplish. Tepals 3–5, persistent, free, rarely connate at base, more or less membranous, lanceolate-linear to obovate or spatulate. Stamens as many as tepals, free; filaments subulate; anthers 4-celled. Ovary 1-ovuled; style 0; stigmas 2–4. Fruit a somewhat fleshy or membranous 1-seeded utricle or capsule, the latter dehiscent by an operculum. Seeds more or less lenticular, brown to black, with annular embryo and central endosperm.

About 50 species, widespread in all tropical and temperate regions.

Literature: A. Thellung, *Amarantus,* in: Aschers. u. Graebn., *Syn. Mitteleur. Fl.* 5, 1 : 225–356 (1914). P. Aellen, Die Amaranthaceen Mitteleuropas mit besonderer Berücksichtigung der adventiven Arten, in: Hegi, *Ill. Fl. Mitteleur.* ed. 2, 3, 2 : 465–516 (1961).

1. Leaf axils with 2 strong spines.   **3. A. spinosus**
– Leaf axils not spiny   2
2. Monoecious or sometimes polygamous plants   3
– Dioecious plants   9

3. Fruit somewhat fleshy, indehiscent. Flowers in axillary glomerules forming terminal, simple or paniculate, leafless spikes   4
– Fruit membranous, opening by an operculum   5
4. Leaves ovate to almost rhombic. Fruit wrinkled. Common annuals.   **10. A. gracilis**
– Leaves linear to lanceolate. Fruit muricate. Perennials; very rare.   **7. A. muricatus**
5(3). Tepals and stamens (2–)3   6
– Tepals and stamens (4–)5 (rarely tepals 3–4 in staminate flowers)   7
6. Bracteoles rigid, somewhat spinescent, twice as long as tepals. Tepals unequal. Seeds up to 0.9 mm. Whitish-green plants.   **8. A. albus**
– Bracteoles not spinescent at apex, tapering or mucronulate, shorter than tepals. Tepals subequal. Seeds 1 mm. or more. Green or reddish plants.   **9. A. graecizans**
7(5). Stems prostrate, glabrescent. Leaves up to 3 × 1 cm. Glomerules axillary, overtopped by leaves. Bracteoles distinctly shorter than tepals.   **6. A. blitoides**
– Stems erect, mostly hairy. Leaves up to 12 cm. Inflorescences spike-like. Bracteoles longer than tepals   8
8. Tepals as long as or longer than fruit. Leaves lighter in colour beneath. Inflorescences often silvery-shining, composed of rather short, branched spikes, the terminal spike thick. Bracteoles one and a third to twice as long as tepals.   **2. A. retroflexus**
– Tepals shorter than fruit. Leaves usually green on both surfaces. Inflorescences greenish, composed of numerous long spikes, the lateral ascending and crowded, the terminal erect, slender, much elongated and tapering. Bracteoles twice as long as tepals.   **1. A. hybridus**
9(2). Leaves ovate to rhombic-lanceolate. Inflorescences with many or few axillary spikes at base. Bracteoles 4–7 mm., up to 3 times as long as tepals. Outer tepals of staminate flowers acute, with median nerve terminating in a sharp prickle. Stamens about as long as tepals. Fruit 2-beaked. Seeds about 1.2 mm. in diam.   **4. A. palmeri**
– Leaves somewhat narrower than in above, obtuse. Inflorescences of almost simple spikes or paniculate. Bracteoles 3–5 mm., mostly as long as or shorter than tepals. Outer tepals not as in above. Stamens shorter than tepals. Fruit 2–3-beaked. Seeds about 1 mm. in diam.   **5. A. arenicola**

Sect. AMARANTHUS. Sect. *Amaranthotypus* Dumort., Florula Belg. 19 (1827). Tepals of pistillate flowers 5(4). Stamens 5(4). Glomerules forming a simple or branched, spike-like, leafless inflorescence. Tepals usually different from bracteoles, the latter spiny-tipped, longer than tepals.

1. Amaranthus hybridus L., Sp. Pl. 990 (1753). *A. chlorostachys* Willd., Hist. Amaranth. 34, t. 10, f. 19 (1790); Boiss., Fl. 4 : 988 (1879). *A. hybridus* L. ssp. *hypochondriacus* (L.) Thell. var. *chlorostachys* (Willd.) Thell., Mém. Soc. Sci. Cherbourg 38 : 205 (1912) et in Aschers. u. Graebn., Syn. 5, 1 : 236 (1914); Aellen in Hegi, Ill. Fl. Mitteleur. ed. 2, 3, 2 : 480 (1961). [Plate 264]

Monoecious annual, greenish, crisp-pubescent to glabrescent, 50–100 cm. Stems erect, almost simple or shortly branching in upper part, greenish or reddish. Leaves up to 12 × 8 cm., long-petioled, rhombic-ovate, cuneate at base, somewhat obtuse or nearly acute or obsoletely emarginate and mucronulate at apex, scabrous-puberulent. Inflorescences erect, more or less paniculate, usually composed of numerous long spikes, the lateral ascending, crowded, the terminal erect, slender, much elongated.

Bracteoles twice as long as tepals, lanceolate-subulate, spiny-tipped. Tepals 5 (sometimes 3–4 in staminate flowers), 1.5–3 mm., shorter than fruit, white-green, oblong-linear to obovate-elliptical, acute or obtuse, mucronate at apex. Fruit dehiscent, somewhat exserted, 2–3-beaked, rugulose. Seeds about 1 mm., black. Fl. April–November.

Hab.: Roadsides, fields and waste places; a noxious weed in irrigated vegetable crops and elsewhere. Acco Plain, Sharon Plain, Philistean Plain, Judean Mts., Lower Jordan Valley.

Area: Origin Trop. America; recently introduced into most temperate and tropical regions.

A neophyte, spreading at a considerable pace throughout the country, hybridizing mainly with *A. retroflexus* and producing introgressed forms.

2. Amaranthus retroflexus L., Sp. Pl. 991 (1753); Boiss., Fl. 4:989 (1879); Aellen in Hegi, Ill. Fl. Mitteleur. ed. 2, 3, 2:485 (1961). [Plate 265]

Monoecious annual, pubescent, 20–80 cm. Stems erect, simple or branched, more or less angular or terete or obsoletely striate, scabrous, puberulent to pilose especially in upper part, generally greenish, sometimes reddish. Leaves up to 12 cm., ovate or ovate-oblong to lanceolate, somewhat rhombic, more or less cuneate or truncate-cuneate at base, obtuse or slightly notched at apex, frequently wavy-margined, sparsely pubescent, green, lighter beneath, prominently nerved, midrib ending in a short mucro; petiole about as long as or shorter than blade, pubescent. Inflorescences composed of many-flowered glomerules arranged in dense, branched, green or reddish thick spikes, often silvery-shining. Bracteoles 3–6 mm., one and a third to twice as long as tepals, rigid, lanceolate-subulate, scarious, keeled below, spiny-tipped. Tepals 5, about 3 mm., as long as or longer than fruit, whitish, oblong-lanceolate to subspatulate, obtuse to retuse, mucronulate at apex. Stamens exserted. Fruit dehiscent, about 2 mm., ellipsoidal or subglobular, compressed, 2–3-beaked, rugulose above. Seeds about 1 mm., more or less lenticular, slightly keeled, almost black, smooth, shining. Fl. May–November.

Var. **retroflexus.** Stems more or less angular. Leaves cuneate. Bracteoles 4–6 mm., twice as long as tepals, stiff and long-spiny.

Hab.: Waste places and irrigated fields and gardens. Common everywhere, notably Acco Plain, Sharon Plain, Philistean Plain, Upper and Lower Galilee, Esdraelon Plain, Samaria, Judean Mts., Dan Valley.

Var. **delilei** (Richter et Loret) Thell., Viert. Naturf. Ges. Zürich 52:442 (1907). *A. delilei* Richter et Loret, Bull. Soc. Bot. Fr. 13:316 (1866). Stems terete, sometimes very low. Leaves truncate-cuneate at base. Bracteoles 3–4 mm., four thirds to one and a half times as long as tepals, less spiny than in preceding variety.

Hab.: Waste places and irrigated fields. Sharon Plain, Philistean Plain, Lower Galilee, Judean Mts., Dan Valley, Edom.

There are transitions between this and the preceding variety.

Area of species: Origin N. America; at present pluriregional.

One of the most common species of *Amaranthus* and one of the dominant weeds of irrigated summer crops and orchards; abundantly hybridizing with *A. hybridus*.

3. Amaranthus spinosus L., Sp. Pl. 991 (1753); Thell. in Aschers. u. Graebn., Syn. 5, 1 : 267 (1914); Aellen in Hegi, Ill. Fl. Mitteleur. ed. 2, 3, 2 : 477 (1961). [Plate 266]

Monoecious perennial, glabrous or almost so, mostly reddish-brown, up to 1 m. Stems erect, branched. Leaves long-petioled, rhombic-ovate or rhombic-oblong or oblong-lanceolate, tapering on both ends, obtuse, notched and mucronate to awned at apex; leaf axils usually with 2, rather strong spines. Inflorescences axillary and terminal, spike-like, erect, slender and elongated, often with a few smaller, remote, axillary spikes at base; lower clusters pistillate, upper staminate. Bracteoles as long as or longer than tepals, scarious, ovate, weakly spiny-tipped. Staminate flowers with 3 oblong-ovate, obtuse or acute or shortly spiny-tipped tepals. Pistillate flowers with 5, 1–1.5 mm. long, white-membranous, acute or shortly spiny-tipped tepals. Fruit as long as tepals, mostly dehiscent, compressed, ellipsoidal, acute or obtuse, 3-beaked. Seeds about 0.75 mm. in diam. Fl. March–August.

Hab.: Waste fields. Sharon Plain. Very rare (casual).

Area: Origin presumably Trop. America; recently introduced into many countries of the eastern hemisphere.

Eaten as pot herb.

4. Amaranthus palmeri S. Wats., Proc. Am. Acad. 12 : 274 (1877). [Plate 267]

Dioecious annual, glabrous or almost so, pale green, up to 1.5 m. Stems erect, simple or short-branched, sulcate-angular. Leaves up to 7 cm., petiolate, ovate or rhombic to lanceolate, more or less obtuse, usually mucronate, prominently nerved. Inflorescences usually unisexual, sometimes polygamous, terminal, spike-like, more or less erect, much elongated, continuous or interrupted below, often with many or few smaller, axillary spikes at base. Bracteoles up to 7 mm., one and a half to 3 times as long as tepals, more or less spreading, lanceolate-subulate, spiny-tipped, rigid in pistillate flowers, shorter and weaker in staminate ones. Staminate flowers with 5 oblong, acute-mucronulate, about 2–3 mm. long tepals, and 5 stamens about as long as tepals. Pistillate flowers with tepals as long as or longer than capsule, whitish-green, oblong or obovate to spatulate, mucronate or mucronulate or emarginate at apex. Fruit dehiscent, sub-globular, 2-beaked, rugose at top. Seeds about 1.2 mm. in diam., dark brown, shining. Fl. June–September.

Hab.: Irrigated fields and waste places. Recently introduced but extensively spreading among summer crops. Sharon Plain, Esdraelon Plain, Judean Mts.

Area: Origin presumably S.W. United States; recently introduced into many countries of the eastern hemisphere.

5. Amaranthus arenicola J. M. Johnston, Journ. Arn. Arb. 29 : 193 (1948); Aellen in Hegi, Ill. Fl. Mitteleur. ed. 2, 3, 2 : 511 (1961). [Plate 268]

Dioecious annual, glabrous or nearly so, pale green, up to 1.5 m. Stems erect, almost simple or much branched above. Leaves up to 9 cm., petiolate, oblong-lanceolate to lanceolate or rhombic-lanceolate to oblanceolate, more or less rounded, usually mucro-nulate, prominently nerved. Inflorescences almost simple or paniculate, with dense spikes, much elongated, interrupted below. Bracteoles up to 5 mm., mostly as long as or shorter than tepals, spreading to erect, lanceolate-subulate, shortly spiny-tipped, rigid in pistillate flowers, weaker and sometimes slightly longer than tepals in staminate ones. Staminate flowers greenish, with 5 ovate-lanceolate, acute, usually mucronulate tepals and 5 stamens. Pistillate flowers greenish-white; their tepals unequal, as long as or longer than fruit, obovate or spatulate, obtuse, truncate or slightly notched, shortly mucronulate. Fruit dehiscent, globular, 2–3-beaked, wrinkled above. Seeds about 1 mm. in diam., dark brown, shining. Fl. June–November.

Hab. : Waste places and fields. Recently introduced. Philistean Plain.

Area : South-western parts of N. America.

Amaranthus caudatus L. and *A. tricolor* L., cultivated as ornamental plants are some-times met with as garden escapes. Both are marked, among others, by their red colour (*A. caudatus* with red to purple bracteoles and sepals; *A. tricolor* with reddish foliage).

Sect. BLITOPSIS Dumort., Florula Belg. 19 (1827). Tepals and stamens mostly (2–)3, sometimes 4 or 5, and then fruit indehiscent, rarely dehiscent, but then all glomerules axillary.

6. Amaranthus blitoides S. Wats., Proc. Am. Acad. 12 : 273 (1877); Aellen in Hegi, Ill. Fl. Mitteleur. ed. 2, 3, 2 : 489 (1961). [Plate 269]

Monoecious, glabrescent or hairy annual, 20–60 cm. Stems prostrate, branching, flexuous, densely leafy, whitish-green or sometimes purplish-red. Leaves usually small, up to 3 × 1 cm., petiolate, often crowded, obovate or oblong-lanceolate, cuneate at base, almost obtuse and shortly mucronate at apex, with whitish nerves. Glomerules axillary, dense, green or red, overtopped by leaves. Bracteoles about 1.5 mm., distinctly shorter than perianth, almost scarious, ovate-lanceolate, tapering-mucronulate. Tepals 4–5, about 2.5 mm., all or some as long as or longer than fruit, oblong-lanceolate, acute or acuminate, those of pistillate flowers unequal. Fruit dehiscent, about 2 mm. long, more or less ovoid, shortly 3-beaked, almost smooth, greenish, sometimes reddish. Seeds about 1.5 mm., sublenticular, acute at margin, black, shining. Fl. May–November.

Hab. : Roadsides and waste places. Very common. Sharon Plain, Philistean Plain, Lower Galilee, Judean Mts., rapidly spreading also into other districts. Observed for the first time in Jerusalem in 1948 and has since spread widely within the country.

Area : N. America; introduced into the Euro-Siberian and Mediterranean countries, probably also elsewhere.

7. Amaranthus muricatus Gillies ex Moq. in DC., Prodr. 13, 2 : 276 (1849) *pro syn.* *Exolus muricatus* Moq. in DC., l.c.

Perennial, usually glabrous, decumbent, up to 60 cm. Leaves 3–6 cm., long-petioled, linear to lanceolate. Inflorescence a long panicle, branched at base. Bracteoles half as long as perianth, ovate, acute. Tepals 5, about 2 × 0.5–0.7 mm., spatulate. Stamens 5. Fruit indehiscent, strongly muricate.

Hab. : Waste places. Esdraelon Plain. Rare (casual).

Area : Origin temperate S. America; naturalized and casual in several other regions.

8. Amaranthus albus L., Syst. ed. 10, 2 : 1268 (1759); Boiss., Fl. 4 : 990 (1879). [Plate 270]

Monoecious annual, glabrous or nearly so, whitish-green, up to 60 cm. Stems erect or ascending, rigid, paniculate. Branches more or less spreading, glabrous or sometimes sparingly pilose. Leaves up to 3.5 (–5) cm., petiolate, ovate-rhombic to oblong-obovate, attenuate-cuneate at base, more or less obtuse or obsoletely emarginate at apex, wavy-margined, punctate and prominently nerved beneath with midrib ending in a short mucro. Glomerules few-flowered, sessile, overtopped by leaves, forming loose, leafy inflorescences. Bracteoles about 3 mm., twice as long as flowers, lanceolate-subulate, greenish, rigid, spinescent at apex. Tepals 3, about 1.5 mm., shorter than fruit, unequal, ovate to lanceolate, tapering at apex, whitish, glabrous or sparingly ciliate at margins, veinless or with a very faint midrib. Stamens mostly 3. Fruit dehiscent, a little longer than perianth, minutely 3-beaked, rugose. Seeds up to 0.9 mm., sublenticular, acute at margin, black, smooth, shining. Fl. June–September.

Hab. : Fields, mostly among irrigated crops and in waste places. Acco Plain, Sharon Plain, Philistean Plain, Esdraelon Plain.

Area : Mainly Trop. and Subtrop. America, probably naturalized in the Mediterranean and adventive in the Euro-Siberian region, also in Africa and Australasia.

9. Amaranthus graecizans L., Sp. Pl. 990 (1753). *A. sylvestris* Desf. var. *graecizans* (L.) Boiss., Fl. 4 : 990 (1879). *A. blitum* L., Sp. Pl. 990 (1753) p.p.; Moq. in DC., Prodr. 13, 2 : 263 (1849). [Plate 271]

Monoecious annual, glabrous or nearly so, 10–75 cm. Stems erect, usually branched, sulcate-angular, green or often reddish. Branches ascending. Leaves up to 6 (–7) × 3 cm., petiolate, rhombic-elliptical to ovate or obovate to ovate-orbicular, rarely linear-lanceolate or oblanceolate, attenuate or cuneate at base, obtuse and obsoletely emarginate with a weak, short mucro at apex, more or less wavy-margined, prominently nerved beneath. Glomerules 2–5 mm., axillary, loose or dense, overtopped by leaves. Bracteoles 1.2–2 mm., shorter than tepals, more or less membranous, ovate-lanceolate to lanceolate-linear, tapering or mucronulate, greenish-white. Tepals 3, up to 2 mm., shorter than fruit or subequal to it, obovate or ovate-lanceolate to oblong-linear, acute or acuminate, sometimes more or less obtuse and mucronulate at apex, greenish-white. Stamens 3. Fruit dehiscent, subspherical to ellipsoidal, 2–3-beaked, wrinkled, green, sometimes becoming reddish. Seeds about 1.5 mm. in diam., lenticular, with more or less acute margins, brown-black, shining.

Var. **graecizans**. *A. angustifolius* Lam., Encycl. 1 : 115 (1783) s.s. *A. angustifolius* Lam. var. *graecizans* (L.) Thell. in Aschers. et Graebn., Syn. 5, 1 : 306 (1914). Leaves narrow, up to 1.5 cm. broad, more or less lanceolate to oblanceolate. Bracteoles lanceolate. Tepals about 1.5 mm., mucronulate or not. Fl. February–November.

Hab. : Waste places and roadsides. Acco Plain, Sharon Plain, Philistean Plain, Upper and Lower Galilee, Esdraelon Plain, Samaria, Judean Mts., Upper and Lower Jordan Valley, Dead Sea area, Gilead, Moav.

Var. **sylvestris** (Vill.) Aschers. in Schweinf., Beitr. Fl. Aethiop. 176 (1867). *A. silvestris* Vill., Cat. Pl. Jard. Strasb. 111 (1807). Leaves rather large, up to 3 cm. broad, ovate-

orbicular to rhombic-elliptical. Bracteoles ovate-lanceolate. Tepals up to 2 mm., mucronulate. Fl. April–August.

Hab. and distrib. : As above.

Area of species : Pluriregional.

10. Amaranthus gracilis Desf., Tabl. 43 (1804). *A. lividus* auct. Fl. Palaest. non L.,
Sp. Pl. 990 (1753). *Albersia blitum* auct. Fl. Palaest. non Kunth, Fl. Berol. 2, 2 : 144
(1838). [Plate 272]

Monoecious annual, glabrous, up to 70 cm. Stems more or less ascending, branching
from base, green, whitish or reddish. Leaves 1–7 × up to 4 cm., petiolate, ovate to
almost rhombic, more or less cuneate at base, obtuse and obsolutely emarginate, mucronulate at apex, more or less wavy-margined, green or reddish, with light or dark spots
on the upper surface, prominently white-nerved beneath. Flowers arranged in axillary,
many-flowered · glomerules and in terminal ones forming leafless spikes. Bracteoles about 1 mm., distinctly shorter than perianth, ovate, acute, white-membranous,
with green midrib. Tepals usually 3, about 1.5 mm., ovate to lanceolate or oblong-
oblanceolate to spatulate, somewhat acute to submucronate, greenish-white, rarely
with very few cilia at base. Stamens 3. Fruit a little longer than tepals, indehiscent,
somewhat fleshy, globular-ovoid, subcompressed, obsoletely beaked, rugulose, green or
reddish. Seeds about 1 mm. in diam., lenticular, black, smooth. Fl. Mainly April–
October.

Hab. : Waste places, roadsides and among irrigated crops. Almost everywhere,
notably in Acco Plain, Sharon Plain, Philistean Plain, Esdraelon Plain, Samaria,
Shefela, Judean Mts., Hula Plain, Upper Jordan Valley, Gilead, Moav.

Area : Pluriregional; in most tropical and subtropical and also in some temperate
regions.

Sometimes grown as a vegetable.

2. AERVA Forssk.

Woolly herbs or shrubs. Leaves mostly alternate, simple, entire. Inflorescences spike-
like or paniculate. Bracts 3 (1 bract and 2 bracteoles). Flowers minute, usually hermaphrodite, rarely unisexual, white or rust-coloured. Tepals (4–)5, almost equal, rather
membranous, the inner 3 or also others woolly-fleecy. Stamens (4–)5, with filaments
connate at base into a short cupule; anthers 2-celled; staminodes 5, subulate-triangu-
lar or tooth-like. Ovary 1-ovuled; style short, with 2 minute stigmas. Fruit an indehiscent, membranous utricle, included in perianth. Seeds vertical, compressed, ovoid
or reniform; endosperm central; embryo annular.

Some 10–12 species, mainly in Trop. Asia and Africa.

1. Aerva persica (Burm. f.) Merr., Philipp. Journ. Sci. 19 : 348 (1921). *Iresine persica*
Burm. f., Fl. Ind. 212, t. 65, f. 1 (1768). *A. tomentosa* Forssk., Fl. Aeg.-Arab. CXXII,
170 (1775). *A. javanica* auct. Fl. Palaest. non (Burm. f.) Juss., Ann. Mus. Paris 2 : 131
(1803); Boiss., Fl. 4 : 992 (1879). [Plate 273]

Densely tomentose dwarf-shrub, 15–80 cm. Stems erect, branched, terete, stiff.
Leaves 0.5–6 (–8) × 0.3–2.5 (–3.5) cm., alternate, short-petioled, ovate or oblong, some-

times lanceolate to linear, obtuse. Inflorescences many, varying in length, up to 8 mm. broad, terminal and axillary, spike-like, sometimes forming leafless panicles. Bracts almost as long as tepals, membranous, ovate, mucronulate, the outer ones hairy. Tepals up to 3 mm., hyaline, lanceolate, 1-nerved, white-villose or woolly. Stamens 5. Utricle about 1.5 mm., ovoid-oblong. Seeds brown, shining. Fl. December–June.

Var. **persica.** Stems branched, stiff, 20–60 cm. or higher. Leaves up to 6 (–8) × 2.5 (–3.5) cm., ovate, oblong or oblanceolate. Spikes about 8 mm. broad. Flowers 2–3 mm. Outer 2 tepals without nerves, somewhat longer than the nerved 3 inner ones.

Hab.: Hot deserts. Judean Desert, C. and S. Negev, Lower Jordan Valley, Dead Sea area, Arava Valley, Edom.

Var. **bovei** (Webb) Chiov., Fl. Somala 286 (1929). *A. javanica* (Burm. f.) Juss. var. *bovei* Webb in Hook. f., Niger Fl. 173 (1849); Boiss., Fl. 4 : 993 (1879). *A. tomentosa* Forssk. var. *bovei* (Webb) C. B. Clarke in This.-Dyer, Fl. Trop. Afr. 6, 1 : 38 (1909). Lower plant, 15–45 cm. Leaves smaller, 0.5–3 (–4) × 0.3–1.2 cm., ovate-oblong to lanceolate or linear, often revolute, sometimes recurved. Spikes shorter, more slender. Flowers smaller, up to 2 mm. All tepals nerved, broader and shorter than in var. *persica*.

Hab.: As above. C. and S. Negev, Dead Sea area, Arava Valley, deserts of Gilead, Moav and Edom.

Area or species : Paleotropical.

Woolly spikes formerly used for stuffing saddles and pillows. Plant also utilized in popular medicine.

3. Digera Forssk.

Annuals. Stems erect. Leaves alternate, petiolate, simple and entire. Inflorescences composed of axillary long spikes. Bracts and bracteoles 3. Flowers in clusters of 3, the middle hermaphrodite, fertile, the lateral sterile, rudimentary, in the form of a lobed crest, all white, greenish or pinkish. Perianth divided into 4–5 almost membranous segments erect in fruit, the outer broader. Stamens 5, free; anthers 2-celled; staminodes 0. Ovary superior, more or less compressed, 1-ovuled; style filiform with 2 short, recurved stigmas. Fruit a subglobular, hard, wrinkled utricle. Seeds more or less lenticular, with membranous testa; endosperm central; embryo annular.

One species in Trop. Africa and Asia.

1. Digera muricata (L.) Mart., Nov. Act. Nat. Cur. 13 : 285 (1826). *Achyranthes muricata* L., Sp. Pl. ed. 2, 295 (1762). *D. alternifolia* (L.) Aschers. in Schweinf., Beitr. Fl. Aethiop. 180 (1867). *A. alternifolia* L., Mant. 50 (1767). *D. arvensis* Forssk., Fl. Aeg.-Arab. 65 (1775). [Plate 274]

Annual, nearly glabrous, up to 50 cm. Stems more or less ascending, branched below, striate-angular. Leaves up to 6 × 3 cm., lanceolate or ovate, cuneate to rounded at base, acute at apex; petiole shorter than blade, somewhat pilose. Inflorescences up to 15 cm., spike-like, slender. Peduncles axillary, almost erect, much longer than leaves. Bracts shorter than flowers, membranous, greenish. Pedicels very short, bearing 2 sterile and 1 fertile flower. Perianth segments 5 (4), 3–4 mm., elliptical to ovate-lanceolate, somewhat acute, rather green. Style exceeding perianth. Fruit longer than crests

but included in perianth, subglobular, nut-like, with hard pericarp. Seeds lenticular.
Fl. March–April.

Hab.: Among crops. Dead Sea area, Arava Valley (Recorded by Hart, 1891).

Area: Trop. Africa and Asia.

Eaten as pot herb.

4. ACHYRANTHES L.

Herbaceous or suffruticose plants. Leaves opposite, petiolate, simple, entire or nearly
so. Inflorescences spike-like or paniculate. Bract 1, persistent, membranous; bracteoles
2, usually spinescent. Flowers hermaphrodite, greenish. Tepals 1–5, almost equal,
subulate, acuminate-aristate, later indurated. Stamens 2, 4 or 5, united at base into a
membranous cupule with 5 staminodes alternating with filaments. Ovary 1-ovuled;
stigma simple, capitate. Fruit an ovoid to almost globular, indehiscent, membranous
utricle, included in perianth. Seeds brownish; endosperm central; embryo annular,
peripheral.

About 12 tropical and subtropical species.

1. Achyranthes aspera L., Sp. Pl. 204 (1753); Boiss., Fl. 4:993 (1879). [Plate 275]

Perennial, appressed-pubescent, 30–100 cm. Stems erect or ascending, herbaceous,
frutescent at base. Branches elongated, ascending, opposite. Leaves 1.5–5.5(–6) ×0.6–
2.5(–3) cm., short-petioled, ovate to oblong-lanceolate, acute or acuminate, densely
appressed-hairy on lower surface. Inflorescences erect, terminal or arising from upper
axils, elongated, slender. Bracts almost as long as flowers, hyaline, aristate; bracteoles
with awn as long or twice as long as membranous blade. Flowers greenish, rarely whitish
or purplish. Tepals as long as to twice as long as bracteoles, sometimes somewhat longer,
persistent, lanceolate, acute, 3-nerved. Stamens 5; staminodes membranous, laciniate.
Fruit more or less ovoid.

Var. **sicula** L., Sp. Pl. 204 (1753). *A. aspera* L. var. *argentea* (Lam.) Boiss., Fl. 4:994
(1879). *A. argentea* Lam., Encycl. 1:545 (1785). Flowers 3–4 mm., somewhat longer
than bracts and bracteoles, whitish or purplish. Awn of bracteoles almost twice as
long as blade. Fl. February–May (–June).

Hab.: Among bushes, hedges and in fields. Sharon Plain, Upper Galilee, Mt. Carmel, Upper and Lower Jordan Valley, Ammon.

Area of species: Tropical regions of Asia and Africa (naturalized in the Mediterranean region, Australia and S. Africa).

Plant ashes employed as a source of alkali in dyeing.

5. ALTERNANTHERA Forssk.

Annual or perennial, prostrate or decumbent herbs. Leaves opposite, sessile or short-
petioled, simple, entire or nearly so. Inflorescences mostly small, axillary glomerules or
spikes. Flowers small, usually hermaphrodite, each in an axil of a membranous bract,
supported by 2 membranous bracteoles. Tepals 5, slightly unequal, glabrous or
minutely hairy, stellate in fruit. Stamens 2–3 or 5; filaments linear, united at base to a

tube; anthers 1-celled; staminodes 5, very small, entire or laciniate. Ovary obovoid or almost globular, 1-ovuled; style with capitate stigma. Fruit a membranous utricle, sometimes winged at margin. Seeds lenticular; endosperm central; embryo annular.

About 70 species in the tropical and subtropical regions of both hemispheres.

1. Leaves oblong-lanceolate to oblanceolate-obovate, acute. Bracts and bracteoles half as long as tepals.   **1. A. sessilis**
– Leaves obovate-spatulate, more or less obtuse. Bracts and bracteoles about as long as the prickly-mucronate tepals.   **2. A. pungens**

1. Alternanthera sessilis (L.) DC., Cat. Hort. Monsp. 77 (1813); Boiss., Fl. 4 : 996 (1879) excl. syn. R. Br.; Melv., Kew Bull. 1958–59 : 171–172 (1958). *Gomphrena sessilis* L., Sp. Pl. 225 (1753). [Plate 276]

Annual or perennial, glabrous or slightly pilose, about 25 cm. Stems branching from base, prostrate, elongated. Leaves 1.5–6(–8.5) × 0.5–1.5 cm., short-petioled, oblong-lanceolate or oblanceolate to almost obovate, tapering at base, acute at apex. Glomerules 3–8 mm., clustered at nodes, sessile, few-flowered, subglobular; rhachis more or less pilose. Bracts and bracteoles almost half as long as tepals, persistent, white-membranous. Tepals about 1.5–2 mm., membranous, ovate-lanceolate, white-shining. Stamens 3, connate into a cupule, alternating with subulate staminodes. Fruit as long as or some-what longer than tepals, lenticular, notched at apex, winged, dark brown. Seeds lenticular, brown, shining. Fl. April–January.

Hab.: Swampy places. Acco Plain, Sharon Plain, Hula Plain, Upper Jordan Valley.
Area: Mediterranean, Irano-Turanian; also in many tropical regions.

2. Alternanthera pungens Kunth in H.B.K., Nov. Gen. et Sp. 2 : 206 (1818); Melv., Kew Bull. 1958–59 : 174 (1958). [Plate 277]

Perennial, more or less pilose-tomentose. Stems procumbent, 30–50 cm. Leaves 1.5–4.5 × 0.5–2.5 cm., obovate-spatulate, tapering at base, rather obtuse at apex, subglabrous. Spikes ovoid, up to 9 × 6 mm., sessile at nodes, caducous, whitish-brown. Bracts and bracteoles about as long as tepals, ovate-lanceolate, acute-mucronate, glabrous, prickly in fruit. Tepals up to 6 mm., distinctly unequal, lanceolate, mucro-nate-prickly in fruit, glabrous, the inner 3 with tufts of white small hairs, glochidiate at tip. Staminal tube and filaments short; staminodes reduced to short teeth. Utricles small, shorter than tepals. Seeds about 1 mm. in diam., brown. Fl. September.

Hab.: Wet places. N. Negev. Very rare (casual).
Area: S. America; adventive in many countries.

Magnoliales

34. LAURACEAE

Trees or shrubs, mostly aromatic, rarely parasitic leafless herbs. Leaves usually alternate, rarely opposite, fasciculate or whorled, persistent and leathery or deciduous, simple and mostly entire, exstipulate. Inflorescences usually axillary, sometimes terminal or lateral. Flowers usually hermaphrodite, sometimes unisexual (plants then dioecious), actinomorphic, mostly 3-merous, small, greenish, yellow or white. Perianth mostly 2-whorled, of (4–)6 sepaloid (rarely petaloid) segments. Stamens adnate to perianth, mostly in 3–4 whorls, the innermost or all reduced to staminodes; filaments usually free; anthers 2- or 4-celled, usually opening by 2 or 4 uplifting valves. Ovary usually superior and free, sometimes entirely or partly included within a concave, woody or fleshy receptacle, mostly 1-celled; ovule solitary, anatropous, pendulous; style 1; stigma entire or rarely 2–3-lobed. Fruit a drupe or berry. Seed solitary; endosperm 0; embryo large, straight; cotyledons fleshy.

Thirty one genera and 2, 250 species mostly in tropical regions of Asia and America.

1. Laurus L.

Evergreen trees or shrubs. Leaves alternate, leathery, simple. Flowers hermaphrodite or unisexual, in axillary or terminal umbels or cymes. Perianth more or less petaloid, with 4 equal, caducous segments. Staminate flowers with (8–)12(–14) stamens in 2–3 whorls; filaments with 2 nectar glands in the middle; anthers 2-celled, dehiscing by a valvar lid; connective broad. Pistillate flowers with 4 filament-like staminodes, each with 2 hastate glands at apex; ovary superior, with 1 ovule; style short; stigma 3-lobed, obtuse. Fruit a 1-seeded berry. Seed pendulous; embryo erect, large.

Two species in the Mediterranean and Macaronesian regions.

1. Laurus nobilis L., Sp. Pl. 369 (1753); Boiss., Fl. 4 : 1057 (1879). [Plate 278]

Tall, glabrous, dioecious tree, 3–6 m. Leaves 5–11 × 2.5–6 cm., short-petioled, leathery, lanceolate to obovate-oblong, tapering below, acuminate or obtuse, more or less wavy-margined, shining on upper surface. Flowers pedicellate, in axillary or terminal umbels or cymes, staminate inflorescences many-flowered, pistillate few-flowered. Bracts involucrate, glabrescent. Perianth segments about 4 mm., obovate, obtuse, greenish-white. Fruit 1.2–1.7 × 0.7–1.2 cm., ellipsoidal to almost globular, black, with crustaceous pericarp. Seeds 5–6 mm., oblong, black. Fl. March–May.

Hab.: Maquis. Upper and Lower Galilee, Mt. Carmel, Samaria, Judean Mts., Dan Valley.

Area: Mediterranean.

One of the most characteristic components of the mesic maquis variants.

Leaves widely used as condiment; leaves and oil used in medicine (folia lauri and oleum lauri).

Ranunculales

35. RANUNCULACEAE

Herbaceous annuals or perennials, very rarely shrubs or woody climbers. Leaves alternate, rarely opposite or in whorls, mostly exstipulate; blade often deeply divided, sometimes compound. Inflorescences single-flowered to paniculate. Flowers usually hermaphrodite, actinomorphic or zygomorphic, with whorled or spiral sepals, petals, stamens and carpels. Perianth double or simple and then petaloid or sepaloid, sometimes accompanied by petal-like nectariferous organs. Stamens indefinite in number, free, with longitudinally dehiscing anthers. Ovary superior, with 1 to many, free or partly or completely united carpels; ovules 1 or numerous, anatropous, inserted in 2 rows along ventral suture. Fruit mostly apocarpous with 1 to many follicles or achenes, sometimes a septicidal capsule or a many-seeded berry. Seeds with copious, oily endosperm; embryo small; cotyledons 2, rarely 1.

About 50 genera and 2,000 species, mostly in temperate regions of the northern hemisphere. Some genera and many species also occur in mountainous regions of the tropics and in the southern hemisphere.

Literature: E. Janchen, Die systematische Gliederung der Ranunculaceen und Berberidaceen. *Denkschr. Akad. Wiss. Wien Math.-Nat. Kl.* 108 : 1–82 (1949).

1. Flowers zygomorphic, spurred	2
– Flowers actinomorphic, not spurred	3
2. Follicles 3 or more.	**2. Delphinium**
– Follicle 1.	**3. Consolida**
3 (1). Perianth simple, petaloid	4
– Perianth consisting of calyx and corolla	5
4. Climbing woody perennials. Tepals 4.	**5. Clematis**
– Herbs, not climbing. Tepals 5–13.	**4. Anemone**
5 (3). Fruit a many-seeded capsule or a group of 3 to numerous, partly connate and many-seeded carpels.	**1. Nigella**
– Fruit a head- or spike-like group of 1-seeded achenes	6
6. Petals without nectariferous pit at base.	**8. Adonis**
– Petals with a nectariferous pit at base, often covered by a scale	7
7. Achenes with 2 empty cells at base.	**7. Ceratocephala**
– Achenes without empty cells at base.	**6. Ranunculus**

Trib. HELLEBOREAE. Flowers solitary or racemose. Fruit mostly of several, many-seeded, dehiscent, or 1-seeded, indehiscent carpels; rarely fruit a berry or a capsule.

1. NIGELLA L.

Annuals, erect or decumbent, glabrous or pubescent, with branching or simple hairs. Leaves alternate, mostly pinnatisect. Flowers hermaphrodite, solitary at ends of branches, actinomorphic, pink, white, bluish to yellow. Calyx of 5 sepals, often petaloid, persistent or deciduous. Corolla of 5 (–8) petals with strongly bent claw and simple or 2-lipped nectariferous limb. Stamens many, first erect then bent downwards. Ovary of 2–10 or more, partly or entirely connate carpels. Fruit a many-seeded capsule or a group of partly connate, folded or inflated, few- or many-seeded carpels, often with long, free, beak-like styles. Seeds angular or discoid, often black.

About 15 species, mainly in Mediterranean and Irano-Turanian regions.

Literature: A. Brand, Monographie der Gattung *Nigella. Helios* 13 : 8–15, 22–28, 33–38 (1896). A. Terraciano, Revisione monografica delle specie del genere *Nigella. Boll. R. Orto Bot. Palermo* 1 : 122–153 (1897); 2 : 19–42 (1898).

1. Calyx greenish, deciduous. Petals with claw about twice as long as 5 mm. long limb.
   Fruit a capsule consisting of 2–3 united, short-beaked carpels.   **1. N. unguicularis**
 − Calyx petaloid, persistent. Petals with claw much shorter than limb. Fruit a capsule,
   consisting of 3 to numerous, long-beaked carpels free in their upper part   **2**
2. Flowers yellow. Seeds compressed.   **3. N. ciliaris**
 − Flowers pink, bluish or whitish. Seeds angular.   **2. N. arvensis**

Sect. GARIDELLA Prantl, Bot. Jahrb. 9 : 244 (1887). Sepals oblong, greenish to white, much shorter than petals, deciduous. Fruit of 2–3 connate carpels with short beaks.

1. Nigella unguicularis (Poir.) Spenn., Monogr. Nig. 12 (1829). *Garidella unguicularis* Poir. in Lam., Encycl. Suppl. 2 : 709 (1812) et in Lam., Tabl. Encycl. t. 379, f. 2 (1793); Boiss., Fl. 1 : 64 (1867). [Plate 279]

Annual, glabrous or sparingly hairy, 15–60 cm. Stems often simple, erect, prominently ribbed, with a few erect branches in upper part. Lower leaves petiolate, ovate, pinnately dissected into linear to capillary, 1–5 cm. long lobes; upper leaves sessile, few-lobed, the uppermost small, subulate. Flowers terminal, 1.2–1.5 cm., yellowish or yellowish-green, on long erect pedicels. Sepals 3–4 mm., much shorter than petals, deciduous, green, oblong. Petals 5, long-clawed, with a nectariferous pore at top of claw; limb 2-lipped, lips 2-fid, the lower much longer, wavy-margined. Fruit a 2–3-celled capsule, 0.6–1 cm. (incl. beak); carpels united all along (except beak). Seeds oblong, netted-pitted and transversally sulcate. Fl. May–June.

Hab.: Batha, stony hillsides and weedy places. Upper and Lower Galilee, Mt. Carmel, Esdraelon Plain, Samaria, Judean Mts., Judean Desert, Gilead.

Area : E. Mediterranean.

Sect. NIGELLA. Sepals white or bluish, longer than petals, stalked, persistent. Carpels 3–5 (–10), partly or entirely connate, rounded at back, with a long beak. Seeds usually angular.

2. Nigella arvensis L., Sp. Pl. 534 (1753); Boiss., Fl. 1 : 65 (1867). [Plate 280]

Annual, erect or decumbent, 10–50 cm. Stems simple or branching, often angular or furrowed, glaucous or green, smooth or scabrous. Lower leaves petiolate, dissected into linear, lanceolate or elliptical segments, the upper sessile, usually short, 3-fid or simple, with smooth or scabrous margins. Flowers 1–2.5 cm. in diam., pink, bluish or white, long-pedicelled, solitary at ends of stems and branches. Sepals 0.5–1 cm., persistent, abruptly stipitate, white or bluish, ovate, more or less cordate at base, acuminate, glabrous. Petals 0.8–1.2 cm., short-clawed; lower lip often hairy, with dark-coloured stripes and with 2, rather long, linear to oblong lobes callous at tip; upper lip about half as long as lower, broadly linear at base. Anthers apiculate or aristate or muticous. Ovary of 3–5 carpels, connate to about one to two thirds of their length; carpels furrowed on back with 3 prominent longitudinal ribs. Capsule obconical, with 3–5, partly connate, long-beaked, smooth or tuberculate carpels. Seeds angular, granular-scabrous. $2n = 12$. Fl. April–July.

Very polymorphic both in its vegetative and reproductive characters. In Palestine which is one of the variation centres of this species, the following varieties have been distinguished and tested for constancy.

1. Plants erect. Carpels smooth   2
 – Plants decumbent or erect. Carpels tuberculate   3
2. Anthers with very short appendages.   var. **arabica**
 – Anthers with long to very long (awn-like) appendages.   var. **beershevensis**
3(1). Plants of the coastal spray zone, usually decumbent. Branches divaricately spread. Lower leaves sparingly divided, with middle lobes 3–8 mm. broad; upper leaves undivided, up to 1 cm. broad.   var. **palaestina**
 – Plants not as above   4
4. Plants of coastal zone, usually erect. Leaves numerous, 1–2-pinnatifid, the lower crowded, rosulate; lobes of upper leaves longer than those of lower.   var. **daucifolia**
 – Plants not as above   5
5. Stems numerous, arising from base. Flowers 2–2.5 cm. across. Lobes of lower petal lip broadly ovate (2 × 1 mm.), with short, ovate to club-shaped tip. Carpels sparingly tuberculate.   var. **multicaulis**
 – Stems few. Flowers smaller. Lobes of lower petal lip linear to oblong   6
6. Mediterranean plants up to 50 cm. or more. Flowers bluish, 2 cm. or more across. Petals 5–7 mm. broad; lobes of lower lip oblong, more or less straight.   var. **submutica**
 – Desert plants, 10–25 cm. Flowers whitish, up to 1.5 cm. across. Petals about 3 mm. broad; lobes of lower lip linear, filiform, 0.3–0.5 mm. broad.   var. **negevensis**

Above varieties have been grouped under 3 subspecies as follows :

Subsp. **arvensis**. Plant mostly erect. Carpels smooth. The typical variety of this subspecies (var. *arvensis*) is one of the common European forms not occurring in Palestine.

Var. **arabica** (Boiss.) Zoh. (comb. nov.). *N. deserti* Boiss. var. *arabica* Boiss., Ann. Sci. Nat. Bot. ser. 2, 16 : 359 (1841). *N. deserti* Boiss., Fl. 1 : 67 (1867) p.p. Carpels smooth, divergent. Lobes of the lower petal lip straight. Anthers with very short appendages.
     Hab. : Steppes; sandy soils. N. Negev.
     Area : W. Irano-Turanian (endemic in Palestine).

Var. **beershevensis** Zoh. * Slender plant, 15–25 cm. Leaves finely dissected into many filiform rigid lobes. Lobes of lower petal lip strongly incurved. Anthers more or less aristate (with long appendages).
     Hab. : Steppes; sandy and loess soils. N. Negev.
     Area : As above (Palestine and probably also Egypt).

Subsp. **tuberculata** (Griseb.) Bornm., Verh. Zool.-Bot. Ges. Wien 48 : 547 (1898). *N. tuberculata* Griseb., Spicil. Fl. Rumel. 1 : 318 (1843). *N. arvensis* L. var. *glauca* Boiss., Fl. 1 : 66 (1867) p.p. Plant erect. Stems rigid. Carpels tuberculate. The typical variety of this species (var. *tuberculata*) occurs in Anatolia, Thrace and elsewhere, but not in Palestine.

* For Latin descriptions of this and the subsequent varieties
see Appendix at end of this volume.

Var. **submutica** Bornm., l.c. Stems rigid, up to 50 cm. Leaves with few divisions. Flowers bluish. Anthers short-appendiculate or muticous.

Hab.: Sandy loams. Acco Plain, Sharon Plain. Common especially in the *Desmostachya bipinnata* community.

Area: E. Mediterranean (endemic in Palestine).

Var. **negevensis** Zoh. Stems poorly branching. Flowers small (up to 1.5 cm. in diam.). Petals 3 mm. broad; lobes of lower lip linear, 0.3–0.5 mm. broad, with slightly club-shaped tips. Anthers muticous.

Hab.: Sands. N. E. Negev (Mishor Yemin).

Area: Saharo-Arabian (endemic in Palestine).

Var. **multicaulis** Zoh. Plant much branching. Flowers rather large (2–2.5 cm. in diam.). Sepals whitish-green. Lobes of lower petal lip 2 mm. long and about 1 mm. broad, with a short, ovate to club-shaped tip. Carpels sparingly tuberculate.

Hab.: Sandy soils. N. and Coastal Negev.

Area: Saharo-Arabian (probably endemic in Palestine).

Subsp. **divaricata** (Beaupré ex DC.) Zoh. (stat. nov.). *N. divaricata* Beaupré ex DC., Syst. 1 : 329 (1818). Low or procumbent-decumbent, often stemless plants, divaricately branching from base into rather thick, glaucous and often scabrous branches. Leaves rather thick. Flowers 1.5–2 cm. in diam. Carpels densely tuberculate. The typical form (var. *divaricata*) does not occur in Palestine.

Var. **palaestina** Zoh. Leaves sparsely divided, divisions oblong-elliptical, the middle one 3–8 mm. or more broad; upper leaves undivided, broadly oblong or elliptical.

Hab.: Spray zone of the Mediterranean Sea; sandy soils. Sharon Plain. Common.

Area: E. Mediterranean (probably endemic in Palestine).

Var. **daucifolia** Zoh. Plants up to 20(–30) cm., more or less erect. Leaves up to 6 cm. long or more, 1–2-pinnatifid, the lower ones forming a dense rosette; upper leaves dissected into linear lobes often longer than those of the lower leaves.

Hab.: Sandy soils, at some distance from spray zone. Coastal Galilee and Philistean Plain.

Area: E. Mediterranean (Palestine and probably also Lebanon).

Area of species: Euro-Siberian, Mediterranean and W. Irano-Turanian.

Sect. NIGELLASTRUM Moench, Meth. 311 (1794). Sepals yellow, stalked, longer than petals. Carpels 5 to many, folded lengthwise, with long beak. Seeds compressed.

3. Nigella ciliaris DC., Syst. 1 : 327 (1817); Boiss., Fl. 1 : 70 (1867). [Plate 281]

Annual, very leafy and hairy, 30–40 cm. Stems sulcate-angulate, patulous-hirsute, strictly branching from below middle. Leaves all along the stem, cut into linear or oblong, stiff, acute to acuminate, 1–3 cm. long lobes; lower leaves petiolate, the upper sessile. Flowers long-pedicelled, 2–3 cm. in diam., yellow. Sepals persistent, spreading, ovate, tapering at base, acuminate, white-hispid. Petals clawed; lower lip cuneate, 2-fid, with linear, hispid lobes provided with a short tooth near base; upper lip very short. Carpels 5–15, folded, connate to about middle and divergent, parallelly and prominently nerved, hairy or glabrous; free portion of carpel flat, ending in an erect or spreading beak. Seeds about 5 mm. in diam., compressed-discoid, brown or black, with wavy margin. 2n = 12. Fl. March–June.

Hab.: Fields and grassy patches of destroyed batha. Coastal Galilee, Acco Plain, Sharon Plain, Philistean Plain, Upper and Lower Galilee, Mt. Carmel, Esdraelon Plain, Samaria, Shefela, Judean Mts., Dan Valley, Hula Plain, Upper Jordan Valley, Gilead.

Area: E. Mediterranean.

The cultivated species *Nigella sativa* L. is the קֶצַח (nutmeg flower, fitches) of the Bible (e.g. Is. xxviii : 25). Formerly widely cultivated as a condiment; at present very rare, also occurring as a garden escape.

2. DELPHINIUM L.

Annual or perennial herbs with simple or tuberous roots. Leaves alternate, palmately lobed or divided. Flowers hermaphrodite, mostly showy, in racemes or panicles, zygomorphic. Sepals 5, petaloid, the upper one prolonged into a conical or cylindrical hollow spur. Petals smaller than sepals, glabrous or hairy, free, the 2 upper ones with nectar secreting spurs included within the sepal spur; the 2 lateral petals with broad limb and narrow claw. Stamens numerous, partly enclosing the ovary with their broad filaments. Gynoecium of 3–9 separate carpels, each with many ovules. Fruit of 3 or more, free, many-seeded follicles. Seeds in 2 rows, mostly black, covered with minute round scales.

About 200 species almost exclusively in the northern hemisphere.

Literature: E. Huth, Monographie der Gattung *Delphinium. Bot. Jahrb.* 20 : 322–499 (1895). P. K. Chowdhuri, P. H. Davis and M. Hossain, Materials for a Flora of Turkey: III. Ranunculaceae: I, *Delphinium* L. *Not. Roy. Bot. Gard. Edinb.* 22 : 403–425 (1958).

1. Flowers violet. Petals all glabrous. Annuals. **2. D. peregrinum**
– Flowers bluish-white. The 2 lateral petals long-hairy. Perennials. **1. D. ithaburense**

1. Delphinium ithaburense Boiss., Diagn. ser. 1, 8 : 9 (1849) et Fl. 1 : 90 (1867). [Plate 282]

Perennial, with tuberous rootstock and thick fibrous roots, 40–80 cm. Stems erect, sparingly branching, striate, glabrous or sparingly patulous-hirsute. Leaves up to 8 cm. across, long-petioled, almost orbicular in outline, palmately 5–7-partite, patulous-hairy; segments cuneate at base, cut into ovate to oblong-linear lobes, with a callous mucro at tip; upper leaves strongly reduced. Racemes simple, dense, many-flowered, elongating in fruit, up to 20 cm. Pedicels half or two thirds as long as spur, with 3 unequal membranous bracts. Flowers 2.5 cm. or more long (incl. spur), bluish-white. Sepals as long as or somewhat shorter than petals, elliptical, hairy or glabrous; spur erect or spreading, somewhat longer than corolla, more or less cylindrical, obtuse, glabrous or hairy. Upper petals glabrous, obliquely truncate, somewhat shorter than the deeply 2-fid and long-clawed, barbate or hairy lateral petals. Follicles 3, 1.2–1.6 cm., mostly erect or somewhat divergent, ellipsoidal, hairy or glabrous, beak one fifth to one sixth the length of the carpel. Seeds triangular, densely covered with membranous crisp tubercles. Fl. April–May.

Hab.: Batha and shady places. Upper and Lower Galilee, Mt. Carmel, Esdraelon Plain, Mt. Gilboa, Judean Mts., Judean Desert, N. Negev.

Area: E. Mediterranean (Palestine-Turkey).

This species varies greatly in indumentum, especially in that of flowers and fruit. There are specimens with glabrous and hairy sepals, with glabrous and hairy carpels, with glabrous sepals and hairy carpels and vice versa. Often these forms also intergrade. There is, therefore, no justification for maintaining either *D. chodati* Opphr., Bull. Soc. Bot. Genève ser. 2, 20.94 (1936), or the varieties separated by Zohary, Palest. Journ. Bot. Jerusalem ser. 2:154 (1941).

2. Delphinium peregrinum L., Sp. Pl. 531 (1753); Boiss., Fl. 1 : 87 (1867). [Plate 283]

Annual, appressed-hairy to glabrescent, 30–60 cm. Stems single or few, erect, with stiff branches. Leaves crowded at base of stem and sparse in upper part; the lower long-petioled, 1–2-palmatisect, with segments cut into linear-lanceolate, acute lobes, 1–3 cm., the upper ones sessile, entire or few-lobed, grading into linear-subulate bracts. Flowers 1.5–2 cm. long (incl. spur), pedicellate, violet, showy, arranged in a long, spike-like raceme. Sepals hairy. Lateral petals with ovate or elliptical limb a little shorter than or as long as claw; spur ascending, up to twice as long as corolla. Filaments glabrous. Follicles 3, 1–1.5 cm. long (incl. slender short beak), many-seeded, glabrous or hairy. Seeds globular, with ovate, membranous scales. Fl. May–July (–August).

Hab.: Batha, steppes and fallow fields. Acco Plain, Sharon Plain, Philistean Plain, Upper and Lower Galilee, Esdraelon Plain, Samaria, Judean Mts., W. and N. Negev, Dan Valley, Upper and Lower Jordan Valley, Gilead, Moav.

Area: Mediterranean and W. Irano-Turanian.

Rather variable in indumentum of flowers and fruit. *D. peregrinum* as conceived here also includes a whole series of forms such as *D. halteratum* Sibth. et Sm., *D. eriocarpum* (Boiss.) Hal., *D. bovei* Decne. and *D. virgatum* Poir. The taxonomic value of these forms can be decided upon only on a basis of experimental studies.

3. CONSOLIDA (DC.) S. F. Gray

Annuals. Leaves alternate, narrowly divided. Flowers hermaphrodite, zygomorphic. Sepals 5, petaloid, the posterior one prolonged into a conical or cylindrical spur. Petals connate, variously lobed, the upper forming a single spur. Stamens many. Carpel 1. Fruit a single, many-seeded follicle. Seeds as in *Delphinium*.

About 100 species in north-temperate regions.

Literature: E. Huth, Monographie der Gattung *Delphinium*. Sect. 1 *Consolida*. Bot. *Jahrb.* 20 : 365–391 (1895). R. V. de Soó, Ueber die mitteleuropäischen Arten und Formen der Gattung *Consolida* (DC.) S. F. Gray. *Oesterr. Bot. Zeitschr.* 71 : 233–246 (1922).

1. Flowers yellow, 0.7–1 cm., solitary or very few in each raceme. Low desert annuals.
2. C. flava

– Flowers white to pink. Plants not as above   2

2. Corolla 5-lobed; spur boat-shaped with a crozier-shaped extremity.   **3. C. scleroclada**
– Corolla 3-lobed; spur conical or cylindrical.   **1. C. rigida**

1. Consolida rigida (DC.) Bornm., Beih. Bot. Centralbl. 31, 2 : 181 (1914). *Delphinium rigidum* DC., Syst. 1 : 244 (1817); Boiss., Fl. 1 : 82 (1867). [Plate 284]

Annual, erect or procumbent, much branched, 15–50 cm. Stems branching dichotomously from base and at top, slightly furrowed, velvety to patulous-pubescent, mostly glandular and viscid above. Branches rigid, almost leafless. Leaves crowded at base of stems and at ramifications, the lower long-petioled, palmately 3–5-parted into segments, each with linear to oblong, acute, hairy to glabrescent lobes; the upper sessile, reduced to few lobes. Racemes few-flowered. Flowers about 1 cm. or less (incl. spur), deflexed, borne on long spreading pedicels, subtended by 3-fid or entire linear bracts. Corolla white to pale pink, 3-lobed, the middle lobe 2-fid, shorter than the obtuse lateral ones; spur almost as long as hairy petal. Follicle 1–1.5 cm., erect or deflexed, somewhat compressed, oblong, slightly curved, tapering at apex, patulous- and glandular-hairy, with beak about half as long as follicle. Seeds covered with appressed, rather distinct scales. Fl. May–July.

Hab.: Among irrigated summer crops and in abandoned fields. Acco Plain, Sharon Plain, Philistean Plain, Upper and Lower Galilee, Mt. Carmel, Esdraelon Plain, Samaria, Judean Mts., Hula Plain.

Area : E. Mediterranean.

Varies considerably in dimensions, indumentum (from almost glabrescent to densely velvety and patulous-hairy) and in size of follicles.

2. Consolida flava (DC.) Schrödgr., Ann. Naturh. Mus. Wien 27 : 43 (1913). *Delphinium flavum* DC., Syst. 1 : 346 (1817); Boiss., Fl. 1 : 83 (1867). *D. deserti* Boiss., l.c. [Plate 285]

Annual, appressed- to patulous-pubescent, glandular above, 10–25 cm. Stems rigid, erect or ascending, divaricately branched. Lower leaves 2–5 cm., crowded, long-petioled, palmately 3-parted into segments, each with linear to linear-lanceolate, acute to mucronate, pubescent lobes; the upper leaves sessile and largely reduced, gradually turning into linear bracts. Racemes 1- to few-flowered. Flowers small, 0.7–0.8 (–1) cm. (incl. spur), deflexed, borne on long, bracteolate, ascending, patulous-hairy and glandular pedicels. Sepals about 5 mm., oblong-elliptical. Corolla yellow with purple spots, shallowly 3-lobed; lateral lobes semiovate, longer than the denticulate middle one; spur about as long as petal, bent upwards. Filaments hirsute. Follicle 1–1.5 cm., many-seeded, oblong-linear, glandular-hairy, with short beak. Seeds minute, angular, covered with distinct scales. Fl. May–June.

Hab.: Deserts and steppes, on sandy, loessy and gravelly ground. Philistean Plain, Judean Desert, W., N. and C. Negev, Lower Jordan Valley, Moav.

Area : E. Saharo-Arabian.

C. flava is extremely variable and includes a whole series of forms (or varieties), among them also that named *Delphinium deserti* by Boissier (l.c.) which, in our opinion, can hardly be considered even a variety. It is particularly variable in the division of the corolla and dentation of the petals, size of the fruit and type of the indumentum. There is also some variation in the length of the leaf lobes. As yet the constancy of these forms has not been tested.

3. Consolida scleroclada (Boiss.) Schrödgr., Ann. Naturh. Mus. Wien 27 : 44 (1913). *Delphinium sclerocladum* Boiss., Diagn. ser. 1, 8.: 8 (1849); Chowdhuri, Davis and Hossain, Not. Roy. Bot. Gard. Edinb. 22 : 414 (1958). *D. anthoroideum* Boiss. var. *sclerocladum* (Boiss.) Boiss., Fl. 1 : 85 (1867). [Plate 286]

Annual, appressed-puberulent to glabrescent, 20–50 cm. Stems single or few, divaricately and often intricately branching, ultimate branches dichotomous. Leaves soon vanishing, often already before flowering, comparatively small, crowded mainly at base; lower leaves appressed-pilose, 2–3 cm., broadly ovate in outline, palmately partite into 3 or more deeply lobed segments, lobes linear, obtuse, with thickened callous mucro; petiole about as long as blade; upper leaves strongly reduced, gradually merging into sessile, linear, acute, mucronate, scattered bracts; bracteoles triangular-lanceolate, often opposite. Flowers 1.5–2 cm., solitary at ends of ultimate branches on appressed-hirsute pedicels. Calyx hairy; spur horizontal, boat-shaped, with crozier-shaped extremity; lower sepals oblong-elliptical, with broad, green, hairy stripes. Corolla white to pink with a 2-fid upper lobe, 2 shorter and obliquely triangular lateral ones and 2 ovate or triangular lower lobes. Follicle about 1.2 cm., compressed, oblong-linear, gradually tapering at apex; beak one quarter to one fifth the length of the follicle. Seeds triquetrous, brown, covered with dense, very narrow, scale-like protuberances. Fl. June–July.

Hab. : Steppes, stony ground. Mountains of C. Negev, Ammon. Rare.

Var. moabitica Zoh.* Differs from the typical form of *C. scleroclada* by the shorter lateral and lower corolla lobes, which is due to the very shallow sinus between upper and lateral lobes.

Hab. : As above. Moav.

Area of species : W. Irano-Turanian.

4. ANEMONE L.

Perennial herbs with tuberous rootstock. Leaves radical, compound or lobed. Scapes leafless, except for an involucre of 3 leaves usually at a considerable distance from the flowers. Flowers conspicuous, hermaphrodite, solitary, terminal or in few-flowered umbel- or head-like clusters, actinomorphic. Perianth usually of many petaloid tepals. Stamens and carpels numerous, inserted on a conical torus. Ovule 1 in each carpel, pendulous. Fruit of numerous achenes, inserted on the elongated torus; style of achene persistent, acuminate or caudate, glabrous or bearded.

About 120 species mostly in temperate regions of the northern hemisphere.

1. Anemone coronaria L., Sp. Pl. 539 (1753); Boiss., Fl. 1 : 11 (1867). [Plate 287]

Perennial herb with tuberous brown rootstock. Scapes 10–40 cm., hairy, 1-, rarely

---

* See Appendix at end of this volume.

2-flowered. Radical leaves long-petioled, ternately or 2-ternately compound, with petiolate segments deeply dissected into coarsely toothed, acute or mucronate lobes. Involucre leaves 3 (–4), sessile, variously divided, often resembling. the radical leaves. Flowers 5–9 cm. in diam., variously coloured. Tepals varying from 5 to 13, usually 6 (very rarely 4), obovate to suborbicular or oblong-ovate, obtuse or apiculate, pubescent outside. Anthers blue or purple. Receptacle conical or oblong. Achenes more or less spherical, densely woolly, with persistent filiform styles. 2n=16. Fl. January–March.

Hab.: Batha, fallow fields and grassy patches. Acco Plain, Sharon Plain, Philistean Plain, Upper and Lower Galilee, Mt. Carmel, Esdraelon Plain, Shefela, Judean Mts., Judean Desert, N. Negev, Dan Valley, Hula Plain, Upper and Lower Jordan Valley, Gilead, Ammon, Moav.

Area: Mediterranean.

This is one of the most common, showy plants of the country's open habitats. It is remarkable for its genetic polymorphism in flower colour. The main colour types are scarlet-red, white, pink, purple to violet (vars. *coccinea, albiflora, rosea, cyanea – parviflora – incisa,* respectively, in Post, Fl. Syr. Pal. Sin. ed. 2, 1 : 5, 1932).

In the northern part of the country mixed polymorphic populations are common. In the south the red colour predominates.

5. CLEMATIS L.

Perennial woody shrubs or vines, climbing by their long coiling petioles. Leaves opposite, usually compound, exstipulate. Flowers hermaphrodite, solitary or in paniculate inflorescences, actinomorphic. Perianth usually of 4 petaloid tepals. Stamens and carpels numerous; staminodes sometimes present. Ovule solitary, pendulous. Achenes numerous in head-like clusters, mostly sessile, dry, with persistent, mostly long-plumose style.

About 230 species in temperate regions, especially of the northern hemisphere.

Literature: O. Kuntze, Monographie der Gattung *Clematis. Verh. Bot. Ver. Prov. Brandenb.* 26 : 83–202 (1885).

1. Flowers solitary, about 5 cm. or more across. Leaves mostly undivided, clustered, coarsely toothed or lobed. **1. C. cirrhosa**
- Flowers in loose panicles, 2–2.5 cm. across. Leaves compound, leaflets entire or lobed or dentate. **2. C. flammula**

Sect. CHEIROPSIS DC., Syst. 1 : 162 (1817). Flowers solitary in leaf axils. Staminodes absent. Style plumose in fruit. Bracteoles united into a 2-lipped involucre.

1. Clematis cirrhosa L., Sp. Pl. 544 (1753); Boiss., Fl. 1 : 2 (1867). [Plate 288]

Evergreen. Stems climbing, almost glabrous, 2–5 m. Leaves 2–6 cm., clustered, long-petioled, mostly undivided, cordate-ovate or ovate-oblong, coarsely toothed or lobed, rarely leaves ternate. Flowers about 5 cm. or more across, solitary, deflexed, subtended near base by a cup-shaped, 2-lipped involucre. Tepals 4, white or yellowish-white, elliptical to obovate, obtuse to apiculate, densely pubescent outside. Achenes flat, ovate, hairy, with a 3–6 cm. long, feathery style. Fl. January–February.

Hab.: Maquis and maquis remnants. Sharon Plain, Philistean Plain, Upper and Lower Galilee, Mt. Carmel, Samaria, Judean Mts.
Area: Mediterranean.

Sect. CLEMATIS. Flowers usually numerous in panicles. Staminodes absent. Style plumose in fruit. Bracteoles free.

2. Clematis flammula L., Sp. Pl. 544 (1753); Boiss., Fl. 1 : 4 (1867). [Plate 289]
Evergreen. Stems climbing, 2–5 m. Leaves mostly 2-pinnate, pubescent when young, later glabrous; pinnae long-petiolulate, mostly 3-foliolate, leaflets ovate-oblong to elliptical, entire, lobed or dentate. Flowers sweet-scented, in loose, much branched panicles. Perianth 2–2.5 cm. across. Tepals 4, 1–1.3 cm., white, oblong-elliptical, obtuse, pubescent on both surfaces. Stamens numerous, as long as carpels, yellow; filaments glabrous. Achenes flattened, narrowly ovoid, brown, puberulent, with prominent margin and 2–4 cm. long, feathery style. Fl. April–May.
Hab.: Maquis. Coastal, Upper and Lower Galilee, Mt. Carmel, Samaria, Dan Valley.
Area: Mediterranean, with slight extensions into W. Irano-Turanian countries.
Used in medicine as diuretic, etc.

6. RANUNCULUS L.

Perennial or annual, hairy or glabrous, sometimes aquatic herbs. Leaves often palmately lobed or otherwise divided or deeply cut, rarely undivided, entire or dentate. Flowers hermaphrodite, solitary and terminal or in branched cymes or panicles, often yellow, rarely white or red, actinomorphic, all parts of flower spirally arranged. Sepals 3–5. Petals 5–8 (–20), with a nectariferous pit near base on upper side. Stamens numerous. Gynoecium superior, composed of many free, 1-ovuled carpels; ovules ascending. Fruit a head of many distinct, 1-seeded achenes with persistent styles.
Over 400 widely distributed species; mainly in the northern extratropical hemisphere.
Literature: P. Graebner f., *Ranunculus* subgen. *Batrachium*, in: Aschers. u. Graebn., *Syn. Mitteleur. Fl.* 5, 3 : 74–98 (1935). M. Zohary, Taxonomical studies in the flora of Palestine and neighbouring countries. *Palest. Journ. Bot. Jerusalem* ser. 2 : 151–153 (1941). R. D. Meikle, The Batrachian Ranunculi of the Orient. *Not. Roy. Bot. Gard. Edinb.* 23 : 13–22 (1959). P. H. Davis, Materials for a Flora of Turkey: IV. Ranunculaceae: II, *Ranunculus* L., *Not. Roy. Bot. Gard. Edinb.* 23 : 103–161 (1960).

1. Aquatic plants with white flowers. All or part of leaves dissected into capillary segments, submerged   2
 – Terrestrial or swamp plants, with yellow or red flowers   5
2. Upper leaves floating on water surface, not finely dissected.   **16. R. saniculifolius**
 – All leaves finely dissected   3
3. Leaves 0.5–2.5 cm. in diam., circular in outline, segments rigid.   **17. R. sphaerospermus**
 – Leaves usually longer, not circular in outline   4
4. Flowers 0.8–1 cm. across. Nectary lunate. Stamens 6–15.   **18. R. trichophyllus**
 – Flowers about 2 cm. across. Nectary circular. Stamens 12 or more.
                                                                                        15. R. aquatilis ssp. **heleophilus**

5 (1). Sepals 3 (–4). Petals 8–12. Leaves ovate-cordate, undivided.   **1. R. ficaria**
– Sepals and petals usually 5   6
6. Corolla red, rarely yellow. Petals 2–4 cm. or more long, 1–1.5 cm. broad. **6. R. asiaticus**
– Corolla yellow, usually much smaller than in above   7
7. Leaves undivided; the lowermost ovate-cordate, the upper oblong-lanceolate. Mature
   head of achenes globular, about 4 mm. or less in diam.   **14. R. ophioglossifolius**
– Leaves distinctly lobed or divided into entire or incised lobes   8
8. Perennials with dimorphic roots: thick, ovoid or cylindrical and thin, fibrous ones.
   Flowers mostly solitary at the end of each branch, mostly 1.5–3 cm. across. Fruiting
   heads mostly ovoid to cylindrical   9
– Annuals, rarely perennials with monomorphic roots – all fibrous. Flowers in branched
   inflorescences, axillary, (0.6–)0.8–2(–2.5) cm. across. Fruiting heads mostly globular,
   rarely oblong or cylindrical but then achenes globular   12
9. Lower leaves 1–2(–3)-pinnately or palmately dissected into numerous, lanceolate to
   linear, 1–2 mm. broad lobes   10
– Lower leaves flabellately or ternately lobed; lobes ovate-oblong, 3–9 mm. broad   11
10. Fruiting heads ovoid, scarcely 1 cm. long; beak of achene hooked, 1–3 times as long
    as disk. Plants occurring in Edom.   **5. R. macrorhynchus**
— Fruiting heads cylindrical, 1 cm. long or more; beak of achene shorter than in previous.
    Common plants.   **4. R. millefolius**
11 (9). Fleshy roots short, ovoid. Achenes 1–2 mm. Flowers 2–3 cm. across.   **3. R. paludosus**
— Fleshy roots oblong-cylindrical. Achenes 4–6 mm. Flowers 1.5–2 cm. across.
                                                                   2. R. damascenus
12 (8). Achenes ovoid or globular, smooth, wrinkled or punctulate   13
— Achenes flattened, tuberculate, muricate or spiny   14
13. Perennials. Flowers 1.5–2.5 cm. across. Fruiting heads globular.
                                                              7. R. constantinopolitanus
— Annuals. Flowers 0.8–1 cm. across. Fruiting heads oblong to short-cylindrical.
                                                                 13. R. sceleratus
14 (12). All leaves cordate at base. Pedicels thickened and deflexed in fruit. Petals almost
    as long as sepals. Tubercles of achenes ending in a hooked hair.   **11. R. chius**
— Leaves not cordate at base. Pedicels not as in above. Petals longer than sepals.
    Tubercles of achenes without hooked hairs   15
15. Achenes spiny or sharply muricate   16
— Achenes tuberculate, neither spiny nor muricate   17
16. Lower leaves orbicular or reniform, crenate or with short lobes. Sepals deflexed.
    Achenes muricate.   **10. R. muricatus**
— Lower leaves cuneate-obovate, dentate or cut into 3 lobes. Sepals spreading. Achenes
    mostly spiny.   **12. R. arvensis**
17 (15). Fruiting heads 1.2–1.5 cm. in diam.; beak of achene 2–3 mm. long, recurved, disk
    3–4 mm. long, with very broad, almost wing-like margin.   **8. R. cornutus**
— Fruiting heads usually 0.7–1 cm. in diam. or shorter; achene and beak not as in above
    and usually much shorter.   **9. R. marginatus**

Subgen. FICARIA (Huds.) Benson. Calyx of 3 sepals. Petals 8–12, yellow, with nectariferous pit covered by a scale.

1. Ranunculus ficaria L. ssp. **ficariiformis** (F. W. Schultz) Rouy et Fouc., Fl. Fr. 1 : 73 (1893). *R. ficariaeformis* F. W. Schultz in F. W. Schultz et Billot, Arch. Fl. 1 : 123 (1848). *R. calthaefolius* (Guss.) Jord., Obs. Pl. Crit. 6 : 2 (1847). [Plate 290]

Perennial, glabrous, 10–20 cm., with dimorphic roots : long, thin and fibrous roots together with long or short, fleshy or tuberous ones. Stems herbaceous, rather short. Leaves (2–)3–6 cm., very long-petioled, ascending, ovate-cordate, entire or obscurely sinuate, glabrous; base of petiole widened and membranous. Scapes 5–15 cm. Flowers golden-yellow, up to 2.5 (–4) cm. across. Sepals 3, ovate, concave, yellowish-white. Petals 8–12, oblong, with a nectary scale at base. Stamens numerous. Fruiting heads globular, about 5 mm. across; achenes globular, 2–2.5 mm., hirsute, with a very short acute beak. Fl. January–April.

Hab. : Fields, deep heavy soil. Sharon Plain, Philistean Plain, Upper Galilee, Samaria, Judean Mts., Gilead, Ammon.

Area : Mediterranean and Euro-Siberian.

Used in medicine (herba, radix ficariae).

Subgen. RANUNCULUS. Calyx with 5 sepals. Petals yellow, rarely crimson, often 5; nectariferous pit mostly covered by a scale.

Sect. RANUNCULASTRUM DC., Prodr. 1 : 27 (1824). Perennials. Roots dimorphic : fleshy, unbranched and filiform, branched. Nectary scale free or largely adnate to petal, rarely 0. Fruiting head and torus mostly elongated. Achenes smooth or tuberculate, with conspicuous beaks.

2. Ranunculus damascenus Boiss. et Gaill. in Boiss., Diagn. ser. 2, 6 : 5 (1859) et Fl. 1 : 30 (1867). *R. oxyspermus* M. B. in Willd., Sp. Pl. 2 : 1328 (1800) ssp. *damascenus* (Boiss. et Gaill.) P. H. Davis, Not. Roy. Bot. Gard. Edinb. 23 : 143 (1960). *R. oxyspermus* M. B. var. *damascenus* (Boiss. et Gaill.) Post, Fl. Syr. Pal. Sin. ed. 2, 1 : 11 (1932). [Plate 291]

Perennial, patulous-hirsute, 10–30 cm. Roots dimorphic : oblong-cylindrical, tuberous and thin, fibrous. Stems erect, simple, branching above, whitish. Lower leaves long-petioled with membranous base; blade 3–5 cm. long, ovate or orbicular in outline, greenish, with cordate or cuneate base, cut into 3 ovate-cuneate, incised or obscurely ternately lobed segments, densely silky-canescent; the uppermost leaves short-petioled or sessile but more deeply incised, with few linear segments, mostly 3-lobed at tip. Flowers few, 1.5–2 cm. across, long-pedicelled, yellow. Sepals 5, later deflexed, silky-hairy. Petals 5, very broadly obovate, as long as or much longer than sepals; obovate-cuneate nectary scale at base of petal free in upper part. Stamens many. Fruiting heads 0.8–1 × 0.7–0.8 cm., ovoid; achenes 4–6 mm. long (incl. beak), triangular-ovate; disk with tubercles ending in a hair; beak slightly arcuate, as long as to much longer than disk. Fl. March–April.

Hab. : Steppes and among rocks. Ammon, Moav, Edom.

Area : W. Irano-Turanian.

3. Ranunculus paludosus Poir., Voy. Barb. 2 : 184 (1789). *R. chaerophyllus* auct. non L., Sp. Pl. 555 (1753). *R. flabellatus* Desf., Fl. Atl. 1 : 438, t. 114 (1798). [Plate 292]

Perennial, appressed- and silky-hairy, 10–35 cm. Roots dimorphic : short, ovoid, tuberous and long, thin, fibrous; root neck covered with a dense net of fibres. Radical leaves orbicular or cuneate-flabellate, coarsely toothed or 3-lobed; lower stem leaves divided into 3 broadly cuneate, deeply incised or lobed or coarsely dentate segments. Scape often terminating in a single flower, rarely with 1–2 additional long-pedicelled flowers. Flowers 2–3 cm. across, yellow, dark-veined. Sepals 5, hairy, spreading or slightly reflexed. Petals 5, broadly obovate, considerably broader and longer than sepals, with a nectary scale at base. Stamens numerous. Fruiting heads about 5 mm. broad, spike-like, ovoid-oblong; achenes 1–2 mm., ovate, punctulate, tapering to a spreading-recurved beak longer than disk. Fl. April–May.

Hab. : Fields and grassy sites on heavy soil. Upper Galilee, Ammon, Moav, Edom.
Area : Mediterranean, slightly extending into the Euro-Siberian region.

4. Ranunculus millefolius Banks et Sol. in Russ., Nat. Hist. Aleppo, ed. 2, 2 : 254 (1794); Eig, Journ. Bot. Lond. 1937 : 187 (1937). *R. orientalis* L., Sp. Pl. 555 (1753) non auct., *nom. confusum. R. orientalis* L. ssp. *hierosolymitanus* (Boiss.) P. H. Davis, Not. Roy. Bot. Gard. Edinb. 23 : 150 (1960). *R. hierosolymitanus* Boiss., Fl. 1 : 36 (1867). *R. myriophyllus* Russ. ex Schrad., Journ. Bot. 1 : 424 (1799) et DC., Syst. 1 : 257 (1817). [Plate 293]

Perennial, divaricately branched, leafy, canescent, 10–20 cm. Roots dimorphic : thick, cylindrical, tuberous and thin, fibrous. Stems thickened, somewhat angular. Lower leaves 6–10(–20) cm., long-petioled, membranously sheathed at base; blade oblong in outline, pinnatisect into oblong to linear, acute lobes; upper leaves sessile, with fewer but longer lobes. Flowers few to many, solitary at the end of branches, 2–2.5(–3) cm. across, long-pedicelled, yellow; pedicels thickened in fruit. Sepals 5, much shorter than petals, appressed or spreading, oblong, hairy. Petals 5, obovate, with a free, triangular-cuneate nectary scale. Fruiting heads 1–1.5 cm., spicate, oblong-cylindrical; achenes 2–3 mm., flat, obovate, glabrous (rarely hairy), smooth or punctulate, very rarely muriculate; beak lanceolate, slightly incurved, much shorter than disk. Fl. February–April.

Hab. : Batha and steppes. Shefela, Upper and Lower Galilee, Mt. Carmel, Samaria, Judean Mts., N. Negev, Gilead, Ammon, Moav, Edom.
Area : E. Mediterranean and W. Irano-Turanian.

5. Ranunculus macrorhynchus Boiss., Diagn. ser. 1, 6 : 5 (1846) et Fl. 1 : 28 (1867). *R. dasycarpus* (Stev.) Boiss. var. *macrorhynchus* (Boiss.) Zoh., Palest. Journ. Bot. Jerusalem ser. 2 : 152 (1941). *R. macrorhynchus* Boiss. ssp. *trigonocarpus* (Boiss.) P. H. Davis, Not. Roy. Bot. Gard. Edinb. 23 : 154 (1960). *R. trigonocarpus* Boiss., Diagn. ser. 1, 8 : 2 (1849) et Fl. 1 : 36 (1867). *R. dasycarpus* (Stev.) Boiss., Fl. 1 : 28 (1867). *R. dasycarpus* (Stev.) Boiss. var. *leiocarpus* Zoh. et var. *edumeus* Zoh., l.c. [Plate 294]

Perennial, densely subappressed and silky-canescent, 8–20 cm. Roots dimorphic : thicker, cylindrical and thin, fibrous. Lower leaves 5–10 cm., long-petioled, ovate in outline, 2–3-palmately dissected into linear lobes; base of petiole sheathed, with membranous margin; upper leaves short-petioled, few-lobed. Pedicels axillary, divaricate in upper part of scape, 3–6 cm., not thickened. Flowers 1.6–2 cm. in diam., yellow. Sepals 5, appressed or spreading, ovate, silky, much shorter than obovate-oblong petals. Fruiting heads ovoid, with few achenes; achenes flat, obtriangular, hirtellous or glabrous at margin; disk smooth or tuberculate; beak slender, incurved, scarcely flattened, 1 to 3 times as long as disk. Fl. March.

Hab.: Steppes; compact soil covered with gravel. Edom.

Area: W. Irano-Turanian.

6. **Ranunculus asiaticus** L., Sp. Pl. 552 (1753); Boiss., Fl. 1 : 31 (1867). [Plate 295]

Perennial, hirsute, 10–30 cm. Roots dimorphic: short, cylindrical, tuberous and long, thin, fibrous. Subterranean short stem often with thin, horizontal runners. Lower leaves long-petioled, broadly ovate to orbicular, crenate or 3-lobed, rarely dissected; the others cuneate or more or less dissected into oblong, toothed or incised lobes, or 3-partite into cuneate ones; the uppermost cut into oblong to linear lobes. Scape solitary, simple or slightly branching, 1–4-flowered. Flowers (2–)3–6(–8) cm. in diam., crimson (rarely orange or yellow). Sepals 5, spreading or deflexed. Petals 5, obovate-cuneate, without nectary scale. Stamens numerous. Receptacle (rhachis) glabrous, elongating in fruit. Heads 2–4 cm., cylindrical, spike-like; achenes 2–3 mm., compressed, papery, ovate-orbicular with a hooked beak about half as long as disk. Fl. February–May.

Hab.: Devastated maquis, batha, stony hillsides, abandoned fields and steppes. Coastal Galilee, Acco Plain, Sharon Plain, Philistean Plain, Upper and Lower Galilee, Mt. Carmel, Mt. Gilboa, Samaria, Judean Mts., Judean Desert, N. and N. E. Negev, Dan Valley, Upper Jordan Valley, Gilead, Ammon, Moav, Edom.

Var. **asiaticus**. Lower leaves undivided or only 3-fid into broad lobes.

Hab.: Throughout the Mediterranean territory.

Var. **tenuilobus** Boiss., l.c. All the leaves much dissected into narrow lobes.

Hab.: Predominantly in steppes and deserts.

Area of species: Mediterranean, with extensions into W. Irano-Turanian territories.

R. asiaticus is one of the most common showy flowers of the late winter and spring flora. It occurs in a variety of habitats and penetrates deeply into steppes in the south and east of the country. Sometimes the two varieties mentioned meet and intermediates between them occur. In Palestine the yellow-flowered form is very rare.

Also grown as ornamental plant.

Sect. CHRYSANTHAE Sp., Hist. Nat. Veg. Phaner. 7 : 212 (1839). Perennials or annuals. Roots monomorphic. Nectary scale free nearly to base. Fruiting heads subglobular; torus not elongating; achenes smooth. Leaves deeply cut or divided.

7. **Ranunculus constantinopolitanus** (DC.) Urv., Mém. Soc. Linn. Par. 1 : 317 (1822). *R. lanuginosus* L. var. *constantinopolitanus* DC., Syst. 1 : 281 (1817). *R. constantinopolitanus* var. *palaestinus* (Boiss.) Boiss., Fl. 1 : 49 (1867). [Plate 296]

Perennial, erect, villose-hispid, up to 1 m., with a bundle of long, cylindrical fibrous roots. Stems striate, leafy mainly at base and paniculately branching above. Lower leaves with 5–15(–20) cm. long, white-hispid petioles and appressed-hirsute blades, almost orbicular in outline, cordate at base, 3-parted, with lobed lateral segments, middle lobe cuneate at base, coarsely dentate-lobulate at apex; upper leaves short-petioled; the uppermost and floral ones reduced to linear-lanceolate segments. Inflorescences branched. Pedicels long, slender. Flowers yellow, 1.5–2.5 cm. in diam. Sepals 5, reflexed, villose. Petals obovate to obovate-oblong, much longer and broader than sepals, with a conspicuous nectary scale. Fruiting heads globular, about 1 cm. across; receptacle glabrous; achenes 4–5 mm. long, ovoid, smooth or finely punctulate, with very narrow margins, ending abruptly in a short (1 mm. long), hooked beak. Fl. March–April.

Hab.: Swamps and damp places on heavy soil. Upper Galilee.

Area: E. Mediterranean, extending into adjacent territories of the Irano-Turanian region.

Sect. ECHINELLA DC., Prodr. 1 : 41 (1824). Annuals. Nectary scale free above or nearly to base. Fruiting heads subglobular; torus not elongating. Achenes tuberculate or muricate or spiny.

8. Ranunculus cornutus DC., Syst. 1 : 300 (1817); Boiss., Fl. 1 : 56 (1867); P. H. Davis, Not. Roy. Bot. Gard. Edinb. 23 : 156 (1960). *R. lomatocarpus* Fisch. et Mey., Ind. Sem. Hort. Petrop. 1 : 36 (1835). [Plate 297]

Annual, sparingly hirsute, tuberculate at base, 30–60 cm. Stems almost hollow, grooved, somewhat flexuous, branching from middle. Lower leaves long-petioled; blades orbicular in outline, the lowermost dentate, not dissected, the next-lowest divided into 3 often 3-sect segments, the middle segment often long-petioled, deeply incised or toothed; stem leaves long-petioled, cut into oblong-lanceolate, 1–3 cm. long segments; the uppermost sessile, with few linear-lanceolate acute lobes. Flowers many, 1.5–2 cm. across, yellow. Sepals 5, shorter than broadly ovate petals, ovate, reflexed, hairy outside; nectary scale cuneate, free in upper part. Fruiting heads globular in outline, 1.2–1.5 cm. across, loose, stellate; achenes 12–15 in number, about 7 mm. long (incl. beak), flat, ovate-orbicular, with a broad, wing-like margin, latter without furrows on either side; disk densely tuberculate, tubercles ending in a thick and short bristle; beak (1–)2–5 mm., broadly lanceolate, slightly recurved. Fl. March–April.

Hab.: Heavy soil. Sharon Plain, Lower Galilee, Lower Jordan Valley. Rare.

Area: E. Mediterranean, extending into W. Irano-Turanian territories.

9. Ranunculus marginatus Urv., Mém. Soc. Linn. Par. 1 : 318 (1822); P. H. Davis, Not. Roy. Bot. Gard. Edinb. 23 : 158 (1960). *R. trachycarpus* Fisch. et Mey., Ind. Sem. Hort. Petrop. 3 : 46 (1837); Boiss., Fl. 1 : 55 (1867). [Plate 298]

Leafy annual, sparingly pubescent or glabrous, 15–30 cm. Stems erect, grooved, mostly branching from middle. Lower leaves long-petioled with petioles broadened at base; blade 2–5 cm., orbicular in outline, much dissected or 3-lobed or 3-sect into broad, cuneate-ovate to orbicular segments, the middle ones often broadly stalked; stem leaves with ovate, orbicular or oblong lobes (or cut into linear segments); the uppermost few-lobed, lobes oblong-linear. Flowers many, 0.8–1.5 cm. across, yellow. Pedicels not thickened in fruit. Sepals 5, ovate-oblong, hairy outside, reflexed, much shorter than broadly obovate petals; nectary scale almost free. Fruiting heads 0.7–1 cm. in diam. or shorter, globular; achenes 10–20 in number, 3–5 mm., flat, ovate to orbicular; disk 2–3 mm., densely tuberculate, with very narrow keeled margin and a narrow furrow on either side; beak 1 mm., triangular to lanceolate, usually not curved. Fl. March–April.

Var. **marginatus.** Leaves moderately divided or parted. Beak triangular-lanceolate.

Hab · Fields and roadsides, Sharon Plain, Philistean Plain, Upper Galilee, Esdraelon Plain, Lower Jordan Valley, Moav. Rather rare.

Var. **scandicinus** (Boiss.) Zoh. (comb. nov.). *R. trachycarpus* Fisch. et Mey. var. *scandicinus* Boiss., Fl. 1 : 55 (1867). [Plate 299] All leaves much dissected, the upper ones cut into linear, acute, somewhat denticulate and sparingly setulose lobes. Beak triangular.

Hab.: Wet fields and along ditches. Acco Plain, Sharon Plain, Philistean Plain, Upper and Lower Galilee, Mt. Carmel, Esdraelon Plain, Samaria, Shefela, Judean Mts., Hula Plain, Beit Shean Valley, Moav. Common.

Area of species: Mediterranean, slightly extending into W. Irano-Turanian and S. Euro-Siberian territories.

A form with minute flowers and fruiting heads of about 5 mm. across, described as *R. trachycarpus* Fisch. et Mey. var. *minor* Zoh., Palest. Journ. Bot. Jerusalem ser. 2 : 152 (1941), from the Sharon Plain, seems to be a "hunger form" of var. *scandicinus*.

R. guilelmi-jordani Aschers., Verh. Bot. Ver. Prov. Brandenb. 21 : 64 (1880) is another form distinguished by the reflexed calyx and by the petals scarcely longer than sepals. Boissier, Fl. Suppl. 14 (1888) was right to reduce this taxon to the rank of a variety. It is recorded by Dinsmore in Post, Fl. Syr. Pal. Sin. ed. 2, 1 : 15 (1832) from Moav; we have not collected this form.

10. Ranunculus muricatus L., Sp. Pl. 555 (1753); Boiss., Fl. 1 : 56 (1867). [Plate 300]

Annual, almost glabrous, 10–20 cm. Stems erect or procumbent, much branching, especially above. Lower leaves long-petioled, with almost orbicular-reniform blade, either undivided and coarsely crenate-lobed or, like upper ones, ternately or 2-ternately dissected; segments cuneate, crenate or lobed; uppermost leaves little divided, short-petioled. Pedicels axillary, opposite leaves. Flowers 1–1.5 cm. in diam., bright yellow. Sepals 5, erect or deflexed, sparingly setulose. Petals 5, obovate-oblong, longer than calyx, with nectary scale. Stamens many. Fruiting heads globular, 1.2–1.5(–2) cm. across; receptacle hairy or glabrous; achenes 7–8 mm. (incl. beak), stellately spreading, compressed, ovate, with 2-furrowed margin; disk studded with long, straight or hooked, prickly tubercles; beak sword-like, erect or slightly curved, triangular-lanceolate, almost as long as or shorter than disk. Mature head becoming detached from axis as a whole. Fl. March–April.

Hab.: Damp places and fields on heavy soils. Sharon Plain, Philistean Plain, Upper Galilee, Esdraelon Plain, Judean Mts., Hula Plain, Lower Jordan Valley, Gilead.

Area: E. Mediterranean and W. Irano-Turanian.

11. Ranunculus chius DC., Syst. 1 : 299 (1817); Boiss., Fl. 1 : 54 (1867). [Plate 301]

Annual, branched, softly and sparingly villose or glabrescent, 10–30 cm. Stems erect to ascending, hollow, striate, branching at sharp angles. Leaves cordate-orbicular, with 3–5 entire or lobed divisions; the upper ones deeply 3-lobed, middle lobe often entire, longer than the 2–4-lobed lateral ones; the uppermost and floral leaves oblong-elliptical, often undivided. Pedicels axillary, opposite leaves, short and thick, the fruiting ones arcuate-deflexed and inflated above, 1.5–2 cm. long, 1.5–2.5 mm. thick. Flowers minute, 6–8 mm. in diam., whitish-yellow. Sepals 5, reflexed, villose. Petals 5, almost as long as sepals, with nectary scale. Fruiting heads 6–8 mm. in diam., globular; receptacle glabrous; achenes 3–4 mm., compressed, ovate, tapering into a hooked beak half as long as disk or shorter; disk studded with tubercles ending in a hooked hair (hair often caducous in ripe achenes); margin smooth, acute, not grooved. Fl. March–April.

Hab. : Grassy sites, woods, batha and among rocks. Sharon Plain, Philistean Plain, Upper and Lower Galilee, Mt. Carmel, Samaria, Judean Mts., Lower Jordan Valley, Gilead.

Area : N. Mediterranean, with slight penetrations into adjacent Irano-Turanian and Euro-Siberian territories.

12. Ranunculus arvensis L., Sp. Pl. 555 (1753); Boiss., Fl. 1 : 57 (1867). [Plate 302]

Annual, hairy or glabrescent, 10–30 cm. Stems erect, branching. Lower leaves long-petioled, obovate-cuneate, dentate or cut into 3 lobes, the remainder often 2-ternate, becoming more dissected towards apex of stem; lobes of uppermost ones 2–5 cm., linear to lanceolate. Flowers axillary and in terminal cymes, borne on hairy pedicels. Sepals 5, spreading, yellowish-green, hairy, shorter than petals. Petals 5, bright lemon-yellow, obovate, spatulate; nectary covered by a free scale somewhat broader than base of petal. Stamens 10 or more (rarely fewer). Receptacle hairy; achenes 4–8 in number, 5–8 mm. (incl. beak), stellately spreading, compressed; disk muricate or prickly, margin elevated at right angle to surface of disk, fringed at both edges with stiff spines or prickles longer than those of disk; spines sometimes reduced to tubercles or teeth; beak 2–3 mm., straight or hooked. Fl. March–May.

Var. **arvensis.** Disk and margin thick, covered with hooked, moderately long spines or prickles.

Hab. : Mainly a weed in cereal crops but also elsewhere; very common. Acco Plain, Sharon Plain, Philistean Plain, Upper and Lower Galilee, Esdraelon Plain, Mt. Gilboa, Samaria, Mt. Carmel, Judean Mts., Judean Desert, N. Negev, Upper and Lower Jordan Valley, Gilead, Ammon, Moav, Edom.

Var. **longispinus** Post, Fl. Syr. Pal. Sin. 42 (1883–1896); Fiori, Nuov. Fl. Anal. It. 1 : 677 (1924). Spines of margin longer than disk and much longer than beak or beak 0.

Hab. : As above. Acco Plain, Sharon Plain, Upper and Lower Galilee, Judean Mts., Upper Jordan Valley, Gilead.

Var. **tuberculatus** (DC.) Fiori, Nuov. Fl. Anal. It. 1 : 677 (1924). *R. tuberculatus* DC. var. *prostratus* Post, l.c. Spines reduced to tubercles or teeth.

Hab.: As above. Samaria (Post, ed. 2, 1 : 17, 1932).

Area of species : Mediterranean, Irano-Turanian and Euro-Siberian.

Sect. HECATONIA (Lour.) DC., Prodr. 1 : 30 (1824). Swamp or water plants. Nectary scale often forked or completely surrounding nectary. Achenes scarcely compressed; beak usually very short.

13. Ranunculus sceleratus L., Sp. Pl. 551 (1753); Boiss., Fl. 1 : 52 (1867). [Plate 303]

Annual, glabrous, leafy, 20–60 cm. Stems erect, hollow, furrowed, branching from base and all along. Lower leaves long-petioled, reniform or 5-angled, palmately 3-lobed, all or the lateral lobes often 2–3-lobed, all lobes crenate or coarsely toothed; upper leaves short-petioled, with fewer but deeply divided lobes; uppermost leaves sessile, with 1–3 lobes. Flowers 0.8–1 cm. across, long-pedicelled, pale yellow, numerous, in paniculate inflorescences. Sepals 5, reflexed, hairy. Petals 5, narrowly obovate-elliptical, slightly longer than sepals; nectariferous cavity almost open, with short scale. Fruiting heads 0.6–1.2 cm., oblong to short-cylindrical; receptacle somewhat hairy; achenes numerous, minute, less than 1 mm., ovoid to almost globular, glabrous, slightly wrinkled on inner side, very short-beaked. Fl. April–June.

Hab.: Swamps. Sharon Plain, Hula Plain.

Area : Euro-Siberian, Mediterranean and Irano-Turanian.

Strongly poisonous; used in medicine (herba ranunculi scelerati).

Sect. FLAMMULA (Webb) Ovcz., in Fl. URSS 7 : 361 (1937). Roots fibrous. Leaves undivided. Petals caducous. Receptacle glabrous; acheres slightly compressed; beak very short.

14. Ranunculus ophioglossifolius Vill., Hist. Pl. Dauph. 3 : 731 (1789); Boiss., Fl. 1 : 53 (1867). [Plate 304]

Annual, glabrous or slightly hairy, 15–40 cm. Stems hollow, furrowed, erect or ascending, branching from base and at top. Lowermost leaves long-petioled, orbicular-cordate to ovate; lower stem leaves long-petioled, with ovate-oblong blades, the upper short-petioled with oblong to elliptical blades, the uppermost linear-lanceolate, sessile, all except uppermost remotely dentate. Flowers numerous, 0.6–1 cm. across, long-pedicelled, pale yellow. Sepals 5, glabrous. Petals 5, slightly exceeding sepals in length, with a nectary scale. Fruiting heads minute, 3–4 mm. across, globular; receptacle glabrous; achenes 1–1.2 mm., slightly compressed, obovoid, narrowly margined, finely tuberculate, very short-beaked. Fl. March–April.

Hab.: Swamps. Sharon Plain, Philistean Plain, Hula Plain.

Area : Mediterranean and W. Irano-Turanian, with extension into the Euro-Siberian region.

Subgen. BATRACHIUM (DC.) A. Gray. All or part of leaves divided into capillary segments. Petals white, nectariferous pit at base of petals naked. Fruiting pedicels arcuate. Achenes transversally wrinkled. Aquatic plants.

15. Ranunculus aquatilis L. ssp. **heleophilus** (Arv.-Touv.) Rikli in Schinz et Keller, Fl. Schweiz ed. 2, 2 : 80 (1905). *R. heleophilus* Arv.-Touv., Essai Pl. Dauph. 19 (1871). *R. aquatilis* L. var. *submersus* Gren. et Godr., Fl. Fr. 1 : 23 (1847) p.p.; Boiss., Fl. 1 : 23 (1867) p.p.; Meikle, Not. Roy. Bot. Gard. Edinb. 23 : 18 (1959). [Plate 305]

Aquatic plant with submerged, up to 4 cm. long leaves; blade 3-fid and repeatedly forked into filiform lobes, collapsing outside water; petiole conspicuous. Flowers medium-sized, white with yellow ground. Petals about 1 cm., contiguous or not, obovate-cuneate, with a circular nectariferous pit at base. Stamens numerous. Fruiting pedicels often recurved. Receptacle pilose. Achenes 1.2–1.8 mm., glabrous or hairy, with convex, often distinctly keeled dorsal margin and acute, slightly beaked apex. Fl. February–April.

Hab.: Swamps, ponds and shallow water bodies. Sharon Plain, Samaria, Judean Mts., Hula Plain, Upper Jordan Valley, Gilead and elsewhere. Fairly common.

Area : Temperate and tropical regions.

Locally occurring plants belong mainly to var. **microcarpus** Meikle, l.c. with 1–1.5 mm. long, mostly glabrous or subglabrous achenes. The typical heterophyllous variety of *R. aquatilis* has so far not been observed in Palestine.

16. Ranunculus saniculifolius Viv., Fl. Lib. Spec. 29 (1824). *R. peltatus* sensu Meikle, Not. Roy. Bot. Gard. Edinb. 23 : 14 (1959). *R. aquatilis* sensu Post, Fl. Syr. Pal. Sin. ed. 2, 1 : 8 (1932). [Plate 306]

Aquatic robust plant. Stems 1 m. or longer. Submerged leaves 1–5 cm. or longer, usually short-petioled or subsessile, with slender, filiform, diverging, rather rigid segments; transitional leaves rarely present; floating leaves usually numerous, 1–2.5 cm. broad or more, long-petioled, generally reniform or semicircular in outline, more or less hairy beneath, divided almost to middle into 3 (–5)-lobed segments, basal margin of blade truncate or forming a wide sinus. Pedicels usually much longer than the submerged leaves, tapering slightly from base to apex. Flowers often 0.8–1 cm., with contiguous or overlapping, obovate-cuneate petals. Receptacle pilose. Achenes 1.5–1.8 mm., usually more or less pilose, sometimes glabrous, with distinctly convex dorsal margin and rounded or obscurely beaked apex. (Description after Meikle, l.c.). Fl. May.

Hab.: In water. Mt. Carmel, Moav (after Dinsmore in Post, l.c.).

Area : Euro-Siberian and Mediterranean; also in other regions.

17. Ranunculus sphaerospermus Boiss. et Bl. in Boiss., Diagn. ser. 2, 5 : 6 (1856). *R. aquatilis* L. var. *sphaerospermus* (Boiss. et Bl.) Boiss., Fl. 1 : 23 (1867). *R. peltatus* Schrank ssp. *sphaerospermus* (Boiss. et Bl.) Meikle, Not. Roy. Bot. Gard. Edinb. 23 : 16 (1959). *R. circinatus* Dinsmore in Post, Fl. Syr. Pal. Sin. ed. 2, 1 : 9 (1932) non Sibth., Fl. Oxon. 175 (1794). [Plate 307]

Leaves all submerged, (0.5–)1.5–2.5 cm., very short-petioled, with numerous, diverging, rather rigid segments, forming more or less dense spherical tufts, not collapsing outside water. Pedicels (2.5–)3–6(–8) cm., often greatly exceeding leaves. Flowers large, 1.5–2.5(–3) cm. in diam., with broadly obovate petals. Achenes minute, about 0.7–1 mm., subglobular, glabrous or pilose at apex. Fl. March–April.

Hab.: In stagnant shallow waters. Hula Plain, Upper Jordan Valley.

Area: E. Mediterranean and probably other regions.

18. Ranunculus trichophyllus Chaix in Vill., Hist. Pl. Dauph. 1 : 335 (1786); Boiss., Fl. 1 : 23 (1867) p.p. excl. var.; Meikle, Not. Roy. Bot. Gard. Edinb. 23 : 20 (1959). [Plate 308]

Leaves all submerged, up to about 6(–9) cm., subsessile or short-stalked, with slender diverging segments collapsing outside water. Pedicels usually shorter than leaves, not distinctly tapering. Petals 4–5 mm., generally not contiguous, narrowly obovate-cuneate, with lunate, hirsute nectariferous pit at base. Stamens 6–15. Fruiting pedicels distinctly recurved. Receptacle pilose. Achenes about 0.7–1.8 mm., glabrous or hairy, with convex, often distinctly keeled dorsal margin and acute short-beaked apex. Fl. February–April.

Hab.: Recorded by Boissier (l.c.) and Nabelek (Publ. Fac. Sci. Univ. Masaryk 35 : 14, 1923) from Upper Jordan Valley. The specimens collected by us in the Upper Jordan Valley are referable to *R. sphaerospermus*. The locality recorded by Dinsmore "Kurretra to Susa" (Meikle, l.c.) lies outside Palestine. Plate 308 was drawn from a Swiss specimen.

Area: Mainly Mediterranean, Irano-Turanian and Euro-Siberian; probably also elsewhere.

7. CERATOCEPHALA Moench

Dwarf annuals with simple or branched stems. Leaves divided into linear segments. Flowers small, pale yellow. Sepals 5. Petals 5 with nectariferous pit covered by a scale and with a narrow claw. Stamens 5 to numerous. Fruit a spike-like head of numerous, erect achenes, each with a pair of empty lateral cells at base and a long, sickle-shaped beak. Plants with aspect of *Ranunculus*.

Seven species, mostly in Euro-Siberian, Irano-Turanian and Mediterranean regions.

1. Ceratocephala falcata (L.) Pers., Syn. 1 : 341 (1805); Boiss., Fl. 1 : 58 (1867). *Ranunculus falcatus* L., Sp. Pl. 556 (1753). [Plate 309]

Dwarf annual, hairy or glabrescent, 3–10 cm. Stems 0 or 1 to several, often unequally long. Leaves all radical, more or less long-petioled; blade 0.5–2 cm., oblong to obovate in outline, palmately dissected into linear lobes, white-hairy to glabrescent. Flowers 0.8–1(–1.4) cm. across, terminal, solitary. Sepals oblong-elliptical, membranous-margined, hairy, shorter than whitish-yellow, oblong, clawed petals; nectary scale 2–2.5 mm., on elongated claw. Fruiting heads spicate, oblong to cylindrical. Achenes numerous, 0.8–1.2 cm., often persistent, each with 2 empty lateral cells at its base; beak falcate or incurved, 4–5 times as long as achene proper. Fl. March–April.

Hab.: Batha, fields and steppes. Upper and Lower Galilee, Samaria, Judean Mts., Judean Desert, N. and C. Negev, Upper and Lower Jordan Valley, Gilead, Ammon, Moav, Edom.

Area: E. Mediterranean and Irano-Turanian, also in some provinces of the Euro-Siberian region.

C. falcata has been divided by Boissier and others into a series of varieties; these are:

Var. **vulgaris** Boiss., l.c.: Plants with scapes equalling or exceeding the leaves in length. Achenes with broad beak, falcate from base.

Var. **exscapa** Boiss., l.c.: Dwarf plants with sessile or subsessile flowers.

Var. **incurva** (Stev.) Boiss., l.c.: Scapes longer than leaves. Beak of achene narrow, more or less erect, with incurved-hooked apex. Tomentose plants.

Since all these varieties may occur together in the same site and are linked by inter-mediates, their constancy is still to be tested.

8. ADONIS L.

Annual or perennial herbs. Leaves alternate, 2–3-pinnately dissected into linear or filiform segments. Flowers mostly terminal and solitary, actinomorphic. Sepals 5 (–8), erect or spreading, glabrous or hairy. Petals 5–20, showy, without nectariferous pit at base. Stamens numerous. Gynoecium superior, composed of many free carpels inserted on a cylindrical or conical receptacle; carpel with erect or pendulous ovule; styles short, persistent, becoming indurated in fruit. Fruit an elongated head of wrinkled, short-beaked achenes.

About 20 species in temperate Eurasia and the Mediterranean region.

Literature: E. Huth, Revision der Arten von *Adonis* und *Knowltonia. Helios, Monatl. Mitt. Gesamtgeb. Nat.* 8 : 61–73 (1890). H. Riedl, Revision der einjährigen Arten von *Adonis* L., *Ann. Naturh. Mus. Wien* 66 : 51–90 (1963).

1. Achene (incl. beak) 5–6 mm., oblong, longitudinally wrinkled-ribbed; beak erect, as long as body of achene. Flowers (2–)5(–6) cm. in diam.   **1. A. aleppica**
– Achene and beak considerably smaller. Flowers generally (1–)1.5–3(–4) cm. in diam.   2
2. Inner (upper) margin of achene not toothed, achene rounded all over.   **5. A. annua**
– Inner (upper) margin of achene with a tooth close to or at some distance from beak   3
3. Achene 3–5 mm. broad; tooth on inner (upper) margin at considerable distance from beak. Flowers 2–4 cm. in diam. Transverse ridge passing around middle of achene. Head up to 8 mm. broad.   **2. A. aestivalis**
– Achene 2–3(–4) mm. broad; tooth or gibbosity on inner margin very close to beak. Flowers smaller   4
4. Flowers yellow or orange-coloured. Tooth on inner (upper) margin of achene narrow, dorsal crest of achene mostly prominent and toothed.   **4. A. dentata**
– Flowers red (very rarely yellow). Gibbosity on inner margin of achene broad, dorsal crest rather obsolete.   **3. A. cupaniana**

1. Adonis aleppica Boiss., Ann. Sci. Nat. Bot. ser. 2, 16 : 350 (1841) et Fl. 1 : 16 (1867). [Plate 310]

Annual, glabrous or slightly hairy, 15–40 cm. Stems erect, striate, branching from base. Leaves cut into linear, setaceous, acute lobes. Flowers long-pedicelled. Sepals ovate, glabrous, half as long as 1.6–2.5(–3) × 0.6–0.8 cm., red or orange-coloured, flat, obovate petals. Fruiting heads cylindrical, more or less dense. Achenes 5–6 mm., oblong, longitudinally wrinkled-ribbed, tapering to an erect beak of same length as broad body of achene; beak acute, canaliculate at base. Fl. February–April.

Hab.: Fields. Sharon Plain, Samaria, Upper Jordan Valley, Gilead.

Area: E. Mediterranean.

The most showy species among local representatives of the genus.

2. Adonis aestivalis L., Sp. Pl. ed. 2, 771 (1762); Boiss., Fl. 1 : 17 (1867). [Plate 311]

Annual, almost glabrous, 20–60 cm. Stems erect, furrowed, simple or branching. Leaves 3–4-pinnatisect; segments linear, acute or mucronate. Flowers long-pedicelled. Sepals appressed to petals, broadly ovate, glabrous. Petals 6–8 in number, 1–1.7 cm., about twice as long as sepals, spreading, scarlet (rarely yellow), sometimes with a dark basal spot, oblanceolate, entire or emarginate. Stamens much shorter than petals, dark violet. Fruiting heads 7–8 mm. broad or more, dense, ovoid-oblong, then cylindrical. Achenes obliquely ovoid, wrinkled-pitted, wih a circular ridge, usually crested-dentate around middle; tooth on inner (upper) margin obtuse or acute, at some distance from the green, short, ascending, recurved or horizontal beak. Fl. April–May.

Var. **aestivalis.** Beak ascending; tooth on inner (upper) margin mostly separated from beak by a sinus.

Hab.: Fields and bathas. Upper and Lower Galilee, Shefela, Judean Mts., Judean Desert, Upper Jordan Valley, Ammon.

Var. **palaestina** (Boiss.) Zoh. (comb. et stat. nov.). A. palaestina Boiss., Diagn. ser. 1, 8 : 1 (1849) et Fl. 1 : 16 (1867). Beak horizontal or recurved; tooth not separated from beak by a sinus.

Hab.: Fields and batha. Sharon Plain, Philistean Plain, Upper and Lower Galilee, Mt. Carmel, Esdraelon Plain, Samaria, Shefela, Judean Mts., Judean Desert, Hula Plain, Upper Jordan Valley, Ammon.

Area of species: Mediterranean, Irano-Turanian and Euro-Siberian.

Dried plants used as a heart tonic.

The record by Dinsmore in Post, ed. 2, 1 : 7 (1932) on the occurrence of A. flammea Jacq., Fl. Austr. 4 : 29 (1776) in Palestine is erroneous.

3. Adonis cupaniana Guss., Fl. Sic. Syn. 2 : 36 (1845). A. aestivalis L. var. cupaniana (Guss.) Huth, Rev. Adon. 64 (1890). A. microcarpa Boiss., Fl. 1 : 18 (1867) non DC., Syst. 1 : 223 (1817). [Plate 312]

Annual, 10–30 cm. Stems erect. Leaves 3–4-pinnatisect; segments linear, acute or mucronate. Flowers about 1 cm. in diam. Sepals mostly glabrous. Petals almost twice as long as sepals, red or rarely yellow, with a dark basal spot, oblong or elliptical, obtuse, mostly irregularly dentate at apex. Stamens violet. Fruiting heads about 1.5 cm., dense, oblong. Achenes about 3 mm., almost rhomboidal, strongly wrinkled, without marked dorsal crest; inner (upper) margin of achene gibbous, the outer (lower) margin with an obtuse tooth; beak short, ascending or slightly incurved, green. Fl. February–April.

Hab.: Fields. Acco Plain, Sharon Plain, Philistean Plain, Upper and Lower Galilee, Mt. Carmel, Esdraelon Plain, Samaria, Shefela, Judean Mts., Judean Desert, Hula

Plain, Upper and Lower Jordan Valley, Dead Sea area, Gilead, Ammon, Moav, Edom.

Area : Mediterranean, extending into some Euro-Siberian and W. Irano-Turanian territories.

4. Adonis dentata Del., Fl. Eg. 287, t. 53 (1813) ; Boiss., Fl. 1 : 18 (1867). [Plate 313]

Annual, 8–25 cm. Stems mostly erect, usually branched. Leaves 2–3-pinnatisect, with filiform, acute segments. Flowers up to 1.5 (–2) cm. across. Sepals glabrous. Petals over twice as long as sepals, oblong or obovate, yellow or orange-coloured. Stamens violet. Fruiting heads 1–3 cm. or shorter, dense, cylindrical. Achenes small, about 2 (rarely 3–4) mm., globular to rhomboidal, wrinkled, dorsal crest prominent (rarely obscure), strongly toothed, inner (upper) margin with an obtuse tooth, very close to and reaching or exceeding beak in height; beak blackish, ascending, sometimes slightly incurved. Fl. February–April.

Var. **dentata.** Flowers up to 1 (–1.5) cm. across. Fruiting heads dense, mostly 1–2 cm. Achenes 2 (–3) mm.

Hab. : Steppes and deserts. Sharon Plain, Philistean Plain, Mt. Carmel, Mt. Gilboa, Samaria, Judean Mts., Judean Desert, N. and C. Negev, Upper and Lower Jordan Valley, Gilead, Ammon, Moav, Edom.

Var. **subinermis** Boiss., Fl. 1 : 19 (1867). Dorsal crest not prominent. Fruiting heads less dense and flowers smaller than in type.

Hab. : As above. Judean Desert, N. Negev, Dan Valley, Upper Jordan Valley, Ammon.

Area of species : Saharo-Arabian and Irano-Turanian, slightly extending into the Mediterranean borderland.

5. Adonis annua L., Sp. Pl. 547 (1753). *A. autumnalis* L., Sp. Pl. ed. 2, 771 (1762); Boiss., Fl. 1 : 16 (1867). [Plate 314]

Annual, usually glabrous, 20–60 cm. Stems erect, striate, usually branched. Leaves 3–4-pinnatisect; segments linear, acute. Flowers 1.5–2.5 cm. in diam. Sepals 5, ovate, spreading or deflexed, green or purplish. Petals (0.5–)0.6–0.8 (–1) cm., slightly longer than sepals, erect to connivent, ovate, deep red with a dark basal spot. Stamens violet. Fruiting heads 1.5–1.8 cm., rather loose, ovoid-oblong. Achenes about 3 mm., obliquely ovoid, glabrous, wrinkled, destitute of a tooth on inner (upper) margin; beak short, straight, green. Fl. February–April.

Hab. : Batha and fallow fields. Acco Plain, Sharon Plain, Upper and Lower Galilee, Mt. Carmel, Esdraelon Plain, Samaria, Judean Mts., Hula Plain, Upper Jordan Valley.

Area : Mediterranean, extending into Euro-Siberian territories.

Formerly used as a heart tonic.

36. BERBERIDACEAE

Perennial herbs or shrubs. Leaves deciduous or sometimes persistent, alternate or rosu-late, simple or compound, stipulate or exstipulate. Flowers hermaphrodite, solitary or in axillary or terminal cymes, racemes or panicles, actinomorphic. Calyx of 3–6 free sepals. Corolla of 4–6 petals, sometimes not clearly differentiated from calyx; petaloid nectaries in 1–2 whorls, often replacing corolla, sometimes 0. Stamens 4 or 6 or more, generally in 2 whorls, those of the outer whorl opposite petals; filaments free; anthers mostly dehiscing by valves. Ovary superior, 1-, rarely 2-celled, 2–3-carpelled; ovules 1 to few on a basal placenta, or many on a parietal placenta, anatropous; style short or 0. Fruit a berry, follicle or a vesicular capsule rupturing at maturity. Seeds with endosperm; embryo mostly small, sometimes with connate cotyledons.

About 12 genera and 200 species, mainly in the north-temperate regions (some also in the southern hemisphere).

1. Leaves pinnatisect. Nectary scales 0.   **2. Bongardia**
– Leaves 2- to 3-ternate. Nectary scales 6.   **1. Leontice**

1. Leontice L.

Perennial herbs with subterranean tuberous stem. Leaves 2- or 3-ternate, mostly cauline. Flowers in richly branching panicles. Sepals 6, petaloid. Petals 6, yellow appearing as clawed nectary scales. Stamens 6, free; anthers dehiscing by valves. Ovary 1-celled; ovules 2–4, basal; style short; stigma small, entire or shortly 2-lobed. Capsule membranous, inflated, irregularly torn at apex. Seeds 1–4, globular.

Four to 6 species, mainly in the Mediterranean and Irano-Turanian regions.

1. Leontice leontopetalum L., Sp. Pl. 312 (1753); Boiss., Fl. 1 : 99 (1867). [Plate 315]
Perennial, 15–50 cm. Aerial stem erect, simple or branching, thick, mostly hollow, leafy, smooth and glabrous. Subterranean tuber 4–8 cm. across, globular or somewhat depressed, warty or lobed, not bearing leaf buds. Leaves 2–3-ternate, with 2–6 cm., ovate or obovate, rarely subcordate leaflets, the uppermost sometimes 2–3-partite. Inflorescence a densely branching conical panicle. Pedicels patulous-erect, much elongated in fruit, up to 7 cm. Flowers yellow, subtended by short, oblong bracts. Sepals petaloid, about 8 mm.; nectary scales orbicular, retuse. Capsule 1.5–2(–3) cm., ovoid or globular, rounded at base, mostly 1-seeded. Dispersal by detachment of the entire aerial fruit-bearing stem from subterranean tuber (tumble weed). Fl. February–March (–April).

Hab.: Fields and steppes. Sharon Plain, Philistean Plain, Upper and Lower Galilee, Mt. Carmel, Esdraelon Plain, Samaria, Judean Mts., Judean Desert, W., N. and C. Negev, Lower Jordan Valley, Dead Sea area, Gilead, Ammon, Moav, Edom.

Area: E. Mediterranean, slightly extending into adjacent Irano-Turanian areas.

Occurring as a primary component of the vegetation in steppes, but exclusively segetal under Mediterranean conditions. One of the most striking examples of a steppe species from which a segetal ecotype has been derived.

Most of the specimens growing in desert parts of the country could be referred to *L. minor* Boiss., Fl. 1 : 100 (1867) [*L. leontopetalum* var. *minor* Ky. ap. Sam. ex Rech. f., Ark. Bot. ser. 2, 5, 1 : 144 (1959)], but there are all transitions between the typical form and this variety.

The tuber contains saponine and is used in the Middle East for cleansing purposes. Also used in S. and E. Europe as a remedy for epilepsy.

2. BONGARDIA C. A. Mey.

Perennial, glabrous herbs with a large subterranean bulbous stem bearing several scattered leaf buds. Flowers hermaphrodite, paniculate on leafless scapes, actinomorphic. Sepals 3–6, petaloid. Petals 6, with a nectariferous pore at base. Stamens 6, free; anthers muticous, opening by valves. Ovary 1-celled; ovules few, basal, with erect funicle; style short; stigma broadened. Capsule membranous, inflated, torn at apex into irregular lobes. Seeds almost globular.

A single species in W. and C. Asia.

1. Bongardia chrysogonum (L.) Sp., Hist. Nat. Veget. Phaner. 8 : 65 (1839); Boiss., Fl. 1 : 99 (1867). *Leontice chrysogonum* L., Sp. Pl. 312 (1753). [Plate 316]

Perennial, with erect, scapose, glabrous, white stem, 20–45 cm. Subterranean tuber (4–)6–8 cm. across, globular or somewhat depressed, with bud-like protuberances. Leaves 10–20 cm., all radical, scattered, ascending or decumbent, pinnatisect into (1–)2–4 cm. long, opposite, sessile, dentate to 3–6-lobed, obovate-cuneate segments with a dark blot at base. Flowers long-pedicelled, in loose, divaricately branching panicles with scale-like leaves at base of branches. Sepals 3. Petals about 1 cm., yellow. Capsule about 1(–1.5) cm., ovoid-oblong. Seeds globular, rather large. Dispersal by detachment of the entire aerial fruit-bearing scape from subterranean tuber (tumble weed). Fl. February–March.

Hab.: Fields; weed on deep heavy soils. Sharon Plain, Philistean Plain, Upper and Lower Galilee, Mt. Carmel, Esdraelon Plain, Samaria, Shefela, Judean Mts., Hula Plain, Lower Jordan Valley, Ammon.

Area : E. Mediterranean and W. Irano-Turanian.

Tuber used to counteract epilepsy (Post). Leaves edible, but in small quantities only.

37. MENISPERMACEAE

Woody or herbaceous dioecious climbers or rhizomatous perennials, rarely trees. Leaves alternate, simple or trifoliate, entire or sometimes palmately lobed, exstipulate. Flowers unisexual, 2–3-merous, generally actinomorphic, greenish. Whorls of sepals, petals and stamens 1 to several, mostly 2. Sepals 3–12 or numerous, rarely 1–2, free or connate. Petals 1–6, rarely numerous or 0, free, sometimes connate. Staminate flowers usually with 6(9) stamens (sometimes less), distinct or connate; anthers 4-celled or seemingly so. Pistillate flowers with or without staminodes, mostly with 3–6 free, 1-celled carpels; ovary superior; ovules 1–2, amphitropous; placentation parietal; style short, mostly curved, sometimes 0. Fruit a drupe or achene. Seed curved, with or without endosperm.

About 67 genera and 425 species mostly in the tropics of the Old World, a few also in the temperate regions.

1. Cocculus DC.

Dioecious shrubs, mostly woody climbers. Leaves simple. Flowers axillary, solitary, cymose or paniculate. Sepals 6, pubescent, in 2 whorls, inner larger, concave. Petals 6, shorter than inner sepals, entire or 2-fid. Staminate flowers with 6–9 free stamens, bearing 1 lobed anthers. Pistillate flowers with 6 (or 0) staminodes and 3 (or 6) carpels, the latter with cylindrical, erect or recurved, undivided style. Drupe laterally compressed, obovoid or spherical, with scar of style near the base. Seed horseshoe-shaped; endosperm scanty; cotyledons linear, appressed.

About 12 species in the tropics of Africa, Asia and also in N. America and the Pacific Islands.

1. Cocculus pendulus (J. R. et G. Forst.) Diels, in Pflznr. 46 : 237 (1910). *Epibaterium pendulum* J. R. et G. Forst., Char. Gen. 108 (1776). *C. laeba* (Del.) DC., Syst. 1 : 529 (1817); Boiss., Fl. 4 : 1201 (1879). [Plate 317]

Woody, much branched, puberulent, glaucous climber. Branches elongated, slender, striate, twining. Leaves small, 1.2–4 × 0.5–1.8 cm., short-petioled to subsessile, somewhat leathery, lanceolate-oblong or ovate-oblong, cuneate or rounded at base, usually obtuse, rarely retuse, mucronate, entire or sometimes obscurely lobed, mostly puberulent at first, then glabrous. Flowers small; the staminate short-pedicelled or sessile in few-flowered cymes; the pistillate solitary or in pairs on 4–8 mm. long pedicels. Drupe about 5 mm. across. Fl. March–July.

Hab.: Climbing on rocks, cliffs and trees. E. and S. Negev, Lower Jordan Valley, Arava Valley, Edom.

Area: Mainly Sudanian. Also in other Paleotropical regions.

Used in medicine as diuretic. Leaves edible.

38. NYMPHAEACEAE

Perennial water plants, usually with thick creeping rhizomes and floating, submerged or rarely aerial, alternate leaves. Floating leaves simple, mostly entire, peltate or cordate, long-petioled. Flowers hermaphrodite, solitary, very long-petioled, with a rudimentary bract at base, actinomorphic, often large, emerging from water. Perianth usually of 3–6 or more sepals and 3 to many petals sometimes showing transition to the numerous stamens; the latter hypogynous, epigynous or inserted on wall of ovary; anthers introrse. Carpels 3 or more, usually free, rarely coalescent into a many-celled, superior or inferior ovary; ovules 1 to many in each cell on parietal placentae. Fruit a group of achenes sunk in the receptacle or a spongy capsule dehiscing by swelling of the internal mucilage. Seeds often arillate, with straight embryo, with or without endosperm.

About 8 genera and 90 species in tropical and extratropical regions.

1. Petals white, blue, rarely rose-coloured. Ovary half-inferior. **1. Nymphaea**
  – Petals yellow. Ovary superior. **2. Nuphar**

1. NYMPHAEA L.

Perennial herbs with thick creeping rhizomes and 1-flowered scapes. Leaves floating or submerged, orbicular or ovate to oblong, more or less cordate or peltate at base, stipulate, lateral nerves of leaves branching at right angles. Flowers large, white, blue pink or red. Sepals 4, hypogynous (sometimes 8 in 2 whorls), mostly caducous. Petals numerous, large, imbricated in several whorls, inserted at base or on walls of ovary, the inner narrower, showing gradual transition into stamens. Stamens numerous, inserted on wall of ovary, the outer ones with petaloid filaments and small anthers, the inner with linear filaments and elongated introrse anthers. Carpels numerous, concrescent with the receptacle into a many-celled, half-inferior ovary with concave top and umbilicate centre; ovules many in each cell, pendulous; styles free, exserted; stigmas radially connate into a discoid or funnel-shaped body. Fruit a spongy, berry-like capsule, globular or ovoid, maturing under water and made to split by the swelling mucilage. Seeds arillate, often with endosperm.

About 40 species widely dispersed in many regions.

1. Flowers white or rose-coloured. **1. N. alba**
  – Flowers light blue. **2. N. caerulea**

1. Nymphaea alba L., Sp. Pl. 510 (1753); Boiss., Fl. 1 : 104 (1867). [Plate 318]

Perennial, with thick, creeping rootstock. Leaves all floating, 10–30 cm., ovate-orbicular, cordate, with sinus reaching up to middle of blade, entire. Flowers 8–15 (–20) cm. in diam., on stalk up to 2 m. long. Sepals lanceolate, green beneath, white above. Petals 20–25, white, rarely rose-coloured, the outer exceeding the sepals. Filaments of inner stamens more or less linear, those of outer ones petaloid. Ovary globular to ovoid; stigmatic rays usually 15–20, yellow. Fruit 1.8–4 cm. in diam., obovoid, with stamen scars almost to the top. Seeds ellipsoidal. Fl. May–July.

Hab.: Lakes and ponds. Sharon Plain, Philistean Plain, Hula Plain. Extremely rare.
Area: Euro-Siberian and Mediterranean.

This species has been preserved only in one or two localities and is in danger of disappearing altogether through the drying up of the swamps.

Rootstock rich in tannins; flowers and roots used variously in medicine (flores et radix nymphaeae albae).

2. Nymphaea caerulea Savigny, Dec. Pl. Eg. 3 : 74 (1779). [Plate 319]

Perennial, with thick creeping rhizome. Leaves all floating, 20–40 cm. in diam., orbicular or ovate-orbicular with a deep basal sinus, entire or slightly wavy at base, green above, purplish all around near margin. Flowers 7–15 cm. in diam., on 1 m. or longer stalks. Calyx somewhat longer than corolla; sepals lanceolate, acuminate, often marked with black lines and dots. Petals 10–30, light blue, linear-oblong or lanceolate. Stamens 50–70; anthers with up to 5 mm. long appendages. Stigmatic rays

10–30, short. Fruit subglobular. Seeds ellipsoidal-globular, minutely and longitudinally striate or smooth. Fl. May–September.

Hab.: Ponds and rivers. Acco Plain, Sharon Plain, Philistean Plain. Very rare. Area: Trop. African.

On the verge of extinction in Palestine; there are only two sites in which this plant still occurs here.

2. NUPHAR Sm.

Perennial herbs with thick creeping rhizomes. Leaves all floating or part of them submerged, cordate with deep sinus, exstipulate, lateral nerves of floating leaves repeatedly forked. Flowers large, semiglobular, yellow, borne on long stalks. Sepals 5 (rarely up to 12), persistent. Petals numerous (sometimes 0), smaller than sepals, spatulate, with nectary on back. Stamens numerous, hypogynous; anthers introrse. Ovary and fruit superior, flask-shaped, adnate to receptacle, 10–16-celled, with many ovules in each cell; stigmas on top of ovary, united into a peltate disk with 5–24 rays. Fruit berry-like, irregularly splitting. Seeds many, large, not arillate, with little endosperm and much perisperm.

About 15 species in the northern extratropical regions.

1. Nuphar lutea (L.) Sm. in Sibth. et Sm., Fl. Gr. Prodr. 1 : 361 (1809); Boiss., Fl. 1 : 104 (1867). *Nymphaea lutea* L., Sp. Pl. 510 (1753). [Plate 320]

Rhizome thick, branched. Leaves 10–30 cm., mostly floating or some submerged, ovate-oblong, deeply cordate with a deep sinus reaching the insertion point of the petiole. Flowers yellow, 4(–6) cm. in diam., on up to 2 m. long stalks. Sepals 5, (1–)2–3 cm., broadly obovate to orbicular, concave, bright yellow within. Petals many, about one third as long as sepals, broadly spatulate. Stamens shorter than ovary. Stigmatic disk 5–24-rayed. Fruit a 3–4 cm. long, flask-shaped, irregularly dehiscing berry. Fl. May–July.

Hab.: Lakes and rivers (in stagnant or slowly streaming water). Acco Plain, Sharon Plain, Philistean Plain, Upper Galilee, Dan Valley, Hula Plain, Upper Jordan Valley.

Area: Euro-Siberian, Mediterranean and Irano-Turanian.

With the drying up of the natural lakes and swamps this water lily has disappeared from several localities.

Rhizome used as source of starch food in famine; formerly used in medicine (flores, radix nuphari lutei).

39. CERATOPHYLLACEAE

Perennial, aquatic, submerged, rootless, monoecious plants. Leaves whorled, sessile, dichotomously divided into filiform or linear serrulate segments; stipules 0. Flowers minute, unisexual, solitary at nodes, actinomorphic. Staminate flowers with a whitish perianth of 9–15 tepals, connate at base; stamens 10–20, filaments short, anthers 2-celled with a thickened, often coloured connective. Pistillate flowers with a green perianth of 9–15 tepals and 1 pistil; ovary superior, 1-carpelled, 1-celled, 1-ovuled, ovule pending from top of ovary; style 1, slender, persistent; stigma 1. Fruit a nutlet. Seeds with large embryo; endosperm 0.

One genus and 3 species, almost cosmopolitan.

1. CERATOPHYLLUM L.

See description of family.

1. Leaves once or twice dichotomously divided; segments linear, densely denticulate. Fruit with 2 spines at base, somewhat shorter than or equalling persistent style.
1. C. demersum

– Leaves divided dichotomously 3–4 times; segments filiform, remotely denticulate. Fruit without spines at base, much longer than persistent style.   **2. C. submersum**

1. Ceratophyllum demersum L., Sp. Pl. 992 (1753); Boiss., Fl. 4:1202 (1879). [Plate 321]

Perennial, submerged, rather stiff herb. Stems slender, branching, glabrous, 20–100 cm. or longer. Leaves 1–2 cm., 4–12 in a whorl, filiform, once to twice divided dichotomously into 2–4 segments; segments dark green, linear, with irregular, short, dense, erect teeth. Flowers almost sessile. Staminate flowers with mostly 12 tepals, 3-denticulate at apex; stamens 10–16, crowded. Pistillate flowers with 9–10 tepals; ovary long-ovoid, style very long, awl-shaped. Fruit up to 5 mm., ovoid, black, rough or somewhat warty, stiff, with 2 spines at base and with persistent style as long as or longer than nutlet. Fl. June–August.

Hab.: Lakes, streams and ponds. In standing or slowly flowing water. Sharon Plain, Mt. Carmel, Hula Plain, Upper Jordan Valley.

Area: Euro-Siberian, Mediterranean and Tropical African.

2. Ceratophyllum submersum L., Sp. Pl. ed. 2, 1409 (1763). [Plate 322]

Perennial, submerged, glabrous, tender herb. Stems filiform, branching, 20–25 cm. Leaves long, many in a whorl, very delicate, almost capillary, light green, mostly 3 times dichotomously lobed with 5–8 ultimate segments, sparsely dentate-prickly. Staminate flowers sessile; tepals mostly 12; stamens 6–16. Pistillate flowers very short-stalked; tepals 9–10. Fruit 4–5 mm., ovoid, black, with fine warts, without basal spines; persistent style much shorter than nutlet.

Hab.: Ponds. Hula Plain. Rather rare.

Area: Euro-Siberian, Mediterranean and Irano-Turanian.

Guttiferales

40. PAEONIACEAE

Perennial herbs or shrubs. Leaves alternate, divided, exstipulate. Flowers hermaphrodite, usually solitary, actinomorphic, very conspicuous. Sepals 5 or more, free, green or showing transition into petals. Petals 5–10(–13), free. Stamens numerous. **Ovary** superior; carpels 2–8, free, borne on a fleshy disk, with short styles and broad, almost sessile stigmas. Fruit a group of 2–8 follicles with several anatropous seeds in 2 rows along the ventral suture of carpel. Endosperm copious; embryo small.

One genus in temperate regions of the northern hemisphere.

1. Paeonia L.

Perennial showy herbs with herbaceous or somewhat woody stems. Rootstock thickened or tuberous. Leaves divided. Flowers solitary, terminal. Sepals mostly 5, imbricated. Petals 5–8, larger than sepals. Stamens numerous, united at base into a fleshy ring. Ovary of a few carpels, each with a short style and broad stigma. Fruit composed of a few large, tomentose or glabrous follicles, dehiscent along their inner (ventral) side. Seeds large, numerous, at first red then dark blue and shining.

About 50 species especially in C. and E. Asia and S. Europe, a few also in N. America.

Literature : F. C. Stern, *A Study of the Genus Paeonia.* London (1946).

1. Paeonia mascula (L.) Mill., Gard. Dict. ed. 8, no. 1 (1768). *P. officinalis* L. var. *mascula* L., Sp. Pl. 530 (1753). *P. corallina* Retz., Obs. Bot. 3 : 34 (1783); Boiss., Fl. 1 : 97 (1867). [Plate 323]

Perennial, 60–90 cm. Rootstock tuberous, roots thick. Stems simple, erect, glabrous. Leaves 2-ternate, glabrous or sparingly hairy and light green beneath; segments up to 10 cm., ovate or broadly elliptical, entire, rounded to cuneate at base, tapering to acuminate at apex. Flowers about 10 cm. across. Sepals 5, broadly ovate, imbricated, the lowermost sometimes leaf-like. Petals 5–8, purple-red, rarely whitish, ovate, somewhat erose or wavy. Stamens with crimson filaments and yellow anthers. Follicles usually 3–5, 2–3 cm., woolly, first horizontal, then recurved; stigmas red. Seeds globular, smooth, shining, at first red then blue and black. Fl. April–May.

Hab. : Clearings in maquis. Upper Galilee.

Area : Mediterranean and W. Euro-Siberian.

41. HYPERICACEAE

Herbs, shrubs or trees. Leaves mostly opposite or whorled, simple, often persistent, mostly exstipulate, often glandular-dotted. Flowers hermaphrodite or sometimes unisexual, solitary or in cymes, often conspicuous. Sepals and petals (2–)4–5 (–14), imbricated in bud. Stamens numerous, often connate in bundles or in a single tube, sometimes partly staminodal. Ovary superior; carpels (1–)3–5 (–15), forming 1 or more cells; placentation axile or parietal; styles free or connate, with peltate, lobed or divided stigma. Fruit a capsule, berry or drupe. Seeds often arillate, with large embryo and without endosperm.

Forty nine genera and 900 species in tropical and subtropical regions, only *Hypericum* also in temperate regions.

Literature : R. Keller, *Hypericum* L., in : Engl. u. Prantl, *Nat. Pflznfam*. ed. 2, 21 : 175–183 (1925).

1. HYPERICUM L.

Perennial herbs or shrubs, rarely trees. Leaves opposite or whorled, sessile, entire, usually glandular-dotted. Flowers hermaphrodite, solitary or in cymes, yellow or purple. Sepals 5, often unequal. Petals 5, twisted in bud, glandless or with a few glands at tip. Stamens numerous, arranged in 3–5 bundles opposite petals, rarely all free. Ovary superior of 3–5 carpels, 1- or 3–5-celled; ovules mostly numerous; styles 3–5, free of connate. Fruit a many-seeded septicidal capsule, rarely a berry. Seeds winged or wingless.

About 200 species in subtropical and temperate regions; many in the Mediterranean region.

1. Stamens 5-adelphous, i.e. arranged in 5 bundles. Seeds winged. Corolla 3–5 cm. across.
   1. H. hircinum
 – Stamens 3-adelphous, i.e. arranged in 3 bundles. Seeds not winged. Corolla much smaller than in above 2
2. Whitish woolly plants.   **7. H. lanuginosum**
 – Glabrous plants 3
3. Stems decussately branched. Leaves mostly crisp-undulate. Calyx 1.5–2 mm.
   6. H. triquetrifolium
 – Plants not as above 4
4. Stems terete. Leaves ovate to orbicular or elliptical 5
 – Stems winged or prominently 2-ridged. Leaves mostly linear-lanceolate or ovate-oblong 6
5. Low, dense, hemispherical shrubs, with very short branches. Sepals acuminate, without glands.   **2. H. nanum**
 – Erect, loosely branching shrubs with long branches. Sepals obtuse, with numerous glands at margin.   **3. H. serpyllifolium**
6 (4). Stems 4-winged. Sepals entire, half as long as petals.   **5. H. acutum**
 – Stems with 2 prominent, decussately alternating lines. Sepals denticulate, one third to one fifth as long as petals.   **4. H. hyssopifolium**

Sect. ANDROSAEMUM (Tourn. ex Adans.) Gren. et Godr., Fl. Fr. 1 : 320 (1847). Stamens 5-adelphous. Capsule baccate, indehiscent or opening only at top.

1. Hypericum hircinum L., Sp. Pl. 1103 (1753); Boiss., Fl. 1 : 788 (1867). [Plate 324]
Glabrous shrub, 1–1.5 m. Stems erect, 4-angled. Leaves 2–7 cm., sessile, oblong-lanceolate, dotted with minute glands. Cymes terminal, few-flowered. Flowers 3–5 cm. across. Sepals lanceolate, acute, entire, eglandular, soon deciduous. Petals large, 2–2.5 cm., obovate-oblong. Stamens 5-adelphous at base, longer than corolla. Styles 5–6 times as long as ovary, longer than petals. Capsule dehiscent at apex. Seeds winged, reticulate-foveolate. Fl. July–October.
Hab. : Moist places. Upper Galilee, Samaria, Dan Valley.
Area : Mainly N. Mediterranean.

Sect. HYPERICUM. Sect. *Euhypericum* Gren. et Godr., Fl. Fr. 1 : 314 (1847). Stamens 3-adelphous. Capsule 3-celled, dehiscing by 3 valves.

2. Hypericum nanum Poir. in Lam., Encycl. Suppl. 3 : 699 (1814); Jaub. et Sp., Ill. Pl. Or. 1 : 46, t. 23 (1842); Boiss., Fl. 1 : 792 (1867).

Hemispherical dwarf-shrub, glabrous, 15–30 cm. Stems tortuous, terete. Branches dense, 4–6-leaved. Leaves 0.7–1.5 × 0.5–0.9 cm., ovate to orbicular, narrowing at base, pellucid-dotted, persistent. Inflorescences 5–9-flowered. Pedicels 4.5–9 mm. Calyx leathery; sepals almost equal, oblong-lanceolate, acuminate. Petals 1.2 cm., yellow, not punctate. Stamens 20–30, 3 adelphous, the longest almost as long as petals; anthers with a glandular point at tip. Styles slender, filiform; stigmas truncate. Capsule 3 celled. Seeds cylindrical. Fl. June–July.

Hab. : Samaria (recorded by Boissier, not observed by us).

Area : E. Mediterranean and W. Irano-Turanian.

3. Hypericum serpyllifolium Lam., Encycl. 4 : 176 (1797); Boiss., Fl. 1 : 793 (1867). [Plate 325]

Dwarf-shrub, glabrous, 40–80 cm. Stems erect. Branches long, diffuse, reddish-brown. Leaves 5–8 × 3 mm., almost sessile, ovate to elliptical, obtuse, with revolute margin, glaucous above, sparsely glandular beneath, persistent. Cymes in densely crowded terminal corymbs. Flowers 1–1.5 cm. across. Sepals oblong, 1 or 2 obtuse, the rest somewhat acute, with black subsessile glands at margin, one third to one quarter as long as petals. Petals yellow, more or less glandular at upper part. Stamens 3-adelphous at base, shorter than corolla. Capsule 3-celled. Seeds cylindrical, minutely pustulate, inconspicuously keeled. Fl. April–July.

Hab. : Maquis and forest. Coastal Galilee, Sharon Plain, Upper and Lower Galilee, Mt. Carmel, Samaria, Judean Mts., Gilead.

Area : E. Mediterranean.

4. Hypericum hyssopifolium Chaix in Vill., Hist. Pl. Dauph. 3 : 505 (1788); Boiss., Fl. 1 : 799 (1867). [Plate 326]

Perennial, glabrous, often suffrutescent, 30–60 cm. Stems slender, with 2 prominent, decussately alternating lines, often paniculately branched. Leaves usually 1–4 × 0.2–1.5 cm., sessile, linear or lanceolate-linear, obtuse or acuminate, pellucid-dotted, with revolute margin, persistent. Cymes peduncled, forming a narrow, elongated panicle. Bracts glandular or eglandular at margin. Flowers up to 2 cm. across. Sepals nearly equal, with stipitate black glands at margin. Petals yellow, with stipitate black glands at margin of upper part only. Stamens 3-adelphous. Capsule ovoid-oblong, 3-celled, 3–4 times as long as calyx. Seeds cylindrical, more or less minutely papillose, foveolate. Fl. April–May.

Var. **latifolium** Boiss., Fl. 1 : 799 (1867). Differs from type by more or less subsessile flowers, eglandular bracts and broader leaves (up to 1.5 cm. broad), scarcely revolute at margin.

Hab. : Maquis and batha. Coastal Galilee, Upper Galilee, Mt. Carmel.

Area of species : Mediterranean and W. Irano-Turanian.

5. Hypericum acutum Moench, Meth. 128 (1794). *H. tetrapterum* Fries, Novit. Fl. Suec. 6 : 94 (1823); Boiss., Fl. 1 : 805 (1867). [Plate 327]

Perennial, stoloniferous, suffrutescent, glabrous, 20–30 cm. Stems 4-winged, loose, much branched. Leaves (1–)1.5–3 cm., half-clasping, ovate-oblong, obtuse, with pellucid glands on upper surface and black glands near margin. Inflorescences terminal and lateral, the former forming a corymbose panicle; rarely flowers few or solitary. Flowers about 1.5 cm. across. Calyx about half as long as corolla; sepals somewhat connate at base, lanceolate, acute, more or less remotely glandular especially near margin, sometimes with a black gland at tip. Petals yellow, with sparse black glands, sometimes intermixed with pellucid ones. Stamens 3-adelphous, connate to one quarter of their length; anthers with 1 black gland at top of connective. Seeds delicately foveolate-reticulate with square pits. Fl. July–September.

Var. **anagallidioides** (Jaub. et Sp.) Boiss. (orth. mut. "anagallidifolium"), l.c. 806. *H. anagallidioides* Jaub. et Sp., Ill. Pl. Or. 1 : 47, t. 24 (1842). This is a much branched plant with shorter (1–1.8 cm.) leaves and with 1–3-flowered terminal cymes.

Hab. : Damp ground. Upper Galilee.

Area of species : Mediterranean and Euro-Siberian.

6. Hypericum triquetrifolium Turra, Farset. Nov. Gen. 12 (1765). *H. crispum* L., Mant. 106 (1767); Boiss., Fl. 1 : 806 (1867). [Plate 328]

Perennial herb, glabrous, glaucescent, 20–40 cm. Stems terete, branching decussately almost from base. Leaves 0.3–1.5 cm., lanceolate, cordate at base, acute, crisp-undulate, pellucid, black-dotted at margin. Inflorescence a pyramidal panicle as broad at base as long, formed by branches with 1–5-flowered terminal cymes. Flowers 1.5–1.8 cm. across. Sepals 1.5–2 mm., ovate-oblong, obtuse or acute, one fifth to one quarter as long as petals, more or less denticulate at upper part. Petals yellow, linear-oblong, not glandular. Stamens 3-adelphous, apiculate, ending with a black gland. Capsule 3-celled. Seeds cylindrical, punctate-pitted. Fl. May–September.

Hab. : Fields. Acco Plain, Sharon Plain, Philistean Plain, Upper and Lower Galilee, Mt. Carmel, Esdraelon Plain, Shefela, Judean Mts., W. and N. Negev, Dan Valley, Hula Plain, Upper Jordan Valley, Transjordan.

Area : E. Mediterranean and W. Irano-Turanian.

7. Hypericum lanuginosum Lam., Encycl. 4 : 171 (1797); Boiss., Fl. 1 : 807 (1867). [Plate 329]

Perennial, tomentellous to canescent, except for glabrous inflorescence, 20–60 cm. Stems terete. Leaves 1–5 cm., ovate-oblong, cordate-amplexicaul, somewhat longer than internodes, with pellucid glands and with sparse black glands near margin, lower leaves obtuse, the upper tapering, acute. Cymes paniculate. Bracts with stipitate glands. Flowers about 1.8 cm. across. Sepals ovate, acute or obtuse, toothed, with black glands. Corolla yellow, 2–3 times as long as calyx; petals scarcely glandular at margin. Stamens 3-adelphous at base, with black gland at top of connective. Capsule ovoid.

Seeds cylindrical, minutely reticulate-foveolate. Fl. April–June.

Hab.: Rocks and rock crevices. Upper and Lower Galilee, Mt. Carmel, Mt. Gilboa, Samaria, Judean Mts.

Area: E. Mediterranean.

Papaverales

42. PAPAVERACEAE

Annual or perennial herbs, rarely shrubs and trees, often with milky or coloured juice. Leaves alternate, often rosulate, the uppermost sometimes whorled, variously lobed or divided, without stipules. Flowers hermaphrodite, solitary or rarely in umbel-like clusters, actinomorphic. Sepals 2 (or 3), free, caducous. Petals 4–6(8–12), rarely 0, free, in 1–2, rarely in 3 whorls, often crumpled in bud. Stamens numerous and spirally arranged or few and cyclic, free. Ovary superior, of 2 to several connate carpels, usually 1–2-celled, with parietal placentae which sometimes form false septa through inward intrusion; ovules few to many, campylotropous or anatropous; style very short or 0; stigmas as many as carpels. Fruit a many-seeded capsule dehiscent by valves or lids, rarely fruit indehiscent. Seeds with oleiferous or mealy endosperm and minute embryo.

Twenty eight genera and about 450 species, mostly in warm and temperate regions of the northern hemisphere; a few in S. America.

Literature: F. Fedde, Papaveraceae-Hypecoideae et Papaveraceae-Papaveroideae, in: *Pflznr.* 40 (IV. 104): 1–430 (1909).

1. Fruit an obovoid, obconical or club-shaped (very rarely linear-oblong), many-carpelled capsule opening by lids. **1. Papaver**
- Fruit a linear or cylindrical, pod-like capsule with 2–4 carpels, opening by valves   2
2. Capsule 1-celled, opening mostly by 3 valves; stigma 2–4 lobed. **2. Roemeria**
- Capsule 2-celled, opening by 2 valves; stigma 2-lobed or 2-horned. **3. Glaucium**

1. Papaver L.

Annual or perennial herbs, mostly hispid or sometimes glabrous, with milky or coloured juice. Leaves mostly radical or rosulate, pinnatifid or otherwise lobed. Pedicels elongated. Flower buds deflexed. Flowers showy, mostly solitary, rarely in corymbose racemes, red, rarely violet, yellow or white. Sepals 2(–3), caducous. Petals 4(–5–6), crumpled in bud, those of outer whorl mostly larger. Stamens numerous. Ovary of few or several united carpels with parietal placentae often extending towards the axis so as to form false cells; stigmas sessile, united into a broad, 4–20-rayed disk, crowning the ovary. Capsule obovoid, club-shaped, obconical to oblong-cylindrical, opening by lids beneath stigmas between placentae. Seeds scrobiculate.

About 100 species, chiefly in north-temperate regions, some also in S. Africa, Australia and N. African savannas.

Literature: F. Fedde, *Papaver* L., in Papaveraceae-Hypecoideae et Papaveraceae-Papaveroideae, in : *Pflznr.* 40 (IV. 104) : 288–386 (1909). N. Feinbrun, Taxonomic studies on *Papaver* Sect. *Orthorhoeades* of Palestine and of some other Mediterranean countries, *Israel Journ. Bot.* 12 : 74–96 (1963).

1. Capsule bristly 2
- Capsule glabrous 3
2. Capsule obovoid-globular to broadly ellipsoidal, densely covered with curved bristles.
7. P. hybridum
- Capsule oblong, club-shaped or oblong-cylindrical, sparsely covered with a few, erect bristles. **6. P. argemone**
3 (1). Stigmatic rays tongue-shaped, longer than broad, not overlapping, at maturity usually curved upwards; disk broader than capsule; capsule large, broadly turbinate-obconical. Bristles of pedicels mostly spreading. **2. P. carmeli**
- Stigmatic rays broader than long or rarely as broad as long, often retuse, overlapping 4
4. Capsule obovoid-turbinate, large (1.5–2 × 0.7–1.2 cm.); disk flat at maturity, not umbonate; stigmas 8–16 (mostly 9–13). Bristles on pedicels usually spreading, stiff; bristles on flower buds often arising from a thick tubercle. Tall, erect plants growing in fields in alluvial soils. **1. P. subpiriforme**
- Capsule narrower, and/or shorter. Bristles on flower buds not as above. Plants not as above 5
5. Stigmatic disk umbonate, i.e., with distinct protuberance in centre; stigmas 5–13, mostly 6–9; capsule oblong or club-shaped, narrow, usually 1.3–2 × 0.5 cm. Lower leaves lyrate, with a large terminal lobe and smaller, ovate-triangular, remote lateral lobes; petiole with long, sparse cilia. Stems usually with sparse spreading hairs below. Plants growing on hills. **3. P. syriacum**
- Stigmatic disk flat, rarely with a minute protuberance in its centre; capsule shorter, (0.8–)1–1.3(–1.5) cm., cupuliform to obovoid or turbinate. Lower leaves not as above 6
6. Stems and usually also peduncles covered with stiff, spreading, yellow bristles. Lower leaves short-petioled, oblong, pinnatifid into ovate-oblong obtuse lobes with cartilaginous margins; stem leaves sessile, oblong or lanceolate, with ovate, obtuse lobes. Capsule turbinate; stigmatic rays 6–8, disk flat. Plants of the inland plains.
5. P. polytrichum
- Stems and leaves almost glabrous, or covered with sparse or dense, often dark purple hairs. Lower leaves long-petioled, pinnatifid into triangular lobes, upper leaves short-petioled, pinnatisect into narrowly linear, usually divaricate lobes. Capsule cupuliform to obovoid or turbinate; stigmatic rays 6–12. Plants growing on sandy soils of the coastal plain and the Negev 7
7. Latex sulphur-yellow. Lower part of stem covered with dense, long and spreading hairs; hairs, stems and lower leaves often dark purple; flower buds often densely hispid or canescent with long and delicate hairs. **4. P. humile** ssp. **sharonense**
- Latex white. Lower part of stem and leaves not as above, often almost glabrous; hairs on flower buds usually sparse. **4. P. humile** ssp. **humile**

Sect. ORTHORHOEADES Fedde, in Pflznr. 40 : 290 (1909). Filaments thread-like. Capsule glabrous. Disk almost flat; stigmatic rays free or overlapping, always rounded or obtuse, never acute.

1. Papaver subpiriforme Fedde, Bull. Herb. Boiss. ser. 2, 5 : 169 (1905) et in Pflznr. 40 : 302 (1909); Feinbr., Israel Journ. Bot. 12 : 82 (1963). *P. rhoeas* L. var. *oblongatum* Boiss., Fl. 1 : 113 (1867). [Plate 330]

Annual, leafy, glabrescent in lower parts, bristly in upper, 30–50 cm. Stems erect, much branched, grooved. Branches strict, almost equal. Leaves pinnatifid, lobes obovate-oblong or oblong, entire or irregularly dentate-lobulate with rounded, subacute or bristle-tipped teeth, glabrous except along median nerve. Pedicels with few long, stiff, mostly spreading bristles. Flower buds 1.5–2 cm., obovoid, more or less rounded at apex, sparingly tuberculate-bristly. Flowers large, 6–7 cm. in diam. Petals 2.5–3 × 4–5 cm., more or less broadly triangular, truncate or rounded at apex. Capsule 1.5–2 × 0.7–1.2 cm., somewhat stipitate, obovoid-turbinate, gradually tapering towards base or narrowly obconical, with a flat disk at apex and (8–)9–13(–16) overlapping stigmatic rays reaching or not the margin of the disk. 2n=14. Fl. March–April.

Hab. : Fields. Acco Plain, Sharon Plain, Philistean Plain, Upper Galilee, Esdraelon Plain, Shefela, Dan Valley, Hula Plain, Upper Jordan Valley, Ammon.

Area : E. Mediterranean.

2. Papaver carmeli Feinbr., Israel Journ. Bot. 12 : 83 (1963). [Plate 331]

Annual, almost glabrous below, (30–)50–60 cm. Stems mostly numerous. Leaves petiolate, lyrate, the lower 7.5–12 cm., with remote, mostly dentate lobes; the upper oblong or narrowly ovate, acute, pinnately parted into narrow, acute, mostly dentate lobes, often pointed with a bristle; all glabrous above, sparingly bristly along nerves beneath; petiole winged, sparingly long-ciliate. Pedicels 20–25 cm., scape-like, with few spreading bristles. Flower buds 1.5–2 cm., ovoid to narrowly ovoid. Flowers fairly large, about 5 cm. in diam. Petals unequal, 3.5–6.5 cm., red with a dark but frequently white-rimmed spot, ovate or cuneate-obovate. Capsule 1–1.5 ×0.7–0.8 cm., shortly stipitate, broadly turbinate-obconical, attenuate at base; disk concave at maturity, broader than capsule (up to 1.3 cm. broad); stigmatic rays (5–)7–11 (–15), longer than broad, not overlapping, tongue-shaped, mostly curved upwards at maturity. Seeds light brown. 2n=14. Fl. March–June.

Hab. : Field edges. Upper and Lower Galilee, Mt. Carmel, Esdraelon Plain, Mt. Gilboa.

Area : E. Mediterranean (endemic in Palestine).

3. Papaver syriacum Boiss. et Bl. in Boiss., Diagn. ser. 2, 6 : 8 (1859); Fedde, in Pflznr. 40 : 305 (1909); Feinbr., l.c. 87. *P. rhoeas* L. var. *syriacum* Boiss., Fl. 1 : 113 (1867) p.p. *P. obtusifolium* Desf. var. *barbeyi* Fedde, Bull. Herb. Boiss. ser. 2, 5 : 447 (1905). [Plate 332]

Annual, 15–50 cm. Stems usually erect or ascending, branching from base, with sparse spreading hairs below. Lower leaves 7.5–10 cm., long-petioled, pinnatipartite-lyrate, segments remote, ovate-triangular, the terminal much larger, oblong-obovate; petiole sparingly long-ciliate; upper leaves sessile, pinnatipartite or pinnatisect, the uppermost linear-lanceolate, serrate; all appressed-hairy, mainly along veins. Pedicels with appressed or spreading bristles. Flower buds 1.5–2 cm., oblong-obovoid, appressed-hirsute. Flowers about 4 cm. in diam. Petals broad, scarlet, usually with a black, often white-rimmed spot. Capsule usually 1.3–2 × 0.5 cm., substipitate, clavate-turbinate to linear-oblong, attenuate at base; disk broader than capsule, with a 1–2 mm. long protuberance in centre; stigmatic rays (5–)6–9(–13), broader than long, overlapping. 2n=14. Fl. March–May.

Hab.: Fields and roadsides. Coastal and Upper Galilee, Shefela, Judean Mts., Dan Valley, Ammon, Moav.

Area: E. Mediterranean.

4. Papaver humile Fedde, Bull. Herb. Boiss. ser. 2, 5 : 446 (1905) et in Pflznr. 40 : 306 (1909); Feinbr., l.c. 89. [Plate 333]

Annual, branching mainly below, 10–50 cm. Stems erect or decumbent, nearly glabrous to densely hispid. Lower leaves about 10 cm., long-petioled, pinnately cut into triangular or linear, usually dentate lobes often forming almost a right angle with leaf axis; upper leaves short-petioled, with narrower, usually divaricate lobes. Pedicels erect, appressed to spreading, bristly. Flower buds 1.2–1.5 cm., obovoid, ovoid or oblong. Flowers 3–4 cm. in diam. Petals scarlet, usually with a black, often white-rimmed spot. Capsule 0.8–1.5 × 0.4–0.7 cm., cupuliform to obovoid or turbinate; disk slightly broader than capsule, flat, rarely with a small protuberance; stigmatic rays (5–)6–12(–15), broader than long, overlapping, rounded, obtuse. 2n=14. Fl. March–June.

Subsp. **humile.** *P. strigosum* (Bönningh.) Schur var. *gaillardotii* Fedde, Bull. Herb. Boiss. ser. 2, 5 : 446 (1905). Latex white. Stems and leaves often with sparse, short bristles or almost glabrous, rarely lower part of stem hispid. Flower buds sometimes with sparse short bristles.

Hab.: Sandy or sandy loess soils. Philistean Plain (southern part), W., N. and C. Negev, Dead Sea area, Moav.

Area: Saharo-Arabian.

Subsp. **sharonense** Feinbr., l.c. 90. Latex sulphur-yellow. Lower part of stem usually densely hispid with long, spreading, often purple hairs. Leaves usually appressed-hairy on both sides, lower leaves often purple. Flower buds often densely covered with long whitish hairs.

Hab.: Fields; sandy clay soil. Coastal Galilee, Acco Plain, Coast of Carmel, Sharon Plain, Philistean Plain (northern part), Shefela.

Area: E. Mediterranean.

5. Papaver polytrichum Boiss. et Ky. in Boiss., Diagn. ser. 2, 5 : 14 (1856) et Fl. 1 : 113 (1867); Fedde, in Pflznr. 40 : 304 (1909); Feinbr., l.c. 92 [Plate 334]

Annual, 15–30 cm. Stems many, erect, simple or frequently dichotomously branching below, covered with stiff, yellow bristles. Leaves densely appressed-pilose, canescent; lower leaves 4–6 cm., short-petioled, oblong, lyrate-pinnatifid into ovate-oblong, obtuse lobes with cartilaginous margin; upper leaves sessile, 1–3 cm., usually oblong or lanceolate, pinnatifid into ovate, obtuse lobes. Pedicels with stiff, spreading bristles. Petals 2–3 cm. across, orbicular, brick-red, mostly without a spot. Capsule 1–1.2 × 0.5

cm., turbinate, glabrous; disk flat, broader than capsule; stigmatic rays 6–8, over-lapping. Fl. March–May.

Hab. : Steppes and field borders. Judean Desert, N. and C. Negev, Lower Jordan Valley, Moav.

Area : W. Irano-Turanian.

A specimen with almost entire leaves was found by Aaronsohn in Moav and considered f. *subintegrum* (O. Ktze.) Fedde, l.c. of above species.

Sect. ARGEMONORHOEADES Fedde, in Pflznr, 40 : 326 (1909). Filaments club-shaped. Capsule almost always bristly; disk almost hemispherical; stigmatic rays borne on slightly elevated keels with sinuses not extending to centre of disk and forming more or less acute teeth at margin.

6. Papaver argemone L., Sp. Pl. 506 (1753); Boiss., Fl. 1 : 118 (1867); Fedde, in Pflznr. 40 : 328 (1909) incl. *P. belangeri* Boiss., Fl. 1 : 117 (1867). [Plate 335]

Annual, leafy, appressed-setulose, 10–50 cm. Stems 1 or more, erect or ascending, simple or branching, setose-hispid to glabrescent. Leaves 1–2-pinnatisect into linear-lanceolate to oblong, acute, revolute-margined segments often pointed with a bristle; those of upper leaves much narrower. Pedicels very long, somewhat thickened at apex, appressed-setulose. Flower buds 0.5–1 cm., oblong-obovoid, mostly obtuse. Sepals bristly. Petals 1–2 cm., scarlet with a dark base, obovate, not contiguous. Filaments violet, thickened above; anthers bluish. Capsule 1–2.5 × 3–4 mm., oblong, club-shaped or oblong-cylindrical, ribbed, sparsely covered with a few erect bristles, especially above; stigmatic disk often narrower than capsule, with 4–6 rays. 2n=14, 42. Fl. March–April.

Var. **argemone**. Capsule sparsely bristly throughout.
Hab. : Batha. Judean Mts., Ammon.

Var. **glabratum** (Coss. et Germ.) Rouy et Fouc., Fl. Fr. 1 : 160 (1893); Fedde, in Pflznr. 40 : 328 (1909). Capsule with a few, rarely (0)1–2 bristles only at apex.
Hab. : As above. Judean Mts., Gilead, Ammon, Edom.

Area of species : Mediterranean, extending into Irano-Turanian, and Euro-Siberian territories.

P. belangeri Boiss., Fl. 1 : 117 (1867); Fedde, in Pflznr. 40 : 330 (1909), was separated from *P. argemone* by the smaller size of its petals, the frequent lack of a dark basal spot on them, its smaller capsule (1 cm. or less long, 2–3 mm. broad) and its procumbent habit. The Palestine specimens referred to present, no doubt, a stunted form of *P. argemone*.

7. Papaver hybridum L., Sp. Pl. 506 (1753); Boiss., Fl. 1 : 117 (1867). *P. apulum* Ten. var. *gracillimum* Fedde, in Pflznr. 40 : 332 (1909). [Plate 336]

Annual with white latex, appressed- or patulous-setulose, 10–50 cm. Stems few or many, erect or ascending, strictly or divaricately branching. Leaves ovate to oblong, the lower ones long-petioled, 2–3-pinnatisect into linear-lanceolate, oblong to ovate, obtuse to acuminate, bristly pointed segments, often with ciliate and revolute margins, the upper ones sessile, less divided, with narrower segments. Pedicels very long, somewhat thickened at apex, setose. Flower buds ovoid-globular with rounded apex. Sepals hispid. Petals about 1.5–2 cm., crimson with a blackish spot at base, obovate-orbicular. Filaments dark violet, thickened above; anthers bluish. Capsule 0.8–1.8 cm., obovoid-globular to broadly ellipsoidal, densely covered with stiff white bristles; stigmatic disk narrower than capsule, with 4–8 prominent rays. 2n=14. Fl. March–April.

Hab.: Fields and batha. Acco Plain, Sharon Plain, Philistean Plain, Upper and Lower Galilee, Esdraelon Plain, Shefela, Judean Mts., Judean Desert, N., C. and S. Negev, Lower Jordan Valley, Ammon, Edom.

Attempts to divide this species into varieties have failed because of the continuous variation in it.

Var. *siculum* (Guss.) Arcang., Comp. Fl. It. ed. 2, 247 (1894) recorded by Bornmueller (exsicc.) in the Philistean Plain (Jaffa) does not differ from the type. Var. *grandiflorum* Boiss., l.c., recorded by Dinsmore in Post, Fl. Syr. Pal. Sin. ed. 2, 1 : 36 (1932) for Jerusalem, has not been observed by us.

Area of species: Mediterranean and Irano-Turanian, extending into Euro-Siberian territories.

Papaver divergens Fedde et Bornm. in Fedde, Pflznr. 40 : 337 (1909) (Sect. *Carinata* Fedde, l.c. 334) is reported from the Coastal Negev (Rafah) by Täckholm, p. 370 (1956).

Argemone mexicana L., Sp. Pl. 508 (1753) has been found as a casual plant in the Sharon Plain. It is well distinguished from all other members of the local Papaveraceae by its spiny stem, leaves and capsule and by its yellow to orange flowers.

Originates from S.W. United States and C. America.

2. ROEMERIA Medik.

Low annuals. Leaves petiolate, 1–3-pinnatisect or very sparsely lobed. Flowers terminal, solitary or few, long-pedicelled, mostly violet or crimson. Sepals 2, caducous. Petals 4, crumpled in bud. Stamens numerous. Ovary one-celled, with (2–)3(–4) carpels; placentae rib-like, not extending into false septa, many-ovuled; stigma sessile, capitate, 2–4-lobed, little extended. Capsule 1-celled, linear-cylindrical, dehiscing by 3–4 valves from apex to base. Seeds reniform.

Six species mainly in the Mediterranean, Irano-Turanian and Saharo-Arabian regions.

1. Fruit glabrous. Leaves spatulate, undivided or with few oblong to linear lobes.
   2. R. procumbens
– Fruit setulose. Leaves 1–3-pinnatisect. **1. R. hybrida**

1. Roemeria hybrida (L.) DC., Syst. 2 : 92 (1821); Boiss., Fl. 1 : 118 (1867); Fedde, in Pflznr. 40 : 239 (1909). *R. dodecandra* (Forssk.) Stapf, Denkschr. Akad. Wien 51 : 295 (1886); Fedde, l.c. 242. *Chelidonium hybridum* L., Sp. Pl. 506 (1753). [Plate 337]

Annual, more or less hairy, 10–30 cm. Stems 1 or few, erect or ascending, divaricately branching. Leaves sparingly setulose to glabrescent, 1–3-pinnatisect into 0.5–2 cm. long, linear or oblong, obtuse to acute, mostly bristly tipped lobes with serrulate margins; the radical leaves long-petioled, the upper sessile. Flower buds (0.8–)1–1.3

cm., oblong, obtuse, sparingly hairy. Sepals oblong. Petals 1–2 cm., deep violet-blue with a darker spot at base, obovate. Stamens with black filaments. Capsule 3–6 (–10) × 0.2–0.3 cm., mostly 3-valved, cylindrical, glabrescent or sparsely or densely covered with spreading setae longer or shorter than width of capsule. Fl. February–April.

Hab. : Batha and steppes. Judean Mts., Judean Desert, Negev, Upper Jordan Valley, Dead Sea area, Arava Valley, Gilead, Ammon, Moav, Edom.

This polymorphic species includes among others also a form with more broadly lobed leaves and longer setae spreading along the whole capsule. This form, referred to *R. dodecandra* (Forssk.) Stapf by some authors, is linked with *R. hybrida* by a series of intermediate forms.

Two other varieties are recorded from Palestine by Nabelek, Publ. Fac. Sci. Univ. Masaryk 35 : 22 (1923) :

Var. velutina DC., l.c. 93 et 122. Entire stem soft velvety-hairy.
Hab. : As above. Gilead.

Var. velutino-eriocarpa Fedde, l.c. 241. Stem soft-villose. Capsule bristly-hairy.
Hab. : As above. Ammon.

Area of species : Mediterranean and Irano-Turanian, extending towards Saharo-Arabian and Euro-Siberian territories.

2. Roemeria procumbens Aarons. et Opphr. in Opphr., Bull. Soc. Bot. Genève ser. 2, 22 : 305 (1931). [Plate 338]

Low, procumbent annual, 4–10 cm. Stems numerous, thin, sparingly beset with bristles. Leaves 2.5–6 cm., spatulate, entire or with few oblong-linear irregular lobes, tapering at base to a long petiole, glabrous. Flower buds about 1 cm., subrhomboidal, acute. Sepals about 1.5 cm., obovate. Petals about 3 cm. broad, purple with a dark spot at base, reniform to suborbicular. Capsule about 4 cm. long, much broader than pedicel, cylindrical with conspicuous sutures, glabrous. Fl. March.

Hab. : Steppes. C. Negev, Ammon Desert, Edom.
Area : W. Irano-Turanian.

3. GLAUCIUM Mill.

Annual or perennial, hairy, glaucous herbs with yellow juice. Leaves large, the radical petiolate, pinnatifid or lobed, the cauline amplexicaul. Flowers axillary or terminal, solitary, long-pedicelled, large, showy, yellow or red to orange. Sepals 2, caducous. Petals 4, convolute in bud, caducous. Stamens numerous, free. Ovary elongated, of 2 connate carpels separated by a false septum into 2 cells; placentae rib-shaped; stigma subsessile, 2-lobed or -horned. Capsule linear, often very long, pod-like, with a spongy septum, dehiscing by 2 valves. Seeds numerous, ovoid-reniform, scrobiculate.

About 20 species mainly in the E. Mediterranean region and S.W. Asia.

Literature : F. Fedde, *Glaucium* Adans., in Papaveraceae-Hypecoideae et Papaveraceae-Papaveroideae, in : *Pflznr.* 40 (IV. 104) : 221–238 (1909).

1. Flowers bright yellow. Capsule scabrous, tuberculate. Coastal perennials or biennials.
                                                                        5. G. flavum

− Flowers red or orange-red, rarely yellow. Capsule hairy or glabrous but not tuberculate. Plants not growing on coast **2**

2. Annuals. All leaves more or less regularly and deeply pinnatipartite into narrowly oblong (3–4 times as long as broad), lobulate or coarsely toothed segments; upper leaves longer than pedicels. Flowers light scarlet or orange-red, 1.5–3 cm.   **1. G. corniculatum**

− Perennials (sometimes flowering in the first year). Leaves, at least part of them, lyrate-pinnatipartite into obovate segments, terminal segment much broader and more or less truncate or crested-dentate; upper leaves much shorter than pedicels **3**

3. Lower leaves 3–10 cm. Petals yellow, 1.5–2(–2.5) cm. long. Mature capsule glabrous or sparingly setose. **2. G. arabicum**

− Lower leaves usually much larger than in above. Petals red (very rarely yellow, but then 4–6 cm. long) **4**

4. Petals 1.5–2.5(–3) cm. long. Capsule glabrescent or sparingly hairy. **3. G. grandiflorum** var. **judaicum**

− Petals 3–5 cm. long **5**

5. Plants glabrescent or very sparingly hairy. Sepals and capsules sparingly hairy to glabrous. **4. G. aleppicum**

− Plants densely pubescent-tomentose. Sepals and capsules more or less densely hairy or papillose. **3. G. grandiflorum** var. **grandiflorum**

1. Glaucium corniculatum (L.) J.H. Rud., Fl. Jen. Pl. 13 (1781); Boiss., Fl. 1 : 119 (1867); Fedde, in Pflznr. 40 : 223 (1909). *Chelidonium corniculatum* L., Sp. Pl. 506 (1753). [Plate 339]

Annual, papillose-hairy, 20–50 cm. Stems erect or ascending, simple or branched, leafy. Leaves more or less regularly 1–2-pinnatipartite; segments narrowly oblong, more or less remote, irregularly lobulate or coarsely dentate; the lower segments often entire; lower leaves 8–25 cm., petiolate, intermediate and upper leaves smaller, half-clasping. Pedicels often shorter than subtending leaves. Flower buds 1.5–3 cm., oblong-conical. Sepals papillose-hairy. Petals 1.5–3 cm., light scarlet or orange-red with a blackish spot at base, obovate-orbicular. Capsule 10–20 cm., straight or slightly arcuate, appressed- or spreadingly-hairy, rarely glabrescent at maturity; stigma much broader than capsule, with horizontal or ascending horns. Fl. March–May.

Hab.: Fields; alluvial soil. Sharon Plain, Philistean Plain, Esdraelon Plain, Judean Mts., Judean Desert, W. and C. Negev, Upper and Lower Jordan Valley, Dead Sea area, Arava Valley.

Area : Mediterranean, Irano-Turanian, with extensions towards the Euro-Siberian region.

2. Glaucium arabicum Fresen., Mus. Senckenb. 1 : 174 (1834); Boiss., Fl. 1 : 120 (1867); Fedde, in Pflznr. 40 : 225 (1909). [Plate 340]

Perennial, sparingly pilose, glaucous herb, 10–30 cm. Stems 1 or many, sparingly and strictly branched. Leaves small, densely hairy to tomentose; lower leaves 3–10 cm., rosulate, short-petioled, oblong, lyrate-pinnatipartite into obovate-oblong segments, lower segments remote, almost entire, upper crowded, acutely dentate; terminal segment truncate or retuse, unequally 3-dentate; intermediate and upper leaves 1.5–4 cm., clasping, 3–5-lobed. Pedicels much longer than upper leaves. Flower buds 1.5–2 cm., oblong-conical. Sepals oblong, papillose-hairy. Petals 1.5–2(–2.5) cm., yellow. Capsule 8–15 cm., straight, sparingly appressed-setose to glabrous. Fl. March–May.

Hab.: Steppes and deserts. S. Negev, Lower Jordan Valley, Moav, Edom.

Area: W. Irano-Turanian (Egypt, Palestine, Iraq).

Var. *gracilescens* Fedde, l.c. cannot be separated from the typical taxon.

3. Glaucium grandiflorum Boiss. et Huet in Boiss., Diagn. ser. 2, 5:15 (1856) et Fl. 1 : 121 (1867); Fedde, in Pflznr. 40 : 227 (1909). [Plate 341]

Perennial, tomentellous, glaucous, 30–50 cm. Stems few to several, erect, profusely and divaricately branching. Lower leaves 10–15 cm., petiolate, oblong, lyrate-pin-natisect, segments remote, obovate-oblong, irregularly dentate or lobulate, teeth obtuse, bristle-tipped; intermediate and upper leaves clasping, more remotely and irregularly divided, lower segments linear-oblong, the terminal crest-like; all whitish, papillose-tomentose (rarely glabrescent). Pedicels much longer than upper leaves. Flower buds 1.5–4(–5) cm., oblong to spindle-shaped, attenuate and more or less acute at apex. Sepals oblong, crisp-hairy. Petals 1.5–4(–6) cm., red to dark orange (very rarely yellow) with a violet-spotted base, obovate-orbicular. Capsule 10–20 cm., straight or somewhat arcuate, tapering at apex, densely or loosely covered with appressed or partly spreading long, crisp hairs. Fl. March–May.

Var. **grandiflorum.** Petals 3–4(–5) cm. Capsule fairly densely hairy.

Hab.: Steppes and semisteppic batha. Judean Mts., Judean Desert, N. and C. Negev, Dead Sea area, Ammon.

Var. **judaicum** (Bornm.) Sam. ex Rech. f., Ark. Bot. ser. 2, 2, 5 : 340 (1952). *G. judaicum* Bornm., Beih. Bot. Centralbl. 29, 2 : 12 (1912). Petals 1.5–2.5(–3) cm. Capsule mostly sparingly hairy or glabrescent.

Hab.: As above. Judean Desert, N. and C. Negev, Lower Jordan Valley, Dead Sea area, Moav, Edom.

Area of species : Mainly Irano-Turanian.

G. grandiflorum reaches the southernmost limit of its distribution range in Palestine and splits into a series of forms not encountered in other parts of its area. Remarkable among them are the small-flowered forms which are confined to the Judean Desert and approach very closely the yellow-flowered *G. arabicum* Fresen. On the other hand, in the typical, large-flowered variety (var. *grandiflorum*), one sometimes encounters yellow-flowered specimens which cannot be considered a particular variety.

A form somewhat intermediate between the type and var. *judaicum* is fairly abundant in Transjordan (Gilead). All this renders the taxonomic rank of var. *judaicum* questionable.

4. Glaucium aleppicum Boiss. et Hausskn. ex Boiss., Fl. 1 : 122 (1867); Fedde, in Pflznr. 40 : 228 (1909). [Plate 342]

Perennial, sparingly crisp-papillose, green-glaucous, 20–50 cm. Stems many, erect or ascending, sparingly branched. Leaves lyrate-pinnatipartite into irregular, oblong-obovate, entire or dentate, segments; lower leaves 10–20 cm., short-petioled, lower segments remote, the upper approximate, short-dentate with obtuse mucronulate teeth, the terminal one shallowly 3-lobulate or crested; intermediate and upper leaves with broad clasping base, and 3 to 5 lobes, the upper lobe club-shaped, entire or cristate-dentate at apex. Pedicels longer than leaves, almost glabrous. Flower buds 3–5 cm., conical or fusiform, glabrous or glabrescent. Sepals 3–5 cm., oblong, broadly scarious-margined on one side, sparingly hairy or glabrous. Petals 4–5 cm. or more, scarlet, with darker spot at base, broadly obovate to suborbicular. Capsule 10–18 cm., mostly sparingly hairy or glabrous. Fl. March–May.

Hab.: Batha and steppes; rocky ground. Upper Galilee, Negev, Upper Jordan Valley, Ammon.

Area: W. Irano-Turanian.

5. Glaucium flavum Crantz, Stirp. Austr. ed. 1, 2 : 133 (1763); Fedde, in Pflznr. 40 : 232 (1909). *G. luteum* Scop., Fl. Carn. ed. 2, 1 : 369 (1772); Boiss., Fl. 1 : 122 (1867). [Plate 343]

Perennial or biennial, more or less papillose-hairy to glabrescent, glaucous, 30–60 cm. Stems few or many, erect or ascending, somewhat dichotomously branched, glabrous or sparingly pilose at base. Leaves thick or succulent, densely covered on both sides with short crisp hairs; lower leaves 15–35 cm., long-petioled, oblong, lyrate-pinnatifid, lower segments few-lobulate or entire, upper ones dentate or incised into triangular, acute lobules; intermediate and upper leaves sessile, much shorter, ovate, cordate, clasping at base, pinnately lobed, the uppermost sometimes sinuate-dentate. Pedicels conspicuous, often shorter than leaves. Flowers 5–7 cm. across, bright yellow. Sepals ovate-oblong, acute, crisp papillose-hairy. Petals almost orbicular. Capsule 10–25 cm., tapering at apex, more or less densely covered with papillae or tubercles, sometimes glabrescent at maturity; stigma broad, often curved upwards. Fl. April–August.

Hab.: Coast; sandy or shingle beaches. Coastal Galilee, Acco Plain, Coast of Carmel, Sharon Plain, Philistean Plain.

Area: Mediterranean, extending towards the Euro-Siberian territories.

43. FUMARIACEAE

Herbaceous, sometimes climbing plants with watery sap. Leaves alternate, rarely almost opposite, usually much divided or dissected. Flowers hermaphrodite, racemose or cymose, actinomorphic or zygomorphic. Calyx of 2, often coloured and bract-like caducous sepals. Corolla of 4, more or less coherent and sometimes basally connate petals, in 2 whorls; in the zygomorphic flowers 1–2 petals of the outer whorl saccate or spurred, the inner ones narrower, crested and united above anthers. Stamens 4 or 2, but then each stamen with one 2-celled middle anther and two 1-celled lateral anthers; nectar glands at base of one or both stamens. Ovary superior, 2-carpelled, 1-celled, with parietal placentae; ovules 1–2 to many, anatropous; style 1; stigma 1, sometimes 2-lobed. Fruit a dehiscent septate capsule or an indehiscent 1-seeded nutlet or a loment. Seeds with copious endosperm and minute embryo.

Twenty genera and 450 species, mainly in north-temperate regions, most of them in the Old World.

1. Fruit a loment of many, 1-seeded segments. Flowers actinomorphic. All or inner petals 3-lobed. **1. Hypecoum**
- Fruit a nutlet or a 2-seeded capsule. Flowers zygomorphic. Petals not 3-lobed 2
2. Plants without tendrils. Fruits all alike. **3. Fumaria**
- Plants with tendrils. Fruits of two kinds in each raceme, the lower 1-seeded nutlets, the upper 1–2-seeded capsules. **2. Ceratocapnos**

1. Hypecoum L.

Annuals, mostly glabrous, glaucous. Stems procumbent to erect, branching from base. Leaves 2–3-pinnatifid or pinnatipartite. Flowers mostly yellow, in short cymes. Sepals 2, caducous. Petals 4, 2-whorled, the outer entire or 3 lobed, the inner smaller, 3-partite, middle lobe frequently concave above. Stamens 4, mostly with winged filaments. Ovary 2-carpelled, 1-celled, many-ovuled; placentae 2; style with 2 stigmas. Fruit a loment with several 1-seeded segments, disarticulating after ripening into joints, rarely fruit a pod-like capsule dehiscing by 2 valves. Seeds ovoid or octahedral, more or less tuberculate.

About 15 species, mainly in the Mediterranean and Irano-Turanian regions; a few also in the Euro-Siberian and Saharo-Arabian territories.

Literature: F. Fedde, *Hypecoum* L., in Papaveraceae-Hypecoideae et Papaveraceae-Papaveroideae, in: *Pflznr*. 40 (IV. 104): 85–97 (1909).

1. Loment pendulous on a deflexed pedicel. Outer petals entire, not 3-lobed. **6. H. pendulum**
- Loment erect or suberect, not pendulous 2
2. Outer petals entire, elliptical or elliptical-rhombic. **5. H. geslinii**
- Outer petals more or less 3-lobed 3
3. Middle lobe of inner petal mostly with entire margin. **4. H. aegyptiacum**
- Middle lobe of inner petal with fringed or ciliate margin 4
4. Outer petals distinctly 3-lobed, lateral lobes equalling or exceeding the middle one. Filaments broadly winged below and narrowly winged above. **3. H. imberbe**
- Outer petals more or less 3-lobed with lateral lobes shorter than middle one. Filaments narrowly winged near base only 5
5. Inner petals yellow with violet lateral lobes. **2. H. deuteroparviflorum**
- Inner petals yellow altogether. **1. H. procumbens**

1. Hypecoum procumbens L., Sp. Pl. 124 (1753); Boiss., Fl. 1: 124 (1867); Fedde, in Pflznr. 40: 87 (1909). [Plate 344]

Annual, glabrous, 5–15 cm. Stems mostly procumbent. Leaves spreading to almost erect, 2–3-pinnatisect with oblong-lanceolate to ovate lobes; lower leaves 5–15 cm., crowded. Inflorescences few- or many-flowered. Pedicels mostly longer than leaves, mostly ascending in fruit. Flowers about 1 cm., yellow. Sepals ovate-lanceolate, obsoletely denticulate, acuminate. Outer petals about 8 mm. broad, 3-lobed, frequently short-clawed; inner petals shorter, 3-partite, middle lobe about as long as lateral ones, short-stipitate, subcordate at base, cochleariform to boat-shaped, fringed-ciliate. Stamens with filaments narrowly-winged at base; anthers about 1.5 mm. Fruit 3–5 × 0.2–0.3 cm., mostly erect, arcuate-falcate, thickened at joints, somewhat compressed, striate, often surrounded by a membranous epidermis; joints hardly separating when ripe. Seeds finely tuberculate. Fl. March–April.

Hab.: Roadsides and fallow fields. Judean Mts., Judean Desert, Lower Jordan Valley, Ammon, Moav, Edom. Rather rare.

Area: Mediterranean, extending towards the Irano-Turanian and the Euro-Siberian regions.

2. Hypecoum deuteroparviflorum Fedde, in Pflznr. 40:90 (1909). *H. parviflorum* Barb., Herbor. Levant 115 (1882); Boiss., Fl. Suppl. 24 (1888) non Kar. et Kir., Bull. Soc. Nat. Mosc. 15:141 (1842).

Annual, low, procumbent, glabrous, glaucescent, 7–15 cm. Leaves up to 10 cm., all prostrate, 3-pinnatisect into crowded segments; petiole 5 cm., thickened at base. Flowers few, small, 5–7 mm., yellow. Sepals membranous, acuminate, subpectinate at margin, half as long as petals. Outer petals ovate-subrhombic, with 3 small lobes, clawed; the inner 3-partite, with stipitate and fimbriate, oblong-linear middle lobe and 2 shorter, somewhat violet, linear lateral lobes. Filaments with narrow, membranous wings at base. Style 2-fid; lobes divaricate, subulate. Fruit almost erect, arcuate, thickened at joints, torulose. Fl. March.

Hab.: Sand fields. W. Negev (after Fedde, l.c.).

Area: Saharo-Arabian.

3. Hypecoum imberbe Sm. in Sibth. et Sm., Fl. Gr. Prodr. 1:107 (1806) et Fl. Gr. 2:t. 156 (1813). *H. grandiflorum* Benth., Cat. Pyr. 91 (1826); Boiss., Fl. 1:125 (1867); Fedde, in Pflznr. 40:91 (1909). [Plate 345]

Annual. Stems and leaves ascending to erect, 12–35 cm. Leaves 2–3-pinnatisect, lobes oblong-lanceolate to oblong-linear. Inflorescences few- to many-flowered. Flowers long-pedicelled. Sepals ovate-lanceolate, long-acuminate, rather entire. Outer petals 1–1.3 cm. broad, orange-yellow, 3-lobed, cuneate at base, lateral lobes as long as or longer than middle one; inner petals smaller, 3-partite, middle lobe stipitate, cochleariform, subcordate, ciliate, shorter than the lateral ones. Filaments broadly winged at base, narrowly winged above; anthers about 2 mm. Fruit 4–5 × 0.2–0.4 cm., erect, arcuate on spreading pedicel, jointed, torulose, striate, hardly disarticulating. Fl. March–April.

Hab.: Roadsides and fields. Sharon Plain, Philistean Plain, Samaria, Judean Mts., Judean Desert, W. Negev, Lower Jordan Valley, Dead Sea area, Gilead, Ammon, Moav, Edom.

Area: Mediterranean and W. Irano-Turanian.

4. Hypecoum aegyptiacum (Forssk.) Aschers. et Schweinf., Ill. Fl. Eg. 37 (1887); Fedde, in Pflznr. 40:93 (1909); Boiss., Fl. 1:125 (1867). *Mnemosilla aegyptiaca* Forssk., Fl. Aeg.-Arab. 22 (1775). [Plate 346]

Annual, rather glabrous, 10–25 cm. Stems ascending to erect, branching, especially from base. Leaves 2–3-pinnatipartite with short lobes, mostly lanceolate, mucronulate; lower leaves 3–10 cm., crowded. Inflorescences often dichotomously branching, few- to many-flowered. Pedicels equal to or longer than leaves, thin, spreading in fruit. Flowers

0.7–1.5 cm. Calyx about 5 mm., with lanceolate, dentate sepals, 2-mucronate above. Outer petals yellow, large, more or less 3-lobed, middle lobe as long as or somewhat longer than lateral ones; inner petals violet at base, 3-partite, middle lobe stipitate, concave, boat-shaped above, with entire, rarely short-ciliate margin, lateral lobes growing after flowering. Filaments narrowly winged. Fruit 3 × 0.1–0.2 cm., slender, erect, straight or somewhat curved, striate, readily disarticulating into segments. Seeds minutely tuberculate. Fl. March–April.

Hab.: Steppes. Philistean Plain, W. and N. Negev.

Area: Saharo-Arabian.

5. Hypecoum geslinii Coss. et Kral., Bull. Soc. Bot. Fr. 4 : 522 (1857); Coss., Comp. Fl. Atl. 2 : 74 (1887); Fedde, in Pflznr. 40 : 94 (1909). [Plate 347]

Very similar to *H. aegyptiacum* (Forssk.) Aschers. et Schweinf. Stems 7–20 cm., procumbent-ascending. Leaf segments linear to narrowly lanceolate. Pedicels more or less equalling leaves in length. Outer petals about 2.5 mm. broad, elliptical or sub-rhombic, not lobed; inner petals 3-partite with middle lobe longer than lateral ones, growing after flowering, boat-shaped and shortly ciliate above, somewhat violet at base. Fruit fairly straight or arcuate. Fl. March–April.

Hab.: Sandy deserts and fields. W. Negev.

Area: Saharo-Arabian, extending into Irano-Turanian territories.

Above description and drawing are based on a N. African specimen.

6. Hypecoum pendulum L., Sp. Pl. 124 (1753); Boiss., Fl. 1 : 125 (1867); Fedde, in Pflznr. 40 : 95 (1909). [Plate 348]

Annual, glabrous, 5–20 cm. Stems ascending to erect, simple or branching from base. Leaves 2–3-pinnatisect with linear-setaceous to filiform lobes; lower leaves 1.5–7 cm., crowded. Inflorescences few-flowered. Pedicels mostly longer than leaves (the lower shorter), thickened and deflexed in fruit. Flowers 0.6–1 cm. Calyx about 3 mm., membranous, yellow-green with ovate sepals. Outer petals 4–7 mm. broad, oblong to subrhombic, not lobed; inner petals 3-partite, with middle lobe stipitate, concave-cochleariform, fringed-ciliate above, longer than lateral lobes. Stamens with narrowly winged filaments. Fruit 4–5 × 0.2–0.3 cm., pendulous, almost straight, articulate and ribbed-angular, with a membranous, readily separating, epidermis. Seeds densely and minutely pitted-tuberculate. Fl. April–May.

Hab.: Fields. Judean Desert, Negev, Ammon, Moav, Edom.

Area: Mediterranean and Irano-Turanian.

2. CERATOCAPNOS Dur.

Slender, tendril-bearing annuals. Leaves 1- to 2-ternately divided into entire leaflets. Inflorescences racemose, usually short, opposite leaves. Flowers small, rose-coloured, 1-bracteate. Sepals 2, small, caducous. Corolla zygomorphic; petals 4 in 2 whorls, connivent; upper petal of outer whorl spurred at base, the inner two plane. Stamens 2, each with one middle 2-celled anther and two lateral 1-celled anthers. Upper stamen spurred at base. Ovary with 2 ovules or more; style filiform; stigma minute, flattened.

Fruits of two kinds: the lower ovoid, 4-angled, indehiscent, 1-seeded, grooved on one side, truncate; the upper lanceolate, dehiscent by 2 valves, 1–2-seeded, beaked.

Two species in the Mediterranean region.

1. Ceratocapnos palaestinus Boiss., Diagn. ser. 1, 8 : 12 (1849) et Fl. 1 : 132 (1867). [Plate 349]

Annual, climbing, glabrous, much branched, 60 cm. or more. Stems slender, with long internodes, somewhat grooved. Leaves alternate, (1–)2-ternate; leaflets 1–2 × 0.4–1 cm., petiolulate, ovate, mucronate, entire. Inflorescences long-peduncled, few-flowered, loose. Pedicels much shorter than linear bracts, later elongating. Flowers 0.8–1 cm., pink. Sepals linear-lanceolate. Outer petals broadly winged at apex; the upper one with broad and short saccate spur at base. Fruits of two kinds: the lower 1-seeded nutlets, 3–4 mm., parallelly and deeply grooved, with an umbonate operculum; the upper 1–2-seeded capsules up to 1.5–1.8 cm. long, tapering into a sword-shaped beak. There are transitions between the two fruit forms. Fl. February–May.

Hab.: Shady places behind rocks and in maquis. Acco Plain, Sharon Plain, Upper and Lower Galilee, Mt. Carmel, Esdraelon Plain, Samaria, Judean Mts., Hula Plain, Gilead.

Area: E. Mediterranean.

3. FUMARIA L.

Annuals, often climbing, with finely dissected leaves. Inflorescences racemose, usually terminal. Flowers with short pedicels and bracts of various lengths. Sepals 2, of various sizes and shapes, broader or narrower than corolla. Petals 4, in 2 whorls, connivent; upper petal of outer whorl spurred at base, semi-cylindrical, winged above, the lower narrow, more or less canaliculate, never spurred; inner petals both spatulate. Stamens 2, each with one middle 2-celled anther and two lateral 1-celled ones. Ovary 1-celled; ovule 1; style filiform; stigma 2-lobed. Fruit an indehiscent, 1-seeded nutlet.

About 50 species in the Mediterranean region, Middle Europe, W. to C. Asia.

Literature: H. W. Pugsley, A revision of the genera *Fumaria* and *Rupicapnos. Journ. Linn. Soc. Lond. Bot.* 44 : 233–355 (1919).

1. Flowers at least 9 mm. long. Lower petal (of outer whorl) not spatulate   2
– Flowers less than 9 mm. long. Lower petal more or less spatulate   5
2. Fruit smooth when dry. Fruiting pedicels recurved.   **3. F. capreolata**
– Fruit rugulose or rugose when dry. Fruiting pedicels mostly erect   3
3. Sepals oblong-lanceolate, narrower than corolla. Petals white to pale pink. Fruit slightly compressed.   **1. F. judaica**
– Not as above   4
4. Sepals broader than corolla. Fruit about 2 mm. across, slightly rugulose when dry.
    4. F. thuretii
– Sepals almost linear, not broader than corolla. Fruit 3 mm. or more, coarsely rugose when dry.   **2. F. macrocarpa**
5 (1). Sepals very small, narrower than corolla.   **7. F. parviflora**
– Sepals large and as broad as or broader than corolla   6
6. Fruiting pedicels arcuate-recurved. Fruit 1–1.6 mm., nearly smooth.   **5. F. kralikii**

– Fruiting pedicels usually erect-spreading. Fruit larger than in above, rugose.

6. F. densiflora

1. Fumaria judaica Boiss., Diagn. ser. 1, 8 : 15 (1849) et Fl. 1 : 138 (1867); Pugsley, Journ. Linn. Soc. Lond. Bot. 44 : 267 (1919). [Plate 350]

Glabrous annual, 30–100 cm. Leaves 5–15 cm., 2–3-pinnatisect, with oblong, mucronulate, unequal lobes. Racemes loose in fruit. Bracts about half as long as fruiting pedicels. Sepals oblong-lanceolate, narrower than and about one quarter the length of the 1–1.2 cm. long corolla, acutely denticulate. Petals whitish to pale pink, the inner ones tipped with purple. Fruit (2–)2.5–3 mm. in diam., slightly compressed, more or less keeled, obtuse, tuberculate-rugose. Fl. January–April.

Hab.: Hedges and shady places. Acco Plain, Sharon Plain, Philistean Plain, Mt. Carmel, Samaria, N. Negev.

Area : E. Mediterranean.

Also includes f. *bipunctata* Aarons. ex Opphr. et Evenari, Bull. Soc. Bot. Genève ser. 2, 31 : 245 (1941) with 2 black points at apex of not tuberculate fruit.

2. Fumaria macrocarpa Parl., Pl. Nov. 5 (1842); Boiss., Fl. 1 : 137 (1867); Pugsley, Journ. Linn. Soc. Lond. Bot. 44 : 268 (1919). *F. megalocarpa* Boiss. et Sprunn. in Boiss., Diagn. ser. 1, 1 : 68 (1843). [Plate 351]

Glabrous annual, 40–80 cm. Lower leaves usually 6–12 cm., 2–3-pinnatisect into broad, oblong-lanceolate, obtuse, abruptly mucronulate, flat lobes. Racemes rather dense. Bracts somewhat shorter than the erect-spreading fruiting pedicels. Flowers 0.9–1.1 cm., pinkish-white. Sepals small, up to 3 × 1 mm. Fruit up to 3(–4) mm. in diam., globular or slightly compressed, obscurely keeled, tuberculate-wrinkled. Fl. March–May.

Hab.: Hedges, fields and gardens. Coastal Galilee, Acco Plain, Sharon Plain, Upper Galilee, Samaria, Dan Valley, Hula Plain, Upper Jordan Valley.

Area : E. Mediterranean.

Var. *oxyloba* (Boiss.) Hamm., Monogr. Fum. 45 (1857); Boiss., l.c. 138, has not been separated from the typical form.

3. Fumaria capreolata L., Sp. Pl. 701 (1753); Boiss., Fl. 1 : 136 (1867); Pugsley, Journ. Linn. Soc. Lond. Bot. 44 : 269 (1919). [Plate 352]

Glabrous annual, 30–120 cm. Leaves 10–20 cm., 2–3-pinnatisect with oblong, mucronulate, flat lobes. Flowering racemes rather dense. Bracts slightly shorter than more or less recurved fruiting pedicels. Flowers 1–1.6 cm. Sepals ovate, acute, more or less dentate, broader than and about half as long as corolla. Wings of upper petal and tips of inner ones dark purple. Fruit subglobular, more or less smooth. Fl. March–June.

Hab.: Humid, shady places. Coastal Galilee, Sharon Plain, Philistean Plain, Upper and Lower Galilee, Mt. Carmel, Samaria, Shefela, Dan Valley.

Area : Mediterranean and Euro-Siberian.

Varies in flower colour and dentation of sepals.

4. Fumaria thuretii Boiss., Diagn. ser. 2, 1 : 15 (1853) et Fl. 1 : 137 (1867); Pugsley, Journ. Linn. Soc. Lond. Bot. 44 : 292 (1919). [Plate 353]

Glabrous and rather robust annual, branching from base. Leaves 4–9 cm., 2–3-pinnatisect into minute, oblong-linear, acute, flat lobes. Racemes mostly many-flowered. Bracts a little shorter than recurved or erect fruiting pedicels. Flowers about 1 cm. Sepals ovate, obscurely repand-dentate to almost entire, pale rose to white, fairly large, broader than pink corolla tube. Wings of upper petal purple, margins of lower one very narrow. Fruit small, about 2 mm. across, ovoid, somewhat rugulose when dry. Fl. March.

Hab. : Field borders. Upper Galilee.

Area : E. Mediterranean.

In our specimens the fruiting pedicels are erect or spreading, not recurved.

5. Fumaria kralikii Jord., Cat. Dijon 19 (1848); Pugsley, Journ. Linn. Soc. Lond. Bot. 44 : 298 (1919). *F. anatolica* Boiss., Diagn. ser. 1, 8 : 14 (1849) et Fl. 1 : 136 (1867). [Plate 354]

Glabrous annual, 20–30 cm. Stems erect, branching. Leaves 4–10 cm., 2–3-pinnatisect, with oblong to linear, flat lobes. Racemes rather loose. Bracts oblong or linear-oblong, somewhat longer than arcuate-recurved fruiting pedicels. Flowers 5–6 mm. Sepals broader than corolla, elliptical, acuminate, denticulate. Petals pink, purple-tipped. Fruit small, 1–1.6 mm. in diam., subglobular, obtuse, almost smooth. Fl. April–May.

Hab. : Fields and among shrubs. Sharon Plain, Upper and Lower Galilee, Mt. Carmel, Samaria, Judean Mts.

Area : Mainly E. Mediterranean.

6. Fumaria densiflora DC., Cat. Hort. Monsp. 113 (1813). *F. micrantha* Lag., Elench. Hort. Matrit. 21 (1816); Boiss., Fl. 1 : 136 (1867); Pugsley, Journ. Linn. Soc. Lond. Bot. 44 : 299 (1919). [Plate 355]

Erect or spreading annual, 10–25 cm. Leaves 4–10 cm., 3–4-pinnatisect into narrowly linear, canaliculate lobes. Inflorescences dense. Bracts linear-oblong, acuminate, a little longer than erect-spreading fruiting pedicels. Flowers (4–)5–7 mm., pink, with pink or purple tip on inner petals and on wings of the upper one. Sepals nearly orbicular-elliptical, as broad as or broader than corolla, dentate. Fruit 2–2.5 mm. in diam., somewhat keeled, obtuse and finely rugose. Fl. January–May.

Var. **densiflora.** Leaf segments linear. Flowers about 6–7 mm. long. Sepals broader than corolla.

Hab. : Fields. Coastal Galilee, Sharon Plain, Philistean Plain, Upper and Lower Galilee, Mt. Carmel, Esdraelon Plain, Samaria, Shefela, Judean Mts., Judean Desert, N. Negev, Dan Valley, Hula Plain, Upper and Lower Jordan Valley, Ammon.

Var. **parlatoriana** (Kral. ex Boiss.) Zoh. (comb. nov.). *F. micrantha* Lag. var. *parlatoriana* Kral. ex Boiss., Fl. 1 : 136 (1867). *F. bracteosa* Pomel, Nouv. Mat. Fl. Atl. 239 (1874); Pugsley, Journ. Linn. Soc. Lond. Bot. 44 : 300 (1919). Leaf segments setaceous.

Flowers minute, 4–5 mm., with much diminished spur. Sepals as broad as and half as long as corolla.

Hab.: Fields. Sharon Plain, Shefela, Judean Mts., Judean Desert, Negev, Upper and Lower Jordan Valley.

Area of species: Mediterranean, extending towards the Euro-Siberian and Irano-Turanian regions.

7. Fumaria parviflora Lam., Encycl. 2 : 567 (1788); Boiss., Fl. 1 : 135 (1867); Pugsley, Journ. Linn. Soc. Lond. Bot. 44 : 322 (1919). [Plate 356]

Glabrous annual, 5–25 cm. Leaves 3–4-pinnatisect, with linear-spatulate, rather acute, usually caniculate lobes. Racemes short-peduncled, dense, up to 20-flowered. Bracts about as long as or longer than erect-spreading fruiting pedicels. Flowers 5–7 mm., white. Sepals minute, as broad or narrower than pedicels, dentate. Corolla white altogether or tinged with pink or purple at top; upper petals entire or slightly emarginate. Fruit subglobular, distinctly keeled, obtuse or apiculate, rugose. Fl. February–May.

Hab.: Fields and roadsides. Coastal Galilee, Acco Plain, Sharon Plain, Philistean Plain, Upper and Lower Galilee, Mt. Carmel, Esdraelon Plain, Samaria, Shefela, Judean Mts., Judean Desert, W. and C. Negev, Dan Valley, Hula Plain, Upper and Lower Jordan Valley, Dead Sea area, Arava Valley, Gilead, Ammon, Moav, Edom.

Area: Mediterranean, Euro-Siberian and Irano-Turanian.

According to Rechinger f. (Ark. Bot. ser. 2, 2, 5 : 341, 1952) and Dinsmore in Post (Fl. Syr. Pal. Sin. ed. 2, 1 : 46, 1932) *Fumaria gaillardotii* Boiss., Fl. 1 : 139 (1867) occurs in the Hula Plain, the Philistean Plain and Samaria. We have not observed this species in Palestine.

The data by Dinsmore (l.c.) on *F. asepala* Boiss., Fl. 1 : 135 (1867) and *F. agraria* Lag., Gen. et Sp. Nov. 21 (1816) from Palestine are most probably erroneous.

44. CAPPARACEAE

Herbs or shrubs, sometimes vines. Leaves alternate, rarely opposite, simple or compound, with unarmed or spiny stipules. Flowers hermaphrodite, rarely unisexual, solitary or in compound, bracteate inflorescences, zygomorphic or actinomorphic. Sepals 4 (–5), free or connate. Petals 4, rarely 0 (very rarely 5 or 8), mostly free. Torus short or elongated, often with a symmetrical or asymmetrical disk. Stamens 4, 6 or many, free or sometimes adnate to torus, often borne on an androphore; anthers 2–4-celled. Ovary superior, sessile or borne on a gynophore, 1-celled, sometimes many-celled by false septa growing from placentae; ovules numerous, rarely 2, campylotropous, inserted on parietal placentae; style 1; stigma 2-lobed or capitate. Sometimes androecium and gynoecium borne on an androgynophore. Fruit a capsule, often 2-valved, or rarely a drupe or berry, often many-seeded (rarely 1–2-seeded). Seeds reniform, with smooth testa and mostly without endosperm; embryo curved.

About 50 genera and 800 species, almost all tropical; a few of them in the Mediterranean and Saharo-Arabian regions.

1. Corolla 0 (in ours). Plants neither spiny nor glandular.   **1. Maerua**
 – Corolla present. Plants spiny or glandular-hairy   **2**
2. Stamens numerous. Shrubs, glabrous or hairy, with spiny stipules.   **2. Capparis**
 – Stamens 4–6. Herbs or shrubs, glandular-hairy, with unarmed stipules.   **3. Cleome**

Subfam. CAPPAROIDEAE. Fruit a berry or drupe, without replum, not regularly dehiscent. Stamens many or 4–16.

1. MAERUA Forssk.

Trees or shrubs, unarmed, glabrous or pubescent. Leaves simple or trifoliate, often with small setaceous stipules; petiole often jointed with branch. Flowers hermaphrodite, solitary or in corymbs or racemes. Sepals 4, caducous, connate below into a cylindrical or infundibuliform tube. Petals 0 or 4. Corona mostly present, formed from the prolonged disk (torus), with or without a toothed, fimbriate or entire margin. Stamens many, free or connate at base and adnate to gynophore. Ovary on a long gynophore, 1–2-celled, with 2–4 parietal placentae forming sometimes false septa; ovules numerous. Fruit an ovoid or cylindrical berry often transversally many-celled, the cells 1-seeded.

Over 80 species, mostly in savannas of Trop. Africa and Asia.

1. Maerua crassifolia Forssk., Fl. Aeg.-Arab. CXIII et 104 (1775). *M. uniflora* Vahl, Symb. Bot. 1 : 36 (1790); Boiss., Fl. 1 : 419 (1867). [Plate 357]

Tree or shrub with ascending or divaricate branches. Young twigs smooth, pubescent. Bark grey, longitudinally fissured. Leaves about 1 cm. or less, succulent, cuneate-obovate, entire, mostly retuse, pubescent or puberulent, with short, non-jointed petioles. Flowers few to numerous on long, hairy peduncles, growing from older branches. Calyx 0.8–1 cm.; sepals 4, oblong, obtuse, reflexed; tube urceolate. Corolla 0. Corona fringed. Stamens about 30, adnate to gynophore. Ovary cylindrical, glabrous, borne on a long gynophore. Fruit a berry, about 1 cm. in diam. Fl. April.

Var. **maris-mortui** Zoh., Imp. For. Inst. Univ. Oxf., Inst. Pap. 26 : 21 (1951). Peduncles 2–4-flowered, appressed-pubescent. Calyx densely hirsute. Ovary glabrous.

Hab.: Wadis and oases in the Dead Sea area. Rare.

Area: Sudanian.

2. CAPPARIS L.

Trees or shrubs, sometimes prostrate or climbing, usually with stipular spines and simple leaves, sometimes almost leafless. Flowers usually hermaphrodite, axillary or terminal, solitary or racemose, corymbose or umbellate, more or less zygomorphic. Sepals 4, the posterior often larger and more concave than others. Petals 4, the 2 posterior ones coherent, forming a nectariferous fleshy protuberance or cavity at the thickened base. Stamens numerous (or 8–16), inserted on torus; filaments sometimes somewhat connate and hairy at base. Ovary borne on a long (rarely short) gynophore, 2–8-carpelled, 1-celled, with 2 or more parietal placentae; ovules numerous; stigma sessile. Fruit a fleshy berry, indehiscent or separating into 3–4 valves. Seeds numerous.

About 350 species mainly in tropical regions of both hemispheres, a few in the Mediterranean and S. W. Asian countries.

Literature : M. Zohary, The species of *Capparis* in the Mediterranean and the Near Eastern countries. *Bull. Res. Counc. Israel* D, 8 : 49–64 (1960). M. Jacobs, The genus *Capparis* (Capparaceae) from the Indus to the Pacific. *Blumea* 12 : 385–541 (1965).

1. Spartoid, almost leafless erect shrubs. Leaves, when present, linear, soon deciduous. Corolla red, small.   **4. C. decidua**
  – Leafy plants, with white, large flowers   **2**
 2. Leaves thick, cartilaginous-fleshy, mostly ovate. Posterior sepal large, up to 4 cm., strongly incurved and helmet-shaped. Evergreen robust shrubs of hot deserts.
                                                                      3. C. cartilaginea
  – Plants not as above   **3**
 3. Leaves orbicular or ovate-orbicular, not hairy-fleecy. Stems glabrous.   **1. C. spinosa**
  – Leaves obovate, ovate or ovate-oblong, often hairy or fleecy. Stems hairy.   **2. C. ovata**

Sect. CAPPARIS. Sepals rather large, the posterior boat- or helmet-shaped. Stamens many. Leafy shrubs or trees.

1. Capparis spinosa L., Sp. Pl. 503 (1753); Boiss., Fl. 1 : 420 (1867); Zoh., Bull. Res. Counc. Israel D, 8 : 50 (1960). [Plate 358]

Shrub with long, decumbent or ascending, smooth, glabrous, often purplish or glaucous branches. Leaves long-petioled, orbicular or slightly ovate, entire, tipped with a spiny mucro, usually glabrous; stipular spines varying in length, often hooked and strong. Flower buds more or less symmetrical. Flowers axillary, solitary, long-pedicelled, showy, white. Sepals glabrous, all more or less equally long, the posterior boat-shaped. Petals obovate-orbicular, almost equal in size, the 2 posterior ones with thickened margins coherent in lower part and forming a nectariferous protuberance. Stamens numerous, with pink or bluish filaments, often shorter than glabrous gynophore. Berry 2.5–4 cm., pyriform, dehiscing by valves.

Var. **aegyptia** (Lam.) Boiss., Fl. 1 : 420 (1867). *C. aegyptia* Lam., Encycl. 1 : 605 (1785); Del., Fl. Eg. 237 (1813). Shrub with aërial shoots decaying at end of summer. Stems and branches often purplish in upper part. Leaves usually 1.5–2.5 cm., deciduous, not fleshy, orbicular or ovate-orbicular to orbicular-cuneiform, rounded or retuse at apex with a small mucro or prickle. Anterior petals 2 × 1.5 cm., obovate or orbicular, elongating after anthesis. Fl. May–August.

Hab.: Walls, fences and grassland, often also in the vicinity of human dwellings. Acco Plain, Sharon Plain, Philistean Plain, Upper and Lower Galilee, Mt. Carmel, Esdraelon Plain, Mt. Gilboa, Samaria, Judean Mts., W. and C. Negev, Gilead, Ammon.

Area : Mediterranean, slightly extending into the Irano-Turanian region.

This variety is the common caper in Palestine. It is a frequent associate in the *Hyparrhenia hirta* community.

Var. **aravensis** Zoh., l.c. 53. Evergreen shrub with branches dying back every year only in their upper part, the rest persisting. Leaves 2–3 cm. in diam., thick, fleshy, brittle, orbicular or broadly ovate, rounded or rarely obcordate at base, truncate or emarginate at apex, surface covered with blue-green, waxy bloom. Petals as in var. *aegyptia*. Fl. March–June.

Hab.: Wadis and cliffs in hot deserts. Judean Desert, E. Negev, Dead Sea area, Arava Valley.

Area: E. Sudanian; probably endemic in Palestine.

Typical *C. spinosa* has large, round-ovate glabrous leaves and its flowers are much larger than in above varieties. *C. spinosa* L. var. *spinosa* has obviously been cultivated in W. Mediterranean countries and Europe and has since escaped from cultivation. Although recorded by Duhamel (Nouv. 1 : 137, 1801) from Palestine, this variety has so far not been observed by us.

The Hebrew names צָלָף (Talmudic) and אֲבִיּוֹנָה (Eccl. xii : 5) are believed to refer to *C. spinosa* and its varieties. Flower buds widely collected and consumed as pickles.

2. **Capparis ovata** Desf., Fl. Atl. 1 : 404 (1798); Zoh., Bull. Res. Counc. Israel D, 8 : 54 (1960). *C. spinosa* L. var. *canescens* Coss., Not. Pl. Crit. 2 : 28 (1849) p.p.; Boiss., Fl. 1 : 420 (1867) p.p. [Plate 359]

Hemicryptophytic shrub, 30–100 cm. Stems erect, ascending or procumbent, whitish, with hairy branches. Leaves 0.8–4 cm., long-petioled, ovate or obovate to elliptical or oblong, tapering or rounded or slightly cordate at base, acute or acuminate, usually pubescent or fleecy, bright green or greyish-green; stipular spines strong, setaceous, straight or curved, rarely 0. Flower buds greyish-canescent, unilaterally gibbous. Flowers medium-sized, long-pedicelled. Posterior sepal boat-shaped, much longer and broader than others. Petals white, the anterior ones slightly to much longer than posterior, often strongly divergent, similar to wings of a butterfly, the posterior coherent at base, their thickened and elevated margins forming a deep nectariferous cavity. Stamens numerous, anterior longer, posterior shorter than petals; filaments purple or white; anthers pink. Gynophore long, pilose at base. Berry pear-shaped. Fl. April–August.

Var. **palaestina** Zoh., l.c. 56. Greyish-green canescent shrub, with white stem, 70–100 cm. Branches often erect or spreading, white, hairy. Leaves usually 1.5–2.5 × 1–2 cm., elliptical to oblong, rarely ovate or obovate, acute and mucronate at apex, hairy or fleecy; stipular spines strong, curved. Anterior petals 1.5–2.5 cm., oblong, strongly divergent and much elongating, up to one and a half times to twice as long as posterior ones. Filaments white. Nectariferous cavity of posterior petals bordered by strongly elevated rim. Berry 2.5–3 × 1.5–2 cm., broadly pyriform.

Hab.: Stony places and abandoned fields in non-typical Mediterranean sites. Upper Galilee, Esdraelon Plain, Mt. Gilboa, Judean Mts., Judean Desert, C. and S. Negev, Hula Plain, Upper Jordan Valley, Beit Shean Valley, Lower Jordan Valley, Dead Sea area, Arava Valley, Ammon.

This variety differs markedly from the varieties of *C. spinosa* by its larger posterior sepals, its longer and more divergent anterior petals, its soft-villose or cobwebby indumentum, its whitish filaments, etc.

Var. **microphylla** (Ledeb.) Zoh., l.c. 59. *C. herbacea* Willd. var. *microphylla* Ledeb., Fl. Ross. 1 : 235 (1841). Very close to var. *palaestina* from which it differs by the prostrate habit of its branches, by smaller (0.8–1.5 cm.) leaves and by smaller (1–1.5 cm.) flowers.

Hab. : Steppes and deserts. Negev. Rare.

Area of species : Arid E. Mediterranean, Irano-Turanian and Saharo-Arabian.

Use as in *C. spinosa*.

3. Capparis cartilaginea Decne., Ann. Sci. Nat. Bot. ser. 2, 3 : 273 (1834); Zoh., Bull. Res. Counc. Israel D, 8 : 60 (1960). *C. galeata* Fresen., Mus. Senckenb. 2 : 111 (1836); Boiss., Fl. 1 : 421 (1867). [Plate 360]

Evergreen shrub, 1–2 m. Stems many, divaricately and intricately branching. Branches very flexuous and thickened at nodes, glaucous, puberulent, longitudinally striatulate. Leaves 2–4 (–6) × 1–4 cm., long-petioled, thick, cartilaginous, ovate to oblong, rounded or somewhat tapering at base, glaucous or green, tipped with a spiny, straight or curved mucro; stipular spines short, recurved, sometimes abortive. Flower buds strongly gibbous. Pedicels stout, recurved in fruit. Flowers 8–9 cm., axillary, solitary, strongly zygomorphic, white. Sepals very unequal, the anterior and the lateral ones 1.5–2 cm., oblong, concave; the posterior about 4 cm., helmet-shaped and semi-circular, curved, with a 1–2 cm. deep and wide cavity. Petals very unequal, the anterior pair free, spreading, broadly ovate, the posterior pair coherent, short, partly enclosed within the "helmet". Ovary borne on a long gynophore. Fruit a 6–9 cm., dehiscent, ovoid-pyriform, ribbed berry, borne on 3–4 cm. long stipe. Seeds reniform. Fl. February–April.

Hab. : Ravines and rocks of hot deserts. C. Negev, Lower Jordan Valley, Arava Valley, Edom.

Area : Sudanian.

This species, as partly also others, exhibits a clear trend towards unisexuality by abortion of the ovary.

Use as in *C. spinosa*.

Sect. SODADA (Forssk.) Endl., Gen. 893 (1839). Sepals small, the posterior concave. Stamens 8–15. Almost leafless shrubs.

4. Capparis decidua (Forssk.) Edgew., Journ. Linn. Soc. Lond. Bot. 6 : 184 (1862); Zoh., Bull. Res. Counc. Israel D, 8 : 61 (1960). *Sodada decidua* Forssk., Fl. Aeg.-Arab. 81 (1775). *C. sodada* R. Br. in Denh. et Clapp., Narr. Trav. Afr. App. 225 (1826); Boiss., Fl. 1 : 419 (1867). [Plate 361]

Spartoid shrub up to 4 m., with bright green, patulous, smooth, glabrous branches and twigs. Leaves few on the young twigs, about 4 mm., linear-convolute, spiny at apex, soon deciduous; stipular spines short, patulous, straight or recurved. Flowers in racemes or in axillary fascicles of (1–)3–5, borne on very short branches. Pedicels up to 1 cm. or more. Sepals unequal, the posterior larger and deeply concave, 2 to 3 times as broad as the rest. Petals about 1 cm., red, ovate, ciliolate, the 2 posterior ones partly enclosed within the posterior sepal. Stamens 8–15, longer than petals. Ovary on a 1–2 cm. long gynophore; style about half as long as ovary. Fruit 0.5–1 cm., long-stipitate, ovoid-subglobular, apiculate. Seeds few to many. Fl. May–March.

Hab. : Wadis and oases in hot deserts, among savanna-like vegetation. Upper Jordan Valley, Dead Sea area, Arava Valley. Rare.

Area : Sudanian, extending into other Paleotropical regions.

Young shoots used in folk medicine.

Subfam. CLEOMOIDEAE. Fruit a capsule with persistent replum. Stamens few or many.

3. CLEOME L.

Annual or perennial herbs or half-shrubs, glabrous or glandular. Leaves simple or composed of 3–7 entire or serrate leaflets. Flowers hermaphrodite, solitary or in racemes, actinomorphic or zygomorphic, white, yellow or purple. Sepals 4, free or connate, sometimes caducous. Petals 4, with or without claw. Torus ring-like, cupular or elongated. Stamens 6 (rarely 4 or many), inserted on the torus or on the short androphore, free or very shortly connate at base. Ovary more or less short-stipitate, 2-carpelled, 1-celled, with a replum; ovules many, on two parietal placentae; styles very short or 0. Fruit a capsule, opening by 2 membranous valves separating from replum. Seeds reniform, asperulous or woolly.

About 200 species, mainly of tropical regions, especially in Africa and Asia but also in America.

1. Leaves, at least the lower ones, trifoliate. Glandular-pubescent annuals.   **3. C. arabica**
− Leaves simple   2
2. Low shrubs, woody at base. Leaves orbicular, cordate or reniform. Fruit 1–1.5 cm. long.
                                                                                    1. C. droserifolia
− Perennials, not shrubby. Leaves ovate or oblong. Fruit 4–6 cm. long.   **2. C. trinervia**

1. Cleome droserifolia (Forssk.) Del., Fl. Eg. 106 (1813); Boiss., Fl. 1 : 415 (1867). *Roridula droserifolia* Forssk., Fl. Aeg.-Arab. LXII et 35 (1775). [Plate 362]

Compact, cushion-like, green-yellowish, glandular-viscid half-shrub, profusely and intricately branching, densely leafy, 25–50 cm. Leaves 0.5–1.5 cm. in diam., long-petioled, simple, orbicular to reniform or subcordate, 3-nerved, densely covered with short simple hairs and long-stipitate glands. Flowers about 1–1.2 cm., solitary in axils of upper leaves, long-pedicelled, somewhat zygomorphic. Sepals glandular, oblong. Petals reddish-yellow, with a thickened scale-like appendage at base, ciliate-glandular at apex. Stamens 4, shorter than style. Ovary subsessile, oblong or ellipsoidal, shortly and densely glandular. Fruit 1–1.5 cm., on erect or spreading pedicel, setose-glandular. Seeds rather compressed, reniform-globular, glabrous, minutely granular. Fl. March–May.

Hab. : Hot deserts, rocks and wadis. Judean Desert, Lower Jordan Valley, Dead Sea area, Arava Valley.

Area : E. Sudanian.

2. Cleome trinervia Fresen., Mus. Senckenb. 1 : 177, t. 11 (1834); Boiss., Fl. 1 : 414 (1867). *C. arabica* Botsch., in Novit. Syst. Pl. Vasc. 130 (1964) non L., Cent. Pl. I. 20 (1755). [Plate 363]

Perennial, erect, glandular-setulose herb, 30–60 cm. Stems numerous, little branching. Leaves 2.5–5 cm., petiolate, simple, ovate to oblong, somewhat acute, setulose-glandular, more or less prominently 3-nerved. Racemes terminal, many-flowered, elongated in fruit. Flowers long-pedicelled, with elliptical to oblong bracts shorter than pedicel. Sepals oblong to ovate, densely glandular-setulose, one third as long as long-clawed, oblong, obtuse, red-veined petals. Stamens 6. Fruit 4–6 × 0.6–0.8 cm., pendulous on spreading pedicels, linear to narrowly-elliptical, glandular-scabrous; gynophore very short or almost 0; styles very short or 0. Seeds velvety to woolly. Fl. January–May.

Hab. : Desert wadis and roadsides. Judean Desert, W. and C. Negev, Lower Jordan Valley, Dead Sea area, Arava Valley.

Area : E. Sudanian and Saharo-Arabian.

3. Cleome arabica L., Cent. Pl. I. 20 (1755). *C. africana* Botsch., in Novit. Syst. Pl. Vasc. 130 (1964). [Plate 364]

Annual, erect, simple or branched, glandular-pubescent, 30–50 cm. Leaves petiolate, all or only lower ones trifoliate; leaflets 1–3 cm., petiolulate, oblong-linear, obtuse, entire, scabrous; upper leaves simple, merging into bracts. Flowers in terminal racemes leafy below, long-pedicelled, the lower ones remote, the upper crowded. Sepals ovate-oblong, scabrous-glandular, one third to one half the length of the clawed, oblong-spatulate, obtuse, yellowish-brown and purple-veined petals. Stamens 6 (5). Ovary borne on a short gynophore. Capsule 4–7 × 0.5–0.8 cm., mostly pendulous, compressed, linear-elliptical, glandular-scabridulous, longitudinally netted-veined. Seeds woolly. Fl. March–July.

Hab. : Deserts, in ruderal sites and wadis. W., N. and C. Negev, Lower Jordan Valley, Dead Sea area, Arava Valley, Edom.

Area : Species probably of Sudanian origin extending into the Saharo-Arabian and adjacent Paleotropical regions.

45. CRUCIFERAE

Annual or perennial herbs, sometimes shrubby or woody at base. Leaves mostly alternate, exstipulate. Inflorescences racemose, usually ebracteate. Flowers hermaphrodite, actinomorphic. Sepals 4, in 2 decussate whorls, erect or spreading, equal or 2 of them saccate at base. Petals 4, free, clawed, mostly yellow, white, pink or purple, alternating with the sepals. Stamens 6 (rarely 4 or 2), 2 of them shorter, forming the outer whorl and 4 longer, forming the inner one; filaments sometimes winged or with tooth-like appendages; nectaries various in size and form, at base of stamens. Ovary superior, syncarpous, consisting of 2 carpels and 2 parietal placentae, usually 2-celled through formation of a false septum by the union of inward intrusions of the placentae; ovules 2 to many, rarely 1, campylotropous; style simple or 0; stigma entire or 2-lobed. Fruit a siliqua (at least 3 times as long as broad) or a silicle (length less than 3 times of width), dehiscent from below by 2 valves or upper part indehiscent or altogether indehiscent; sometimes fruit a loment (breaking transversally into 1-seeded indehiscent joints). Seeds usually in 1 or 2 rows in each cell; endosperm almost 0; embryo folded.

About 350 genera and 3,000 species, chiefly in the north-temperate regions of both hemispheres, but also in mountains of tropical regions; some genera cosmopolitan.

The family has recently been divided into 15 tribes, 6 of which are represented in the local flora.

Literature : A. v. Hayek, Entwurf eines Cruciferen-Systems auf phylogenetischer Grundlage. *Beih. Bot. Centralbl.* 27, 1 : 127–335 (1911). O. E. Schulz, Cruciferae-Brassiceae, I, II. in : *Pflznr.* 70 (IV. 105) : 1–290 (1919); 84 (IV. 105) : 1–100 (1923). O. E. Schulz, Cruciferae-Sisymbrieae, in : *Pflznr.* 86 (IV. 105) : 1–388 (1924). O. E. Schulz, Cruciferae – *Draba* et *Erophila*, in : *Pflznr.* 89 (IV. 105) : 1–384 (1927). O. E. Schulz, Cruciferae, in : Engl. u. Prantl, *Nat. Pflznfam.* ed. 2, 17b : 227–658 (1936). E. Janchen, Das System der Cruciferen, *Oesterr. Bot. Zeitschr.* 91 : 1–28 (1942).

1. Fruit a siliqua (i.e., at least 3 times as long as broad)   2
 – Fruit a silicle (i.e., at most twice, rarely two and half times as long as broad)   50
2. Fruit 1-seeded, pendulous, indehiscent, less than 3 cm. long, surrounded by a wing-like extension. Flowers yellow.   **8. Isatis**
 – Fruit not as above   **3**
3. Fruit consisting of 2 joints; the lower seed-bearing or seedless, dehiscing or not but always consisting of 2 valves (rarely valves obsolete); upper joint always containing seed, sometimes necklace-shaped or beak-like, never opening by valves   4
 – Fruit not jointed, dehiscing by 2 valves or not dehiscing; beak if present seedless   13
4. Flowers white or pink or purple, rarely yellow but then fruit pendulous and very long (5–8 cm.)   5
 – Flowers yellow or pale yellow. Fruit not pendulous   8
5. Lower joint 4–6 mm. or more thick, 1–2-seeded, indehiscent. Littoral succulent plants.
                                                                                        59. Cakile
 – Lower joint thin, seedless or 1- to many-seeded. Plants not as above   6
6. Lower joint inconspicuous (or rudimentary), seedless, the upper over 2 cm. long.
                                                                                      64. Raphanus
 – Lower joint 2–4- to many-seeded, upper shorter than in above   7
7. Fruit (incl. beak) 0.7–1.3 cm.; lower joint somewhat flattened, torulose, usually indehiscent.   **57. Reboudia**
 – Fruit (incl. beak) usually 1.5–3 cm. long; lower joint cylindrical or almost so, not torulose, readily or tardily dehiscing.   **58. Erucaria**
8(4). Upper joint many-seeded, lomentaceous or necklace-shaped.   **65. Enarthrocarpus**
 – Upper joint (or beak of fruit) 1–2-seeded   9
9. Lower joint dehiscent by 2 valves, 4- to many-seeded   10
 – Lower joint 1–2(–3–4)-seeded, indehiscent   12
10. Fruit 1.5–2.5 cm. long, appressed to stem; beak bent sidewards, 1-seeded, 3–7 mm. long.
                                                                                      53. Hirschfeldia
 — Fruit mostly not appressed to stem; beak never bent sidewards, often longer than in above   11
11. Flowers 6–8 mm. across, pale yellow. Fruit (incl. beak) 4.5–9 cm.   **51. Brassica**
 — Flowers 1–1.5 cm. across, intensely yellow. Fruit shorter than in above.   **52. Sinapis**
12(9). Lower joint not lomentaceous, upper joint globular or pyriform, without expanding wings.   **60. Rapistrum**

— Lower joint (1–)2–4-seeded, lomentaceous, upper joint muricate, with 2–4 broad wings.   **66. Cordylocarpus**

13(3). Fruit with 2–3 horns at apex; valves mostly hairy or tomentose   14

— Fruit without horns; valves glabrous or hairy   16

14. Flowers about 3 mm. across. Fruit about 1 cm. or less in length, each valve terminating with a horn.   **21. Notoceras**

— Flowers larger than in above; horns formed from stigmas   15

15. Fruit about 2 cm. or shorter; horns of fruit short, consisting of indurated stigmas.
                                                                             20. Morettia

— Fruit (2–)3–10 cm. long; horns consisting of lateral lobes of stigmas.   **19. Matthiola**

16(13). Flowers yellow or rarely cream-coloured   17

— Flowers white, pink, purple, lilac or flesh-coloured   29

17. Petals large, with deep violet veins. Fruit 1.5–2.5 × 0.2–0.6 cm., with beak at least half as long as valves.   **54. Eruca**

— Petals and fruit not as above   18

18. Seeds in 2 rows in each cell   19

— Seeds (in ours) in 1 row in each cell   20

19. Pedicel much shorter than fruit. Fruit not falcate.   **50. Diplotaxis**

— Pedicel half as long to nearly as long as fruit. Fruit falcate.   **12. Nasturtiopsis**

20(18). Leaves glabrous, glaucous, entire, obovate or oblong, with clasping base. Fruit 4-angled, 6–12 cm. long.   **47. Conringia**

— Leaves and fruit not as above   21

21. Fruit 1-seeded, with an oblique beak as long as or longer than indehiscent, crustaceous and tuberculate seed-bearing part.   **9. Schimpera**

— Fruit many-seeded and different from above   22

22. Leaves finely dissected into linear or filiform or terete lobes. Fruit beakless.
                                                                             2. Descurainia

— Leaves not as above   23

23. Seeds broadly winged.   **27. Arabis**

— Seeds not winged   24

24. Fruit falcate, with pedicels at least half as long as fruit. Low desert plants.
                                                                             12. Nasturtiopsis

— Fruit and pedicels not as above   25

25. Leaves simple, entire or toothed, the cauline ones not clasping. Fruit hairy, on thick pedicels.   **10. Erysimum**

— Leaves pinnatifid or otherwise lobed, and if entire or toothed then the cauline ones clasping   26

26. Fruit straight, appressed to stem, not over 2.5 cm. long, with distinct beak. Flowers 6 mm. or longer. Plants 0.6–2 m.   **51. Brassica**

— Fruit straight or falcate, not appressed to stem, and if appressed then flowers much smaller   27

27. Fruit divaricate or pendulous, 6–9 cm. long, glabrous. Leaves entire or dentate, the upper clasping. Flowers up to 1.5 cm. Segetal plants up to 10 cm.   **14. Malcolmia**

— Plants not as above   28

28. Plants about 10–25 cm. Flowers 2–3 mm., not bracteate. Fruit 1.5–3 cm., spreading, hairy, with 1-nerved valves. Cauline leaves not long-auriculate. Very rare.
                                                                             3. Arabidopsis

— Plants usually taller. Fruit usually with 3-nerved valves, glabrous or hairy, longer than

3 cm., sometimes fruit shorter but then cauline leaves long-hastate or flowers bracteate.
1. Sisymbrium

29 (16). Fruit woody or crustaceous, 2–5 mm. thick, with 2 rows of numerous 1-seeded cells generally alternating with sterile ones. Flowers purple.   **22. Chorispora**

— Fruit not as above   30

30. Seeds in 2 rows in each cell   31

— Seeds in 1 row in each cell   35

31. Glabrous aquatic plants. Leaves pinnately divided into ovate or elliptical leaflets. Flowers white, 4–6 mm. long. Fruit 1–2 cm. long.   **25. Nasturtium**

— Terrestrial plants. Leaves not as in above   32

32. Perennials with entire, broad-ovate, clasping, succulent leaves.   **49. Moricandia**

— Annuals. Leaves not as above   33

33. Leaves or lobes of leaves filiform. Fruit indehiscent.   **18. Leptaleum**

— Leaves or lobes not filiform. Fruit readily dehiscing   34

34. Stems 3–10 cm. Leaves 1–2 cm. long. Style much narrower than valves.
13. Stigmatella

— Stems (15–)20–60 cm. Leaves 3–15 cm. long. Style nearly as broad as valves.
50. Diplotaxis

35 (30). Fruit pubescent or tomentose (rarely glabrescent but then fruit 2-horned)   36

— Fruit glabrous   43

36. Fruit not over 2.5 cm. long, tipped with 2 thickened (horn-like) stigmatic lobes. Procumbent, canescent and stellate-hairy desert perennials with entire or toothed leaves.
20. Morettia

— Plants not as above   37

37. Flowers up to 3 mm., white. Fruit 1–3 cm., subsessile, straight, curved, contorted or spirally twisted, tardily or not at all dehiscent.   **16. Torularia**

— Plants not as above   38

38. Fruit pendulous, 5–8 cm. Strigulose-pubescent and glandular perennials with large flowers.   **11. Hesperis**

— Fruit erect or spreading. Plants not glandular-strigulose   39

39. Seeds flattened, narrowly winged   40

— Seeds globular, ovoid or oblong, not winged   41

40. Fruit (1.5–)2–2.5 cm. Flowers 4–8 mm.   **15. Eremobium**

— Fruit 3–5 cm. Flowers 1 cm. or more.   **19. Matthiola**

41 (39). Caespitose, grey-canescent perennials with almost entire, linear or filiform leaves and densely canescent fruit. Desert plants.   **10. Erysimum**

— Annuals. Leaves oblong or obovate, toothed or lobed, rarely entire   42

42. Style in fruit tapering into a conical pointed stigma. Fruiting pedicel thick, as broad as fruit or almost so.   **14. Malcolmia**

— Style cylindrical in fruit, terminating with a 2-lobed or capitate stigma. Pedicel thinner than fruit.   **17. Maresia**

43 (35). Leaves divided into 2–4 pairs of leaflets, or finely dissected into linear, terete or filiform lobes   44

— Leaves entire or dentate, rarely pinnatifid or lyrate   45

44. Flowers minute (about 3 mm.), white.   **26. Cardamine**

— Flowers large, more than 1 cm., violet or pink.   **48. Pseuderucaria**

45 (43). Desert shrubs with glabrous, glaucous, entire or somewhat repand leaves. Flowers pale violet. Fruit on erect or spreading pedicel, (3–)4–7 cm. long.   **49. Moricandia**

— Plants not as above   46

46. Fruit 6–15 cm., spreading or deflexed   47
— Fruit much shorter   48
47. Lower leaves lyrate. Petals up to 8 mm. Fruit mostly deflexed. Plants 40–80 cm.
                                                                    28. Turritis
— Lower leaves entire to dentate. Petals 1.5–2.5 cm. Fruit spreading. Plants much smaller.
                                                                    14. Malcolmia
48(46). Fruiting pedicel 0 or very short, as thick as fruit. Cauline leaves not clasping.
   Valves often torulose. Flowers white, minute. Desert plants.   **16. Torularia**
— Plants with a different set of characters   49
49. Cauline leaves not clasping and not cordately rounded at base. Flowers often lilac.
   Fruiting pedicel much thinner than fruit. Fruit more or less torulose. Desert or sandy
   soil plants.   **17. Maresia**
— Cauline leaves ovate, orbicular or oblong, clasping or broadly cordate at base. Flowers
   white or pink but then fruiting pedicel thick. Fruit not torulose. Plants not occurring
   in deserts or on sandy soils.   **27. Arabis**
50(1). Fruit with 2 disk-like, flattened, smooth, 1-seeded valves, separating from replum.
   Flowers yellow.   **43. Biscutella**
— Fruit not as above   51
51. Fruits of two forms : flat-discoid and urn-shaped.   **41. Aethionema**
— Fruits uniform   52
52. Spiny desert shrubs. Fruit almost globular, with spiny beak. Flowers large, pink or
   lilac.   **63. Zilla**
— Plants not as above   53
53. Fruit 2-jointed, the lower joint terete or linear, seedless or 1–4-seeded, the upper one
   globular, ovoid or ellipsoidal or oblong, 1-seeded   54
— Fruit not jointed, 1- or many-seeded or if jointed in appearance then upper joint
   seedless   56
54. Fruit cylindrical, compressed; lower joint with few seeds. Flowers white or pink.
                                                                    57. Reboudia
— Fruit not cylindrical; upper joint globular or ovoid or conical   55
55. Flowers white to cream-coloured. Stigma sessile.   **61. Crambe**
— Flowers yellow. Stigma borne on a long style.   **60. Rapistrum**
56(53). Fruit with coarse and very prominent tubercles or protuberances   57
— Fruit not tuberculate as above (but sometimes with rounded knobs at apex or prominently 3-ribbed)   59
57. Flowers white. Fruit made of 2 subglobular, verrucose valves.   **46. Coronopus**
— Flowers yellow. Fruit simple   58
58. Fruit 1-seeded, with oblique flattened beak longer than fruit proper. **9. Schimpera**
— Fruit 2-seeded, beakless or almost so.   **4. Ochthodium**
59(56). Fruit with rounded knobs at apex or keeled and 3-ribbed between keels, glabrous,
   indehiscent, 1-seeded. Leaves ovate or oblong, undivided or auriculate at base. Glabrous, glaucous plants   60
— Fruit smooth or nerved, glabrous or variously hairy but neither gibbous nor ribbed
   as above   61
60. Fruit single-celled; pedicel thin, spreading.   **6. Boreava**
— Fruit 3-celled, 2 lateral cells sterile; pedicel thick, club-shaped, appressed to stem.
                                                                    5. Myagrum
61(59). Fruit ovate or broadly elliptical, (0.9–)1–3(–4) × 0.5–1.5 cm. Seeds many   62
— Fruit not as above   65

62. Flowers yellow. Perennial tomentose plants.   **31. Fibigia**
— Flowers pink or purple or flesh-coloured   63
63. Fruit appressed-pubescent. Seeds winged. Shrubby perennials.   **29. Farsetia**
— Fruit glabrous. Annuals   64
64. Petals deeply notched or 2-fid. Fruit 1.5–3(–4) cm. long. Seeds wingless.   **30. Ricotia**
— Petals entire. Fruit 0.9–1.2 cm. long. Seeds winged.   **56. Savignya**
65(61). Fruit 1-seeded, indehiscent   66
— Fruit 2- or many-seeded   70
66. Fruit globular, obovoid or pear-shaped, glabrous   67
— Fruit not as above   69
67. Fruit pendulous, 5–8 mm. across, not apiculate.   **7. Texiera**
— Fruit smaller, apiculate   68
68. Flowers white.   **62. Calepina**
— Flowers yellow.   **37. Neslia**
69(66). Fruit discoid, orbicular or elliptical, about 5 mm. or less in length. Leaves entire,
   stellate-hairy.   **34. Clypeola**
— Fruit elliptical, obovate or oblong, (6–)8 mm. or more long. Leaves not as above.
                                                                              8. Isatis
70(65). Fruit with a spoon-like or lingulate seedless beak (upper joint) as long as seed-
   bearing part; valves readily dehiscing. Leaves 2–3-pinnatisect. Petals purple-veined.
                                                                              55. Carrichtera
— Fruit not as above   71
71. Fruit with lignified valves; each valve ending with a concave, somewhat spoon-shaped
   appendage; septum thick; cells 1–2-seeded.   **23. Anastatica**
— Fruit not as above   72
72. Fruit more or less dorsally compressed, rarely almost globular; valves flattened, disk-
   shaped or convex; septum parallel to and as broad as valves   73
— Fruit laterally compressed; valves boat-shaped; septum perpendicular to and much
   narrower than width of valves   77
73. Leaves all radical, rosulate. Petals 2-fid, white. Seeds not winged.   **35. Erophila**
— Leaves radical and cauline. Seeds winged or not   74
74. Aquatic plants, 30–45 cm. or more tall. Leaves pinnate or lobed or dentate. Fruit
   globular or ellipsoidal. Flowers yellow.   **24. Rorippa**
— Plants not as above   75
75. Leaves clasping to sagittate at base. Fruit ovoid or obovoid; valves convex or inflated.
   Pedicels as long as or longer than fruit. Hairs of plant not stellate.   **36. Camelina**
— Plants not as above   76
76. Seeds broadly winged. Fruit covered with simple or 2-armed hairs. Petals white.
                                                                              33. Lobularia
— Seeds wingless or very narrowly winged. Fruit mostly with stellate hairs or glabrous.
   Petals yellow or cream-coloured, very rarely white.   **32. Alyssum**
77(72). Cells of fruit 1(–2)-seeded   78
— Cells of fruit (2–)4- or many-seeded   80
78. Leaves clasping. Perennials. Fruit indehiscent, inflated, usually cordate at base; valves
   keeled, not winged.   **45. Cardaria**
— Leaves not clasping. Annuals or rarely perennials. Valves more or less winged, rarely
   not winged but then leaves entire.   79
79. Inflorescence racemose. Petals equal.   **44. Lepidium**
— Inflorescence corymbose. Outer petals longer than inner.   **42. Iberis**

80 (77). Fruit winged.   **40. Thlaspi**
— Fruit not winged   81
81. Fruit obtriangular-obcordate.   **38. Capsella**
— Fruit ovoid or ellipsoidal, obtuse or truncate.   **39. Hymenolobus**

Trib. SISYMBRIEAE. Ovary mostly sessile. Nectar glands generally forming a closed ring around the base of filaments. Fruit a siliqua, rarely a silicle, usually not dorsally flattened.

1. SISYMBRIUM L.

Annual, biennial or perennial, glabrous or pubescent herbs. Stems erect. Leaves pinnatifid-lyrate, sometimes runcinate or entire. Racemes many-flowered, bracteate or ebracteate. Flowers conspicuous or minute. Calyx erect-spreading, scarcely 2-saccate. Petals as long as or longer than calyx, generally yellow, sometimes white or pink, more or less clawed. Stamens 6, free, edentate. Style short; stigma undivided or 2-lobed. Fruit a dehiscent, terete or slightly compressed, narrowly linear, long siliqua; valves convex, with distinct midrib and 2 lateral veins; septum thickened or membranous. Seeds usually in 1 row, oblong or ellipsoidal, yellowish-brown, pendulous on short free funicles; cotyledons conduplicate.

About 80 species mainly in temperate regions of the northern and southern hemispheres.

Literature: O. E. Schulz, *Sisymbrium* L., in Cruciferae-Sisymbrieae, in: *Pflznr.* 86 (IV. 105):46–157 (1924).

1. Flowers solitary, in axils of bracts or leaves. Petals about 3 mm. Fruit 1.5–2.5 cm. long, curved, on very short (up to 2 mm.) pedicels.   **1. S. runcinatum**
– Flowers in bractless racemes   2
2. Fruit 1–1.8 cm., tapering, strongly appressed to stem.   **7. S. officinale**
– Fruit not as above   3
3. Fruiting pedicel 2–4 mm. long, as thick as 2.5–4 cm. long fruit.   **6. S. erysimoides**
– Pedicel and fruit usually longer   4
4. Fruit glabrous. Sepals 2–3 mm. Petals 3–4 mm.   **2. S. irio**
– Fruit hairy, at least when young. Sepals and/or petals larger than above   5
5. Petals 1–1.5 cm. Outer sepals prominently horned at apex.   **4. S. bilobum**
– Petals smaller. Outer sepals not horned   6
6. Flowers 0.8–1 cm. long. Fruit (6–)8–10(–15) cm. long. Pedicel as thick as fruit.
                                                                   5. S. orientale

– Flowers and fruit smaller. Pedicel much thinner.   **3. S. damascenum**

Sect. KIBERA (Adans.) DC., Syst. 2:477 (1821). Racemes leafy. Petals white or pale yellow. Stigma 2-lobed.

1. Sisymbrium runcinatum Lag. in DC., Syst. 2:478 (1821); Boiss., Fl. 1:220 (1867); O. E. Schulz, in Pflznr. 86:131 (1924). [Plate 365]

Glabrous or puberulent annual. Stems erect, branching, 15–25 cm. Radical leaves 3–8 cm., congested in rosettes, petiolate, oblong-linear, pinnatipartite, with about 6 pairs of alternate, acute, runcinate lateral lobes; cauline leaves subsessile, lyrate-pinnatifid, with oblong, acute terminal lobe. Racemes dense, with leaf-like bracts. Flowers minute, solitary, short-pedicelled. Sepals 2 mm., oblong. Petals 3 mm., narrowly obovate-spatulate. Fruiting racemes elongated. Fruit 1.5–2.5 cm., curved, borne on a thick, erect, 1–2 mm. long pedicel; valves thick, prominently nerved, nerves anastomosing; septum thickened; style conspicuous, with 2-lobed stigma. Seeds oblong, yellow-orange. Fl. March–April.

Var. **runcinatum.** Plant and fruit glabrous or glabrescent.
   Hab.: Steppes. Judean Desert, Gilead.

Var. **hirsutum** (Lag.) Coss., Not. Pl. Crit. 4 : 95 (1851); O. E. Schulz, l.c. *S. hirsutum* Lag. in DC., Syst. 2 : 478 (1821). Plant and fruit puberulent throughout.
   Hab.: Steppes. N. Negev, Lower Jordan Valley, Ammon.

Area of species : W. Irano-Turanian, extending into adjacent territories of the Mediterranean and Saharo-Arabian regions.

Sect. IRIO DC., Syst. 2 : 463 (1821). Racemes ebracteate. Petals yellow. Stigma more or less 2-lobed. Fruiting pedicel filiform or thickened. Valves 3-nerved.

2. Sisymbrium irio L., Sp. Pl. 659 (1753); Boiss., Fl. 1 : 217 (1867); O. E. Schulz, in Pflznr. 86 : 89 (1924). [Plate 366]
   Annual, glabrous, 20–45(–90) cm. Stems erect, branched. Leaves (except upper ones) petiolate; the lower 3–15 cm., runcinate-pinnatisect to sublyrate, lobes oblong-linear, toothed to entire; upper with fewer lateral lobes or simple, hastate. Flowering racemes compact, ebracteate, overtopped by young fruit. Flowers pedicellate. Sepals 2–3 mm. Petals 3–4 mm., pale yellow, narrowly spatulate. Fruiting racemes elongated and dense. Fruit 4–6 × 0.1 cm., erect or incurved, terete, glabrous; fruiting pedicels filiform, 0.5–1.5 cm., spreading; valves membranous, torulose, 3-nerved; septum white, membranous; style short; stigma 2-lobed. Seeds minute, yellowish-brown. Fl. January–May.
   Hab.: Roadsides and waste places. Coastal Galilee, Acco Plain, Sharon Plain, Philistean Plain, Upper and Lower Galilee, Judean Mts., Judean Desert, W., N. and C. Negev, Hula Plain, Upper and Lower Jordan Valley, Dead Sea area, Arava Valley, Gilead, Ammon, Moav, Edom.
   Area : Mediterranean and Irano-Turanian extending into the Euro-Siberian, Saharo-Arabian and some tropical regions.

3. Sisymbrium damascenum Boiss. et Gaill. in Boiss., Diagn. ser. 2, 6 : 11 (1859) et Fl. 1 : 218 (1867); O. E. Schulz, in Pflznr. 86 : 94 (1924). [Plate 367]
   Annual or biennial, hirsute with spreading or subretrorse hairs, (10–)30–60 cm. Stems erect, branching from base. Leaves up to 12 cm., petiolate, lyrate, runcinate, terminal lobe triangular-ovate or oblong, acute or obtuse, lateral lobes 3–4 pairs, triangular, acutely dentate; the upper leaves smaller, with fewer pairs of lateral lobes. Racemes compact, ebracteate, overtopped by young fruit; the fruiting ones elongated. Flowers

pedicellate. Sepals 2–4 mm., linear or oblong, obtuse, densely pilose. Petals 3–7 mm., yellow, obovate. Fruit 5–7 cm., terete, sparingly pilose also when mature, on erect or spreading, 0.6–1 cm. long pedicel; valves 3-nerved; septum membranous; stigma truncate. Seeds minute, in 1 row, oblong or ellipsoidal, yellowish-brown. Fl. February–April.

Hab.: Fields. Judean Mts., Judean Desert, Ammon, Edom. Rather rare.

Area: W. Irano-Turanian.

Sect. *Sisymbrium*, Sect. *Pachypodium* (Webb et Berth.) Fourn., Rech. Cruc. 86 (1865). Racemes ebracteate. Petals yellow. Stigma often deeply 2-lobed. Fruiting pedicels straight, thick. Fruit straight, spreading, with very thick style.

4. Sisymbrium bilobum (C. Koch) Grossh., Fl. Cauc. 2 : 163 (1930). *Diplotaxis biloba* C. Koch, Linnaea 15 : 252 (1841). *S. septulatum* DC., Syst. 2 : 471 (1821) prol. *bilobum* (C. Koch) O. E. Schulz, Repert. Sp. Nov. 15 : 372 (1918) et in Pflznr. 86 : 122 (1924). [Plate 368]

Annual, glabrous or pilose, especially at base, 20–30 cm. Stems erect, branching from base. Radical leaves 5–10 cm., rosulate, petiolate, lyrate, terminal lobe obtuse, obsoletely lobulate, lateral lobes oblong-ovate; upper leaves with narrow, elongated, lanceolate, entire lobes. Flowers conspicuous, pedicellate, in loose, ebracteate racemes elongating in fruit. Sepals about 4 mm., linear, pilose, the inner more or less saccate, the outer with a prominent horn at apex. Petals 1–1.5 cm., yellow, whitish when dry, obovate. Pedicels 4–5 mm., erect-spreading, as thick as fruit. Fruit 4.5–6.5 cm., hairy – at least when young; septum white, membranous; stigma 2-lobed. Seeds oblong, brown. Fl. March–April.

Hab.: Steppes. Ammon, Moav, Edom.

Area: W. Irano-Turanian.

5. Sisymbrium orientale L., Cent. Pl. II. 24 (1756); O. E. Schulz, in Pflznr. 86 : 122 (1924). *S. columnae* Jacq., Fl. Austr. 4 : 12, t. 323 (1776). [Plate 369]

Annual, shortly retrorse-hispid, 25–90 cm. Stems erect, branched, sometimes violet at base. Radical leaves 3.5–9(–20) cm., subrosulate, petiolate, pinnatisect to lyrate, terminal lobe triangular-ovate, obtuse, repand-dentate to crenate, confluent with proximal lateral lobes; cauline leaves long-petioled, lyrate with hastate terminal lobe and few small lateral lobes; upper leaves hastate or oblong, entire. Racemes ebracteate, dense, sometimes overtopped by the young lower fruits. Flowers pedicellate. Sepals 4–5 mm. Petals 0.8–1 cm., yellow or whitish-yellow, abruptly clawed; limb obovate. Fruit (6–)8–10(–15) × 0.1–0.2 cm., usually hirsute, rarely glabrous, as broad as thickened, 4–9 mm. long pedicel, tapering into short style; valves 3-nerved; septum thick. Seeds oblong, orange-brown. Fl. March–April.

Hab.: Fields. Philistean Plain, Upper Galilee, Mt. Carmel, Judean Mts., Ammon, Moav, Edom.

Area: Mediterranean and Irano-Turanian, extending into the Euro-Siberian region.

Varies greatly in pubescence and length of fruit. Hardly separable into varieties.

Sect. OXYCARPUS Paol. in Fiori et Paol., Fl. Anal. It. 1 : 434 (1898). Racemes ebracteate. Petals yellow. Stigma depressed, slightly 2-lobed. Fruiting pedicels short, as thick as fruit. Fruit spreading, subulate, tapering; style very short.

6. Sisymbrium erysimoides Desf., Fl. Atl. 2 : 84 (1798); Boiss., Fl. 1 : 217 (1867); O. E. Schulz, in Pflznr. 86 : 134 (1924). [Plate 370]

Annual, pilose or glabrous, 15–60 cm. Stems erect, terete, somewhat flexuous, sparingly branched. Radical leaves up to 12 cm., petiolate, lyrate-pinnatisect, terminal lobe obovate-obtuse, lateral lobes in 2–4 pairs, opposite or alternate, obovate to oblong, acute, irregularly dentate; cauline leaves short-petioled, lyrate-pinnatifid, with subhastate to truncate terminal lobe and 2–3 pairs of lateral lobes; all leaves glabrous or sparingly pilose. Racemes ebracteate, dense. Flowers minute, short-pedicelled. Sepals about 2 mm., erect, oblong, obtuse. Petals yellow, as long as or slightly longer than sepals, spatulate. Fruiting racemes elongated; pedicels 2–4 mm. long, as thick as fruit. Fruit 2.5–4 cm., horizontal or spreading, rigid, terete, more or less glabrous, tapering gradually into the short style and terminating with a slightly 2-lobed stigma; valves membranous, torulose, 3-nerved; septum membranous. Seeds oblong, orange-yellow. Fl. January–May.

Hab.: Waste places and shady sites, mainly in steppes and deserts. Judean Desert, C. Negev, Upper and Lower Jordan Valley, Dead Sea area, Gilead, Ammon, Edom. A camp-follower in deserts.

Area: Mediterranean and Saharo-Arabian, extending into the Sudanian and Macaronesian regions.

Sect. VELARUM DC., Syst. 2 : 459 (1821). Racemes long, ebracteate. Petals yellow to pale yellow. Stigma slightly 2-lobed. Pedicels thick, short. Fruit appressed, tapering towards apex.

7. Sisymbrium officinale (L.) Scop., Fl. Carn. ed. 2, 2 : 26 (1772); Boiss., Fl. 1 : 220 (1867); O. E. Schulz, in Pflznr. 86 : 137 (1924). *Erysimum officinale* L., Sp. Pl. 660 (1753). [Plate 371]

Annual, green or more or less canescent or retrorsely hispid, 30–60(–90) cm. Stems erect, branched above. Radical leaves 6–10 cm., subrosulate, petiolate, obovate, pinnatisect, terminal lobe short, broad, obtuse, irregularly dentate-crenate, confluent with proximal lateral lobes or subhastate, lateral lobes triangular or oblong, acute; cauline leaves petiolate, pinnatipartite to lyrate, with long, hastate terminal lobe; uppermost leaves sessile, hastate, entire. Racemes many-flowered, ebracteate. Flowers minute, short-pedicelled. Sepals 2–2.5 mm., oblong, rounded at apex. Petals 3–4 mm., pale yellow, obtuse-truncate. Fruiting racemes elongated. Fruit 1–1.8 × 0.1–0.15 cm., hirsute or glabrous, borne on a 1–1.5 mm. long, thick, erect pedicel and appressed to stem, tapering gradually into a short style and capitate stigma; valves firm, 3-nerved; septum submembranous, nerveless. Seeds small, oblong, orange-brown. Fl. March–April.

Var. **officinale**. More or less canescent plant. Fruit hirsute.

Hab.: Waste places. Acco Plain, Mt. Carmel, Upper Galilee, Esdraelon Plain, Dead Sea area, Gilead, Ammon.

Var. **leiocarpum** DC., Syst. 2 : 460 (1821); O. E. Schulz, l.c. 141. Greenish plant, less hairy than var. *officinale*. Fruit glabrous.

Hab.: Waste places on heavy soils. Upper Galilee, Samaria, Dan Valley.

Area of species : Euro-Siberian, Mediterranean and Irano-Turanian (also adventive in America and Australia).

2. Descurainia Webb et Berth.

Annual or biennial, sometimes perennial herbs, with branched or simple or glandular hairs, rarely glabrous. Leaves finely 2- or 3-pinnatisect; the lower ones petiolate, the upper subsessile. Racemes mostly ebracteate. Flowers small. Sepals spreading to erect, equal. Petals about as long as sepals or longer, yellow to pale yellow, spatulate. Stamens 6, exserted, without appendages. Style short; stigma capitate. Fruit dehiscent, 2-celled, erect, terete, borne on a filiform pedicel; valves torulose, slightly convex, with distinct midrib and lateral veins; septum membranous. Seeds numerous, in 1 or 2 rows in each cell, oblong or ellipsoidal; cotyledons oblong.

About 45 species in cold and temperate regions, mainly in America.

1. Descurainia sophia (L.) Webb ex Prantl in Engl. et Prantl, Nat. Pflznfam. 3, 2 : 192 (1890); O. E. Schulz, in Pflznr. 86 : 309 (1924). *Sisymbrium sophia* L., Sp. Pl. 659 (1753); Boiss., Fl. 1 : 216 (1867). [Plate 372]

Annual or biennial, 10–50 cm., densely canescent with branched and simple hairs. Stems erect, leafy, terete, branched above. Leaves greyish-green, finely 2- or 3-pinnatisect into narrowly oblong-elliptical to linear, acute lobes. Inflorescences many-flowered, corymbose, elongated in fruit, overtopped by young fruit. Flowers up to 3 mm., shorter than pedicels. Sepals suberect, rather glabrous. Petals almost as long as sepals, pale yellow, narrowly spatulate, long-clawed. Fruit 1.5–4 × 0.1 cm., erect on spreading pedicel, linear-filiform, straight or slightly incurved; valves torulose with distinct midrib, glabrous or with sparse substellate hairs. Seeds small, ovoid, in 1 row in each cell. Fl. February–April.

Var. **sophia**. Fruit glabrous, 2–4 cm. Pedicels 6–8 mm.

Hab.: Fields and steppes. Arava Valley, Ammon, Moav, Edom.

Var. **brachycarpa** (Boiss.) O. E. Schulz, l.c. 315. *Sisymbrium sophia* L. var. *brachycarpum* Boiss., l.c. Fruit short, 1.5–2.5 cm., somewhat broader than in type. Pedicels slender, sometimes slightly longer than in type.

Hab.: Fields and steppes. Judean Mts. (Jerusalem).

Area of species : Euro-Siberian, Mediterranean, Irano-Turanian and Sino-Japanese; probably introduced elsewhere.

3. ARABIDOPSIS (DC.) Heynh.

Annual or perennial tender herbs, covered with simple, stellate or otherwise branched hairs or rarely glabrous. Leaves oblong, entire to pinnatifid. Racemes sometimes bracteate. Flowers small or conspicuous, white or pink, rarely yellowish. Sepals erect, oblong, obtuse, almost not saccate at base. Petals obovate-cuneate or spatulate, clawed, longer than sepals. Style short, as broad as ovary; stigma depressed-capitate or 2-lobed. Fruit dehiscent, flat, linear; valves 1-nerved. Seeds in 1, rarely in 2 rows. (Distinguished from *Arabis* mainly by the incumbent cotyledons.)

Thirteen species, mainly in the Euro-Siberian, Mediterranean and Irano-Turanian regions.

1. Arabidopsis pumila (Steph. ex Willd.) Busch in Kusn., Busch et Fomin, Fl. Cauc. Crit. 3, 4 : 457 (1909). *Sisymbrium pumilum* Steph. ex Willd., Sp. Pl. 3 : 507 (1800); Boiss., Fl. 1 : 213 (1867). [Plate 373]

Low annual, scabrous, with short branching hairs, 10–25 cm. Stems erect. Radical leaves subrosulate, obovate-oblong, sinuate-dentate or pinnatifid, rounded at apex; stem leaves oblong, sagittate, acute, subentire. Racemes subcorymbose. Pedicels 2–3 mm. Flowers minute. Sepals 1.5 mm., oblong, obtuse. Petals about 2 mm., cream-coloured to yellow, narrowly spatulate, rounded at apex. Fruit 1.5–3 × 0.01 cm., flat, linear, curved, pubescent, longitudinally 1-nerved, borne on spreading, 2–5 mm. long pedicels. Seeds in 1 row in each cell, oblong-ellipsoidal. Fl. February–March.

Hab. : Walls. Judean Mts. (after Post, ed. 2, 1 : 68, 1932). Very rare.

Area : Mainly Irano-Turanian, with extensions into adjacent regions (usually in saline habitats).

4. OCHTHODIUM DC.

Annuals. Stems branched, erect or ascending. Leaves variable in shape and margin, the lower pinnatipartite or lyrate, the uppermost lanceolate, undivided. Inflorescences paniculate, ebracteate, elongating in fruit. Pedicels slender. Flowers small, yellow. Sepals spreading, more or less equal at base. Petals short-clawed. Stamens free, edentate. Fruit an indehiscent, 2-celled, globular or obscurely tetragonous, 2-valved silicle with thick septum; style very short, pyramidal; stigma shortly 2-lobed; valves thick, knobby along angles. Seed 1 in each cell, pendulous; cotyledons incumbent or obliquely accumbent.

One species in the E. Mediterranean.

1. Ochthodium aegyptiacum (L.) DC., Syst. 2 : 423 (1821); Boiss., Fl. 1 : 369 (1867). *Bunias aegyptiaca* L., Syst. ed. 12, 3 : 231 (1768). [Plate 374]

Annual, sparingly hirsute, 30–80(–100) cm. Stems erect or ascending, usually with spreading branches. Radical leaves 5–20 cm., petiolate, oblong-lanceolate, lyrate or pinnatipartite; cauline leaves 2–6 cm., sessile or short-petioled, lanceolate or lanceolate-oblong, undivided, dentate or crenate, rarely lyrate with few lateral lobes and a hastate terminal lobe. Flowering racemes dense, paniculate, elongating in fruit. Sepals about

2 mm., spreading. Petals 3–3.5 mm., oblong-spatulate. Fruiting pedicels 4–6(–8) mm., thick, stiff, erect or spreading. Fruit about 4 × 3 mm., densely tuberculate-knobby, with a short pyramidal style. Fl. February–May.

Hab.: Fields and waste places. Acco Plain, Sharon Plain, Philistean Plain, Upper and Lower Galilee, Mt. Carmel, Esdraelon Plain, Samaria, Judean Mts., Dan Valley, Hula Plain, Upper and Lower Jordan Valley, Gilead, Ammon, Moav, Edom.

Area: E. Mediterranean (recorded from Egypt by Linné, but not found there by others).

5. Myagrum L.

Annual, glabrous, glaucous. Stems branched. Leaves entire; the lower ones petiolate, the upper sessile or auriculate. Racemes ebracteate, erect, elongated. Pedicels short. Flowers small. Sepals suberect, equal or the inner somewhat saccate. Petals yellow, oblong. Stamens 6, the longer ones sometimes connate at base in pairs. Fruit an indehiscent, leathery-corky, obcordate or obpyramidal silicle, spuriously 3-celled but only central cell containing seed, broadening at apex into 2 rounded knobs; style short-conical, tetragonous. Seed solitary, obovoid or oblong, pendulous; cotyledons incumbent.

One species in the Mediterranean and W. Irano-Turanian regions; adventive in the Euro-Siberian region.

1. Myagrum perfoliatum L., Sp. Pl. 640 (1753); Boiss., Fl. 1 : 371 (1867). [Plate 375]

Annual, about 30 cm. Stems branching from base. Leaves 2–4 cm., entire, the lower oblong, tapering into a petiole; the upper oblong to lanceolate, more or less acute, sessile or auriculate. Pedicels thick, club-shaped, appressed to stem and about 4 mm. in fruit. Sepals about 2 mm. Petals 3–4 mm. Fruit about 6 mm.; valves longitudinally ridged, broadening towards apex, with rounded knobs. Seeds solitary in the lower cell and abortive in the lateral cells. Fl. March–April.

Hab.: Fields. Mt. Carmel, Gilead, Edom. Rare.

Area: As for the genus.

6. Boreava Jaub. et Sp.

Annual, glaucous, glabrous, erect and branching. Leaves ovate to oblong and lanceolate, amplexicaul, entire. Inflorescences paniculate-corymbose, ebracteate. Flowers conspicuous. Sepals spreading, equal or the inner somewhat saccate. Petals pale yellow, narrowly elliptical with rounded limb, tapering into a claw. Stamens 6, free, edentate. Fruit indehiscent, 1-celled, 1-seeded, nut-like, hard, ovoid or tetragonous, wingless or winged, with a conical style and a 2-lobed stigma. Seeds large, pendulous, ovoid; cotyledons incumbent.

Some 2 or 3 species, mainly in the E. Mediterranean.

1. Boreava aptera Boiss. et Heldr. in Boiss., Diagn. ser. 1, 8 : 49 (1849) et Fl. 1 : 372 (1867).

Annual, glaucous, glabrous, 30–40 cm. Stems erect, terete, branched. Leaves ovate-oblong, cordate-amplexicaul and auriculate at base, obtuse, entire. Racemes elongating in fruit. Calyx somewhat 2-saccate. Petals spatulate, long-clawed. Pedicels erect-patulous, somewhat shorter than fruit. Fruit 5–7 mm., 2-keeled, ovoid, without wings, 3-ribbed on either side between keels, and with a short, compressed, acute beak. Fl. April.

Hab.: Fields. Judean Mts. (Dinsmore in Post, ed. 2, 1 : 111, 1932; no specimen deposited in HUJ).

Area : E. Mediterranean.

7. TEXIERA Jaub. et Sp.

Annuals, glabrous, glaucous. Stems erect, branched. Leaves sessile, cordate-sagittate, entire or dentate. Racemes many-flowered, ebracteate, elongating in fruit. Pedicels long, deflexed. Flowers inconspicuous. Sepals spreading, equal at base. Petals pale yellow or whitish, oblong, clawed. Stamens 6, free, toothless. Stigma sessile, peltate, slightly 2-lobed. Fruit an indehiscent, 1-celled, globular-ellipsoidal silicle with membranous epicarp, thick-spongy mesocarp and crustaceous, unequally crested endocarp. Seed 1, pendulous; cotyledons longitudinally concave or plicate.

One species in the western part of the Irano-Turanian region.

1. Texiera glastifolia (DC.) Jaub. et Sp., Ill. Pl. Or. 1 : t. 1 (1842); Boiss., Fl. 1 : 373 (1867). *Peltaria glastifolia* DC., Syst. 2 : 330 (1821). [Plate 376]

Annual, glabrous, about 30–50 cm. Stems erect, branched from base or above. Leaves 2–6.5 cm., with prominent white midrib; the radical ones oblong, sinuate-dentate at margin, upper oblong-lanceolate, sagittate. Flowering racemes dense. Flowers inconspicuous. Calyx about 1 cm. Petals much longer than sepals, oblong, tapering into a claw. Fruiting racemes elongated. Pedicels 1.2–1.5 cm., deflexed, twisted to one side of rhachis. Fruit 6–8 mm., nut-like, indehiscent, 1-celled, glabrous, smooth, white. Seeds large. Fl. April–May.

Hab.: Fields. Edom.

Area : W. Irano-Turanian.

8. ISATIS L.

Annual or perennial, glabrous or hairy herbs. Radical leaves petiolate, more or less entire; cauline leaves sessile, auriculate. Racemes corymbose-paniculate, ebracteate, elongating in fruit. Flowers small. Sepals erect-spreading, equal at base. Petals exserted, yellow, entire, clawed. Stamens free, edentate. Stigma 2-lobed, more or less sessile. Fruit an indehiscent, 1-celled, 1-seeded, pendulous, flat, linear-oblong to obovate or orbicular silicle; valves glabrescent or pubescent, more or less winged. Seeds pendulous; cotyledons mostly incumbent.

About 30 species; most of them in the Mediterranean and Irano-Turanian, some in the Saharo-Arabian and Euro-Siberian regions.

1. Fruit linear-oblong, 1.2–2.2 × 0.25–0.5 cm., on about 1 cm. long pedicel. Petals 4–8 mm. Subglabrous common herbs.   **2. I. lusitanica**
– Fruit elliptical-obovate, 0.6–1 × 0.2–0.5 cm., on shorter pedicel. Petals shorter. Glabrous desert herbs.   **1. I. microcarpa**

1. Isatis microcarpa J. Gay ex Boiss., Ann. Sci. Nat. Bot. ser. 2, 17 : 201 (1842); Boiss., Fl. 1 : 382 (1867). [Plate 377]

Annual, somewhat glaucous, glabrous or almost glabrous, up to 40 cm. Stems fairly erect, branched. Radical leaves petiolate, oblong-obovate, subentire or slightly crenate-dentate; cauline leaves sessile, triangular-lanceolate, sagittate. Racemes paniculate, elongating in fruit. Flowers 3–4 mm., yellow. Calyx glabrous. Fruiting pedicels less than 1 cm., deflexed, sparingly pilose. Fruit 0.6–1 × 0.2–0.5 cm., elliptical-obovate or oblong, obtuse at apex; valves spongy-membranous, glabrous, pubescent or ciliate, longitudinally ridged. 2n=14. Fl. March–April.

Var. **microcarpa.** Fruit glabrous.
Recorded from El-Arish (Sinai).

Var. **blepharocarpa** Aschers. in Aschers. et Schweinf., Ill. Fl. Eg. Suppl. 747 (1889). Fruit short-pubescent, ciliate.
Hab. : Deserts. Negev, Arava Valley.
Area of species : E. Saharo-Arabian.

2. Isatis lusitanica L., Sp. Pl. 670 (1753). *I. aleppica* Scop., Del. Insubr. 2 : 31, t. 16 (1787); Boiss., Fl. 1 : 382 (1867). *I. aegyptica* L., Sp. Pl. 671 (1753). [Plate 378]

Annual, glabrescent above, with short spreading hairs below, 15–70 cm. Stems erect, with ascending branches. Radical leaves up to 14 × 4 cm., subsessile or petiolate, spatulate-obovate, acute or obtuse, entire or somewhat denticulate; cauline leaves smaller, sessile, lanceolate-triangular, sagittate-auriculate, acute, dentate or entire. Racemes many-flowered, paniculate, elongating in fruit. Flowers 4–8 mm., yellow. Calyx more or less pilose. Fruiting pedicels about 1.2 cm., deflexed, terete-filiform, slightly thickened towards apex. Fruit 1.2–2.2 × 0.25–0.5 cm., linear-oblong, winged, apex truncate or retuse; valves spongy-membranous, pubescent to subglabrous, ciliate-hairy at margins, sometimes black, with distinct ridge. 2n=14. Fl. March–May.

Hab. : Fields and waste places. Sharon Plain, Philistean Plain, Upper and Lower Galilee, Mt. Carmel, Esdraelon Plain, Mt. Gilboa, Samaria, Shefela, Judean Mts., W. and N. Negev, Hula Plain, Upper and Lower Jordan Valley, Gilead, Ammon, Moav, Edom.
Area : E. Mediterranean and W. Irano-Turanian.

As there are transitions between the typical form and that with somewhat shorter and less pubescent fruit (var. *pamphylica* Boiss., l.c. 383), the latter is not separated here.

9. SCHIMPERA Hochst. et Steud.

Desert annuals. Stems decumbent or ascending, branched, yellowish-green, leafy. Racemes ebracteate, dense, elongating in fruit. Flowers small. Sepals spreading, equal at base. Petals yellow, slightly emarginate, somewhat longer than sepals. Stamens 6, free, edentate. Stigma 2-lobed. Fruit an indehiscent, 1-celled, 1-seeded, ovoid, compressed, strongly tuberculate silicle with a 2-nerved, compressed, oblique or horizontal beak. Seed pendulous; cotyledons subconduplicate.

Two species in the Saharo-Arabian and E. Sudanian regions.

1. Schimpera arabica Hochst. et Steud. ex Boiss., Fl. 1 : 384 (1867); Steud., Nom. ed. 2, 2 : 530 (1841). [Plate 379]

Desert annual, sparingly papillose-pubescent, 10–50 cm. Stems branched in upper part or from base, erect or decumbent, becoming hollow, straw-like. Radical leaves 5–9(–18) cm., numerous, rosulate, petiolate, oblong, entire or dentate or lyrate-runcinate, with 3–7 broadly triangular, acute or obtuse, dentate-crenate or entire lateral lobes; cauline leaves sessile, oblong-lanceolate, auriculate-sagittate at base, entire or dentate. Racemes dense. Calyx about 2 mm. Petals 3 mm., yellow, obovate, tapering into a short claw. Fruiting racemes rigid, elongated. Pedicels 2–3 mm., thickened, erect, appressed to rhachis. Fruit 0.7–1 cm. (incl. beak), glabrous or papillose, lower part erect or somewhat spreading, strongly tuberculate; beak 4–7 mm., about 2–3 times as long as fruit proper, rigid, usually compressed, horizontal or oblique. Seeds brown. Fl. March–April.

Var. **arabica.** Fruit glabrous.

Hab. : Deserts, mostly in sandy depressions. W., N. and C. Negev, Dead Sea area, Arava Valley.

Var. **lasiocarpa** Boiss., l.c. Fruit densely papillose-woolly.

Hab. : As above. Negev, Edom.

Area of species : E. Saharo-Arabian.

Trib. HESPERIDEAE. Median nectar glands separated from the transversal ones or the former lacking altogether. Fruit a terete or compressed siliqua, rarely a loment or an indehiscent silicle.

10. ERYSIMUM L.

Annual or perennial pubescent herbs, with simple or branching hairs. Leaves oblong-lanceolate to linear, entire or sinuate-dentate, rarely pinnatifid. Inflorescences racemose, ebracteate. Flowers yellow, sometimes lilac, violet or white. Calyx erect, 2-saccate or with sepals equal at base. Petals clawed. Stamens free, edentate. Stigma capitate or emarginate or 2-lobed. Fruit a dehiscent siliqua, elongated, terete or tetragonous or more or less compressed; valves linear, with midrib or keel. Seeds in 1 row in each cell, not winged or only slightly marginate at apex, oblong; cotyledons mostly incumbent.

About 80 species in the temperate regions of the northern hemisphere, mainly in the Old World.

1. Fruit erect, up to 2.5 × 0.2 cm., tuberculate-scabrous and covered with 4–5-fid hairs.
   2. E. scabrum
   – Fruit not as above   2
2. Annuals. Leaves mostly repand-dentate. Flowers yellow, about 7 mm. Hairs partly 3-partite. Fruit 3–7.5 cm., torulose, tetragonous.   **1. E. repandum**
   – Perennials. Stem leaves or all leaves entire; radical leaves numerous, crowded. Flowers violet or yellow, mostly larger than in above. Hairs all 2-partite. Fruit shorter than in above, not torulose   3
3. Flowers pink-violet, 0.7–1 cm., subsessile. Style distinct. Fruit twisted or straight, 1.5–3 cm., almost terete.   **4. E. oleifolium**

– Flowers yellow, 0.8–1.2 cm., short-pedicelled. Style inconspicuous. Fruit 2–5 cm., terete-tetragonous.   **3. E. crassipes**

1. Erysimum repandum L., Demonstr. Pl. Hort. Upsal. 17 (1753); Boiss., Fl. 1 : 189 (1867) incl. *E. rigidum* DC., Syst. 2 : 505 (1821). [Plate 380]

Annual, appressed-pubescent, 8–30 cm.; hairs mostly 2-partite, few 3-partite. Stems usually simple, sometimes branching from base. Leaves usually up to 6 × 0.6 cm., sometimes larger, linear-lanceolate, repand-dentate, some more or less recurved at apex. Racemes dense, elongating in fruit. Flowers about 7 mm., short-pedicelled. Calyx slightly 2-saccate, appressed-pubescent. Petals yellow. Fruit 3–7.5 × 0.1–0.15 cm., spreading on thick short pedicels, tetragonous, torulose, appressed-hairy, with a distinct style and retuse stigma. Seeds oblong, slightly marginate at apex. Fl. March–April.

Hab.: Fields and steppes. Judean Mts., Lower Jordan Valley, Gilead, Ammon, Moav, Edom.

Area: Irano-Turanian, extending into adjacent Mediterranean and Euro-Siberian territories.

2. Erysimum scabrum DC., Syst. 2 : 505 (1821); Boiss., Fl. 1 : 195 (1867).

Biennial, canescent, 15–20 cm. Stems simple, erect. Radical leaves 2.5 cm., oblong, tapering to a petiole, obtuse, entire or denticulate; stem leaves linear-lanceolate, acute. Flowers 8 mm., yellow. Fruiting racemes short. Fruit 2–2.5 × 0.2 cm., subsessile, erect, tetragonous, acute, minutely tuberculate-scabrous and covered with 4–5-fid hairs; style somewhat longer than width of fruit, conical-filiform. Fl. March–April.

Hab.: Mountains. Gilead (after Post, ed. 2, 1 : 65, 1932).

Area: E. Mediterranean.

3. Erysimum crassipes Fisch. et Mey., Ind. Sem. Hort. Petrop. 1 : 27 (1835); Boiss., Fl. 1 : 206 (1867). [Plate 381]

Perennial, appressed-canescent with forked hairs, 20–65 cm. Stems numerous, erect, almost simple, thin and rigid. Leaves 6–12 × 0.05–0.2(–0.25) cm., linear or narrowly oblanceolate, acute, entire, the radical ones densely crowded. Racemes dense, elongating in fruit. Flowers 0.8–1.2 cm., short-pedicelled. Calyx slightly 2-saccate, appressed-hairy. Petals yellow, with few hairs on outer side. Style very short, with retuse stigma. Fruit 2–5 (–6) × about 0.1 cm., more or less spreading on short thickened pedicels, usually straight, terete-tetragonous, appressed-pubescent. Seeds oblong, brown, slightly marginate at apex. Fl. April–June.

Hab.: Batha and steppes; stony places. Upper Galilee, Mt. Gilboa, Samaria, Shefela, Judean Mts., Judean Desert, Gilead, Ammon, Moav, Edom.

Area: W. Irano-Turanian, extending into adjacent Mediterranean territories.

4. Erysimum oleifolium J. Gay, Erysim. Nov. 6 (1842); Boiss., Fl. 1 : 208 (1867). [Plate 382]

Perennial, caespitose, appressed-canescent with forked hairs, 10–30 cm. Stems many, thin, erect, often simple. Leaves up to 5–10 × 0.05–0.2 cm., narrowly oblanceolate-spatulate to linear, entire, radical leaves densely crowded, longer than cauline ones, tapering towards base. Racemes dense, elongating in fruit. Flowers 0.7 × 1 cm., subsessile. Calyx 2-saccate, appressed-hairy. Petals pink-violet, rather glabrous. Fruit 1.5–3.5 × 0.1 cm., on spreading pedicels, almost terete, twisted or more or less straight, appressed-pubescent, with a filiform style and 2-lobed stigma. Seeds small, oblong. Fl. April–May.

Hab.: Steppes. Moav, Edom.

Area: W. Irano-Turanian.

11. HESPERIS L.

Biennial or perennial herbs with branched or simple, often also with glandular hairs. Leaves entire, toothed or lyrate. Inflorescences racemose, sometimes bracteate below. Flowers conspicuous. Calyx erect; inner sepals saccate. Petals large, white, livid or purple, long-clawed. Stamens free, edentate, the inner 4 with dilated filaments. Stigma deeply 2-lobed. Fruit a tardily dehiscing siliqua, cylindrical to somewhat tetragonous, torulose; valves with distinct midrib. Seeds in 1 row in each cell, marginate or not; cotyledons mostly incumbent.

About 25 species in the Mediterranean and Irano-Turanian regions; a few also in the Euro-Siberian region.

1. Hesperis pendula DC., Syst. 2: 457 (1821); Boiss., Fl. 1: 236 (1867). [Plate 383]

Perennial, 20–50(–75) cm., with mixed, short, glandular – and long, retrorse or spreading hairs. Stems erect, branched. Radical leaves up to 10(–20) cm., rosulate, tapering into a narrow petiole, oblong-lanceolate, coarsely dentate or runcinate-lyrate; cauline leaves few, nearly sessile, sometimes subauriculate, triangular-lanceolate, acute, dentate. Racemes rather loose, elongated. Flowers large, short-pedicelled; pedicels suberect, then deflexed to pendulous. Calyx about 1 cm.; sepals erect, the inner saccate, short-hispid, with white-membranous margin. Petals about twice as long as calyx, livid, long-clawed; limb oblong-lanceolate. Fruit 5–8 × about 0.2 cm., on pendulous or twisted-deflexed pedicels, more or less terete, subtorulose, covered with glandular and branched hairs, and tapering to a 2-lobed stigma. Seeds oblong. Fl. March–May.

Hab.: Batha, among rocks. Upper Galilee, Gilead, Ammon, Moav, Edom.

Area: E. Mediterranean.

12. NASTURTIOPSIS Boiss.

Annuals, small, usually pubescent. Stems mostly numerous. Leaves nearly all radical, dentate or pinnately divided. Racemes corymbose, ebracteate. Flowers small, yellow, pedicellate. Sepals equal at base. Petals entire, clawed. Stamens free, dilated at base. Stigma capitate, emarginate or slightly 2-lobed. Fruit a readily dehiscing, falcate, oblong to linear, many-seeded, 2-valved siliqua; septum membranous. Seeds mostly in 2 rows; cotyledons incumbent.

One or 2 species, Saharo-Arabian.

1. Nasturtiopsis arabica Boiss., Fl. 1: 237 (1867). *N. coronopifolia* (Desf.) Boiss. var. *arabica* (Boiss.) O. E. Schulz, in Pflznr. 86: 254 (1924). [Plate 384]

Annual, patulous-pubescent, rarely glabrescent, 10–25 cm. Stems ascending to erect, branched from base. Radical leaves up to 10 cm., more or less rosulate, oblong or spatulate, tapering at base into a short petiole, obtusely dentate or pinnatifid with oblong-linear lobes; cauline leaves sessile, small. Racemes compact, elongating in fruit. Pedicels (0.5–1.2 cm. in fruit) filiform, spreading. Sepals 1.5–2.5 mm., spreading, pubescent. Petals about 3–6 mm., short-clawed; limb obovate-round. Fruit 1–1.8 ×0.1–0.2 cm., torulose, with a short style and a distinct capitate stigma; valves with acute base and obtuse apex. Seeds small, usually in 2 rows in each cell. Fl. March–April.

Hab.: Deserts; sandy, calcareous and gypsaceous ground. Judean Desert, W., N. and C. Negev, Lower Jordan Valley, Dead Sea area.

Area: Saharo-Arabian.

13. STIGMATELLA Eig

Dwarf, annual, canescent herbs with branching hairs. Leaves mostly rosulate, almost entire. Racemes few-flowered, ebracteate, more or less elongated. Flowers rather conspicuous. Calyx erect, 2-saccate. Petals lilac, much longer than calyx, clawed. Stamens free, filaments dilated at base. Ovary substipitate; style very long, with a long 2-lobed stigma. Fruit a terete-compressed siliqua terminating in a very long beak; valves readily dehiscent. Seeds in 2 rows in each cell; cotyledons accumbent.

One species, Saharo-Arabian.

1. Stigmatella longistyla Eig, Palest. Journ. Bot. Jerusalem ser. 1 : 80 (1938). [Plate 385]

Annual dwarf herb, with dense canescent indumentum of branched erect hairs, 3–10 cm. Stems branching at base. Radical leaves small, 1–2 cm., rosulate, petiolate, spatulate, obtuse, entire or subrepand, rarely minutely dentate; cauline leaves few, subsessile, oblong. Pedicels 3–6 mm. Sepals 4–5 mm., linear, tomentose. Petals 0.8–1 cm., lilac; limb obtriangular, obtuse-truncate. Stamens included, 4–5 mm. Fruit 1.5–1.8 cm., erect-spreading, terete-linear, terminating abruptly in a 3–4 mm. long style and a forked stigma; valves torulose; septum membranous, 1-nerved. Seeds ovoid, brown. Fl. April.

Hab.: Sandstone steppe. Edom.

Area: E. Saharo-Arabian (endemic in Palestine).

14. MALCOLMIA R. Br.

Annual, glabrescent, pubescent or canescent herbs. Leaves repand or toothed or pinnatifid, rarely entire, the radical often petiolate, the upper mostly sessile or auriculate. Racemes ebracteate. Flowers conspicuous. Sepals erect, equal or the inner saccate at base. Petals white, yellow, pink or violet, clawed. Longer stamens connate in pairs. Stigma conical, with decurrent lobes. Fruit an elongated, often tardily dehiscing siliqua. Seeds small, mostly in 1 row in each cell, ovoid or oblong, rarely globular; cotyledons incumbent.

About 25 species, mostly in the Irano-Turanian and Mediterranean regions.

1. Flowers yellow. **4. M. exacoides**
- Flowers pink, lilac, violet to white 2
2. Flowers large, 1.5–2.5 cm. long. Fruit 6–14 cm. Leaves crenulate-repand.
3. M. crenulata
- Flowers smaller, up to 1.2 cm. Fruit up to 6(–7) cm. 3
3. Petals entire. Fruit usually spreading, hispid or with scattered patulous hairs. Leaves dentate or repand-toothed. **1. M. africana**
- Petals notched. Fruit erect, appressed-tomentose. Leaves entire or remotely denticulate.
2. M. chia

1. Malcolmia africana (L.) R. Br. in Ait., Hort. Kew. ed. 2, 4 : 121 (1812); Boiss., Fl. 1 : 223 (1867). *Hesperis africana* L., Sp. Pl. 663 (1753). [Plate 386]

Annual, pubescent with scattered, stellate or forked hairs, 10–40 cm. Stems much branched, leafy. Leaves 1.5–5 (–8) × 0.5–1 (–4) cm., oblong or oblong-obovate, tapering into a petiole, obtuse or subacute, dentate or repand, forked-hairy. Racemes few-flowered, elongating in fruit. Flowers pink to pale lilac to violet. Sepals nearly equal at base, 5–6 mm. Petals about 1 cm., entire, long-clawed. Pedicels spreading, thickened, as thick as fruit. Fruit 3–6 × 0.2 cm., straight or curved, more or less tetragonous, densely hispid, tapering into a pointed, 0.5–1.5 mm. long stigma; septum thickened. Fl. March–April.

Hab.: Steppes and deserts. Judean Mts. (rare), Judean Desert, C. Negev, Lower Jordan Valley, Edom.

Area: Irano-Turanian and Saharo-Arabian, also extending into Mediterranean territories.

2. Malcolmia chia (L.) DC., Syst. 2 : 440 (1821); Boiss., Fl. 1 : 228 (1867). *Cheiranthus chius* L., Sp. Pl. 661 (1753). [Plate 387]

Annual, canescent, with appressed, forked hairs, 10–25 cm. Stems erect, simple or branched. Leaves 1.5–4 cm., obovate to oblong and spatulate, tapering into a petiole, obtuse, apiculate, remotely denticulate to entire; the upper ones subsessile, spatulate to elliptical. Racemes few-flowered. Sepals about 5 mm., longer than pedicels, the inner saccate. Petals 0.8–1 (–1.2) cm., clawed; limb pink to violet, sometimes white, notched. Pedicels as thick as fruit, erect to ascending. Fruit 3–6 × 0.1–0.2 cm., cylindrical, erect, straight or curved, appressed-tomentose, gradually tapering from base to apex, terminating in a conical stigma; valves firm; septum submembranous. Fl. February–April.

Hab.: Batha, among rocks. Sharon Plain, Upper and Lower Galilee, Mt. Carmel, Esdraelon Plain, Mt. Gilboa, Samaria, Judean Mts., Judean Desert, Hula Plain, Gilead, Edom.

Area: E. Mediterranean.

3. Malcolmia crenulata (DC.) Boiss., Fl. 1 : 229 (1867). *Hesperis crenulata* DC., Syst. 2 : 456 (1821). [Plate 388]

Annual, scabrous or glabrescent, 8–15 cm. Stems often numerous, branching from base. Leaves 2–5 × 0.4–1.6 cm., the radical ones tapering into a short petiole, obovate, obtuse or acute; the cauline sessile, oblong-lanceolate, subauriculate, acute or obtuse;

all more or less crenulate-repand (to entire). Racemes few-flowered. Flowers large. Sepals 1–1.8 cm., erect, the inner ones saccate, often purple. Petals 1.5–2.5 cm., pink or white. Pedicels erect or spreading, as thick as fruit. Fruit 6–14(–16) × 0.2 cm., straight or somewhat curved, spreading, glabrous or sparingly papillose-hairy, tapering gradually into 1–1.5 cm. long beak (style) and terminating with the decurrent lobes of the stigma; valves convex, firm, tardily dehiscing or indehiscent; septum thickened. Seeds small, brown. Fl. February–March.

Var. **crenulata.** Petals pink.

Hab.: Fields and batha. Acco Plain, Sharon Plain, Philistean Plain, Upper and Lower Galilee, Esdraelon Plain, Samaria, Judean Mts., Judean Desert, N. Negev, Upper and Lower Jordan Valley, Ammon, Edom.

Var. **albiflora** Eig, Bull. Inst. Agr. Nat. Hist. 6: 2 (1927). Petals white.

Hab.: As above. Judean Mts., Judean Desert, Gilead.

Area of species : E. Mediterranean and adjacent W. Irano-Turanian territories.

4. Malcolmia exacoides (DC.) Spreng., Syst. Veg. 2 : 899 (1825). *Sisymbrium exacoides* DC., Syst. 2 : 463 (1821). *M. conringioides* Boiss., Fl. 1 : 230 (1867).

Annual, green, scabrous-papillose, 5–10 cm. Stems simple or branching. Radical leaves 2.5–4 cm., oblong-obovate, tapering at base, repand to entire; stem leaves short-auriculate, acutely denticulate. Flowers up to 1.5 cm., yellow. Petals with short-exserted claw and oblong, entire limb. Fruit 6–9 cm., spreading or deflexed, glabrous, with spongy septum; style terete, obscurely 2-fid. Fl. March–April.

Hab.: Fields. Recorded from the Judean Mts. by Barbey, Herbor. Levant 116 (1882), and from the Lower Galilee by De Saulcy (fide Post, ed. 2, 1 : 76, 1932). Not observed by us.

Area : E. Mediterranean extending into W. Irano-Turanian territories.

15. E r e m o b i u m Boiss.

Annual or perennial herbs with appressed, stellate hairs. Leaves sessile, narrow, oblong to linear. Inflorescences racemose, ebracteate. Calyx erect, slightly 2-saccate, mostly persistent. Petals entire, linear-spatulate. Stamens free. Lobes of 2-fid stigma connivent. Fruit a siliqua, dehiscent, erect or spreading, linear, torulose, hairy. Seeds in 1 row in each cell, compressed, ellipsoidal-globular, narrowly marginate or winged; cotyledons obliquely accumbent.

Two species in the Saharo-Arabian region.

1. Eremobium aegyptiacum (Spreng.) Aschers. et Schweinf. ex Boiss., Fl. Suppl. 30 (1888). *Malcolmia aegyptiaca* Spreng., Syst. Veg. 2 : 898 (1825). [Plate 389]

Annual, sometimes perennial, prostrate to ascending, canescent, with scattered, stellate, appressed hairs, 10–35 cm. Roots thickened. Leaves sessile, up to 3 × 0.2–0.3 cm., linear or oblong-linear, obtuse. Racemes compact or elongated. Flowers pedicellate, 4–8 mm. Petals longer than sepals, exserted, violet to lilac, linear-spatulate, entire. Fruit usually 1.5–2.5 cm., dehiscent, terete or more or less compressed, tapering into a short style. Seeds ellipsoidal, more or less marginate-winged. Fl. March–April.

Var. **aegyptiacum**. *Malcolmia aegyptiaca* Spreng. var. *aegyptiaca* Coss. in Coss. **et** Barr., Ill. Fl. Atl. 1 : 23, t. 15 f. 16–25 (1882). Low, grey plants. Racemes dense. Pedicels about half as long as calyx. Flowers mostly about 5 mm., overtopped by young fruit. Calyx often readily caducous. Fruiting racemes somewhat compact. Fruit many-seeded, twice as broad as short pedicel. Seeds narrowly marginate.

Hab. : Sandy deserts. Philistean Plain, W. and C. Negev, Arava Valley.

Var. **lineare** (Del.) Zoh. (comb. nov.). *E. lineare* (Del.) Aschers. et Schweinf. ex Boiss., Fl. Suppl. 30 (1888). *Matthiola linearis* Del. in Laborde et Linant, Voy. Arab. Petr. 85 (1830). *Malcolmia aegyptiaca* Spreng. var. *linearis* (Del.) Coss., l.c. 22, t. 15, f. 1–15. Usually taller than var. *aegyptiacum,* sometimes viscid, with more or less fleshy leaves. Flowering and fruiting racemes loose, mostly elongated. Flowers larger, about 8 mm. Calyx frequently subpersistent. Fruit compressed, 3 times as broad as pedicel. Seeds distinctly marginate.

Hab. : Deserts; sandy soils. Philistean Plain, W., C. and S. Negev, Arava Valley, Moav, Edom.

Area of species : Saharo-Arabian (extending towards Sudanian territories).

The seeds vary considerably in width of their margin (wing), and there is no basis for retaining *Eremobium diffusum* (Decne.) Botsch., in Novit. Syst. Pl. Vasc. 359 (1964) as **a** separate species. Boissier was, therefore, justified in regarding *Hesperis diffusa* Decne. (1835) as synonymous with *E. lineare* (Del.) Aschers. et Schweinf. It is also almost certain that *Eremobium pyramidatum* (Presl) Botsch., l.c. 359 should be included within the range of variation of *E. aegyptiacum* as conceived here.

16. TORULARIA (Coss.) O. E. Schulz

Annual, low herbs. Stems erect, branched. Leaves oblong-linear, remotely dentate or pinnatifid or entire. Racemes sometimes bracteate. Flowers inconspicuous. Sepals sub-erect, almost equal, caducous. Petals white or pink, oblong-spatulate. Stamens 6, free, edentate. Style short; stigma capitate or slightly 2-lobed. Fruit a tardily dehiscing, 2-valved siliqua, rigid, terete-linear, straight or contorted, hispid or glabrous. Seeds small, in 1 row in each cell; cotyledons incumbent.

Some 12 species, mainly in the Irano-Turanian region.

1. Torularia torulosa (Desf.) O. E. Schulz, in Pflznr. 86 : 214 (1924). *Malcolmia torulosa* (Desf.) Boiss., Fl. 1 : 225 (1867). *Sisymbrium torulosum* Desf., Fl. Atl. 2 : 84 (1798). [Plate 390]

Annual, hispid or glabrescent, with branching hairs, 5–25 cm. Stems more or less erect or ascending or sometimes prostrate, branching at base. Radical leaves 3–10 cm., rosulate, tapering into a long petiole, oblong-linear, remotely dentate, rarely pinnatifid or more or less entire, soon withering; cauline leaves subsessile. Racemes ebracteate, dense, overtopped by young fruits. Flowers up to 3 mm. Sepals subequal. Petals twice as long as calyx, white, spatulate-oblong, truncate. Fruit 1–3 cm., as thick as 1 mm. long erect pedicel, rigid, straight, curved or strongly contorted, hispid or glabrous; valves firm, tardily dehiscing; septum thickened; style short; stigma capitate. Seeds compressed, oblong-ovoid. Fl. February–April.

Var. **torulosa.** Fruit and upper part of plant hispid.

Hab.: Steppes and deserts. Philistean Plain, Judean Mts., Judean Desert, N. and C. Negev, Lower Jordan Valley, Dead Sea area.

Var. **scorpiuroides** (Boiss.) O. E. Schulz, l.c. 217. *Malcolmia torulosa* (Desf.) Boiss. var. *leiocarpa* Boiss., l.c. *Sisymbrium scorpiuroides* Boiss., Ann. Sci. Nat. Bot. ser. 2, 17 : 62 (1842). Fruit and upper part of stem glabrous.

Hab.: As above. Judean Mts., Judean Desert, W. and N. Negev, Lower Jordan Valley, Dead Sea area, Arava Valley, Ammon, Moav, Edom.

Area of species: Irano-Turanian, slightly extending into adjacent territories, mainly into the Saharo-Arabian region.

17. Maresia Pomel

Annual, low, tomentose or glabrous herbs. Leaves ovate, oblong-linear to spatulate, entire or dentate or pinnatifid. Racemes ebracteate. Flowers large or small. Sepals almost erect, unequal at base, obtuse. Petals longer than sepals and stamens, clawed; limb obovate, truncate. Stamens free; filaments barely dilated at base; anthers ovoid to oblong, obtuse. Stigma capitate or 2-lobed. Fruit a dehiscent, torulose, linear siliqua. Seeds in 1 row in each cell, small, brown; cotyledons incumbent.

Some 5 species in the Mediterranean and Saharo-Arabian regions.

1. Flowers 3–5 mm. Tomentose-canescent plants.   **1. M. nana**
– Flowers larger. Green plants   2
2. Leaves almost all in a basal rosette, petiolate, sinuate-toothed or pinnatifid, mostly somewhat tomentose.   **3. M. pygmaea**
– Leaves mostly cauline, the lower long-, the upper short-petioled, entire or subpinnatifid, mostly glabrous.   **2. M. pulchella**

Sect. MARESIA. Sepals suberect, the inner slightly saccate, small (2–3 mm.). Petals small (3–7 mm.). Fruiting pedicels short (3–6 mm.).

1. Maresia nana (DC.) Batt. in Batt. et Trab., Fl. Alg. 1 : 68 (1888). *Sisymbrium nanum* DC., Syst. 2 : 486 (1821). *Malcolmia nana* (DC.) Boiss., Fl. 1 : 222 (1867). *Malcolmia confusa* Boiss., l.c. 221. [Plate 391]

Annual, tomentose-canescent with stellate hairs, 10–25 cm. Stems erect or prostrate, branched. Leaves 0.7–3 × 0.2(–0.4) cm., subsessile, oblong or oblong-linear, entire or repand-dentate. Racemes dense, corymbose, overtopped by fruits, later elongating. Flowers 3–5 mm., about as long as pedicels, white to pink. Calyx about 2 mm., slightly 2-saccate. Fruit 2–2.5 × 0.1 cm., on erect to spreading, thickened pedicels, terete, tomentose or glabrescent, abruptly terminating in a short style and a capitate, truncate or 2-lobed stigma; valves dehiscent, truncate at base, 1-nerved; septum membranous, thickened between the 2 longitudinal veins. Seeds ellipsoidal, dark brown. Fl. January–March.

Hab.: Coastal sands. Acco Plain, Sharon Plain, Philistean Plain.

Area: Mediterranean; slightly extending into coastal areas of adjacent regions.

In Palestine only var. **nana** occurs; it is marked by its tomentose fruit.

Sect. DIBOTHRIUM O. E. Schulz, in Pflznr. 86 : 209 (1924). Sepals erect, inner ones strongly saccate. Petals large (0.7–1.5 cm.). Fruiting pedicels long (0.8–2.3 cm.).

2. Maresia pulchella (DC.) O. E. Schulz, in Pflznr. 86 : 211 (1924). *Hesperis pulchella* DC., Syst. 2 : 455 (1821). *Malcolmia pulchella* (DC.) Boiss., Fl. 1 : 222 (1867). [Plate 392]

Annual, glabrous or sparingly appressed-pubescent below, 10–20 cm. Stems many, erect or prostrate, diffuse. Leaves 1.2–2.5 × 0.3 cm., most or nearly all cauline; the lower long-petioled, obovate-spatulate; upper short-petioled, oblong-ovate or linear; all more or less entire to almost pinnatifid, glabrous or minutely hairy. Calyx about 5 mm., 2-saccate, often lilac. Petals 0.8–1.5 cm.; limb usually violet, lilac or white; claw yellow-white. Fruit 2–3.5 cm., straight or subincurved, on long, erect-spreading pedicels; valves readily dehiscing; septum 2-nerved; stigma 2-lobed. Seeds dark brown. Fl. December–April.

Hab. : Sandy soils. Coastal Galilee, Acco Plain, Sharon Plain, Philistean Plain.

Area : E. Mediterranean (Palestine, Lebanon).

3. Maresia pygmaea (Del.) O. E. Schulz, in Pflznr. 86 : 210 (1924). *Hesperis pygmaea* Del., Fl. Aeg. Ill. 67, n. 596, t. 63, Suppl. f. 15 (1813). *Malcolmia pygmea* Boiss., Fl. 1 : 222 (1867). [Plate 393]

Annual, tomentose or minutely hairy desert herb, 5–15 cm. Stems branched, erect or ascending, almost leafless, sometimes violet. Leaves 0.3–2 × 0.2 cm., almost all in a basal rosette, petiolate, spatulate or oblong-cuneate, sinuate-toothed, pinnatifid, pinnatipartite or lyrate with obtuse lobes, somewhat tomentose, often with minutely branched hairs. Racemes few-flowered, buds overtopped by open flowers. Pedicels longer than calyx, filiform, somewhat thickened in fruit, erect-spreading. Calyx about 3 mm., erect, 2-saccate. Petals 0.7–1.2 cm., limb obovate-rounded or truncate, usually lilac; claw white. Fruit 1.5–2.5 (–3.5) × 0.1 cm., straight or slightly curved, linear, glabrescent; valves readily dehiscing. Seeds ovoid, brown. Fl. March–April.

Hab. : Deserts; sandy soils. C. Negev, Moav, Edom.

Area : E. Saharo-Arabian.

Intermediate forms between *M. pulchella* and *M. pygmaea* have been observed.

18. LEPTALEUM DC.

Small annual herbs with slender stems. Leaves linear, undivided or finely divided into filiform lobes. Inflorescences short, racemose, few-flowered. Flowers bracteate. Sepals erect, linear, not saccate. Petals linear. Stamens 6, the longer ones long-connate in pairs, often reduced by abortion to 2. Fruit a dorsally flattened, indehiscent siliqua, opening follicularly when moistened; lobes of stigma connate into a minute cone; valves 1-nerved, reticulate-veined; septum submembranous. Seeds many, small, in 2 rows in each cell; cotyledons incumbent.

One species, mainly in the Irano-Turanian region.

1. **Leptaleum filifolium** (Willd.) DC., Syst. 2 : 511 (1821); Boiss., Fl. 1 : 243 (1867). *Sisymbrium filifolium* Willd., Sp. Pl. 3 : 495 (1800). [Plate 394]

Tiny annual, glabrescent or pilose, up to 10 cm. Stems erect or prostrate, simple or branching from base. Leaves 2–3 × 0.1 cm., entire or pinnatifid into 3–5 filiform segments. Pedicels short, thickened. Flowers axillary, solitary or in pairs. Sepals 3–5 mm., linear. Petals about 0.5–1 cm., white or pink. Fruit 1.5–2.5 × 0.2–0.3 cm., linear, dorsally flattened, obtuse and asymmetrical at base, gradually tapering at apex, glabrous or with scattered hairs, sometimes violet; valves leathery. Seeds oblong. Fl. February–March.

Hab. : Steppes and deserts Judean Desert, C. Negev, Moav, Edom.

Area : Irano-Turanian, extending into Saharo-Arabian territories.

19. Matthiola R. Br.

Annual or perennial, tomentose or canescent (rarely green) herbs with stellate or forked hairs. Leaves simple, entire or sinuate to pinnatifid. Inflorescences racemose, ebracteate, sometimes with yellowish, glandular scales below flowers. Flowers pink, purple, yellow or livid. Sepals erect, the inner saccate. Petals linear, oblong or obovate, entire or wavy, clawed. Filaments free, the inner broader. Stigma of 2 erect connivent lobes, each lobe generally with a dorsal protuberance or horn-like process. Fruit a many-seeded, linear siliqua; valves indehiscent or tardily dehiscing; septum thickened. Seeds in 1 row in each cell, compressed, mostly narrowly winged; cotyledons accumbent.

About 50 species, mostly Mediterranean, Irano-Turanian and S. African.

Literature : P. Conti, Les espèces du genre *Matthiola*. *Mém. Herb. Boiss.*, 18 : 1–86 (1900). C. Sauvage, Au sujet des *Matthiola* marocains du groupe *Oxyceras*. *Bull. Soc. Sci. Nat. Maroc* 30 : 121–130 (1950).

1. Tall, tomentose-canescent perennials. Horns of fruit 0. Leaves mostly lanceolate-linear, entire. **1. M. arabica**
 – Annuals. Fruit horned (rarely horns very short) 2
2. Petals 0.6–0.8(–1.5) cm., slightly longer than calyx. Inflorescences often overtopped by leaves. **4. M. parviflora**
 – Petals considerably longer than 1 cm. 3
3. Coastal plants. Fruit about 3 mm. thick, with 3 more or less equal horns, formed by the erect, conical stigma and its 2 divergent or horizontal protuberances. **2. M. tricuspidata**
 – Fruit terminating with 2 horns much longer than stigma, sometimes horns strongly reduced in length. Plants not coastal 4
4. All or part of the fruit contorted; horns often short, up to 2(–3) mm., sometimes extremely reduced, then stigma 2-gibbous. Leaves linear to lanceolate. Petals linear with undulate margin. Desert plants. **6. M. livida**
 – Plants different from above. Horns well developed 5
5. Leaves almost entire or obsoletely repand, greenish. Petals obovate-spatulate, margin entire. **3. M. aspera**
 – Leaves, at least the radical ones, sinuate-dentate or pinnatifid, canescent. Petals linear-lanceolate or oblong to spatulate, margin undulate or entire. **5. M. longipetala**

Sect. LUPERIA DC., Syst. 2 : 169 (1821). Fruit not horned, sometimes 2-gibbous. Petals narrow, oblong to linear.

1. Matthiola arabica Boiss., Ann. Sci. Nat. Bot. ser. 2, 17 : 49 (1842); Boiss., Fl. 1 : 152 (1867). [Plate 395]

Perennial, canescent with short, stellate or simple hairs, 50–75 (–100) cm. Stems many, erect or ascending, rigid, branching from base. Leaves 3–6 × 0.3–0.5 cm., crowded at base, linear to lanceolate, entire, more or less pubescent, the lower long-petioled, the cauline short-petioled. Flowers sessile. Sepals about 8 mm., linear, obtuse, scarious-margined. Petals about 1.2 cm. or more, livid, linear, retuse. Fruit up to 5 cm., spreading, curved or twisted; stigma ovoid, somewhat broader than fruit; horns absent. Fl. March–April.

Hab. : Stony places in deserts. S. Negev, Edom.

Area : Saharo-Arabian.

Sect. ACINATUM DC., Syst. 2 : 175 (1821). Fruit 2-horned. Petals obovate.

2. Matthiola tricuspidata (L.) R. Br. in Ait., Hort. Kew. ed. 2, 4 : 120 (1812); Boiss., Fl. 1 : 154 (1867). *Cheiranthus tricuspidatus* L., Sp. Pl. 663 (1753). [Plate 396]

Annual, low, canescent, 10–20 cm. Stems ascending or procumbent, diffusely branching from base. Leaves 2–5 cm., somewhat fleshy, oblong to oblanceolate, sinuate-dentate or pinnatifid, rarely almost entire, lobes rounded. Racemes short. Flowers short-pedicelled. Sepals about 8 mm., oblong, tomentose. Petals 1.5–1.8 (–2) cm., ex-serted, purple or pink, obovate, clawed. Fruit 3–5 × 0.2–0.3 cm., on thick terete pedicels, subtorulose; stigma 2-lobed, erect, conical, with spreading, awl-shaped horns, equalling or somewhat exceeding the stigma in length. Fl. March–June.

Hab. : Sandy soil; spray zone. Coastal Galilee, Acco Plain, Sharon Plain.

Area : Mediterranean.

3. Matthiola aspera Boiss., Diagn. ser. 1, 8 : 16 (1849) et Fl. 1 : 155 (1867). [Plate 397]

Annual, glaucous-green, rough, with scattered branching hairs or almost glabrous, 15–30 cm. Stems erect or ascending. Leaves (3–)5–7 cm., oblong-lanceolate to linear, tapering, obtuse, almost entire or obsoletely repand, slightly tomentose or with spreading stellate hairs, greenish. Racemes many-flowered. Flowers large, sessile. Sepals 0.8–1 cm., scarious-margined, tomentose at base. Petals 1.2–1.5 (–2) cm., pink or purple, obovate-spatulate with entire margin. Fruit 4–6 × 0.2 cm., firm, spreading, glabrous or sparingly stellate-hairy; stigma conical, more or less elongated; horns twice as long as diameter of fruit, awl-shaped, erect or spreading. Fl. February–April.

Var. **aspera.** Stems with branching hairs. Fruit with stellate hairs.

Hab. : Deserts. Judean Mts., Judean Desert, W. Negev, Lower Jordan Valley, Dead Sea area, deserts of Ammon and Moav.

Var. **leiocarpa** Bornm., Verh. Zool.-Bot. Ges. Wien 48 : 550 (1898). Stems almost glabrous. Fruit glabrous.

Hab.: Deserts. Judean Desert, Lower Jordan Valley, Dead Sea area, deserts of Ammon.

Area of species : Saharo-Arabian (endemic in Palestine).

4. Matthiola parviflora (Schousb.) R. Br. in Ait., Hort. Kew. ed. 2, 4 : 121 (1812); DC., Syst. 2 : 176 (1821). *Cheiranthus parviflorus* Schousb., Vextr. Marok. 195 (1800). [Plate 398]

Annual, tomentose with branched hairs, 10–30 cm. Stems almost erect, diffusely branched at base. Leaves 4.5–5 (–7.5) × 0.7–1 (–1.5) cm., crowded at base and along stem, oblanceolate, sinuate-dentate or pinnatifid with acute lobes; lower leaves long-petioled, upper and floral sessile. Racemes short, often overtopped by leaves. Flowers short-pedicelled or sessile. Sepals 5 (–7) mm. Petals 0.6–0.8 (–1.5) cm., purple, obovate. Fruit 4.5–6 cm., erect or spreading; horns 4–5 mm. Fl. March–April.

Var. **maris-mortui** Zoh., Palest. Journ. Bot. Jerusalem ser. 2 : 156 (1941). Densely tomentose plants. Leaves densely clustered at base. Inflorescences overtopped by leaves. Flowers 1–1.2 cm.

Hab. : Deserts. Judean Desert, S. Negev, Dead Sea area.

Area of species : Saharo-Arabian, extending into the Mediterranean region.

Sect. PINARIA DC., Syst. 2 : 172 (1821). Fruit 2-horned (horns sometimes very reduced). Petals narrow, oblong to linear.

5. Matthiola longipetala (Vent.) DC., Syst. 2 : 174 (1821); Zoh., Palest. Journ. Bot. Jerusalem ser. 2 : 156 (1941). *Cheiranthus longipetalus* Vent., Descr. Pl. Jard. Cels t. 93 (1802). [Plate 399]

Annual, pubescent, sparingly glandular or eglandular, 10–40 cm. Stems branching from base, rarely leafy. Leaves 3–8 (–10) cm., crowded at base, linear, linear-lanceolate or lanceolate to oblong, sinuate-dentate or pinnatifid, canescent. Flowers sessile or more or less short-pedicelled. Sepals 0.7–1 cm. Petals 1.4–2 (–3) cm.; limb pink or purple, linear-lanceolate or oblong, margin undulate or entire; claw yellow, livid or purple. Fruit 2–10 × 0.1–0.25 cm., erect or horizontally spreading or rarely recurved, straight, rarely contorted; stigma short, more or less obtuse; horns up to 1.2 cm., sometimes very short and thick, horizontal or recurved or more or less incurved. Fl. February–May.

This species is very variable. The following division into varieties is tentative only because of the occurrence of intermediate forms between the varieties of this species and between them and some varieties of *M. livida*.

1. Fruit 2–3 cm. long.   var. **brachyloma**
 – Fruit 4–10 cm. long   2
2. Fruit thin, contorted.   var. **contorta**
 – Fruit straight   3
3. Petal limb linear-lanceolate. Horns of fruit 1–1.2 cm.   var. **longipetala**
 – Petal limb oblong. Horns of fruit short (4–5 mm.), horizontal or strongly curved upwards.
                                                                                var. **bicornis**

Var. **longipetala**. Zoh., Palest. Journ. Bot. Jerusalem ser. 2 : 157 (1941). *M. longipetala* (Vent.) DC. var. *oxyceras* (DC.) Zoh., l.c. *M. longipetala* (Vent.) DC. var. *macrocarpa* (Eig) Zoh., l.c. 158. *M. oxyceras* DC., Syst. 2 : 173 (1821). *M. oxyceras* DC. var. *lunata* Boiss., Fl. 1 : 156 (1867). Stems branching from base. Leaves 3.5–6.5 cm., often all crowded at base, oblong or lanceolate, sinuate-dentate or pinnatifid, lobes mostly linear, acute; indumentum sparse; plants greenish. Petals longer than calyx, limb linear or linear-lanceolate, more or less undulate. Fruit 6–7.5(–8–10) cm., lateral horns up to 1.2 cm., horizontal or recurved.

Hab. : Batha and steppes. Sharon Plain, Philistean Plain, Judean Mts., Judean Desert, N. and C. Negev, Dead Sea area, Arava Valley, Ammon, Moav, Edom.

Var. **bicornis** (Sibth. et Sm.) Zoh., l.c. *M. bicornis* (Sibth. et Sm.) DC., Syst. 2 : 177 (1821); Boiss., Fl. 1 : 155 (1867). *Cheiranthus bicornis* Sibth. et Sm., Fl. Gr. Prodr. 2 : 26 (1813). *M. oxyceras* DC. var. *forcipifera* Boiss., Fl. 1 : 156 (1867). Leaves 5–8(–10) cm., oblong-lanceolate, the lower sinuate-dentate or pinnatifid, upper linear, entire. Petal limb oblong-spatulate. Horns of fruit 4–5 mm., horizontal or strongly incurved.

Hab. : As above. Sharon Plain, Philistean Plain, Judean Mts., N. Negev, Dead Sea area, Arava Valley.

Var. **brachyloma** Zoh., l.c. (err. "brachypoda"). Leaves mostly lanceolate-linear. Fruit 2–3 cm. long, thick. Horns of fruit short.

Hab. : Steppes. Judean Desert, N. Negev.

Var. **contorta** Zoh., l.c. Fruit thin, 1–2 mm. broad, recurved, contorted.

Hab. : Steppes. Edom.

Area of species : E. Mediterranean, W. Irano-Turanian and some adjacent provinces.

6. **Matthiola livida** (Del.) DC., Syst. 2 : 174 (1821). *Cheiranthus lividus* Del., Fl. Aeg. Ill. 591 (1813). [Plate 400]

Annual, canescent, often also with scattered glandular hairs, (12–)30–40 cm. Stems erect, often branching from base. Leaves 2–5(–6) × 0.3–1.2 cm., linear to lanceolate-linear, obtuse or subacute, entire or sinuate-repand, sparingly to densely stellate-hairy. Flowers sessile. Sepals 1–1.2 cm., linear, obtuse, scarious-margined, sometimes purplish. Petals up to 1.8 cm.; limb livid, cream-coloured to pale pink, linear with undulate margin. Fruits 4–6 × 0.1–0.2 cm., erect-spreading, straight or often part of them or all contorted, sparingly to densely hairy, rarely glabrous; horns up to 3 mm. or almost 0 or stigma 2-gibbous. Fl. March–May.

Var. **livida**. Stems with spreading, branching hairs. Fruit more or less densely covered with branching hairs.

Hab. : Steppes. Philistean Plain, N. and C. Negev, Lower Jordan Valley, Edom.

Var. **leiocarpa** Zoh., Beih. Bot. Centralbl. 50, 1 : 45 (1932). Stems appressed-hairy. Fruit glabrous.

Hab. : As above. N. Negev.

Area of species : Mainly Saharo-Arabian.

The area under review must be regarded as one of the differentiation centres of section *Pinaria* to which the last two species belong.

20. Morettia DC.

Perennials, canescent, stellately hairy or scabrous. Stems erect or procumbent. Leaves all cauline, undivided. Flowers axillary or in terminal racemes, pale pink. Sepals erect, equal at base. Petals oblong-linear, entire. Stamens free, edentate. Fruit a siliqua borne on a thick pedicel, terete to tetragonous, with a short style ending in 2 long persistent stigmas thickened at base; valves with incomplete transversal partitions on their inner surface; septum membranous. Seeds in 1 row in each cell; cotyledons accumbent.

Four species, mainly in the Sudanian region.

1. Flowers about 4 mm. Fruit strongly curved. Leaves more or less long-petioled, ovate-elliptical.   **3. M. parviflora**
– Flowers larger. Fruit straight or slightly curved. Leaves short-petioled, lanceolate or oblong   2
2. Leaves often repand. Fruit tetragonous, 1.2–2.5 cm.; style thick, somewhat hairy.
                                                                                                          1. M. philaeana
– Leaves entire. Fruit cylindrical, 0.8–1.5 cm.; style thin, glabrous.   **2. M. canescens**

1. Morettia philaeana (Del.) DC., Syst. 2 : 427 (1821); Boiss., Fl. 1 : 145 (1867). *Sinapis philaeana* Del., Fl. Aeg. Ill. 99, t. 33, f. 3 (1813). [Plate 401]

Perennial, tomentose with scabrous, brittle, stellate hairs, adhering to fingers. Stems thick, ascending, about 25 cm. Leaves about 1–1.2 (–2) cm., short-petioled, oblong, often repand. Sepals 4–6 mm., oblong. Petals 5–8 mm., with slightly exserted claw. Fruit 1.2–1.5 (–2.5) cm., thick, appressed to branches, tetragonous, straight or slightly curved, abruptly ending in a thick, hairy to glabrescent style; lobes of stigma thick, divergent; valves thick, torulose, longitudinally nerved. Fl. March–April.

Hab. : Hot deserts, on granitic rocks. Arava Valley, Edom.

Area : E. Sudanian.

2. Morettia canescens Boiss., Diagn. ser. 1, 8 : 17 (1849) et Fl. 1 : 145 (1867). [Plate 402]

Perennial, canescent, 10–50 cm. Stems many, prostrate, much branching. Leaves 0.8–1.2 cm., short-petioled, lanceolate or ovate-oblong, acute, entire, folded or revolute-margined, with appressed, stellate hairs. Sepals 4–5 mm. Petals 0.6–1 cm.; claw more or less exserted. Fruit 0.8–1.5 cm., appressed to branches, straight, cylindrical, upper part tapering; valves torulose; style thin, often glabrous; stigma with 2 rather thick divergent lobes. Fl. February–May.

Hab. : Hot deserts, mainly on granitic rocks or sandstone. S. Negev, Dead Sea area, Arava Valley, Edom.

Area : E. Sudanian with extensions into adjacent Saharo-Arabian territories.

3. Morettia parviflora Boiss., Ann. Sci. Nat. Bot. ser. 2, 17 : 60 (1842) et Fl. 1 : 146 (1867). [Plate 403]

Perennial, white-tomentose with stellate hairs. Stems prostrate or ascending, 20–30 cm. Leaves 0.8–2.5 cm., more or less long-petioled, ovate or ovate-elliptical to oblong, acute, subentire to entire. Sepals about 3 mm., oblong, hairy. Petals about 4 mm., subexserted, not clawed, oblong-lanceolate, entire. Fruit 0.8–1.2 cm., thin, somewhat compressed, first straight later strongly curved, gradually tapering above, densely stellate-hairy; valves torulose; style long, woolly; lobes of stigma thick, triangular-lanceolate. Fl. March–April.

Hab.: Hot deserts; rocky places. S. Negev, Arava Valley, Edom.

Area : E. Sudanian.

21. Notoceras R. Br.

Low annuals, appressed-canescent with forked and simple hairs. Stems decumbent or ascending, branched. Leaves lanceolate, tapering into a petiole. Racemes many-flowered. Sepals more or less spreading, equal at base. Petals minute, white. Stamens free, edentate. Style distinct; stigma capitate. Fruit a tardily dehiscing, 2-celled, rigid, tetragonous siliqua; valves with distinct midrib, foveolate inside, each with an apical horn. Seeds in 1 row in each cell, ellipsoidal, wingless, brown; cotyledons accumbent.

One species, mainly in the Saharo-Arabian region.

1. Notoceras bicorne (Sol.) Caruel, Flor. Toscan. 536 (1860). *Erysimum bicorne* Sol. in Ait., Hort. Kew. ed. 1, 2 : 394 (1789). *N. canariense* R. Br. in Ait., Hort. Kew. ed. 2, 4 : 117 (1812); Boiss., Fl. 1 : 314 (1867). [Plate 404]

Low annual with appressed, simple and forked hairs, 5–10(–30) cm. Stems procumbent to ascending, branching from base. Branches often spreading or horizontal. Leaves 2.5–3.5 (–6) cm., oblong-linear or lanceolate, tapering at base. Racemes dense, over-topped by leaves, compact in fruit. Pedicels thickened, appressed to rhachis. Sepals about 1 mm., hirsute, with membranous margin. Petals about 1.5–2 mm., oblong-linear, obtuse. Fruit 0.6–1 × 0.2 cm., tetragonous; valves rigid, distinctly 1-nerved, horned at apex; septum membranous. Seeds 2–3 in each cell, compressed. Fl. January–March.

Hab.: Deserts. Judean Desert, N., C. and S. Negev, Upper and Lower Jordan Valley, Dead Sea area, Arava Valley, Moav, Edom.

Area : Saharo-Arabian, slightly extending into the neighbouring regions.

Most characteristic of the *Anabasis articulata – Notoceras* community in the Plain of Jericho and elsewhere. Also occurring in other ephemeral communities in hot deserts.

A hygrochastic plant : fruit diverge from stem when moistened.

22. Chorispora R. Br.

Annual or perennial, pilose or glandular herbs. Leaves dentate or pinnatifid. Racemes ebracteate. Flowers large, long-pedicelled. Sepals erect, the inner saccate at base. Petals yellow or purple, retuse or entire, clawed. Filaments free, not appendiculate. Stigma indistinctly 2-lobed. Fruit an indehiscent siliqua, constricted between seeds, elongated, cylindrical, curved, composed of many transversal cells arranged in 2 rows and an elongated beak; the 1-seeded cells often alternating with empty ones. Cotyledons frequently accumbent.

Ten species mainly in the Irano-Turanian region.

1. Chorispora purpurascens (Banks et Sol.) Eig, Journ. Bot. Lond. 75 : 189 (1937). *Brassica purpurascens* Banks et Sol. in Russ., Nat. Hist. Aleppo ed. 2, 2 : 258 (1794). *C. syriaca* Boiss., Ann. Sci. Nat. Bot. ser. 2, 17 : 384 (1842) et Fl. 1 : 143 (1867). [Plate 405]

Annual, glandular or hairy, 10–30 cm. Lower leaves petiolate, lanceolate, lyrate to pinnatifid, with dentate lobes; the upper subsessile, lanceolate, pinnatifid to dentate. Racemes elongated. Pedicels long, thickened in fruit. Calyx large, about 1 cm., coloured, hairy. Petals 1.5–2 cm., purple, long-clawed. Stamens exserted; filaments free. Style long; stigma 2-lobed. Fruit 2–4 (–5) × up to 0.5 cm., slightly laterally compressed, torulose, scabrous with retrorse papillae, ending in a very long, terete, gradually tapering beak. Fl. March–April.

Hab.: Fields. Judean Mts., Judean Desert, Ammon, Moav, Edom.
Area: W. Irano-Turanian.

23. ANASTATICA L.

Annual, low, stellate-canescent herbs, becoming woody. Stems dichotomously branched. Leaves simple, petiolate. Inflorescences short, spike-like. Flowers small, white, rather sessile. Sepals erect, almost equal. Petals clawed. Stamens with dilated bases of filaments. Style long, 2-lobed, persistent; stigma capitate. Fruit an indurated, hardly dehiscent silicle; valves convex, broad, with a spoon-shaped appendage at apex; cells transversally divided, forming 1-seeded compartments; septum thick; beak subulate, indurated. Seeds compressed, ovoid; cotyledons incumbent.

One species in the hot deserts of the Saharo-Arabian and adjacent Sudanian territories.

1. Anastatica hierochuntica L., Sp. Pl. 641 (1753); Boiss., Fl. 1 : 316 (1867). [Plate 406]

Annual, stellately hairy-canescent, 3–20 cm. Stems procumbent or ascending, dichotomously branched. Leaves up to 3.5 × 1.5 cm., mostly smaller, spatulate-obovate, tapering to a petiole, obtuse or acute, remotely dentate. Flowering racemes mostly in forks between branches, overtopped by leaves. Flowers more or less sessile. Sepals about 2 mm., stellate-hirsute. Petals about twice as long as calyx, clawed; limb broadly obovate. Fruiting branches elongated, woody, incurved and forming a globular body enclosing fruits; when moistened the branches diverge and spread exposing the fruits to the rain and wind. Fruit 5–6 × 3–4 mm., indehiscent, ovoid, stellate-hirsute; valves extended at apex into broad, spoon-shaped appendages equal to or shorter than the 2–5 mm. long, persistent style; septum thick. Seeds pendulous, wingless. Fl. February–April.

Hab.: In gravelly or sandy depressions or wadis of hot deserts. Judean Desert, Negev, Dead Sea area, Arava Valley, Moav.

Area : As for the genus.
The true *Rose of Jericho*.

Trib. ARABIDEAE. Median nectar glands present, sometimes united with the transversal ones to a ring. Fruit a terete or compressed siliqua, rarely a dehiscent or indehiscent silicle.

24. RORIPPA Scop.

Annual or perennial herbs, glabrous or sparingly hairy. Stems erect or ascending. Leaves simple to pinnate. Inflorescences ebracteate. Inner sepals saccate at base. Petals yellow, clawed. Stamens free, without appendages. Style short; stigma capitate or 2-lobed. Fruit a spherical or ellipsoidal silicle; valves convex, with midrib indistinct or vanishing above middle. Seeds in 2 rows in each cell; cotyledons accumbent.

About 90 species throughout the north-temperate regions.

1. Rorippa amphibia (L.) Bess., Enum. Pl. Volh. 27 (1822). *Sisymbrium amphibium* L., Sp. Pl. 657 (1753). *Nasturtium amphibium* (L.) R. Br. in Ait., Hort. Kew. ed. 2, 4 : 110 (1812). [Plate 407]

Perennial, glabrous or pilose, 30–45 cm. Stems stoloniferous, hollow, furrowed, erect, branching. Leaves 4–12 cm., elliptical or oblanceolate, tapering into a short petiole, entire or dentate, the lower leaves sometimes pinnatifid, obtuse or acute, yellowish-green. Racemes many-flowered, elongating in fruit. Pedicels 1–1.5 cm., terete, filiform, spreading, horizontal or deflexed. Flowers small. Sepals about 2 mm., spreading. Petals about 5–7 mm., yellow. Style filiform, 1–2 mm.; stigma capitate. Fruit small, 2–5 × 1–3 mm., substipitate, ellipsoidal to globular. Fl. April–May.

Hab. : Swamps. Hula Plain. Rare.

Area : Euro-Siberian, extending into some Irano-Turanian and Mediterranean territories.

25. NASTURTIUM R. Br.

Glabrous or sparingly hairly perennial herbs. Stems ascending, leafy. Leaves usually pinnate. Inflorescences usually ebracteate, racemose. Flowers small, numerous. Inner sepals spreading, saccate at base. Petals white or yellow, sometimes turning lilac, entire, clawed. Stamens without appendages. Stigma capitate or indistinctly 2-lobed. Fruit a linear, generally turgid, more or less flattened siliqua; valves nerveless or obsoletely 1-nerved. Seeds usually in 2 rows in each cell, with numerous polygonal depressions on surface; cotyledons accumbent.

About 50 species in both hemispheres.

1. Nasturtium officinale R. Br. in Ait., Hort. Kew. ed. 2, 4 : 110 (1812); Boiss., Fl. 1 : 178 (1867). *Sisymbrium nasturtium-aquaticum* L., Sp. Pl. 657 (1753). *Rorippa nasturtium-aquaticum* (L.) Hay., Sched. Fl. Stir. Exs. 22 (1905). [Plate 408]

Glabrous perennial, 15–70 cm. Stems hollow, angular, procumbent and rooting below. Leaves 4–18 cm., pinnate, mostly with 3–7 pairs of leaflets; lateral leaflets broadly ovate or elliptical, the terminal round-ovate, all repand-toothed. Inflorescences ebracteate, elongating in fruit. Sepals about 2 mm., linear. Petals 4–5 mm., twice as long as sepals, white. Fruit 1–2 × 0.1–0.2 cm., oblong-linear, often incurved, on 0.8–1.2 cm. long, horizontal or somewhat deflexed pedicels; valves with distinct midrib and faint lateral nerves. Seeds in 2 distinct rows in each cell, ovoid, brown, pitted. Fl. February–September.

Hab.: Brooks and springs. Coastal Galilee, Sharon Plain, Philistean Plain, Upper Galilee, Esdraelon Plain, Samaria, Judean Mts., Dan Valley, Hula Plain, Upper and Lower Jordan Valley, Edom.

Area. Pluriregional; in many temperate parts of the Holarctis and also in some tropical regions.

26. CARDAMINE L.

Annual or perennial, glabrous or pubescent herbs. Leaves trifoliate or pinnately compound, sometimes simple. Inflorescences racemose, ebracteate. Flowers white or purple. Sepals erect, equal or inner slightly saccate at base. Petals clawed, exserted. Stamens 6(4–5). Stigma simple or 2-lobed. Fruit an erect or spreading, straight, compressed, linear siliqua; valves obsoletely nerved, dehiscing from base with spiral twist. Seeds in 1 row in each cell, compressed; cotyledons accumbent.

About 130 species, almost cosmopolitan.

1. Cardamine hirsuta L., Sp. Pl. 655 (1753); Boiss., Fl. 1 : 160 (1867). [Plate 409]

Annual, usually glabrous, 10–30 cm. Stems erect, branched from base. Radical leaves many, 2–8 cm., rosulate, pinnate with (1–)2–4 pairs of ovate to obovate-orbicular, petiolulate leaflets, the terminal one larger, suborbicular-reniform; cauline leaves few, compound of shorter and narrower leaflets; all subglabrous or with scattered simple hairs. Inflorescences corymbose-racemose, elongating in fruit. Flowers about 2–3 mm. Sepals equal, purplish. Petals about twice as long as sepals (sometimes 0), white. Stamens 4–6, yellow. Fruit about 1.5–2.5 × 0.1 cm., erect or spreading on short pedicels, glabrous, overtopping flowers. Seeds 1 mm. in diam., narrowly winged. Fl. February–April.

Hab.: Wet soils. Acco Plain, Sharon Plain, Philistean Plain, Upper and Lower Galilee, Hula Plain, Moav.

Area: Mediterranean, Euro-Siberian extending towards the Irano-Turanian region.

27. ARABIS L.

Annual or perennial herbs, glabrous or more frequently pubescent with simple or branched hairs. Radical leaves crowded, more or less petiolate; cauline sessile, often clasping, frequently dentate. Racemes often bracteate at base. Sepals erect, equal or inner saccate. Petals white, yellow, pink or violet, clawed. Stamens without appendages. Style short; stigma entire or 2-lobed. Fruit a long, linear siliqua; valves flattened or convex, 1-nerved. Seeds in 1 row in each cell, compressed, ovoid, wingless, rarely winged; cotyledons accumbent.

Over 100 species, mainly in the north-temperate regions of both hemispheres.

Literature : B. M. G. Jones, *Arabis* L., in : Tutin et al., *Fl. Europaea* 1 : 290–294 (1964).

1. Fruit 7–12(–15) cm., deflexed, in one-sided racemes. Seeds winged. Perennial herbs.
   4. A. turrita
 – Fruit 2.5–6(–7) cm., erect or spreading, in flexuous racemes. Seeds wingless   2
2. Flowers pink-violet, 5–8 mm. long. Leaves dentate-serrate; the cauline amplexicaul.
   Pedicels shorter than hirsute calyx.   **1. A. verna**
 – Flowers white   3
3. Perennials with procumbent stems and leaves arranged in many rosettes, softly and
   densely pubescent. Petals about 1–1.5 cm.   **5. A. caucasica**
 – Annuals with erect stems. Leaves and petals not as above   4
4. Cauline leaves broadly cordate at base but not amplexicaul; all leaves sessile. Pedicels
   shorter than calyx. Valves faintly nerved.   **3. A. aucheri**
 – Cauline leaves subamplexicaul, the radical tapering into a petiole, nearly glabrous.
   Pedicels longer than calyx. Valves with prominent midrib.   **2. A. nova**

Sect. ALOMATIUM DC., Syst. 2 : 214 (1821). Petal limb erect. Fruit erect or spreading. Seeds not winged. Annuals.

1. Arabis verna (L.) R. Br. in Ait., Hort. Kew. ed. 2, 4 : 105 (1812); Boiss., Fl. 1 : 168 (1867). *Hesperis verna* L., Sp. Pl. 664 (1753). [Plate 410]

Annual, with simple or mostly stellately branching hairs, 7–30 cm. Stems erect, usually simple. Leaves dentate-serrate; the radical 1.5–6 cm., more or less obovate, tapering at base, obtuse; cauline leaves few, smaller, ovate, amplexicaul. Racemes few-flowered, ebracteate, with young flowers overtopped by fruit. Pedicels shorter than the hirsute calyx. Petals 5–8 mm., exserted; limb pink-violet, very rarely white; claw white. Fruiting racemes elongated, flexuous. Fruit 2.5–6 × 0.1–0.2 cm., borne on thickened pedicel, erect, more or less compressed, rather glabrous; valves many-nerved. Seeds in 1 row in each cell, wingless. Fl. February–April.

Hab. : Batha, among rocks and on walls. Upper and Lower Galilee, Mt. Carmel, Judean Mts., Gilead, Ammon.

Area : Mediterranean.

2. Arabis nova Vill., Prosp. Pl. Dauph. 39 (1779). *A. auriculata* Lam., Encycl. 1 : 219 (1783); Boiss., Fl. 1 : 169 (1867). [Plate 411]

Annual, with scattered, simple or branched hairs, 10–30 cm. Stems fairly erect, simple, slender. Leaves dentate-crenate or almost entire, glabrous or nearly so; the radical ones obovate-oblong, tapering; the cauline ovate-lanceolate, cordate-sagittate at base, subamplexicaul. Racemes many-flowered, ebracteate, elongated. Pedicels longer than calyx, slender. Flowers 3–5 mm., white. Fruiting racemes flexuous. Fruit about 4–5 × 0.1 cm., erect-spreading; valves convex with prominent midrib. Seeds in 1 row in each cell, wingless. Fl. April.

Hab. : Batha. Upper Galilee, Gilead, Ammon.

Area : Mediterranean and Irano-Turanian, extending into adjacent Euro-Siberian territories.

3. Arabis aucheri Boiss., Ann. Sci. Nat. Bot. ser. 2, 17 : 52 (1842) et Fl. 1 : 170 (1867). [Plate 412]

Annual with branched hairs, 10–35 cm. Stems erect, slightly branched. Leaves 0.5–2 cm., sessile, ovate-oblong, irregularly denticulate or almost entire; cauline similar to radical but smaller, broadly cordate at base, not amplexicaul. Racemes few-flowered, ebracteate, elongated in fruit and flexuous. Pedicels shorter than calyx, thickened in fruit. Flowers 4–6 mm., white. Fruit 3–4 × 0.1 cm., erect, usually glabrous. Seeds in 1 row in each cell, wingless. Fl. April.

Hab.: Batha and fields. Upper Galilee, Gilead.

Area: E. Mediterranean, slightly extending into the adjacent Irano-Turanian borderland.

Sect. TURRITA (Wallr.) Reichb. in Mössler, Handb. Gewächsk. 11–15 (1828). Sepals not saccate. Petal limb spreading. Fruit deflexed. Seeds winged. Perennials.

4. Arabis turrita L., Sp. Pl. 665 (1753); Boiss., Fl. 1 : 177 (1867). [Plate 413]

Perennial or biennial, with a dense cover of branching-stellate hairs, 35–65 cm. Rhizome horizontal. Stems branching at base; flowering stems generally simple. Leaves dentate, the radical up to 12 cm., oblong-elliptical to obovate, tapering at base into a 6–10 cm. long petiole; the cauline 2–9 cm., shorter and narrower, oblong-lanceolate, amplexicaul with round basal lobes. Racemes one-sided, bracteate; bracts small. Petals pale yellow, with patulous limb. Fruit (7–)9–12(–15) × 0.2–0.3 cm., compressed, somewhat torulose, deflexed, on an erect-spreading pedicel; valves obsoletely nerved, glabrous or nearly so. Seeds in 1 row in each cell, flattened, ovate-obtuse with conspicuous marginal wing. Fl. April–August.

Hab.: Maquis and garigue. Upper Galilee.

Area: Mediterranean and W. Euro-Siberian.

Sect. ARABIS. Inner sepals saccate. Petal limb broad, spreading. Fruit often deflexed. Seeds narrowly margined. Perennials, with sterile branches, bearing rosulate leaves.

5. Arabis caucasica Schlecht. in Willd., Enum. Pl. Hort. Berol. Suppl. 45 (1813). *A. albida* Stev. in Fisch., Cat. Hort. Gorenk. 51 (1812) *nom. nud.*; Boiss., Fl. 1 : 174 (1867).

Perennial herb, canescent or tomentose, 15–25 cm. Stems procumbent, simple or branching. Leaves mostly rosulate, the radical obovate-oblong, obtuse; upper leaves ovate-cordate, clasping, all sparingly toothed. Flowers white, conspicuous, in short racemes. Sepals 5–8 mm., the inner saccate. Petals 1–1.5 cm., white, with patulous limb and long claw. Fruit 4–7 × 0.15–0.25 cm., erect-patulous, torulose; valves with a distinct median nerve. Seeds narrowly margined. Fl. April–June.

Hab.: Rock fissures. Upper Galilee (Herb. Aaronsohn). Not collected after 1906.

Area: Subalpine zones. Mediterranean, Irano-Turanian and Euro-Siberian.

28. Turritis L.

Perennial tall herbs with branched and simple hairs only at base. Stems erect, leafy. Radical leaves rosulate, lyrate-pinnate or crenate; the cauline sagittate or cordate at base. Racemes ebracteate. Flowers conspicuous. Inner sepals somewhat saccate. Petals white or yellowish. Stamens free, without appendages. Style short; stigma capitate or slightly 2-lobed. Fruit a dehiscent siliqua, terete or tetragonous; valves convex, with distinct midrib; septum thick. Seeds ovoid, 1–2 rows in each cell.

Three species in temperate regions of the Old and New World.

1. Turritis laxa (Sibth. et Sm.) Hay., Prodr. Fl. Pen. Balc. 1 : 402 (1925). *Arabis laxa* Sibth. et Sm., Fl. Gr. Prodr. 2 : 28 (1813); Boiss., Fl. 1 : 168 (1867). *A. cremocarpa* Boiss. et Bal. in Boiss., Diagn. ser. 2, 5 : 16 (1856). *A. laxa* Sibth. et Sm. var. *cremocarpa* (Boiss. et Bal.) Boiss., Fl. l.c. [Plate 414]

Perennial or biennial, 40–80 cm. Stems rigid, branching, glabrous. Radical leaves 4–15 cm., lyrate, tapering into a petiole, hirsute with branching or stellate hairs; cauline leaves shorter, obovate to lanceolate, subamplexicaul, cordate at base, entire, glabrous or sparsely ciliate. Racemes dense, elongated and mostly one-sided in fruit. Flowers about 5 mm., usually white. Fruit (8–)9–15 × 0.2 cm., spreading or deflexed, slender, more or less tetragonous, with a short, distinct style; valves with prominent midrib. Seeds in 1 row, narrowly winged or wingless. Fl. April–June.

Hab. : Maquis and garigue, among rocks. Upper Galilee.

Area : E. Mediterranean.

29. Farsetia Turra

Perennial herbs or half-shrubs, mostly canescent, with forked, stiff, appressed hairs. Leaves subsessile or tapering at base, linear-oblong, more or less obtuse, entire. Inflorescences racemose, ebracteate. Flowers large, pedicellate. Sepals erect, more or less equal at base. Petals white, pink, yellow, purple or livid, linear-oblong, sometimes undulate, with exserted claw. Shorter stamens sometimes appendiculate. Style short and thick; stigma capitate. Fruit a siliqua or silicle, dorsally flattened, ovate-oblong to linear-elliptical, dehiscing by 2 flat or convex valves. Seeds many, in 1–2 rows in each cell, flat, almost orbicular, brown, broadly winged; cotyledons accumbent.

About 12 species, mainly in Sudanian deserts, also in C. Asia.

1. Farsetia aegyptiaca Turra, Farset. Nov. Gen. 5, t. 1 (1765); Boiss., Fl. 1 : 158 (1867). *Cheiranthus farsetia* L., Mant. 94 (1767). [Plate 415]

Half-shrub, canescent with appressed, forked hairs, sometimes thorny when dry, 25–60 cm. Stems many, with dense, erect-ascending branches. Leaves up to about 5 (–6) × 0.2–0.3 cm., subsessile, linear, obtuse. Racemes loose. Pedicels about half as long as calyx or shorter, elongating in fruit. Flowers conspicuous, spreading to erect. Calyx 0.9–1.3 cm.; sepals subequal at base, appressed-hairy. Petals purple-mauve or livid, often wavy-margined, much longer than calyx. Fruit 1–2.5 × 0.5–0.7 (–0.8) cm., dehiscent, oblong or ovate, more or less obtuse, with a distinct style; septum with a rather distinct midrib. Seeds (4–)6–12, in 2 rows in each cell, broadly winged. Fl. April–May.

Var. **aegyptiaca.** Fruit 1.5–2.5 cm., oblong, 2–4 times as long as broad, about 12-seeded. Flowers mauve or purple.

Hab.: Hot deserts. C. and S. Negev. Dead Sea area, Arava Valley, Edom.

Var. **ovalis** (Boiss.) Post, Fl. Syr. Pal. Sin. 80 (1883–1896). *F. ovalis* Boiss., Diagn. ser. 1, 8 : 32 (1849) et Fl. 1 : 159 (1867). Fruit 1–1.6 cm., ovate-elliptical, one and a half to twice as long as broad, about 4–8-seeded. Branches more thorny when dry. Scarcely a variety.

Hab.: As above. C. and S. Negev, Dead Sea area, Arava Valley, Moav.

Area of species : E. Sudanian, extending into the adjacent Saharo-Arabian territories.

Trib. ALYSSEAE. Ovary sessile or short-stipitate. Nectar glands mostly transversal only. Fruit a silicle, usually dorsally compressed. Hairs mostly forked or stellate.

30. RICOTIA L.

Glabrous or pilose herbs. Leaves entire, trifoliate or pinnate. Inflorescences terminal or axillary, racemose. Flowers large, pink, violet or white, pedicellate. Calyx erect, 2-saccate. Petals long-clawed; limb emarginate or obcordate, rarely entire. Stamens free, edentate. Stigmas 2-lobed or capitate. Fruit a silicle borne on a deflexed pedicel, dehiscent, dorsally strongly flattened, elliptical to oblong-ovate, 2-valved; valves thin, netted-veined; replum thin; septum very delicate, sometimes 0 and then fruit 1-celled. Seeds in 1 or 2 rows in each cell, compressed, orbicular, not winged at margin; cotyledons accumbent.

Eight species in the E. Mediterranean region.

1. Ricotia lunaria (L.) DC., Syst. 2 : 284 (1821); Boiss., Fl. 1 : 254 (1867). *Cardamine lunaria* L., Sp. Pl. 656 (1753). [Plate 416]

Annual, glabrous or sparingly hairy, 15–40 cm. Stems procumbent, often pendulous, branched. Leaves up to 10 cm., petiolate, 1–3-pinnatisect; segments petiolulate, ovate or oblong, acute, dentate-crenate. Racemes few-flowered, elongated, buds overtopped by open flowers, lowermost flower often bracteate. Pedicels long, terete. Flowers large, showy. Calyx 0.5–1 cm.; sepals erect. Petals 1–1.8 cm., long-clawed; limb pink, obcordate to 2-fid. Fruiting pedicels shorter than to as long as fruit, deflexed. Fruit 1.5–3 (–4) × 0.5–1.5 cm., 2-celled, becoming 1-celled when mature, strongly flattened, lenticular to elliptical, with a short apiculate style; valves papery, netted-veined. Seeds 1–7, in 1 row, 5–7 mm. across, compressed, orbicular. Fl. February–April.

Hab.: Batha and rocky ground. Coastal, Upper and Lower Galilee, Mt. Carmel, Mt. Gilboa, Dan Valley, Hula Plain, Upper Jordan Valley.

Area: E. Mediterranean (Palestine, Lebanon).

31. FIBIGIA Medik.

Perennial herbs or shrubs with tomentose-stellate indumentum. Stems erect, branched
or simple. Leaves simple, the lower long-petioled, most of them rosulate. Inflorescences
racemose, ebracteate or bracteate. Flowers conspicuous, pedicellate, often yellow. Sepals
erect, the inner ones sometimes saccate. Petals long-clawed; limb oblong. Stamens free,
the smaller sometimes dentate or appendiculate. Stigma capitate. Fruit a dehiscent,
2-celled, 2-valved silicle, dorsally flattened, elliptical to orbicular, hirsute. Seeds large,
in 2 rows in each cell, compressed, mostly winged; cotyledons accumbent.

About 12 species in E. Mediterranean and W. Irano-Turanian regions.

1. Fibigia clypeata (L.) Medik., Pflanzengatt. 90 (1792); Boiss., Fl. 1 : 257 (1867).
Alyssum clypeatum L., Sp. Pl. 651 (1753). *F. rostrata* (Schenk) Boiss., l.c. *Farsetia
rostrata* Schenk, Pl. Specim. Aeg. 42 (1840). [Plate 417]

Perennial, densely canescent-tomentose herb, woody at base, with stellately branched
hairs, 15–50 cm. Stems few or many, erect, simple. Radical leaves up to 10 cm., obovate, spatulate or oblanceolate, mostly entire, sometimes repand-dentate, tapering into
a long, narrow petiole; cauline leaves numerous, sessile, oblong-linear or oblanceolate.
Racemes simple or branched, bracteate or ebracteate, dense, elongating in fruit. Pedicels short, thickening. Sepals 0.6–1 cm., the inner ones saccate, narrowly membranous
at margins, covered with long, branched hairs. Petals about twice as long as calyx,
yellow; limb obovate-oblong, truncate to emarginate. Fruit about 1.2–2.5 × 0.8–1 cm.,
suberect, elliptical-oblong to ovate or obovate, sometimes somewhat contorted, obtuse,
with a 3–6 mm. long, terete, apiculate style; valves dehiscent, canescent, with short,
stellate or branched long hairs. Seeds 4–8, in 2 rows in each cell, flattened, orbicular,
with broad circular wing. Fl. February–April.

Var. **clypeata.** Racemes ebracteate or nearly so, simple. Sepals 0.8–1 cm. Fruit canescent with short, stellate hairs. Leaves entire or almost so.

Hab.: Batha and garigue, among rocks. Upper Galilee, Samaria, Shefela, Judean
Mts., Gilead, Ammon, Moav.

Var. **eriocarpa** (DC.) Thiéb., Fl. Lib.-Syr. 1 : 67 (1936). *Fibigia eriocarpa* (DC.)
Boiss., l.c. 258. *Farsetia eriocarpa* DC., Syst. 2 : 288 (1821). Sepals about 6 mm.
Fruit covered with short, stellate and velutinous, branched long hairs. Leaves broadly
sinuate-repand or coarsely dentate.

Hab.: Batha and garigue. Upper Galilee.

Area of species : E. Mediterranean, W. Irano-Turanian.

Intermediate forms between the above varieties, with smaller, almost entire leaves,
larger fruit (about 2 × 1 cm.) and less dense cover of long hairs were observed in Upper
Galilee.

32. ALYSSUM L.

Annual, perennial or suffrutescent, hairy-canescent plants, usually with dense, stellate
(rarely simple) hairs. Stems simple or branched. Leaves simple, mostly entire. Inflo-

rescences racemose, ebracteate, elongating in fruit. Flowers small, generally yellow, rarely cream-coloured or whitish. Calyx with fairly erect sepals, not saccate at base. Petals clawed; limb entire to 2-fid. Stamens free; filaments filiform or winged, dentate or edentate. Stigma slightly 2-lobed or entire. Fruit a 2-celled, dehiscent, dorsally flattened, discoid, ovoid or obovoid, ellipsoidal or lenticular, hairy, rarely glabrous silicle. Seeds 1–2 or more in each cell, more or less compressed, with or without membranous margin.

About 120 species mainly in the Mediterranean, Euro-Siberian and Irano-Turanian regions.

Literature : J. Baumgartner, Die ausdauernden Arten der Section *Eualyssum* aus der Gattung *Alyssum. Jahresb. Landes-Lehrersem. Wiener–Neustadt* 34 : 1–35 (1907); 35 : 1– 58 (1908); 36 : 1–38 (1909); *idem, Jahresb. Landes-Gymn. Baden bei Wien* 48 : 1–18 (1911). M. Zohary, Taxonomical studies in the flora of Palestine and neighbouring countries. *Palest. Journ. Bot. Jerusalem* ser. 2 : 160–162 (1941). T. R. Dudley, Studies in *Alyssum*. Near Eastern representatives and their allies, I. *Journ. Arn. Arb.* 45 : 57–95, pls. I–V (1964). T. R. Dudley, Synopsis of the genus *Alyssum. Journ. Arn. Arb.* 45 : 358–373 (1964).

1. Seeds 3–6 in each cell   2
– Seeds 1–2 in each cell   3
2. Petals yellow, almost twice as long as calyx.   **1. A. meniocoides**
– Petals white or cream-coloured, slightly longer than calyx.   **2. A. linifolium**
3(1). Fruit glabrous, with scattered tubercles at margin. Seeds narrowly winged.
                                                                              5. A. homalocarpum
– Fruit hairy   4
4. Half-shrubs with spinescent branches.   **10. A. subspinosum**
– Annuals or not spinescent perennials   5
5. Perennials, woody at base. Petals 0.5–1 cm.   6
– Tiny annuals. Petals shorter than in above   7
6. Flowers about 1 cm. Filaments winged.   **9. A. baumgartnerianum**
– Flowers shorter. Filaments wingless.   **6. A. iranicum**
7(5). Filaments winged or appendiculate   8
– Filaments without wings or appendages   9
8. Fruiting pedicels erect, appressed to rhachis. Valves densely covered with minute stellate hairs, often somewhat glabrescent at centre; style minute, one fifth the length of the fruit or less.   **8. A. marginatum**
– Fruiting pedicels spreading. Stellate hairs conspicuous, sometimes intermixed with simple or 2-armed bristles. Style longer than in above.   **7. A. minus**
9(7). Fruit orbicular; valves compressed at margin, one of them more convex than the other; style filiform-cylindrical, glabrous.   **3. A. damascenum**
– Fruit longer than broad; valves wholly and equally convex; style tapering, hairy.
                                                                               4. A. dasycarpum

Sect. MENIOCUS (Desv.) Hook. f. in Benth. et Hook f., Gen. Pl. 1 : 74 (1862). Fruit flat. Ovules 4–8 in each cell. Filaments dentate or winged. Annuals.

1. Alyssum meniocoides Boiss., Ann. Sci. Nat. Bot. ser. 2, 17 : 158 (1842) et Fl. 1 : 286 (1867). [Plate 418]

Annual, with appressed, stellate hairs, 5–15 cm. Stems erect, branching especially at base. Leaves up to 2 × 0.2 cm., sessile, linear-subulate, entire. Inflorescences racemose to corymbose, elongating in fruit. Flowers yellow, 6–8 mm. Petals twice as long as calyx, tapering into a short claw, apex retuse, truncate or obtuse. Filaments mostly with 2-lobed appendage. Fruit 4–7 mm., shorter than spreading or erect pedicel, flat, obovate-elliptical, glabrous, with a 0.5–1 mm. long style. Seeds 3–6 in each cell, wingless or very narrowly margined. Fl. March–April.

Var. **aureum** (Fenzl) Zoh., Palest. Journ. Bot. Jerusalem ser. 2 : 162 (1941). *A. aureum* (Fenzl) Boiss., Fl. 1 : 286 (1867). *Meniocus aureus* Fenzl, Pugill. 13 (1842). Differs from the typical form as follows : Plant grass-green (scarcely canescent). Petals deep golden-yellow. Flowers and leaves larger than in typical form as described above.

Hab. : Fields. Judean Mts., Ammon, Moav.

Area : W. Irano-Turanian.

2. Alyssum linifolium Steph. ex Willd., Sp. Pl. 3 : 467 (1800); Boiss., Fl. 1 : 286 (1867). [Plate 419]

Annual dwarf herb, shortly stellate-tomentose, 4–15 cm. Stems erect or ascending, branching especially at base. Leaves up to 2 × 0.25 cm., linear or narrowly oblanceolate-subulate, obtuse, entire. Inflorescences racemose to corymbose, overtopped by leaves. Flowers minute, white or cream-coloured. Petals 2–3 mm., slightly longer than or as long as calyx, oblong-obovate. Filaments with 1-toothed wings. Fruiting racemes elongated. Pedicels up to 5 (–7) mm., erect or spreading. Fruit 4–8 × 2–3.5 mm., compressed, elliptical or obovate, glabrous, with a very short, apiculate style or sessile stigma; valves nerveless. Seeds 3–6 in each cell, ellipsoidal, with or without narrow wing. Fl. March–April.

Hab. : Fields and steppes. W. and C. Negev, Ammon, Moav, Edom.

Area : Irano-Turanian, extending into the Saharo-Arabian territories.

Sect. PSILONEMA (C. A. Mey.) Hook. f. in Benth. et Hook. f., Gen. Pl. 1 : 74 (1862). Fruit with convex valves. Ovules 2 in each cell. Filaments subulate, toothless. Annuals.

3. Alyssum damascenum Boiss. et Gaill. in Boiss., Diagn. ser. 2, 6 : 18 (1859); Boiss., Fl. 1 : 285 (1867). [Plate 420]

Annual, canescent, with appressed, stellate hairs, 4–12 cm. Stems erect or ascending, usually branching from base. Leaves up to 2 cm., subsessile, oblong to lanceolate, tapering towards base, acute. Racemes dense. Flowers minute, pale yellow. Petals 2-fid. Filaments toothless. Fruiting pedicels as long as fruit or longer, erect, more or less appressed to rhachis. Fruit in cylindrical racemes, 3–5 mm. in diam., orbicular, biconvex with compressed margins; valves covered with equal stellate hairs; style filiform, about 1 mm., glabrous. Seeds 1–2 in each cell, wingless. Fl. March–April.

Hab. : Steppes and deserts. Judean Desert, Upper Galilee, Lower Jordan Valley, deserts of Gilead, Ammon, Moav, Edom.

Area : W. Irano-Turanian (Palestine-Syria).

A hygrochastic plant : fruiting pedicels spread apart when moistened.

4. Alyssum dasycarpum Steph. ex Willd., Sp. Pl. 3 : 469 (1800); Boiss., Fl. 1 : 285 (1867). [Plate 421]

Annual, canescent, with stellate-tomentose hairs, 5–15 cm. Stems erect or ascending, branching from base. Leaves up to 2 cm., obovate to oblong-lanceolate, more or less acute, tapering into a short petiole. Inflorescences dense, corymbose, not overtopped by young fruit. Petals 2.5–3 mm., white, more or less linear, 2-fid. Fruiting pedicels short, thickened, erect, not appressed to rhachis. Fruit 3–3.5 mm., in dense but elongated racemes, biconvex, more or less ellipsoidal; valves with simple and short-stellate, somewhat spreading hairs; style tapering, somewhat shorter than fruit. Seeds 2 in each cell, wingless. Fl. February–April.

Hab.: Fields and roadsides. Negev, Lower Jordan Valley, Ammon, Moav, Edom.

Area: Irano-Turanian.

5. Alyssum homalocarpum (Fisch. et Mey.) Boiss., Fl. 1 : 285 (1867). *Psilonema homalocarpum* Fisch. et Mey., Ind. Sem. Hort. Petrop. 6 : 21 (1839). [Plate 422]

Dwarf annual, canescent-tomentose with appressed, stellate hairs, up to 15 cm. Stems erect, branching slightly at base. Leaves up to 3 cm., dense, subsessile, lanceolate to oblanceolate, to spatulate below, obtuse. Inflorescences dense, elongating in fruit. Pedicels short, spreading in fruit. Flowers minute. Petals included, yellow, cuneate-subspatulate. Fruit up to 5 mm. in diam., compressed, orbicular to slightly obovate, glabrous, with scattered tubercles at margin and with a short, filiform style. Seed solitary in each cell, narrowly winged. Fl. March–April.

Hab.: Mountain steppe. Edom.

Area: W. Irano-Turanian (Sinai, Palestine, Iraq).

Sect. ALYSSUM. Fruit with convex valves. Ovules 2 in each cell. Filaments winged, dentate or appendiculate, rarely toothless. Annuals and perennials.

6. Alyssum iranicum Hausskn. ex Baumg., Jahresb. Kaiser Franz Josef Land.-Gymn. Baden 48 : 9 (1911) non Czerniak (1924). *A. dimorphosepalum* Eig, Palest. Journ. Bot. Jerusalem ser. 4 : 171 (1948). [Plate 423]

Stellate-scurfy, canescent perennial, 5–10 cm. Stems erect or ascending, leafy, branching and woody at base. Leaves up to 1.5 × 0.4 cm., dense, subsessile, linear to oblanceolate, tapering gradually at base. Inflorescences racemose to corymbose, dense, elongating in fruit, buds overtopped by open flowers. Pedicels spreading, thickened in fruit. Flowers up to 8 mm., the lowermost bracteolate. Calyx erect, sepals unequal in width, hyaline at margin, caducous. Petals pale yellow, purple-veined, long-clawed; limb obovate, entire or somewhat retuse. Filaments wingless and edentate. Fruit 5–6 × 3–4 mm., biconvex, ovate-elliptical; style filiform, somewhat shorter than fruit. Seeds 1 (–2) in each cell, wingless. Fl. March–April.

Hab.: Mountain steppe. Moav, Edom.

Area: Irano-Turanian.

7. Alyssum minus (L.) Rothm., Repert. Sp. Nov. 50 : 77 (1941). *Clypeola minor* L., Fl. Monsp. (Nathhorst, Dissert.) No. 70, 21 (1756) non Fl. Monsp. in Amoen. Acad. 4 (1759). *A. campestre* L., Sp. Pl. ed. 2, 2 : 909 (1763) p.p. [Plate 424]

Annual, canescent or densely tomentose with stellate hairs, 5–20 cm. Stems erect, simple or branched at base. Leaves up to 2.5 cm.; the radical mostly obovate, petiolate; the cauline ones oblanceolate, spatulate or tapering into a short petiole. Inflorescences dense, corymbose. Flowers small, up to 3 mm. Sepals spreading. Petals longer than calyx, whitish to cream-coloured, dentate or 2-fid. Filaments winged. Fruiting raceme elongated, rather dense. Pedicels up to 6 mm., spreading. Fruit (3–)4–6(–7) mm. in diam., biconvex but compressed at margin, orbicular, covered with stellate hairs often intermixed with simple or 2-armed longer bristles; valves veinless; style about 1 mm. Seeds 1–2 in each cell, ovoid-compressed, narrowly-winged. Fl. February–March.

Var. **minus**. *A. parviflorum* M.B., Fl. Taur.-Cauc. Suppl. 3 : 434 (1819). Hairs all stellate, usually short-armed, appressed to valves.

Hab.: Fields, batha and among rocks. Judean Mts., Judean Desert, Negev, Edom. Not very common in Palestine.

Var. **strigosum** (Banks et Sol.) Zoh. (stat. et comb. nov.). *A. strigosum* Banks et Sol. in Russ., Nat. Hist. Aleppo, ed. 2, 2 : 257 (1794). Hairs of fruit of two kinds, short, appressed, stellate and long, erect or spreading, 1–2-armed ones. Very common.

Hab.: As above. Coastal Galilee, Sharon Plain, Philistean Plain, Upper and Lower Galilee, Mt. Carmel, Esdraelon Plain, Samaria, Judean Mts., Judean Desert, Negev, Dan Valley, Hula Plain, Upper Jordan Valley, Dead Sea area, Ammon, Moav, Edom.

Area of species : Mediterranean, Irano-Turanian and Euro-Siberian.

8. Alyssum marginatum Steud. ex Boiss., Ann. Sci. Nat. Bot. ser. 2, 17 : 157 (1842); Boiss., Fl. 1 : 282 (1867). [Plate 425]

Dwarf annual, canescent, with appressed, stellate hairs, 5–15 cm. Stems simple or branching from base. Leaves up to 2.5 cm., oblong, spatulate, tapering into a short petiole, more or less obtuse, uppermost sometimes overtopping the fruiting raceme. Inflorescences dense, simple or branching, corymbose. Flowers minute. Calyx a little shorter than petals, caducous. Petals pale yellow, cuneate, retuse. Filaments with short (dentate) wings. Fruiting racemes short, ovoid to cylindrical, dense. Pedicels erect, as long as or longer than fruit, the lowest 2–3 times as long as upper ones, closely appressed to rhachis. Fruit 2–3 mm. in diam., imbricated, orbicular-ovate, biconvex with narrow compressed margin, very densely covered with short stellate-scurfy hairs, often glabrescent at centre of valve; apex somewhat truncate-retuse, with minute style. Seeds 2 in each cell, wingless. Fl. February–April.

Hab.: Steppes. Judean Desert, Negev, Arava Valley, deserts of Ammon, Moav, Edom.

Area : Irano-Turanian.

A form with larger fruit and more cylindrical fruiting racemes found in the Negev and in Edom might perhaps be considered as var. **macrocarpum** Zoh. (var. nov.). The drawing was made of the latter form.

Sect. GAMOSEPALUM (Hausskn.) Dudl., Journ. Arn. Arb. 45 : 70 (1964). Sepals dimorphic, persistent, often inflated in fruit. Long filaments winged. Cells with 2 ovules. Perennials.

9. Alyssum baumgartnerianum Bornm. in Baumg., Jahresb. Kaiser Franz Josef Land.-Gymn. Baden 48 : 16 (1911). *A. tetrastemon* Boiss. var. *latifolium* Boiss., Fl. 1 : 278 (1867).

Much branched perennial, shrubby at base, densely scurfy-stellate, canescent, 25–50 cm. Stems short, leafy. Leaves 1–2 cm., broadly oblong-elliptical, cuneate and tapering towards base. Inflorescences few-flowered, corymbose. Flowers up to 1 cm., long-pedicelled. Petals entire, sulphur-yellow, not purple-veined, with rather broad, hairy claw constricted above. Longer filaments contiguous, with overlapping wings. Fruit on erect pedicel, somewhat compressed, ovate, retuse, much longer than style, densely scurfy. Seeds 2, narrowly winged. Fl. June.

Hab. : Recorded by Boissier from mountains near Nazareth (Lower Galilee), leg. Roth. Has since not been observed in Palestine. The description was made from a specimen (Bornm. 4405) collected in Jebel Barukh (Lebanon).

Area : E. Mediterranean and W. Irano-Turanian.

Sect. ODONTARRHENA (C. A. Mey.) Koch, Syn. Fl. Germ. Helv. 59 (1835). Long filaments free, winged. Fruit convex or flat, dehiscent or indehiscent, borne on rigid divergent pedicels, with 1-seeded cells. Mostly perennials.

10. Alyssum subspinosum Dudl., Not. Roy. Bot. Gard. Edinb. 24 : 160 (1962). Half-shrub, about 20 cm., with divaricate branches covered up to middle with white-silvery, stellate hairs. Uppermost branches 1.3–2.5 cm., spreading, spinescent. Leaves of sterile branches very dense; others loose, (2–)4–6(–9) × (0.5–)1–2 mm., sessile, oblanceolate, acute. Inflorescences umbellate. Pedicels erect-spreading, 2–2.5 mm. Sepals 1.5–2 mm., elliptical or almost ovate, membranous-margined. Petals 2.5–3 mm., yellow, obovate. Longer filaments winged. Fruit (unripe) 1.5–2.5 mm., 2-seeded, compressed, elliptical, densely covered with stellate hairs. Fl. May.

Hab. : Sandstone. Edom (E. of Naqb Ishtar). Not collected by us. Description taken from Dudley, l.c.

Area : Saharo-Arabian.

Gamosepalum alyssoides (L.) Hausskn., Mitt. Thür. Bot. Ver. N.F. 11 : 75 (1897), has been observed by Père P. Mouterde in Transjordan (oral communication).

33. LOBULARIA Desv.

Annual or perennial canescent herbs, with appressed, forked hairs. Stems branched. Leaves entire. Racemes terminal. Flowers small. Sepals erect, not saccate. Petals white, entire, clawed. Stamens free with edentate filaments, sometimes filaments dilated. Style distinct; stigma capitate. Fruit a dehiscent, dorsally flattened, orbicular or ovate silicle. Seeds 1–6 in each cell, winged; cotyledons accumbent.

Five species in the Mediterranean, Macaronesian and Saharo-Arabian regions.

1. Plants 6–10 cm. Fruit 2–3 mm. in diam., orbicular, with 1–2 seeds in each cell.
2. L. arabica
– Plants 10–40 cm. Fruit 4–6 mm. long, obovate-elliptical, with 4–6 seeds in each cell.
1. L. libyca

1. Lobularia libyca (Viv.) Webb et Berth., Phyt. Canar. 1 : 90 (1837). *Lunaria libyca* Viv., Fl. Lib. Spec. 34, t. 16, f. 1 (1824). *Koniga libyca* (Viv.) R. Br. in Denh. et Clapp., Narr. Trav. Afr. App. 215 (1826); Boiss., Fl. 1 : 289 (1867). [Plate 426]

Annual or perennial canescent herb with appressed, forked hairs, 10–40 cm. Stems erect or prostrate, branched. Leaves 1–2 (–3) cm., oblanceolate, tapering into petiole, obtuse. Racemes dense, bractcatc at base, overtopped by young fruit, elongating in fruit. Flowers about 5 mm. Sepals half as long as petals. Fruiting pedicels shorter than or as long as fruit, erect-spreading, terete. Fruit 4–6 mm., obovate-elliptical, with a short almost capitate style, glabrescent, with scattered forked hairs; valves obsoletely 1-nerved, slightly convex. Seeds 4–6 in each cell, winged. Fl. February–April.

Hab.: Deserts, mainly on sandy soils. Philistean Plain, Judean Desert, W. and C. Negev, Lower Jordan Valley, Dead Sea area, Moav, Edom.

Area: Saharo-Arabian, extending slightly into some Mediterranean localities.

2. Lobularia arabica (Boiss.) Muschl., Man. Fl. Eg. 421 (1912). *Koniga arabica* Boiss., Diagn. ser. 1, 8 : 26 (1849) et Fl. 1 : 290 (1867). [Plate 427]

Prostrate annual, with forked hairs, 6–10 cm. Stems branching from base. Leaves 0.4–0.6 (–1) cm., oblong-spatulate to linear, tapering into a short petiole, obtuse. Racemes compact in flower, bracteate at base, overtopped by young fruit, elongating in fruit. Flowers minute, about 5 mm. Petals twice as long as sepals, obovate, entire, tapering into a claw. Fruiting pedicel as long as fruit, spreading horizontally. Fruit 2–3 mm., orbicular, with apiculate style, glabrescent with scattered forked hairs; valves convex, 1-nerved. Seeds 1–2 in each cell, winged. Fl. February–April.

Hab.: Deserts and sandy places. Sharon Plain, Philistean Plain, W. and C. Negev, Lower Jordan Valley.

Area: Saharo-Arabian.

34. Clypeola L.

Annual, small, stellately hairy herbs. Stems slender, simple or branched. Leaves linear to spatulate, entire. Inflorescences racemose, dense, ebracteate. Flowers minute, white to yellow. Sepals suberect or spreading, not saccate. Petals as long as or longer than sepals. Stamens free, with toothed filaments. Style long to 0. Fruit an indehiscent, dorsally compressed, discoid or ellipsoidal silicle, borne on a recurved pedicel, 1-celled, 1-seeded, with or without marginal wing. Seeds elliptical, wingless; cotyledons accumbent.

Some 8 species in the Mediterranean and the Irano-Turanian regions.

Literature: D. A. Chaytor and W. B. Turrill, The genus *Clypeola* and its intraspecific variation. *Kew Bull.* 1935 : 1–24 (1935). M. Breistroffer, Révision systématique des variations du *Clypeola jonthlaspi* L. *Candollea* 7 : 140–166 (1936). M. Breistroffer, Nouvelles

contributions à l'étude monographique du *Clypeola jonthlaspi* L. *Candollea* 10 : 241–280 (1946).
1. Fruit membranous, orbicular, with light-coloured wing at margin, glabrous or with simple appressed hairs. **1. C. jonthlaspi**
– Fruit coriaceous, wingless, with retrorsely barbed, rigid hairs 2
2. Style very short. Pedicel shorter than fruit. **2. C. aspera**
– Style long. Pedicel twice as long as fruit. **3. C. lappacea**

Sect. CLYPEOLA. Fruit membranous, with almost smooth hairs or glabrous and with entire margin.

1. Clypeola jonthlaspi L., Sp. Pl. 652 (1753); Boiss., Fl. 1 : 308 (1867). *C. microcarpa* Moris, Diar. Riun. Scienz. It. 13 : 7 (1841); Boiss., l.c. [Plate 428]

Tiny annual with appressed, stellate hairs, 5–20 cm. Stems fairly erect, simple or branched. Branches ascending. Leaves up to 2 cm., linear-spatulate or oblanceolate, tapering into a short petiole, obtuse to acute. Flowers inconspicuous. Calyx about 1 mm.; sepals persistent, narrowly membranous at margins. Petals about as long as calyx, linear, entire. Stigma sessile. Fruiting racemes elongated, rather dense. Fruiting pedicels up to 3 mm. erect-spreading but recurved at tip. Fruit 2–5 mm. in diam., discoid-orbicular, with entire, winged margin and slightly retuse apex; valves puberulent or pilose or glabrous. Seeds pendulous. Fl. February–March.

Var. **jonthlaspi**. Disk of fruit pilose, margin glabrous.

Var. **lasiocarpa** Guss., Fl. Sic. Prodr. 2 : 197 (1828). Disk of fruit pilose and puberulent, margins pilose-ciliate.

Var. **glabriuscula** Grun., Bull. Soc. Imp. Nat. Mosc. 40, 4 : 396 (1867); Busch, in Fl. URSS 8 : 366 (1939). Disk of fruit glabrous, wings with dense, short cilia.

Hab. (of species): Batha and stony ground. Philistean Plain, Upper and Lower Galilee, Esdraelon Plain, Samaria, Judean Mts., Judean Desert, C. Negev, Hula Plain, Gilead, Ammon, Moav, Edom.

Area of species: Mediterranean and W. Irano-Turanian, extending into some Euro-Siberian territories.

Sect. BERGERETIA DC., Syst. 2 : 326 (1821). Fruit leathery, denticulate or lobed at margin, with rigid, retrorsely barbed hairs.

2. Clypeola aspera (Grauer) Turrill, Journ. Bot. 60 : 269 (1922). *Peltaria aspera* Grauer, Decuria 6 (1784). *C. echinata* DC., Syst. 2 : 328 (1821); Boiss., Fl. 1 : 309 (1867). [Plate 429]

Tiny annual, stellately hairy, canescent, 5–15 cm. Stems erect, simple or branching. Leaves 2–4 × 0.5 cm., oblong-linear, more or less acute, 1-nerved. Racemes at first dense, then elongated. Flowers 1.5–2.5 mm., pedicellate. Petals slightly exceeding calyx in length, white or cream-coloured, emarginate at apex. Fruiting pedicel shorter than fruit, recurved. Fruit 2–3 mm., leathery, subglobular to ovoid (to ellipsoidal) obtuse, somewhat retuse at apex, denticulate at margin; style very short; stigma capitate; valves wholly covered with stiff, denticulate or retrorsely barbed bristles. Seeds wingless. Fl. February–March.

Hab.: Fields and steppes. Philistean Plain, Júdean Mts., Judean Desert, Gilead, Ammon, Moav, Edom.

Area: W. Irano-Turanian.

3. Clypeola lappacea Boiss., Ann. Sci. Nat. Bot. ser. 2, 17:174 (1842) et Fl. 1:310 (1867). [Plate 430]

Tiny annual, scurfy-canescent, with appressed, stellate hairs, 5–15 cm. Stems erect, branching from base, rarely simple. Leaves 1.5–3.5 × up to 0.5 cm., tapering to a narrow petiole, obtuse, 1-nerved, the radical obovate-spatulate, the upper linear. Inflorescences dense. Sepals spreading, not saccate. Petals one and a half to 3 times as long as calyx, yellow, becoming white, oblong, tapering into a short claw. Fruiting racemes elongated. Pedicels terete, up to 6 mm., deflexed, appressed to rhachis. Fruit 2–3 × 2–2.5 mm., ovoid-ellipsoidal, with a 2 mm. long, filiform, apiculate style; valves leathery, subinflated, densely covered with long, stiff, retrorsely barbed hairs, margins dentate-lobed. Seeds pendulous, wingless. Fl. March–April.

Hab.: Semisteppe. Judean Mts. Very rare.

Area: Irano-Turanian.

35. EROPHILA DC.

Annual tiny herbs. Stems scapose. Leaves rosulate, entire or toothed or lobed. Inflorescence a corymbose raceme, mostly ebracteate, elongating in fruit. Flowers white or rarely reddish. Sepals erect-spreading, equal at base. Petals obovate, emarginate to 2-fid. Stamens free, edentate. Style short. Fruit a silicle, dehiscent into 2 valves, 2-celled, dorsally compressed, linear-oblong or oblong or elliptical or obovate; valves flat or convex, with slender median branched vein. Seeds many, small, in 2 rows, ovoid-compressed, brown.

About 7 polymorphic species, mostly in W. Asia, N. Africa and Europe.

Literature: O. E. Schulz, *Erophila* DC., in Cruciferae – *Draba* et *Erophila,* in: *Pflznr.* 89 (IV. 105): 343–372 (1927). O. Winge, Taxonomic and evolutionary studies in *Erophila,* based on cytogenetic investigations. *Compt. Rend. Trav. Labor. Carls. sér. Physiol.* 23: 41–74 (1940). S. M. Walters, *Erophila* DC., in: Tutin et al., *Fl. Europaea* 1: 312–313 (1964).

The extremely variable local populations of this genus are tentatively classed under the following two species:

1. Leaves obovate-spatulate, rarely elliptical. Seeds up to 0.5 mm. long, 10–20 in each cell. Hairs simple or forked. **1. E. verna**
– Leaves narrowly linear or elliptical. Seeds 0.75–1 mm. in diam., up to 8 in each cell. Hairs simple. **2. E. minima**

1. Erophila verna (L.) Bess., Enum. Pl. Volh. 26 (1822). *Draba verna* L., Sp. Pl. 642 (1753). [Plate 431]

Low annual, pilose at base, with simple or forked hairs, 5–10(–15) cm. Stems usually many, sometimes single, slender, erect, glabrous towards apex. Leaves up to 2(–2.5) cm., petiolate, obovate-spatulate, rarely elliptical, entire or sparsely dentate, ciliate. Racemes rather loose. Pedicels filiform, elongated in fruit. Sepals 1–1.5 mm., glabrous or pilose, white-margined. Petals 2–2.5 mm., 2-fid. Fruit about 5–7(–8) × 2.5 mm., erect-spreading, obovate or elliptical to oblanceolate, rounded, with short style at apex. Seeds minute, about 0.5 mm., 10–20 in each cell. Fl. January–March.

Hab. : Batha, rocky places, stone fences. Upper and Lower Galilee, Mt. Carmel, Mt. Gilboa, Judean Mts., Judean Desert, Lower Jordan Valley, Ammon and probably elsewhere.

Area : Mediterranean, Irano-Turanian and Euro-Siberian.

Very polymorphic. Of the infraspecific taxa into which this species has been subdivided (Walters, Repert. Sp. Nov. 69 : 57, 1964), ssp. **praecox** (Stev.) Walters seems to be the most common one in Palestine. It is distinguished from other subspecific taxa mainly by its elliptical to oblanceolate, 5–6 mm. long silicles.

2. **Erophila minima** C. A. Mey., Verz. Pfl. Cauc. 184 (1831); Boiss., Fl. 1 : 303 (1867); O. E. Schulz, in Pflznr. 89 : 371 (1927). [Plate 432]

Small annual, 2–8 cm., pilose at base with simple hairs, glabrous or nearly so towards apex. Leaves up to 2 × 0.2 cm., narrowly linear or elliptical, entire or more or less dentate. Racemes rather loose, corymbose, elongating in fruit. Pedicels filiform, as long as flowers, elongated in fruit. Flowers about 2 mm. Sepals about 1 mm., glabrous or sparingly pilose. Petals longer than sepals, 2-fid, white. Fruit about 3–5 × 2–2.5 mm., compressed, ovate to obovate to broadly elliptical, obtuse with a short style at apex, somewhat netted-veined. Seeds about 0.75–1 mm., up to 8 in each cell, wingless. Fl. January–March.

Hab. : Batha and rocky places. Sharon Plain, Upper and Lower Galilee, Mt. Carmel, Esdraelon Plain and most probably in other districts.

Area : E. Mediterranean and Irano-Turanian.

Exceedingly variable in leaves.

Trib. LEPIDIEAE. Ovary sessile or short-stipitate. Nectar glands mostly transversal only. Fruit a silicle, usually laterally compressed. Hairs mostly simple.

36. CAMELINA Crantz

Annual or biennial herbs, glabrous or hispid, with simple or forked hairs. Stems erect. Radical leaves oblong, stem leaves oblong to lanceolate, sagittate-auriculate, amplexicaul. Inflorescences ebracteate. Sepals equal at base. Petals yellow or whitish, spatulate, entire. Stamens free; filaments edentate. Stigma capitate. Fruit ovoid, obovoid or globular or linear-cylindrical with keeled margins; valves indurated; style distinct. Seeds numerous, ovoid-angular, in 2, rarely in 1 row; cotyledons incumbent.

About 10 species in the Euro-Siberian, Mediterranean and Irano-Turanian regions.

1. **Camelina hispida** Boiss., Ann. Sci. Nat. Bot. ser. 2, 17 : 176 (1842) et Fl. 1 : 312 (1867). *C. persistens* Rech. f., Ark. Bot. ser. 2, 1, 5 : 304 (1949). [Plate 433]

Annual, patulous-hispid, 10–30 cm. Stems erect, branching from base. Radical leaves about 5 cm., rosulate, oblong-lanceolate, obtuse to acute at apex, repand or dentate; stem leaves gradually decreasing in length, 1–3 cm., lanceolate, clasping-sagittate, acute, almost entire. Racemes dense, elongating in fruit. Sepals 3 mm., hispid. Petals 4–5 mm., yellow. Pedicels about as long as or longer than fruit, rather thick, spreading. Fruit 5–6 mm., obovoid or globular, with turgid valves, glabrous or patulous-hairy; style shorter or longer than fruit. Fl. March–April.

Hab. : Steppes. Moav, Edom.

Area : W. Irano-Turanian.

The local specimens vary greatly in the length of the style, the indumentum of the silicle and the direction and length of the pedicels. The persistence of the calyx on which *C. persistens* Rech. f., l.c. is based seems to be inconsistent and therefore inadequate for separating Rechinger's taxon from *C. hispida*. The above findings also raise some doubts as to the specific rank of *C. lasiocarpa* Boiss. et Bl. in Boiss., Fl. l.c.

37. NESLIA Desv.

Erect annuals with branched hairs. Leaves simple, sagittate, sessile or the basal ones petiolate. Racemes ebracteate, dense in flower, elongating in fruit. Flowers small, yellow. Sepals almost erect, not saccate. Petals entire. Filaments not dilated. Fruit a short-stipitate, indehiscent, 1-celled, 1-seeded, somewhat dorsally compressed, obovoid-spherical silicle; valves reticulate; style jointed; stigma retuse. Seed solitary, pendulous, ovoid; cotyledons incumbent.

Two species in the Mediterranean, Irano-Turanian and Euro-Siberian regions.

1. Neslia apiculata Fisch., Mey. et Avé-Lall. in F. et M., Ind. Sem. Hort. Petrop. 8 : 68 (1842). *Vogelia apiculata* (Fisch., Mey. et Avé-Lall.) Vierh., Oesterr. Bot. Zeitschr. 70 : 167 (1921). *N. paniculata* (L.) Boiss., Fl. 1 : 371 (1867) non Desv., Journ. Bot. Appl. 3 : 162 (1814). [Plate 434]

Annual with spreading, branched hairs, (10–)30–60 cm. Stems erect, branching above. Branches ascending. Leaves 2–6(–8) cm., acute, entire or dentate, hirsute; radical leaves subsessile or petiolate, oblong-lanceolate, sagittate; the cauline sessile, lanceolate to linear, sagittate-auriculate. Inflorescences corymbose-paniculate, elongating in fruit. Sepals 1–2.5 mm., erect-spreading, yellowish, with membranous margin. Petals 2–3.5 mm., linear-oblong, truncate, tapering at base. Fruiting pedicels 0.8–1.3 cm., about 3 times as long as fruit, spreading. Fruit 1.5–3 mm. in diam., short-stipitate, more or less globular or broadly obovoid and somewhat compressed, reticulate, with an apiculate style. Fl. February–April.

Hab. : Fields. Acco Plain, Sharon Plain, Philistean Plain, Upper and Lower Galilee, Mt. Carmel, Judean Mts., Judean Desert, Negev, Lower Jordan Valley, Gilead, Ammon, Edom.

Area : Mediterranean and W. Irano-Turanian, extending into adjacent Euro-Siberian provinces.

38. Capsella Medik.

Annual or perennial · glabrous or pubescent herbs. Stems mostly erect, simple or branched. Leaves entire, dentate or pinnatifid, the radical rosulate, the cauline amplexicaul. Inflorescences racemose, dense, elongating in fruit. Flowers inconspicuous, white, reddish or yellowish. Sepals equal at base. Petals obovate, rounded or truncate, rarely 0. Stamens free; filaments linear, edentate. Style short; stigma capitate. Fruit a laterally compressed silicle, obcordate to obtriangular, glabrous, dehiscent into 2 boat-shaped, wingless, keeled valves. Seeds several in each cell, small, ellipsoidal; cotyledons incumbent, rarely accumbent.

Five species in temperate and subtropical regions.

1. Petals white, distinctly exceeding sepals.   1. C. bursa-pastoris
– Petals reddish, not or scarcely exceeding sepals.   2. C. rubella

1. Capsella bursa-pastoris (L.) Medik., Pflanzengatt. 85 (1792); Boiss., Fl. 1 : 340 (1867). *Thlaspi bursa-pastoris* L., Sp. Pl. 647 (1753). [Plate 435]

Annual, glabrous or sparingly hairy, 10–50 cm. Stems usually erect, simple or branching below. Radical leaves 2–16 cm., rosulate, oblanceolate (rarely linear), tapering into petiole, acute, entire or coarsely dentate to pinnately lobed, sometimes runcinate; cauline 1–7 cm., sessile, linear-oblong, amplexicaul-sagittate at base, serrate-dentate or entire. Inflorescences dense, corymbose. Flowers small, about 2 mm., usually white, sometimes overtopped by young fruits. Petals distinctly longer than sepals. Fruiting racemes elongated. Pedicels spreading to almost horizontal, terete-filiform. Fruit 4–8 mm. (rarely less), obcordate to obtriangular, varying in shape of base and apex; valves keeled; style short but distinct in the broad apical notch. Seeds up to 12 in each cell, not marginate. Fl. December–April.

Var. bursa-pastoris. Fruit mostly 4–8 mm.

Hab. : Fallow fields, roadsides and waste places. Acco Plain, Sharon Plain, Philistean Plain, Upper and Lower Galilee, Mt. Carmel, Esdraelon Plain, Shefela, Judean Mts., Judean Desert, N. Negev, Dan Valley, Hula Plain, Upper and Lower Jordan Valley, Ammon, Moav.

Var. minuta Post, Fl. Syr. Pal. Sin. Add. 4 (1883–1896). Fruit 1.5–2.5 mm. Leaves linear, smaller than in type.

Hab. : Fallow fields and roadsides. Sharon Plain, Upper Galilee, Judean Mts.
Area of species : Pluriregional.

2. Capsella rubella Reut., Compt. Rend. Soc. Hallér. 2 : 18 (1854) is distinguished from the other *Capsella* species by reddish petals, not or scarcely exceeding reddish sepals. It has been recorded from Samaria, the Judean Mts. and Edom by Rechinger f., Ark. Bot. ser. 2, 2, 5 : 346 (1952).

39. HYMENOLOBUS Nutt. ex Torr. et A. Gray

Tiny annuals, glabrous or with simple hairs. Stems simple or branching from base. Leaves spatulate, often pinnatilobed. Inflorescences small, racemose. Flowers inconspicuous. Sepals divergent, not saccate. Petals white, spatulate. Filaments filiform. Stigma sessile. Fruit a laterally compressed, ovoid or ellipsoidal silicle, rounded or truncate at apex, dehiscent into 2 boat-shaped valves. Seeds few to many in each cell, small, ovoid, not marginate; cotyledons incumbent.

Some 3 species, mainly in the Mediterranean, Irano-Turanian and Euro-Siberian regions but also in N. America, Chile, Australia.

1. Hymenolobus procumbens (L.) Nutt. ex Torr. et A. Gray, Fl. N. Amer. 1 : 117 (1838). *Lepidium procumbens* L., Sp. Pl. 643 (1753). *Capsella procumbens* (L.) Fries, Novit. Fl. Suec. ed. 2, Mant. 1 : 14 (1832). *Hutchinsia procumbens* (L.) Desv., Journ. Bot. Appl. 3 : 168 (1814). [Plate 436]

Dwarf annual, glabrous or glabrescent, 5–20 cm. Stems simple or branching below, slender. Leaves pinnately divided into few elliptical-lanceolate lobes, rarely entire; cotyledons persistent, obovate-spatulate. Inflorescences compact, elongating in fruit. Flowers minute, about 1 mm. Petals as long as or slightly longer than oblong, hairy sepals. Fruiting pedicels 2–4 times as long as fruit, spreading-ascending. Fruit 3–4 mm., ovoid or ellipsoidal, obtuse or slightly emarginate. Seeds 2–4 in each cell. Fl. January–February.

Hab. : Saline soils. Acco Plain, Lower Jordan Valley.

Area : Pluriregional; mainly Mediterranean, Irano-Turanian, Saharo-Arabian and partly Euro-Siberian; also introduced elsewhere.

40. THLASPI L.

Annual or perennial, usually glabrous herbs. Stems erect, leafy. Radical leaves petiolate; the cauline amplexicaul, entire or dentate. Inflorescences racemose, ebracteate, dense. Flowers small, white or pink or violet, rarely yellow. Sepals ascending, not saccate. Petals short-clawed, with entire or emarginate limb. Stamens free, edentate. Style short or long; stigma depressed-capitate or slightly 2-lobed. Fruit a dehiscent, laterally compressed, obcordate-orbicular, oblong or obtriangular, glabrous silicle; valves boat-shaped, keeled and winged. Seeds 1–8 in each cell; cotyledons accumbent.

About 60 species in the Mediterranean, Euro-Siberian and Irano-Turanian regions especially in high mountains. A few species also in N. and S. America.

1. Dwarf herbs. Radical leaves rosulate. Fruit 5–7 mm. in diam.   **2. T. perfoliatum**
– Higher plants, up to 60 cm. Radical leaves not rosulate. Fruit 1–1.2 cm. in diam.
                                        1. T. arvense

1. Thlaspi arvense L., Sp. Pl. 646 (1753); Boiss., Fl. 1 : 323 (1867). [Plate 437]

Annual, glabrous, 15–60 cm. Stems erect, branched. Radical leaves not rosulate; lower leaves petiolate, oblanceolate or obovate; the cauline sessile, oblong-lanceolate, amplexicaul-sagittate, all entire or sinuate-dentate. Flowering racemes short. Petals 3–4 mm., white, twice as long as sepals. Fruiting racemes elongated. Pedicels erect-spreading, up to 1 cm. Fruit 1–1.2 cm. across, suborbicular-obovate, glabrous; valves boat-shaped with broad wings narrowing towards base and forming a narrow sinus at apex; style very short. Seeds compressed, ellipsoidal, black-brown, 5–8 in each cell. Fl. March.

Hab.: Fields. Esdraelon Plain. Rare (casual).

Area: Pluriregional; north-temperate regions of the eastern hemisphere; also introduced elsewhere.

2. Thlaspi perfoliatum L., Sp. Pl. 646 (1753); Boiss., Fl. 1 : 325 (1867). [Plate 438]

Dwarf annual, glabrous, 5–20 cm. Stems ascending or erect, usually branching from base, leafy. Radical leaves up to 4 cm., rosulate, obovate or broadly spatulate, tapering into a long petiole, entire or repand-dentate; cauline leaves small, ovate-cordate, long-auriculate. Flowering racemes short, overtopped by fruits. Flowers small. Sepals 1–2 mm., linear. Petals about 3–4 mm., linear, white. Fruiting racemes elongated. Pedicels 3–6 mm., terete, filiform, spreading, or horizontal. Fruit 5–7 mm., broadly obcordate; valves with broad wings narrowing towards base and rounded at apex; style minute. Seeds usually 2 (–3–4) in each cell, yellowish-brown. Fl. January–February.

Hab.: Batha, fields and roadsides. Coastal Galilee, Acco Plain, Sharon Plain, Philistean Plain, Upper and Lower Galilee, Mt. Carmel, Esdraelon Plain, Mt. Gilboa, Samaria, Judean Mts., Judean Desert, Upper and Lower Jordan Valley, Gilead, Ammon, Moav.

Area: Mediterranean and Irano-Turanian, extending into the Euro-Siberian region.

Varies in size of flowers and fruit and in leaf dentation.

41. AETHIONEMA R. Br.

Glabrous annuals or perennials, sometimes woody at base. Leaves entire. Inflorescences racemose or spicate, ebracteolate, dense, elongating in fruit. Flowers small, purple, pink or white, rarely yellow. Calyx erect, slightly 2-saccate at base. Petals clawed; limb entire. Stamens free or connate in pairs, with or without appendages. Fruit mostly a 2-celled silicle, dehiscent into keeled valves, laterally compressed, obovate or suborbicular, cordate at base, emarginate at apex, wingless or with broad, entire or dentate wings; in some species the fruits are of two forms: dehiscent, compressed, discoid, 2-celled, with 1 to few seeds in each cell and indehiscent, urceolate, 1-celled, 1-seeded ones. Seeds ellipsoidal; cotyledons incumbent.

About 40 species, mainly in the Mediterranean and Irano-Turanian regions.

1. Rare shrubby perennials. Cells of all silicles 1-ovuled, 1-seeded.   **1. A. gileadense**
– Annuals. Cells of discoid silicles 2–3-seeded   2
2. Wings of fruit coarsely dentate; style distinctly shorter than sinus.   **2. A. carneum**
– Wings of fruit entire; style almost as long as or longer than sinus.   **3. A. heterocarpum**

1. Aethionema gileadense Post, Journ. Linn. Soc. Lond. Bot. 24 : 421 (1888) et Fl. Syr. **Pal. Sin. ed. 2, 1 : 102 (1932).**

Perennial glabrous herb with shrubby base, up to 20 cm. or more. Stems some-what branched. Lower leaves on sterile branches petiolate, blade about 5 mm., obovate with undulate margin; upper leaves about 1 cm., sessile, linear, acute. Fruiting racemes 10 cm., loose. Lower fruits about 5 mm., 2-celled, obovate-orbicular, retuse at base, emarginate at apex, wings as broad as cells, plicate, involute; upper fruits almost urceolate, crenate, later fimbriate; stigma sessile. Seeds solitary in each cell. Fl. March.

Hab.: Woods. Gilead (after Post, l.c.). Not seen by us.

Area: E. Mediterranean (endemic in Palestine).

2. Aethionema carneum (Banks et Sol.) Fedtsch., Fl. W. Tian-Shan 176 (1905) et Act. Hort. Petrop. 23: 428 (1909). *Thlaspi carneum* Banks et Sol. in Russ., Nat. Hist. Aleppo ed. 2, 2: 257 (1794). *A. cristatum* DC., Syst. 2: 560 (1821); Boiss., Fl. 1: 352 (1867). [Plate 439]

Annual, glabrous, 10–15 cm. Stems simple or branching from base. Branches spreading to erect. Leaves entire; the lower ones short-petioled, oblong-ovate; the cauline sessile, lanceolate, half-clasping. Racemes dense, elongating in fruit. Pedicels erect to ascending. Flowers minute, pink, purple to white. Sepals erect, about half as long as petals. Filaments of longer stamens broadened at base, almost connate, not dentate. Fruits of two types: dehiscent, 2-celled, flattish, with coarsely dentate wings broader than 2–3-seeded cells; and indehiscent, 1-celled, 1-seeded, urceolate ones; apex broadly emarginate, with style shorter than sinus. Fl. February–March.

Hab.: Batha and fields. Judean Mts., Edom.

Area: W. Irano-Turanian.

3. Aethionema heterocarpum J. Gay in Fisch. et Mey., Ind. Sem. Hort. Petrop. 4: 1 (1837); Boiss., Fl. 1: 352 (1867). [Plate 440]

Annual, glabrous, 10–15 cm. Stems simple or branching. Branches erect or spreading. Leaves somewhat glaucous; the lower ones 0.8–1.5 (–2) cm., short-petioled, oblong or ovate, obtuse; the cauline 0.4–1 (–1.5) cm., sessile, ovate-lanceolate, cordate and half-clasping, acute. Racemes dense, elongating in fruit. Pedicels about as long as flowers, erect, terete. Flowers small, pink, purple to white. Calyx erect, sometimes violet. Petals 2–3 mm., almost twice as long as calyx, oblong. Filaments of longer stamens often almost connate in pairs, not dentate. Fruits of two types: dehiscent, flattish, with 2, 2–3-seeded cells and broad, entire wings and indehiscent, urceolate, 1-celled, 1-seeded, with narrow, inrolled wings; apex emarginate, with style longer than sinus. Fl. February–April.

Hab.: Batha and fields. Upper and Lower Galilee, Esdraelon Plain, Judean Mts., Gilead, Ammon, Moav.

Area: W. Irano-Turanian, slightly extending into adjacent E. Mediterranean territories.

42. IBERIS L.

Annual or perennial herbs. Leaves almost entire or pinnatifid. Inflorescences racemose-corymbose, ebracteate. Calyx not saccate. Petals white or pink, the two outer larger than the inner ones. Stamens free, edentate. Style long; stigma capitate, somewhat

2-lobed. Fruit a dehiscent, 2-valved, laterally compressed, glabrous silicle; valves keeled or winged. Seeds large, solitary in each cell, often winged; cotyledons accumbent.

About 30 species, mainly in the Mediterranean region, a few also in the Euro-Siberian region.

1. Iberis odorata L., Sp. Pl. 649 (1753); Boiss., Fl. 1 : 335 (1867). [Plate 441]

Annual, sparsely papillose or short-hairy, 5–10 cm. Stems erect to ascending, branched at base. Leaves up to 3 cm., narrow, spatulate-linear, tapering at base, pinnatifid or coarsely dentate towards apex, sparingly ciliate. Racemes corymbose. Flowers small, short-pedicelled Calyx 2 mm.; sepals persistent, not saccate, becoming purple. Petals longer than calyx, obovate, clawed, inner pair somewhat shorter than outer. Fruiting racemes short, umbellate. Pedicels about 7 mm., short-hairy. Fruit about 7 mm. (including wings), ovate, broadly notched at apex, glabrous, with style shorter than 1 mm.; valves winged; wings acute at apex. Seeds compressed, narrowly margined. Fl. March–May.

Hab.: Damp patches among rocks. Judean Mts.

Area: Mediterranean, extending into adjacent Irano-Turanian territories.

A hygrochastic plant; fruiting pedicels spread apart when moistened.

43. BISCUTELLA L.

Annual or perennial, hirsute, rarely glabrous herbs. Leaves mostly rosulate, few cauline, usually simple. Inflorescences racemose, ebracteate. Flowers small. Sepals erect-spreading, equal or the inner saccate at base. Petals entire, yellow, clawed. Stamens free, edentate. Style long. Fruiting racemes elongated. Fruit laterally compressed, forming 2 discoid and winged valves, separating from replum, each containing 1 seed. Seeds orbicular, wingless; cotyledons accumbent.

About 7 species, mostly in Mediterranean and C. European countries.

In modern floras there is a trend to adopt a narrow species concept for this genus and to split former binomials into several species also on the basis of vegetative characters. The local populations have not been adequately studied as yet and are, therefore, classed under a single species.

Literature: E. Malinowski, Monographie der Gattung *Biscutella* L. *Bull. Int. Acad. Sci. Cracovie, Cl. Math. et Nat.* 1910 : 111–139 (1910).

1. Biscutella didyma L., Sp. Pl. 653 (1753). [Plate 442]

Annual, with dense, erect-spreading simple hairs, or glabrescent, 15–50 cm. Stems erect, simple or branched. Radical leaves 5–10 cm., rosulate, oblanceolate-spatulate, tapering into a petiole, more or less acute, more or less dentate-serrate with apiculate teeth; the cauline sessile, oblong-lanceolate, half-clasping, dentate-serrate to entire; all hirsute and ciliate. Racemes dense. Flowers 3–4 mm. Sepals glabrescent, equal at base. Petals obovate-oblong, truncate, tapering to a claw. Fruiting racemes elongated, generally compact. Pedicels about as long as fruit or longer, filiform. Fruit consisting of 2 compressed, orbicular valves, glabrous or short-pubescent, sometimes ciliate; valves

4–8 mm. in diam., separating from replum; style about 2 mm. Seeds compressed, more or less orbicular, wingless. Fl. February–April.

Var. **columnae** (Ten.) Hal., Consp. Fl. Gr. 1 : 105 (1900). *B. columnae* Ten., Prodr. Fl. Nap. 38 (1811–1815); Boiss., Fl. 1 : 321 (1867). Fruit short-pubescent throughout with unequal hairs.

Hab.: Batha and fallow fields. Coastal Galilee, Sharon Plain, Philistean Plain, Upper and Lower Galilee, Esdraelon Plain, Mt. Carmel, Judean Mts., Judean Desert, C. Negev, Dan Valley, Hula Plain, Upper and Lower Jordan Valley, Ammon, Edom. Common.

Var. **ciliata** (DC.) Hal., Consp. Fl. Gr. 1 : 105 (1900). *B. ciliata* DC., Ann. Mus. Hist. Nat. 18 : 297 (1811). Fruit glabrous with hairs on margins only.

Hab.: Batha, fallow fields and steppes. Acco Plain, Sharon Plain, Philistean Plain, Upper and Lower Galilee, Mt. Carmel, Judean Mts., Judean Desert, N. Negev, Gilead.

Var. **leiocarpa** (DC.) Hal., Consp. Fl. Gr. 1 : 105 (1900). *B. leiocarpa* DC., l.c. Fruit altogether glabrous, 4–6 mm. in diam.

Hab.: Batha, fallow fields, steppes and deserts. Sharon Plain, Philistean Plain, Mt. Carmel, Judean Mts., Judean Desert, N. and C. Negev, Lower Jordan Valley, Edom.

Area of species: Mediterranean and W. Irano-Turanian.

Above varieties represent the most common but, by far, not all forms of this species.

44. LEPIDIUM L.

Annual or perennial herbs or half-shrubs, with simple hairs or glabrous. Leaves entire to pinnately divided, the radical often rosulate. Inflorescences racemose, simple or paniculate, ebracteate. Flowers small, white or pink, rarely yellow. Calyx not saccate. Petals longer or sometimes shorter than sepals or extremely reduced, sometimes 0. Stamens 2, 4 or 6, free, edentate. Style short. Fruit a laterally compressed, 2-celled, 2-valved silicle, dehiscent or not; valves mostly winged, sometimes with keeled or rounded back, often forming a sinus at apex. Seeds solitary (rarely 2) in each cell, pendulous, not marginate; cotyledons mostly incumbent.

About 130 species, mostly in warm-temperate regions, almost cosmopolitan.

Literature: A. Thellung, Die Gattung *Lepidium* (L.) R. Br. Eine monographische Studie. *Mitt. Bot. Mus. Univ. Zürich* 28 : 1–340 (1907).

1. Perennial herbs. Leaves not divided or lobed. Fruit 1–2 mm. long, not winged.
   1. L. latifolium
 - Annuals with pinnatifid or otherwise lobed leaves. Fruit winged (sometimes wings very narrow) 2
2. Fruit 4–6 mm. long. Stems and fruiting pedicels erect 3
 - Fruit 2–3 mm. long 5
3. Style much projecting beyond the apical notch of silicle; silicle broadly elliptical to ovate. **2. L. spinescens**
 - Style not projecting beyond the apical notch of silicle 4
4. Rhachis of fruiting raceme spinescent; fruits appressed to rhachis and diverging when moistened. Pedicels 1–2 mm. **3. L. spinosum**

‒ Rhachis of fruiting raceme not spinescent; fruit not appressed to rhachis. Pedicels
  3–4 mm.   **5. L. sativum**
5(2). Desert prostrate plants. Fruit appressed to rhachis.   **4. L. aucheri**
‒ Non-desert erect plants. Style not exceeding apical notch; fruit on spreading pedicels.
                                                                         6. L. ruderale

Sect. LEPIDIUM. Fruit not or very slightly winged, not or slightly notched. Fruiting
pedicels slender, as long as or longer than fruit. Stamens 6.

1. Lepidium latifolium L., Sp. Pl. 644 (1753); Boiss., Fl. 1 : 359 (1867). [Plate 443]
    Perennial herb, glabrous or nearly so, tall, 40–100 cm. Stems erect, branched above,
arising from a branched rootstock. Leaves dentate-serrate to entire; the radical large,
up to 25 × 5 cm., long-petioled, spatulate to oblong-lanceolate; cauline leaves petiolate
to subsessile, lanceolate-linear, acute or apiculate. Flowering racemes numerous, forming
a dense panicle. Flowers minute, up to 1.5 mm., white. Sepals pilose, white-margined.
Stamens 6. Fruiting racemes dense. Pedicels ascending, terete-filiform, 2–4 times as
long as fruit. Fruit 1–2 mm. in diam., ellipsoidal to ovoid, terminating in a very short
style and capitate stigma; valves pilose, sometimes rather glabrous, not winged. Seeds
minute, solitary in each cell. Fl. May–July.
    Hab.: Banks of streams and ditches. Upper Galilee, Esdraelon Plain, Judean Mts.,
Hula Plain, Lower Jordan Valley, Ammon, Moav, Edom.
    Area: Euro-Siberian, Mediterranean and Irano-Turanian.
    Formerly grown as a condiment.

Sect. LEPIOCARDAMON Thell., Mitt. Bot. Mus. Zürich 28 : 57 (1907). Fruit appressed
to stem, distinctly notched and winged. Style much shorter or longer than sinus of fruit.

2. Lepidium spinescens DC., Syst. 2 : 534 (1821); Boiss., Fl. 1 : 354 (1867). [Plate 444]
    Annual, glabrous, 15–60 cm. Stems thickened, tapering towards apex, with branches
widely spreading. Radical leaves up to 12 cm., somewhat rosulate, finely 3-pinnatisect,
lobes linear, acute; cauline leaves 1–2-pinnatisect, the upper ones entire or lobed.
Flowering racemes crowded, elongating in fruit. Flowers about 2 mm., pink to violet.
Sepals pilose, half as long as petals. Fruiting racemes rigid on terminal and lateral
spreading branches tapering at apex to a spine. Fruiting pedicels shorter than fruit,
appressed to rhachis, terete, hygrochastic (i.e., diverging when moistened). Fruit 4–
6 × 2.5–4.5 mm., compressed, ovate or elliptical; valves narrowly winged, forming a
short sinus at apex; style longer than sinus. Seeds solitary in each cell. Fl. March–
May.
    Hab.: Humid fields. Philistean Plain, Upper and Lower Galilee, Esdraelon Plain,
Judean Mts., Dan Valley, Hula Plain, Upper Jordan Valley, Ammon.
    Area: E. Mediterranean, slightly extending into the Irano-Turanian region.

3. Lepidium spinosum Ard., Animadv. Bot. Spec. Alt. 2 : 34 (1763); L., Mant. Alt.
253 (1771). *L. cornutum* Sm. in Sibth. et Sm., Fl. Gr. Prodr. 2 : 6 (1813); Boiss., Fl.
1 : 354 (1867). [Plate 445]

Annual, rather glabrous, 10–40 cm. Stems erect, with short, ascending to divaricate branches, spiny at apex. Radical leaves up to 8 cm., rosulate, petiolate, 2–3-pinnatisect; middle cauline leaves sessile or short-petioled, long-linear, entire or pinnately divided into linear lobes; the uppermost linear, entire. Flowering racemes many, terminal or axillary, small, dense, elongating in fruit. Flowers about 2 mm., white. Calyx glabrous, about half as long as corolla. Fruiting racemes dense, with rhachis terminating in a short spine. Fruiting pedicels short, appressed, thickened, hygrochastic (i.e., diverging when moistened). Fruit 5–6 × 2–3 mm., compressed, ovate to obtriangular, glabrous or sometimes pilose, veined; valves with narrow wings, broadening towards apex to form a V-shaped sinus much longer than style. Seeds solitary in each cell. Fl. March–May.

Hab.: Humid fields. Upper and Lower Galilee, Dan Valley, Hula Plain.

Area: E. Mediterranean.

4. Lepidium aucheri Boiss., Ann. Sci. Nat. Bot. ser. 2, 17 : 195 (1842); Boiss., Fl. 1 : 354 (1867). [Plate 446]

Annual, subglabrous or sparsely stellate-pubescent, somewhat glaucous. Stems 5–50 cm., prostrate, with unequal branches arising from base. Radical leaves up to 5 cm., rosulate, lyrate-pinnatifid with small, remote, entire or dentate-crenate lobes; cauline leaves smaller, subsessile, oblong-spatulate, lobed or toothed to entire. Inflorescences terminal and axillary, small, dense. Flowers about 2 mm., white to cream-coloured. Fruiting pedicels appressed to rhachis, hygrochastic. Fruit about 3 mm., ovate-sub-cordate to quadrangular, subglabrous to puberulent, shortly and narrowly winged at apex, with small V-shaped notch and style shorter or longer than sinus. Seeds solitary in each cell. Fl. March–April.

Hab.: Deserts; somewhat saline ground. Deserts of Moav, Edom.

Area: W. Irano-Turanian.

5. Lepidium sativum L., Sp. Pl. 644 (1753); Boiss., Fl. 1 : 354 (1867). [Plate 447]

Annual, almost glabrous, 25–60 cm. Stems erect, branched. Radical leaves petiolate, 1–2-pinnately divided or irregularly lobed; cauline short-petioled or sessile, the upper entire, linear. Flowering racemes many, terminal and axillary, paniculate. Flowers about 2 mm., white. Calyx about half as long as petals, more or less pilose. Fruiting pedicels erect, terete, shorter than fruit. Fruit 4–6 × 3–5 mm., elliptical to orbicular, glabrous; valves distinctly winged, wings obtuse, broadening towards apex to form a deep narrow sinus scarcely shorter than or equal to style. Seeds solitary in each cell. Fl. March–May.

Hab.: Fields. Sharon Plain, Philistean Plain (casual), Judean Mts., Lower Jordan Valley.

Area: Pluriregional; in temperate and subtropical zones.

Formerly grown as a salad plant. Believed to be native to the E. Mediterranean countries.

Sect. DILEPTIUM DC., Syst. 2 : 538 (1821). Silicle either not winged or narrowly winged but then style free, not exceeding apical notch. Stamens 2–4.

6. Lepidium ruderale L., Sp. Pl. 645 (1753); Boiss., Fl. 1 : 361 (1867). [Plate 447a]

Annual or biennial, sparingly hairy or glabrous, 10–30 cm. Stems erect or ascending, divaricately branching. Radical leaves 3–7 × 1–2 cm., oblong-linear in outline, 1–2-pinnatifid; the upper linear, entire. Flowering racemes dense, later elongating. Flowers minute. Sepals up to 1 mm., oblong-elliptical. Petals often 0 or very reduced. Stamens 2. Fruit 2–2.5 mm., broadly elliptical, notched at apex, narrowly winged; style not exceeding notch. Fl. February–March.

Hab. : Philistean Plain. Rare (casual).

Area : Euro-Siberian, extending into the Irano-Turanian and Mediterranean regions.

45. Cardaria Desv.

Perennial hairy herbs. Stems erect, leafy and branched. Leaves undivided, the radical petiolate, stem leaves amplexicaul. Flowering racemes corymbose. Flowers small, white. Sepals diverging, the outer oblong, the inner ovate, not saccate. Petals somewhat unequal, white, with notched limb and long, narrow claw. Ovary ovoid, with 2–4 ovules; style elongated; stigma semiglobular. Silicle laterally compressed, ovoid, cordate at base, often broader than long, rounded or notched at apex, with keeled and inflated, indehiscent valves. Seeds solitary in each cell, pendulous, ovoid, not winged; cotyledons incumbent.

Two species in the Mediterranean, Irano-Turanian and Euro-Siberian regions; also adventive elsewhere.

1. Cardaria draba (L.) Desv., Journ. Bot. Appl. 3 : 163 (1814). *Lepidium draba* L., Sp. Pl. 645 (1753); Boiss., Fl. 1 : 356 (1867). [Plate 448]

Perennial, pubescent herb, 10–50 cm. Stems leafy, erect, arising from branching woody stock. Leaves up to 10(–12) × 3(–4.5) cm.; the radical ones more or less petiolate, spatulate or obovate-oblong, entire or dentate-repand, sometimes lyrately lobed; stem leaves sessile, spreading to erect, oblong-lanceolate to broadly elliptical or ovate, sagittate-amplexicaul, acute or obtuse, entire or dentate. Inflorescence a dense, corymbose panicle. Flowers 3–4 mm. Sepals spreading, glabrous, white-margined. Petals twice as long as sepals, white. Fruiting racemes elongated. Fruiting pedicels more or less horizontal, terete-filiform, about 3 times as long as fruit. Fruit (2–)3–4 × 3–5 mm., indehiscent, transversely ovoid-cordate, with a rather long style and capitate stigma, sometimes 1 of the 2 cells rudimentary; valves reticulate, glabrescent, somewhat turgid. Seeds 1–2, ellipsoidal, brown. Fl. March–April (–May).

Hab. : Wastes and roadsides, especially in mountainous regions. Philistean Plain, Upper Galilee, Esdraelon Plain, Judean Mts., Judean Desert, W., N. and C. Negev, Hula Plain, Upper Jordan Valley, Gilead, Ammon, Moav, Edom.

Area : Mainly Mediterranean and Irano-Turanian; adventive in other regions.

46. Coronopus Hall.

Annual or biennial herbs. Leaves alternate, often deeply pinnatisect or irregularly lobed. Inflorescences short, racemose. Flowers inconspicuous, frequently white. Calyx short;

sepals spreading, not saccate. Petals exserted, linear-oblong, entire, sometimes abortive. Stamens 2–6, free, edentate. Fruit an indehiscent, laterally compressed silicle, made of 2 almost globular, verrucose-tuberculate, 1-seeded valves; style pyramidal. Seeds solitary in each cell, globular.

About 10 species in temperate and subtropical regions.

Literature : R. Muschler, Die Gattung *Coronopus* (L.) Gaertn. *Bot. Jahrb*. 41 : 111–147 (1908).

1. Coronopus squamatus (Forssk.) Aschers., Fl. Prov. Brandenb. 1 : 62 (1860). *Lepidium squamatum* Forssk., Fl. Aeg.-Arab. 117 (1775). *Senebiera coronopus* (L.) Poir. in Lam., Encycl. 7 : 76 (1806); Boiss., Fl. 1 : 363 (1867). [Plate 449]

Annual or biennial, procumbent, mostly glabrous, 3–20 cm. Stems spreading, much branched at base. Leaves variable in size, long-petioled, deeply pinnatisect into linear-oblanceolate to oblong, entire or irregularly toothed lobes. Flowering racemes axillary, short-peduncled, dense. Flowers 2 mm. Petals about twice as long as sepals, white. Fruit 3–4 × 5 mm., very short-pedicelled, crowded, indehiscent, broadly ovoid to subreniform, narrowly constricted between the 2 subglobular, verrucose-tuberculate or deeply furrowed and crested valves; style short, apiculate. Seeds solitary in each cell, globular. Fl. February–April.

Hab. : Roadsides and damp sites drying out in summer, also on waste and trampled ground. Acco Plain, Sharon Plain, Philistean Plain, Upper and Lower Galilee, Mt. Carmel, Esdraelon Plain, Judean Mts., Hula Plain, Upper and Lower Jordan Valley, Gilead, Ammon.

Area : Mediterranean and Euro-Siberian extending into W. Irano-Turanian borderland; introduced elsewhere.

Trib. BRASSICEAE. Ovary sessile or stipitate. Lateral nectar glands generally distinct from median ones, mostly limited to inner side of the transversal stamens. Fruit a siliqua with or without seeds in the stylar part or a loment or a dehiscent or indehiscent silicle.

47. CONRINGIA Adans.

Annual, glaucous, rather glabrous herbs. Leaves entire, oblong or obovate, the radical ones tapering at base, cauline auriculate-clasping. Inflorescences racemose, ebracteate. Flowers short-pedicelled, mostly yellow, rarely white or cream-coloured. Sepals erect, the inner pair saccate. Petals with obovate, obtuse limb tapering into a claw. Stamens free, edentate. Style long; stigma capitate or 2-lobed. Fruit a dehiscent, terete to tetragonous siliqua tapering into a cylindrical, sometimes thickened, often seedless beak; valves 1–3-nerved. Seeds in 1 row in each cell, oblong, brown; cotyledons frequently incumbent.

Six species in the Euro-Siberian, Mediterranean and Saharo-Arabian regions.

1. Conringia orientalis (L.) Andrz. in DC., Syst. 2 : 508 (1821); Boiss., Fl. 1 : 210 (1867). *Brassica orientalis* L., Sp. Pl. 666 (1753). [Plate 450]

Annual or biennial, glaucous, glabrous, up to 60 cm. or more. Stems erect, simple or branched. Leaves 2–4.5 cm., the radical subsessile, obovate, the upper broadly oblong, cordate-amplexicaul, auriculate, obtuse, often mucronate, undulate at margins. Inflorescences loose, elongating in fruit. Flowers conspicuous, yellow to yellowish-green. Sepals erect, about 6 mm. Petals about 0.8–1.2 cm. Fruiting pedicels 0.4–1.2 cm., ascending, thickened. Fruit 6–12 cm., rather erect, more or less tetragonous, with a long, tapering style; stigma 2-lobed; valves distinctly 1-nerved, obsoletely keeled. Seeds 2–3 mm., dark brown, oblong, rough. Fl. March–April.

Hab.: Fields. Judean Mts., Transjordan. Rare.

Area: Mainly Mediterranean, extending into adjacent Euro-Siberian and Irano-Turanian territories.

48. PSEUDERUCARIA (Boiss.) O. E. Schulz

Annual, glaucous, glabrous, pruinose herbs. Leaves all petiolate, fleshy, pinnately divided; lobes long, terete. Racemes few-flowered, loose, ebracteate. Flowers conspicuous, pedicellate. Sepals erect, the inner saccate. Petals violet or pink, clawed. Stamens free, edentate. Style long and broad; stigma indistinctly 2-lobed. Fruit a dehiscent, 2-celled, long, linear siliqua; valves with prominent midrib. Seeds small in 1–2 rows, brown, white-margined; cotyledons various.

Some 3 species, mainly in the Saharo-Arabian region.

1. Pseuderucaria clavata (Boiss. et Reut.) O. E. Schulz, Bot. Jahrb. 54, Beibl. 119 : 54 (1916) et in Pflznr. 84 : 80 (1923). *Moricandia clavata* Boiss. et Reut. in Boiss., Diagn. ser. 2, 5 : 25 (1856) et Fl. 1 : 387 (1867). [Plate 451]

Annual, glabrous, 10–35 cm. Stems ascending, terete, branching from base. Leaves 3–9 cm., petiolate, fleshy, 1-, 2- or almost 3-pinnatisect into linear, terete, obtuse lobes; radical leaves with broader 3-dentate lobes and longer petioles. Flowers showy. Calyx 0.6–1 cm., violet. Petals 1.3–2.2 cm., pale violet or pink, long-clawed. Fruiting racemes elongated. Pedicels 3–6 mm., erect. Fruit 3.5–7.5 × 0.2 cm., erect or slightly recurved, compressed, linear, subtorulose, terminating in a conical to ovoid style with a minute stigma. Seeds minute, ovoid-ellipsoidal, compressed, in 1 row in each cell. Fl. February–April.

Hab.: Deserts. Philistean Plain, Judean Desert, S. Negev, Lower Jordan Valley, Dead Sea area, Arava Valley.

Area: E. Saharo-Arabian.

49. MORICANDIA DC.

Annual or perennial herbs or shrubs, glabrous, much branched. Leaves undivided, the lower sessile, the upper amplexicaul, often entire. Inflorescences racemose, ebracteate. Pedicels short. Flowers large. Sepals erect, the inner ones saccate. Petals pinkish-purple to white, clawed, with obovate, truncate limb. Stamens free, edentate. Fruit a dehiscent, terete, more or less tetragonous siliqua; valves keeled, 1-nerved; septum membranous. Seeds ellipsoidal, in 1 or 2 rows; cotyledons conduplicate.

Eight species in the Mediterranean and Saharo-Arabian regions.

1. Seeds in 2 rows. Cauline leaves ovate-orbicular, succulent. Flowers about 1 cm.
   2. M. sinaica
 – Seeds in 1 row. Cauline leaves oblong-oblanceolate to obovate, not succulent. Flowers
   1.5–2 cm.   **1. M. nitens**

1. Moricandia nitens (Viv.) Dur. et Barr., Prodr. Fl. Lib. 15 (1910); O. E. Schulz, in Pflznr. 84 : 70 (1923). *Hesperis nitens* Viv., Fl. Lib. Spec. 38, t. 5, f. 3 (1824). *M. dumosa* Boiss., Diagn. ser. 1, 8 : 25 (1849) et Fl. 1 : 386 (1867). [Plate 452]

Perennial, glabrous dwarf-shrub or herb with woody base, 15–40 cm. Stems erect, much branching. Radical leaves 1–5 cm., sessile or subpetiolate, obovate, narrowing towards base, obtuse or acute, entire or repand; cauline ones subamplexicaul or sessile, oblong or oblanceolate, more or less cordate, acute or obtuse, entire. Flowering racemes few-flowered, later elongating. Flowers up to 2 cm. Calyx 6–8 mm. Petals 1.5–2 cm., pale violet, often with darker veins; claw long, equal to calyx. Fruit (3–)4–7 × 0.1–0.2 cm., on thickened spreading pedicel, erect, narrowly linear to tetragonous, straight to somewhat incurved, tapering to a 3 mm. long beak with decurrent stigmas; valves with distinct midrib. Seeds 1 mm., in 1 row in each cell, brown. Fl. Mainly March–April (–November–December).

Hab. : Deserts, mostly in gravelly wadis. W., N. and S. Negev, Arava Valley.
Area : Saharo-Arabian.

2. Moricandia sinaica (Boiss.) Boiss., Fl. 1 : 386 (1867). *Brassica sinaica* Boiss., Ann. Sci. Nat. Bot. ser. 2, 17 : 85 (1842). [Plate 452a]

Perennial, glaucous, glabrous herb, woody at base, 50–70 cm. Stems erect or ascending, terete, whitish, finely striate, branching all along. Leaves succulent, 5–10 × 3–8 cm., the lower ones oblong, tapering at base, the cauline ones ovate-orbicular, deeply cordate-auriculate, clasping with broad auricles, all entire. Racemes congested, then elongating. Flowers pink, about 1 cm., borne on rather thick pedicels shorter than calyx. Calyx 5–6 mm., glabrous, with erect, oblong sepals. Petals 1 cm., with oblong-obovate limb and long-linear claw. Fruit 3–5 cm., on a 5–7 mm. long pedicel, erect, somewhat compressed, linear, tetragonous; valves prominently 1-nerved, readily dehiscing; style short, inconspicuous; stigma obscurely 2-lobed. Seeds in 2 rows in each cell, 0.5–1 mm., compressed, ovoid-globular. Fl. March–April.

Hab. : Rocky deserts. Negev. Rare.
Area : E. Saharo-Arabian.

50. DIPLOTAXIS DC.

Annual or perennial herbs, hispid with simple hairs, scabrous or sometimes glabrescent. Leaves usually lyrate-pinnatipartite, sometimes dentate to entire. Inflorescences almost ebracteate, elongating in fruit. Flowers conspicuous, yellow, white, pink or lilac. Sepals erect-spreading, the inner equal or saccate at base. Petals generally short-clawed. Stamens free, edentate. Style short; stigma more or less 2-lobed. Fruit a compressed, long, linear, dehiscent siliqua with flattened, somewhat torulose, 1-nerved valves; septum membranous. Seeds in 2 rows in each cell, ellipsoidal or ovoid; cotyledons more or less longitudinally conduplicate.

Thirty five species, mainly in the Euro-Siberian, Mediterranean and Saharo-Arabian regions.

1. Flowers yellow   2
 – Flowers white, pink or lilac   3
2. Glabrescent plants. Flowers up to 5 mm. Fruit erect-spreading. Low annuals.
                                                                                           4. D. viminea
 – Hirsute to hispid plants. Flowers 0.8–1 cm. Fruit deflexed. Biennial or perennial desert herbs.   **2. D. harra**
3(1). Radical leaves mostly lobed. Flowers up to 1.6 cm. Petals usually white, rarely pinkish. Style 2–6 mm. Mediterranean weeds.   **3. D. erucoides**
 – Radical leaves almost entire or coarsely toothed. Flowers usually larger than in above. Petals commonly violet-pink. Style shorter than 2 mm. Desert plants.   **1. D. acris**

Sect. HESPERIDIUM O. E. Schulz, in Pflznr. 70 : 150 (1919). Sepals erect, inner broader than outer. Petals abruptly long-clawed. Ovary 150- or more-ovuled. Gynophore conspicuous.

1. Diplotaxis acris (Forssk.) Boiss., Fl. 1 : 389 (1867); O. E. Schulz, in Pflznr. 70 : 150 (1919). *Hesperis acris* Forssk., Fl. Aeg.-Arab. 118 (1775). [Plate 453]

Annual, glabrescent or sparingly hairy, 15–50 cm. Stems erect, branching from base. Radical leaves 3–10 cm., obovate-oblong, tapering to a petiole, coarsely toothed or almost entire; cauline ones few, shorter and narrower, subsessile, oblong-linear, dentate. Pedicels 0.5–1 cm., patulous-hairy, elongating in fruit. Flowers showy, up to 2 cm. Calyx 5–8 mm.; sepals erect, hairy, the inner subsaccate. Petals more than twice as long as sepals, violet-pink, long-clawed; limb obovate, rather obtuse, veined. Stigma subsessile, slightly 2-lobed. Fruit 2–5 × 0.15–0.3 cm., stipitate, erect-spreading, terminating in a very short beak; valves torulose, with distinct midrib. Seeds in 2 rows, minute, ovoid, orange-brown. Fl. January–April.

Hab.: Steppes and deserts. Judean Desert, C. and S. Negev, Dead Sea area, Arava Valley, deserts of Moav, Edom.

Area : Saharo-Arabian.

Sect. CATOCARPUM DC., Syst. 2 : 629 (1821). Sepals half-spreading, oblong, the inner slightly saccate. Petals tapering into a claw. Ovary 50–150-ovuled. Gynophore conspicuous.

2. Diplotaxis harra (Forssk.) Boiss., Fl. 1 : 388 (1867); O. E. Schulz, in Pflznr. 70 : 152 (1919). *Sinapis harra* Forssk., Fl. Aeg.-Arab. 118 (1775). [Plate 454]

Biennial or perennial, hispid-pilose, rarely glabrescent, 20–60 cm. Stems often woody and branched at base. Leaves mostly radical, 3–10(–14) cm., obovate, tapering to a petiole, coarsely dentate or lobed, sometimes almost entire; cauline lanceolate. Pedicels almost as long as or shorter than flowers, usually glabrous. Flowers conspicuous, yellow. Sepals 4–7 mm., erect-spreading, hispid-pilose, the inner slightly saccate. Petals 0.8–1 cm.; limb obovate, gradually tapering to a claw. Fruit (1.5–)3–4.5 × 0.2–0.3 cm., stipitate, deflexed; valves membranous, flattened, torulose; beak very short; stigma broadly 2-lobed. Seeds in 2 rows in each cell. Fl. February–May.

Var. **harra.** Plant hairy throughout, more densely so at base.

Hab.: Steppes and deserts. Judean Desert, W., C. and S. Negev, Dead Sea area, Arava Valley, deserts of Ammon, Moav, Edom.

Var. **subglabra** (DC.) O. E. Schulz, l.c. 155. *D. harra* (Forssk.) Boiss. var. *glabra* Post, Fl. Syr. Pal. Sin. 77 (1883–1896). *D. hispida* DC. var. *subglabra* DC., Syst. 2 : 630 (1821). Stems hairy only at base. Leaves glabrescent, sometimes almost entire.

Hab.: Steppes and deserts. Judean Desert, C. Negev, Dead Sea area, Arava Valley, deserts of Moav, Edom.

Var. **brevisiliquosa** (O. Ktze.) O. E. Schulz, l.c. 155. *Brassica acris* O. Ktze. f. *brevisiliquosa* O. Ktze., Rev. Gen. 1 : 19 (1891). Fruit only 1.5–2 cm. long.

Hab.: As above. Judean Mts., Judean Desert, W. and C. Negev, Dead Sea area. Area of species : Saharo-Arabian.

The above varieties are probably not constant.

Sect. RHYNCHOCARPUM Prantl in Engl. et Prantl, Nat. Pflznfam. 3, 2 : 176 (1891) pro parte; O. E. Schulz, in Pflznr. 70 : 162 (1919). Sepals erect-spreading. Petals gradually tapering to a claw. Ovary 20–85-ovuled. Beak of fruit more or less attenuate towards apex, mostly 1–2-seeded (not in ours). Gynophore 0 or nearly so.

3. Diplotaxis erucoides (L.) DC., Syst. 2 : 631 (1821); Boiss., Fl. 1 : 388 (1867); O. E. Schulz, in Pflznr. 70 : 162 (1919). *Sinapis erucoides* L., Cent. Pl. II. 24 (1756) et Amoen. Acad. 4 : 322 (1759). [Plate 455]

Annual, 15–50 cm., with scattered, papillose, rather appressed hairs. Stems erect, branching mainly from base. Radical leaves 5–15 cm., tapering to a petiole or subsessile, oblong, pinnatipartite-lyrate or irregularly lobed, sometimes leaves almost entire; cauline leaves smaller, oblong-lanceolate to oblong-linear, subamplexicaul, pinnatifid to almost entire. Pedicels 0.5 (–2) cm., usually hairy, elongating in fruit. Sepals 4–8 mm. (rarely smaller), erect-spreading, sometimes coloured, the inner ones slightly saccate. Petals about twice as long as calyx, white, rarely pinkish, limb obovate-elliptical. Fruit (1–1.5–)2–4 × 0.1–0.25 cm., very short-stipitate at base, erect-spreading, terminating in a 2–6 mm. long style with a broad 2-lobed stigma, base of style almost as broad as fruit; valves torulose, with distinct midrib. Seeds minute in 2 rows in each cell. Fl. (November–) February–May.

Var. **erucoides.** Flowers 0.6–1.5 cm. Fruit 2–4 × 0.1–0.25 cm.; style 3–6 mm. Fruiting pedicels much shorter than fruit.

Hab.: Fields. Acco Plain, Sharon Plain, Philistean Plain, Upper and Lower Galilee, Esdraelon Plain, Judean Mts., Judean Desert, N. Negev, Upper and Lower Jordan Valley, Moav, Edom.

Var. **valentina** (Pau) O. E. Schulz, l.c. 165. *D. valentina* Pau, Not. Bot. 1 : 9 (1887).

Flowers up to 8 mm. Fruit 1–1.5 (–2) cm.; style 2 mm. Fruiting pedicels almost as long as fruit. Lower plants with smaller leaves.

Hab.: Fields. Lower Galilee, Esdraelon Plain, N. Negev, Ammon. Scarcely a variety.

Area of species: Mediterranean, slightly extending into adjacent regions.

A noxious weed on the heavy alluvial soils of the area.

Sect. ANOCARPUM DC., Syst. 2 : 630 (1821). Sepals and petals as in Sect. *Rhynchocarpum*. Ovary 16–74-ovuled. Gynophore 0. Beak of fruit slender, tapering at base, seedless.

4. Diplotaxis viminea (L.) DC., Syst. 2 : 635 (1821); Boiss., Fl. 1 : 387 (1867); O. E. Schulz, in Pflznr. 70 : 177 (1919). *Sisymbrium vimineum* L., Sp. Pl. 658 (1753). [Plate 456]

Annual, glabrescent, 5–40 cm. Stems ascending to procumbent (rarely erect), branched or simple. Leaves 2–12 cm., nearly all radical, tapering to a petiole, dentate, sinuate-lyrate or pinnatipartite; cauline leaves few, small. Pedicels shorter than flowers, elongating in fruit. Flowers 3–5 mm. Sepals erect-spreading, glabrous, the inner ones subsaccate. Petals shortly exserted, yellow. Stigma obsoletely 2-lobed. Fruit 1.5–2.7 (–3) × 0.1–0.2 cm., not stipitate, erect-spreading, subtorulose; valves more or less 1-nerved; style (beak) 0.1–1.5 mm. Seeds minute, in 2 rows (rarely in a single row). Fl. March–May.

Var. **viminea.** Slender plant with few stems, about 5–25 cm. Leaves lyrate-pinnatipartite.

Hab.: Fallow fields. Philistean Plain, Mt. Carmel, Samaria, Judean Mts., Judean Desert, Ammon, Moav.

Var. **integrifolia** Guss., Fl. Sic. Prodr. 2 : 275 (1828); Boiss., Fl. 1 : 388 (1867); O. E. Schulz, l.c. 179. Leaves long-spatulate, not lobed, with 3–6 small teeth on each side or with slightly crenulate margins. Low plants.

Hab.: Fallow fields. Sharon Plain, Philistean Plain, Upper and Lower Galilee, Mt. Carmel, Judean Mts., Judean Desert, Upper and Lower Jordan Valley, Ammon.

Var. **subcaulescens** Lojacono-Poj., Fl. Sic. 1 : 117 (1888); O. E. Schulz, l.c. 179. Plant with numerous erect stems, 20–40 cm. Leaves frequently large, 4–12 cm., tapering to a long petiole, oblong-lanceolate, pinnatipartite or undivided, irregularly dentate.

Hab.: Fallow fields. Philistean Plain, Judean Mts.

Area of species: Mediterranean with extensions into the Euro-Siberian region.

Intermediate forms occur between above varieties and the constancy of the latter has not been proved.

51. BRASSICA L.

Annual, biennial or perennial herbs, or rarely shrubs, glabrous or hairy. Roots more or less fusiform. Stems generally erect and branching. Leaves simple or pinnately lobed; the radical rosulate, the cauline alternate. Inflorescences ebracteate or lower part only bracteate, elongated. Flowers yellow, rarely white. Sepals erect or spreading, equal at base, sometimes the inner ones saccate. Petals entire, clawed. Stamens free, edentate. Ovary sometimes with a short gynophore; style long; stigma capitate or 2-lobed. Fruit a siliqua, more or less terete, tetragonous or somewhat flattened; lower part 2-celled, many-seeded, dehiscent by 2 convex, more or less torulose, 1 (–3)-nerved valves; upper part indehiscent, short, beak-shaped, sometimes 1–2-seeded. Seeds in 1 row, spherical or ovoid, wingless; cotyledons longitudinally folded.

About 50 species, mainly in the Mediterranean and Irano-Turanian regions.

Literature: O. E. Schulz, *Brassica* L., in Cruciferae-Brassiceae I, in: *Pflznr.* 70 (IV. 105): 21–84 (1919).

1. Fruit shorter than 3 cm., on short pedicel appressed to stem; beak seedless, short and narrow. Flowers yellow. **2. B. nigra**
– Fruit longer than 3 cm.; pedicels long, spreading; beak 1–2-seeded, 1–2.5 cm. long. Flowers pale yellow. **1. B. tournefortii**

Sect. BRASSICA. Sepals erect or erect-spreading. Ovary with 9–45 ovules. Fruit 1.5–10 cm. long; beak often 1–2-seeded.

1. Brassica tournefortii Gouan, Ill. Obs. Bot. 44, t. 20A (1773); Boiss., Fl. 1: 393 (1867); O. E. Schulz, in Pflznr. 70: 67 (1919). [Plate 457]

Annual, 20–75 cm., densely hispid below, glabrescent above. Stems more or less branching. Radical leaves 7–20 (–40) cm., rosulate, petiolate, lyrate-pinnatisect, terminal lobe broadly ovate, obtuse or acute, lateral lobes 4–12, oblong, obtuse to acute, all lobes dentate-crenate; cauline leaves sessile, with few lateral lobes; the uppermost small, oblong-linear, entire or denticulate; all leaves more or less hispid, especially on midrib. Inflorescences 10–20-flowered, corymbose. Flowers pale yellow. Calyx 3–4 mm., green or somewhat violet. Petals 6–8 mm., linear or oblong, obtuse, long-clawed. Fruiting racemes elongated, loose. Fruiting pedicels 1–4 cm., spreading. Fruit 3.5–6.5 × 0.2–0.3 cm., erect or ascending, forming an obtuse angle with the pedicel, torulose, glabrous, with a 1–2-seeded beak 1–2.5 cm. long, somewhat narrower than fruit. Seeds 1–2 mm. in diam., brown. Fl. January–April.

Hab.: Sandy soils. Acco Plain, Sharon Plain, Philistean Plain, W. and C. Negev, desert of Edom.

Area: E. Mediterranean and Saharo-Arabian.

The desert specimens are more densely hispid and have smaller flowers; they are probably to be classed as a separate subspecies.

Sect. MELANOSINAPIS (DC.) Boiss., Fl. 1: 390 (1867). Sepals erect-spreading. Ovary usually with 5–16 ovules. Fruit short, 0.5–3 cm., with seedless beak.

2. Brassica nigra (L.) Koch in Roehling, Deutschl. Fl. ed. 3, 4: 713 (1833); Boiss., Fl. 1: 390 (1867); O. E. Schulz, in Pflznr. 70: 75 (1919). *Sinapis nigra* L., Sp. Pl. 668 (1753). [Plate 458]

Annual herb, 0.5–1.5 (–2) m., sparingly hispid to hirsute with sparse stiff hairs, sometimes glabrous. Stems erect, branched above. Lower leaves large, up to 15–40 cm., petiolate, lyrate-pinnatisect, lateral lobes 2–6, oblong-ovate, the terminal one large, ovate, obtuse, all irregularly serrate-dentate; cauline leaves short-petioled, with hastate base and 1–2 pairs of lobes, acute; uppermost leaves petiolate or sessile, linear, entire. Inflorescences much branched, many-flowered, dense. Flowers small, yellow. Sepals 5 mm., spreading, equal at base. Petals twice as long as sepals, long-clawed; limb obovate. Fruiting racemes elongated. Fruiting pedicels 2–6 mm., erect, appressed to rhachis. Fruit 0.8–2.5 × 0.1–0.3 cm., erect, almost cylindrical, torulose, straight; beak narrower, 2–5 mm. long, seedless; stigma capitate, 2-lobed. Seeds small, 1–2 mm., globular, dark-brown. Fl. April–July.

Var. nigra. Sparingly hispid, up to 1.5 m. Uppermost leaves sometimes sessile. Pedicels up to 5 mm. Flowers ebracteate. Fruit short, 0.8–1.5 × 0.1–0.2 cm.; beak 2–3 mm. Seeds about 1 mm. in diam.

Hab.: Fields and roadsides. Acco Plain, Sharon Plain, Philistean Plain, Esdraelon Plain, Upper and Lower Galilee, Upper Jordan Valley.

Especially abundant in E. Galilee on basalt ground and in hemicryptophytic batha, where it attains a height of 2 m.

Some believe this to be the *mustard tree* of the New Testament (Math. xiii : 31–32).

Var. bracteolata (Fisch. et Mey.) Sp. ex Coss., Bull. Soc. Bot. Fr. 6 : 609 (1859); O. E. Schulz, l.c. 78. *B. bracteolata* Fisch. et Mey., Ind. Sem. Hort. Petrop. 4 : 33 (1837); Boiss., Fl. 1 : 391 (1867). Plant glabrous or nearly so. All leaves petiolate; middle leaves 2–5-lobed at base. Pedicels up to 6 mm. Lower flowers subtended by narrowly linear bracts. Fruit up to 2.5 × 0.3 cm.; beak 4–5 mm. Seeds about 2 mm. in diam. Fl. March–May.

Hab.: Irrigated fields. Sharon Plain. Rare.

Area of species: Euro-Siberian, Mediterranean and Macaronesian; adventive in Africa, Asia, Australia, America.

52. Sinapis L.

Annual, rarely perennial, mostly hispid herbs, with simple hairs. Stems erect, branching. Radical leaves petiolate, generally pinnatifid or pinnatisect or lyrate, sometimes almost entire; cauline ones more or less sessile, divided or entire. Racemes ebracteate. Sepals erect-spreading, equal at base, often yellowish. Petals usually yellow, clawed; limb obovate. Stamens free, without appendages. Stigma truncate, somewhat 2-lobed. Fruit a linear or cylindrical siliqua, dehiscent by 2 convex, distinctly 3–7-nerved valves (as distinguished from *Brassica*); beak indehiscent, containing 1–2 seeds, ensiform or conical. Seeds in 1 row, spherical, pendulous or erect; cotyledons conduplicate.

About 10 species mainly in the Mediterranean region, but also in the adjacent Saharo-Arabian and Irano-Turanian territories; some species adventive, widespread.

1. Beak conical, much shorter than valvar part of fruit; the latter 8–16-seeded.
                                                                          2. S. arvensis
– Beak ensiform or flattened, almost as long as or longer than valvar part of fruit; the latter 4–8-seeded.   **1. S. alba**

Sect. sinapis. Sect. *Leucosinapis* DC., Syst. 2 : 607 (1821). Ovary 4–8-ovuled. Beak of fruit strongly compressed, ensiform, 1–2-seeded. Seeds pendulous.

1. Sinapis alba L., Sp. Pl. 668 (1753); Boiss., Fl. 1 : 395 (1867); O. E. Schulz, in Pflznr. 70 : 129 (1919). [Plate 459]

Annual, retrorsely hispid or rarely glabrescent, 25–100 cm. Stems erect, branched. Radical leaves 5–20 cm., petiolate, lyrate-pinnatifid, terminal lobe obovate, rather obtuse, lateral lobes oblong, obtuse; upper leaves with fewer lateral lobes or more or less simple. Flowering racemes dense. Pedicels 5–8 mm. Flowers conspicuous, aromatic. Sepals 5–8 mm., spreading. Petals 0.8–1.5 cm., yellow, short-clawed; limb obovate. Fruit 2.5–4 × 0.3–0.4 cm., on a spreading pedicel, with 2–4 seeds in each cell, readily dehiscing by 2 torulose, 3–5-nerved valves, hispid with antrorse bristles; beak 1-seeded, compressed, ensiform, sparingly hirsute, from almost as long as to longer than valvar part; stigma 2-lobed. Seeds spherical, yellow, pale or dark brown. Fl. February–May.

Var. **alba.** Seeds pale brown or yellow. Fruit hirsute.

Hab. : Waste places and roadsides, especially on grey calcareous soils. Sharon Plain, Philistean Plain, Lower Galilee, Mt. Gilboa, Samaria, Judean Mts., Judean Desert, Upper and Lower Jordan Valley, Gilead, Ammon.

Var. **melanosperma** Alef., Landwirtsch. Fl. 250 (1866); O. E. Schulz, l.c. 131. Seeds dark brown, white around hilum.

Hab. : Waste places. Coastal Galilee, Upper and Lower Galilee, Mt. Carmel, Samaria, Judean Mts., N. Negev, Upper and Lower Jordan Valley, Dead Sea area, Gilead, Ammon.

Area of species : Mediterranean and Irano-Turanian; also in some parts of the Euro-Siberian region and in other regions.

Very abundant in some ruderal roadside communities; rarely segetal.

Sect. ceratosinapis DC., Syst. 2 : 607 (1821). Ovary 8–16-ovuled. Beak of fruit conical, 1–2-seeded. Seeds erect.

2. Sinapis arvensis L., Sp. Pl. 668 (1753); Boiss., Fl. 1 : 394 (1867); O. E. Schulz, in Pflznr. 70 : 119 (1919). [Plate 460]

Annual, hispid or glabrescent, (15–)25–60 cm. Stems erect, branched. Radical leaves large, 5–15 cm., petiolate, pinnatifid or pinnatisect to lyrate, sometimes more or less entire, terminal lobe obovate, irregularly dentate; upper leaves sessile, usually undivided, ovate or oblong-linear, acute, dentate. Flowering racemes dense. Pedicels 3–4 mm. Flowers yellow. Sepals 4–5 mm., more or less erect, equal at base, hirsute or glabrous. Petals 0.6–1.2 cm., clawed; limb obovate, truncate. Fruit (1.5–)2.5–4.5 × 0.2–0.3 cm., on a short, 3–7 mm., thick, erect or spreading pedicel, mostly tardily dehiscing by 2 torulose or plane, 3- or 5-nerved valves, glabrous, glabrescent or hispid with stiff retrorse hairs; beak 1–1.8 cm., 1–2-seeded, conical, tapering, mostly glabrous. Seeds dark-brown. Fl. February–May.

Var. arvensis. Plant often glabrescent. Fruit glabrous or glabrescent.

Hab.: Fields and waste places. Sharon Plain, Philistean Plain, Upper and Lower Galilee, Mt. Carmel, Esdraelon Plain, Samaria, Judean Mts., N. Negev, Hula Plain, Upper Jordan Valley, Gilead, Ammon, Moav.

Var. orientalis (L.) Koch et Ziz, Cat. Fl. Palat. 12 (1814); Boiss., Fl. 1 : 395 (1867); O. E. Schulz, l.c. 123. Fruit hispid with stiff retrorse hairs.

Hab.: As above. Philistean Plain, Upper Galilee, Mt. Carmel, Esdraelon Plain, Judean Mts., Judean Desert, Hula Plain, Upper Jordan Valley, Ammon.

Area of species: Origin probably in the Mediterranean region; secondary in the Irano-Turanian and Euro-Siberian regions; introduced to many other, chiefly temperate regions.

Forms corresponding to var. *stricta* Čelak., Prodr. Fl. Boehm. 3 : 470 (1874), to var. *divaricata* O. E. Schulz ex Thell. in Hegi, Ill. Fl. Mitteleur. 4 : 265 (1918), to var. *schkuhriana* (Reichb.) Hag., Tent. Fl. Basil. Suppl. 138 (1843) and other varieties based on fruit size and shape (see Schulz, l.c.) have been observed. These forms, however, intergrade largely and can hardly be considered as varieties.

One of the most noxious weeds in the area.

53. HIRSCHFELDIA Moench

Annual, pilose-hispid herbs, with simple hairs. Radical leaves lyrate-pinnate; upper leaves linear. Inflorescences racemose, ebracteate. Flowers small. Sepals spreading, almost equal. Petals yellow, mostly short-clawed. Fruit a short, dehiscent, linear siliqua, with an indehiscent, 1–2-seeded, conical beak, somewhat inflated at apex; valves obtuse at base and apex, 1–3-nerved. Seeds in 1 row, ovoid or oblong, pendulous, those of beak erect; cotyledons longitudinally conduplicate.

Two species in the Mediterranean and Sudanian regions.

1. Hirschfeldia incana (L.) Lagrèze-Fossat, Fl. Tarn et Garonne 19 (1847); O. E. Schulz, in Pflznr. 70 : 137 (1919). *Sinapis incana* L., Cent. Pl. I. 19 (1755). *H. adpressa* Moench, Meth. 264 (1794); Boiss., Fl. 1 : 390 (1867). [Plate 461]

Annual or biennial, more or less pilose, 20–75 cm., sometimes higher. Stems erect, branched. Leaves more or less hispid; the radical ones up to 20 cm., rosulate, petiolate, frequently lyrate-pinnatisect with terminal lobe broadly ovate, obtuse, and lateral ones more or less oblong, all lobes dentate-crenate; cauline leaves few, lyrate-pinnate with fewer lobes; uppermost entire or nearly so, linear. Racemes dense, later elongating. Flowers 6–9 mm., short-pedicelled. Sepals 4–5 mm., spreading, sparingly pilose. Petals yellow, one and a half times to twice as long as sepals, short-clawed; limb obovate. Stigma small, capitate. Fruit 1.5–2.5 × 0.1–0.15 cm., appressed to stem, linear, borne on a 2–3 mm. long, club-shaped pedicel and terminating with a 1-seeded, usually geniculate, 3–7 mm. long beak; valves more or less torulose, longitudinally nerved, hairy or glabrous. Seeds small, 3–6 in each cell, ovoid-oblong, brown. Fl. February–May.

Hab.: Fields and roadsides. Coast of Galilee, Acco Plain, Sharon Plain, Philistean Plain, Upper and Lower Galilee, Mt. Carmel, Judean Mts., Judean Desert, Negev,

Upper and Lower Jordan Valley, Gilead, Ammon, Moav, Edom. Very common.
Area: Mediterranean and Irano-Turanian.

The dominant species of the *Hirschfeldia–Carduus* community on roadsides.

There is no justification for separating var. *geniculata* (Desf.) Bonn. (in Bonn. et Barr., Cat. Pl. Tun. 22, 1896) from the typical form.

54. ERUCA Mill.

Annual or perennial herbs, with simple hairs or subglabrous. Leaves usually pinnatifid. Inflorescences racemose, few-flowered, ebracteate. Flowers large. Calyx erect; inner sepals somewhat saccate at base. Petals mostly yellow, dark-veined, clawed. Stamens free, without appendages. Style long; stigma 2-lobed. Fruit somewhat inflated, dorsally compressed, linear or oblong, terminating in a broad and long, seedless, ensiform beak; valves convex, rounded above, midrib distinct; septum hyaline. Seeds spherical or ovoid, often in 2 rows in each cell; cotyledons longitudinally conduplicate.

Five species, mainly in the Mediterranean, but also in the Irano-Turanian and Saharo-Arabian regions.

1. Eruca sativa Mill., Gard. Dict. ed. 8, no. 1 (1768); Boiss., Fl. 1 : 396 (1867); O. E. Schulz, in Pflznr. 70 : 181 (1919). [Plate 462]

Annual, hispid-hirsute, glabrescent above, 15–60 cm. Stems erect, rigid, branched. Lower leaves up to 15 (–30) cm., petiolate, the upper shorter, sessile; all lyrate, pinnatisect or pinnatipartite, rarely dentate or almost entire, terminal lobe obovate to orbicular, lateral ones ovate to oblong, pilose or subglabrous. Inflorescences much elongated in fruit. Pedicels shorter than calyx, thickened in fruit. Flowers 1.5–2 cm. Sepals about 1 cm., glabrous or somewhat pilose, sometimes coloured. Petals about twice as long as calyx, yellow or cream-coloured, mostly violet-veined, obovate, obtuse, tapering to a long-linear, 3-veined claw. Fruit usually 1.5–3.5 (incl. beak) × (0.2–)0.3–0.6 cm., erect, more or less appressed to stem; valves glabrous or setose-hairy, with distinct midrib; beak up to 1 cm. or more, flat, ensiform. Seeds 1.5–3 mm. in diam., brown or reddish-yellow. Fl. February–April.

1. Fruit more or less hairy-setose. var. **eriocarpa**
– Fruit glabrous 2
2. Fruit about 2 × 0.3 cm., beak up to 1 cm., as long as valves. Calyx more or less pilose.
var. **erysimoides**
– Fruit about 2.5 × 0.3–0.6 cm., beak shorter than valves. Calyx more or less glabrous.
var. **sativa**

Var. **sativa.** Fruit glabrous, about 2.5 × 0.3–0.6 cm., beak about half as long as valves. Petals yellowish-whitish with violet veins. Leaves pinnately divided, lyrate. Fairly tall plants.

Hab.: Ruins, fences, roadsides, on soils rich in nitrogenous matter. Sharon Plain, Philistean Plain, Upper Galilee, Samaria, Judean Mts., Judean Desert, Upper Jordan Valley, Beit Shean Valley, Lower Jordan Valley, Dead Sea area, Gilead, Moav.

Var. **eriocarpa** (Boiss.) Post, Fl. Syr. Pal. Sin. 79 (1883–1896); O. E. Schulz, l.c. 185. *E. cappadocica* Reut. var. *eriocarpa* Boiss., Fl. 1 : 396 (1867). Fruit medium-sized, with broad-based retrorse hairs.

Hab. : As above. Sharon Plain, Judean Mts., Judean Desert, Upper and Lower Jordan Valley, Dead Sea area, Arava Valley, Gilead, Ammon, Moav, Edom.

Var. **erysimoides** (Sieb. ex Spreng.) Fiori, Nuov. Fl. Anal. It. 1 : 587 (1924). *Brassica erysimoides* Sieb. ex Spreng., Syst. Veg. 2 : 912 (1825). *E. sativa* Mill. prol. *longirostris* (v. Uechtritz) Rouy, Exc. Bot. Esp. 2 : 52 (1883); O. E. Schulz, l.c. 185. Plants usually lower. Sepals mostly pilose. Fruit short, about 3 mm. broad, with up to 1 cm. long beak as long as valves.

Hab. : As above. Sharon Plain, Mt. Carmel, Beit Shean Valley, Ammon.

Area of species : Mediterranean and Irano-Turanian, extending into adjacent areas of the Saharo-Arabian and S. Euro-Siberian regions.

Formerly cultivated as a salad plant. Believed by Löw, Flora der Juden 4 : 28 (1934), to be the אלת of the Bible (2 Ki. iv : 39).

55. CARRICHTERA Adans.

Glabrous or somewhat hairy herbs. Stems leafy, branching diffusely from base. Leaves petiolate, 2–3-pinnatisect. Inflorescences racemose, ebracteate. Flowers small, short-pedicelled, ebracteate. Sepals erect, mostly equal. Petals longer than sepals, yellow, white or cream-coloured, sometimes dark purple-veined, entire, clawed. Stamens free, edentate. Ovary on a short gynophore; stigma sessile, shortly 2-lobed. Fruit a silicle, terminating with a seedless, foliaceous, broad, spoon-shaped or lingulate, 5-nerved beak; lower, seed-bearing part 2-celled, short-ellipsoidal or globular, dehiscent by 2 coriaceous, convex, 3-nerved valves. Seeds 2–4 in each cell, subglobular, brown; cotyledons longitudinally conduplicate.

One species, mainly Saharo-Arabian with extensions into adjacent regions.

1. Carrichtera annua (L.) DC., Mém. Mus. Hist. Nat. 7 : 250 (1821); O. E. Schulz, in Pflznr. 84 : 41 (1923). *Vella annua* L., Sp. Pl. 641 (1753). *C. vellae* DC., Syst. 2 : 642 (1821); Boiss., Fl. 1 : 397 (1867). [Plate 463]

Small annual, glabrous or hairy, 5–25 cm. Stems branched, especially at base. Leaves 1.5–5(–9) cm., tapering to a petiole, frequently 2- to 3-pinnatisect with narrowly linear to linear-oblong, obtuse lobes. Inflorescences rather loose, much elongated in fruit. Flowers 0.6–1 cm., pedicellate. Sepals equal, hairy or glabrous. Petals much longer than sepals, white or cream-coloured, purple-veined, long-clawed, with obovate limb. Fruit about 0.8–1 cm., on deflexed pedicel; lower part 2-celled, almost globular; valves with 3 nerves, bearing stiff bristles; beak spoon-shaped, seedless, 5-nerved, glabrous. Seeds more or less globular, somewhat compressed. Fl. March–May.

Hab. : Steppes, deserts and fallow fields. Sharon Plain, Lower Galilee, Mt. Carmel, Samaria, Judean Mts., Judean Desert, W. and C. Negev, Upper and Lower Jordan Valley, Dead Sea area, Arava Valley, Gilead, Ammon, Moav, Edom.

Area : As for the genus.

56. SAVIGNYA DC.

Annual low herbs. Stems branched. Leaves somewhat fleshy, pinnately lobed or dentate. Racemes dense, then elongating. Flowers small, long-pedicelled, sometimes opposite. Calyx erect-spreading, somewhat saccate. Petals longer than calyx, pale rose or pale violet, rarely white, entire, clawed. Stamens free. Style short; stigma depressed-capitate, or 2-lobed. Fruit a stipitate silicle, pendulous on long filiform pedicel, 2-valved and 2-celled, compressed, elliptical, tipped with an erect style. Seeds numerous, in 2 rows, compressed, with a distinct membranous wing; cotyledons conduplicate.

Two species in the Saharo-Arabian region.

1. Savignya parviflora (Del.) Webb in Parl., Giorn. Bot. It. 2 : 215 (1847); O. E. Schulz, in Pflznr. 84 : 57 (1923). *Lunaria parviflora* Del., Fl. Aeg. Ill. 19 n. 584 (1813). *S. aegyptiaca* DC., Syst. 2 : 283 (1821); Boiss., Fl. 1 : 397 (1867). [Plate 464]

Annual, glandular or hairy below, more or less glabrous above, 15–30 cm. Stems branching from base. Branches erect or ascending, rarely more or less procumbent. Leaves fleshy, the radical ones 3–5 cm., rosulate, petiolate, obovate, obtuse-truncate, crenately lobed or sometimes pinnatifid; cauline irregularly pinnatifid, lobes long, linear, obtuse; upper leaves sessile, linear, pinnatifid to entire. Racemes ebracteate. Pedicels filiform, much longer than flowers. Calyx 2–3 mm.; sepals spreading to sub-erect, caducous, pale purple. Petals 4–5 mm.; limb truncate or slightly emarginate. Fruit 0.9–1.2 × 0.5–0.8 cm., short-stipitate, compressed, on a 1.2–2.5 cm. long, filiform, horizontal or deflexed pedicel; valves yellowish-green, sometimes violet at margins; septum membranous; style 1–2 mm.; stigma 2-lobed. Seeds 0.5–1 mm., flat, disk-like, with circular broad wing. Fl. February–April.

Hab.: Steppes and deserts. W., N. and C. Negev, Arava Valley.

Area: Saharo-Arabian.

Variation in fruit size has been observed; among others there is a form classed as var. *oblonga* (Boiss.) O. E. Schulz, l.c. 59, which intergrades with the typical form and should not be maintained as a separate variety.

57. REBOUDIA Coss. et Dur.

Glabrous, pubescent-strigulose or hirsute annuals. Leaves pinnately divided. Inflorescences racemose, ebracteate, dense in flower, elongated in fruit. Pedicels short. Calyx erect, the inner sepals subsaccate at base. Petals white, pink or lilac with filiform claw. Stamens free, edentate. Stigma capitate. Fruit short, 2-jointed and tipped with a long filiform style; lower joint 2-valved, (1–)2–8(–12)-seeded, compressed or cylindrical; upper joint 1–2-seeded, rarely sterile, as broad as or broader than lower joint. Seeds subglobular to short-ellipsoidal; cotyledons conduplicate.

Two species in the Saharo-Arabian region.

1. Reboudia pinnata (Viv.) O. E. Schulz, Bot. Jahrb. 54, Beibl. 119 : 56 (1916) et in Pflznr. 70 : 144 (1919). *Raphanus pinnatus* Viv., Fl. Lib. Spec. 37 (1824). *Erucaria microcarpa* Boiss., Diagn. ser. 1, 8 : 47 (1849) et Fl. 1 : 366 (1867). [Plate 465]

Annual, glabrous, pubescent-strigulose or hirsute, 10–35(–45) cm. Stems erect, ascending or prostrate, branching from base. Leaves 3–10 cm., the lower oblong to elliptical, 2- or 3-pinnatipartite, lobes usually at more or less right angles to axis, oblong, entire or 3-partite; cauline leaves pinnatifid, lobes linear, entire. Sepals 3–4 mm., linear or oblong. Petals 0.8–1 cm., white to pink, with filiform claw as long as obovate-truncate limb. Fruiting racemes much elongated. Pedicels 2–2.5 mm., thickened, appressed to rhachis or spreading. Fruit 0.5–0.8(–1) cm., glabrous or hairy; lower joint 2-valved, (1–)2–4-seeded, compressed-cylindrical to obovate-elliptical; valves 5-nerved and often torulose, tardily or not at all dehiscent; upper joint usually 1-seeded, ovoid or subglobular, apex compressed, terminating in a 2–3 mm. long filiform style, often longer than upper joint; stigma usually capitate. Seeds about 1 mm., compressed-globular, brown. Fl. February–April.

1. Lower joint of fruit short, obovate-elliptical.   var. **absalomii**
– Lower joint of fruit cylindrical or linear   2
2. Fruit glabrous.   var. **liocarpa**
– Fruit hirsute, scabrous.   var. **pinnata**

Var. **pinnata.** Fruit hirsute, scabrous, lower joint 2–4-seeded.

Hab.: Deserts and steppes. Judean Desert, W. and N. Negev, Lower Jordan Valley, Arava Valley.

Var. **liocarpa** O. E. Schulz, l.c. 145. Fruit glabrous; lower joint 2–4-seeded.

Hab.: As above. Philistean Plain, Judean Desert, W., N. and C. Negev, Lower Jordan Valley, Dead Sea area, Moav.

Var. **absalomii** Opphr. in Opphr. et Evenari, Bull. Soc. Bot. Genève ser. 2, 31 : 258 (1941). Lower joint (1–)2-seeded, short, obovate-elliptical.

Hab.: As above. Coastal Negev, Lower Jordan Valley.

Area of species : E. Saharo-Arabian.

The majority of our plants belongs to var. *liocarpa*. Plants to be referred to var. *major* (Post) O. E. Schulz, l.c., with very long styles and to var. *boissieri* O. E. Schulz, l.c. with larger flowers, were also observed in Palestine, but these forms can scarcely be considered as varieties.

58. ERUCARIA Gaertn.

Annual, glabrous or somewhat pilose herbs. Stems erect. Leaves 1–2-pinnatifid or -pinnatisect. Inflorescences racemose, ebracteate. Sepals erect, almost equal at base. Petals pink or lilac-violet, rarely white, obovate, clawed. Stamens edentate. Fruit 2-jointed, longitudinally nerved; lower joint cylindrical, dehiscent by 2 valves, with 2–8 seeds in each of the 2 cells; upper joint indehiscent, ovoid or ensiform, rostrate, with 1 seed in each of the 1–4 transversal cells. Seeds more or less compressed, oblong-ellipsoidal to subglobular, pale brown; cotyledons conduplicate.

Sixteen species, mainly in steppes and deserts of the Saharo-Arabian and Irano-Turanian regions.

Literature : O. E. Schulz, *Erucaria* Gaertn., in Cruciferae-Brassiceae II, in : *Pflznr.* 84 (IV. 105) : 8–18 (1923).

1. Fruit curved or hook-shaped, up to 2.8 cm. long; upper joint much longer than lower. Leaves pinnatisect. Cotyledons oblong. Flowers up to 8 mm.   **3. E. uncata**
– Fruit straight, shorter, about (1–)2–2.2 cm. Flowers longer than 8 mm. Leaves frequently 2-pinnatisect. Cotyledons linear   2
2. Fruit tapering into a slender filiform style. Seeds oblong or linear; cotyledons usually straight, as long as or slightly longer than radicle. Desert plants.   **2. E. boveana**
– Fruit abruptly terminating in a subulate style. Seeds ellipsoidal-globular; cotyledons curved, usually twice as long as radicle. Non-desert plant.   **1. E. hispanica**

Sect. ERUCARIA. Sect. *Euerucaria* O. E. Schulz, in Pflznr. 84 : 9 (1923). Fruit straight; upper joint obconical or obovoid, tipped with a filiform style. Cotyledons linear.

1. Erucaria hispanica (L.) Druce, Repert. Bot. 3 : 418 (1914). *Sinapis hispanica* L., Sp. Pl. 669 (1753). *E. myagroides* (L.) Hal., Consp. Fl. Gr. 1 : 123 (1900); O. E. Schulz, in Pflznr. 84 : 9 (1923). *Bunias myagroides* L., Mant. 96 (1767). *E. aleppica* Gaertn., Fruct. 2 : 298 (1791); Boiss., Fl. 1 : 365 (1867). [Plate 466]

Annual, glabrous or somewhat puberulent, 20–80 cm. Stems branched. Leaves variable in size and shape, up to 12 (–16) cm., petiolate; lower leaves 2-pinnatisect into linear lobes, sometimes pinnatifid into oblong-obovate, lobed or dentate segments; upper leaves more or less pinnatilobed with narrow, dentate or entire lobes. Racemes elongated in fruit. Pedicels distinctly shorter than calyx. Flowers 0.9–1.5 cm. Sepals subequal at base, sparingly pilose, sometimes coloured. Petals twice as long as calyx, pink to lilac, rarely white, long-clawed. Fruit 2-jointed; lower joint 1–1.7 × 0.2–0.3 cm., frequently as long as or longer than upper, dehiscent, somewhat compressed-cylindrical, obtuse, glabrous, longitudinally nerved with (1–)2–4 seeds in each cell; the upper one 6 mm. or more long, indehiscent, 1–3-seeded, often torulose. Seeds compressed, ellipsoidal-globular; cotyledons curved, twice as long as radicle. Fl. March–June.

The range of variability of this species is only partly reflected in the following varieties whose constancy is still to be verified.

1. Lower leaves 1-pinnatisect; lobes dentate or slightly lobulate.   var. **tanacetifolia**
– Lower leaves 2-pinnatisect   2
2. Fruit more or less spreading.   var. **patula**
– Fruit appressed or nearly so   3
3. Upper joint 2–3-seeded, torulose.   var. **polysperma**
– Upper joint 1 (–2–3)-seeded.   var. **hispanica**

Var. **hispanica**. *E. myagroides* (L.) Hal. var. *tel-avivensis* Zoh., Palest. Journ. Bot. Jerusalem ser. 2 : 162 (1941). Leaves with linear lobes. Fruit erect, appressed; upper joint 1 (–2–3)-seeded, shorter than or about as long as lower joint.

Hab. : Fallow fields and roadsides. Sharon Plain, Philistean Plain, Upper and Lower Galilee, Mt. Carmel, Esdraelon Plain, Mt. Gilboa, Samaria, Judean Mts., N. Negev, Upper Jordan Valley.

Var. **patula** Zoh. (comb. nov.). *E. myagroides* (L.) var. *patula* Zoh., l.c. Fruit more or less spreading; upper joint keeled.

Hab.: As above. N. Negev.

Var. **polysperma** (Boiss.) Zoh. (comb. nov.). *E. myagroides* (L.) Hal. var. *polysperma* (Boiss.) O. E. Schulz, l.c. 10. *E. aleppica* Gaertn. var. *polysperma* Boiss., Fl. 1 : 366 (1867). Upper joint of fruit as long as or longer than the lower, 2–3-seeded, more or less torulose; lower joint 2–3-seeded.

Hab.: As above. Sharon Plain, Philistean Plain, Esdraelon Plain, N. Negev.

Var. **tanacetifolia** Opphr. in Opphr. et Evenari, Bull. Soc. Bot. Genève ser. 2, 31 : 257 (1941). Leaves mostly 1-pinnatisect, lobes dentate-incised or coarsely dentate in distal part.

Hab.: As above. Samaria.

Area of species : Mainly E. Mediterranean.

2. Erucaria boveana Coss., Ill. Fl. Atl. 1 : 45 (1884); O. E. Schulz, in Pflznr. 84 : 11 (1923). *E. lineariloba* Post, Fl. Syr. Pal. Sin. 104 (1883–1896). *Didesmus rostratus* Boiss., Fl. Suppl. 67 (1888). [Plate 467]

Annual, subglabrous to puberulent, (15–)20–50 cm. Stems branched. Radical leaves up to 15 cm., variable in size and shape, petiolate, 2–3-pinnatisect into ovate to narrowly linear, acute or obtuse, almost entire or dentate lobes; cauline leaves shorter, pinnatifid into narrower lobes. Racemes short, much elongating in fruit. Pedicels shorter than calyx. Flowers 0.8–1.2 cm. Calyx slightly 2-saccate. Petals almost twice as long as calyx, pink-violet to nearly white. Fruit 1–2 cm., erect or sometimes spreading, glabrous or puberulent; lower joint often (2–)4–6-seeded, 1–2 mm. broad, subcylindrical; upper joint 1–2(–3)-seeded, frequently somewhat shorter and wider than lower joint, sometimes winged, tapering into a slender filiform style. Seeds linear-oblong; cotyledons usually more or less straight, as long as or slightly longer than radicle. Fl. February–April.

Very polymorphic. The following division into varieties is tentative only. The varieties are recorded here to illustrate the array of forms included in this species. Their constancy is still to be confirmed experimentally.

Variability in fruit :

1. Fruit (mainly upper joint) manifestly winged 2
– Fruit wingless 4
2. Fruit pubescent. .... var. **maris-mortui**
– Fruit glabrous 3
3. Fruit spreading. .... var. **patula**
– Fruit erect. .... var. **alata**
4(1). Upper fruits spreading. .... var. **horizontalis**
– Fruit erect 5
5. Upper joint longer than lower. .... var. **torulosa**
– Upper joint as long as or shorter than lower 6
6. Fruit puberulent. .... var. **puberula**
– Fruit glabrous. .... var. **boveana**

Var. **boveana.** Fruit more or less erect, wingless, glabrous, with upper joint as long as or shorter than lower.

Hab.: Steppes and fields. Judean Desert, W., N. and C. Negev, Upper Jordan Valley, Dead Sea area, Arava Valley, Moav, Edom.

Var. **patula** Zoh., Palest. Journ. Bot. Jerusalem ser. 2:162 (1941). Fruit more or less spreading, winged or prominently keeled, glabrous; upper joint as long as or shorter than lower.

Hab.: As above. Judean Desert, Lower Jordan Valley, Dead Sea area.

Var. **maris-mortui** Zoh., l.c. Fruit frequently spreading, winged or dentate, pubescent; upper joint as long as or shorter than lower.

Hab.: Fields, steppes and deserts. Judean Desert, Lower Jordan Valley, Dead Sea area.

Var. **alata** Zoh., l.c. 163. Fruit erect, more or less winged, glabrous; upper joint as long as or longer than lower.

Hab.: As above. Judean Desert, Negev, Lower Jordan Valley, Dead Sea area, Edom.

Var. **torulosa** Zoh., l.c. 163. Fruit up to 2 cm. (with style), erect, wingless, glabrous; upper joint mostly longer than lower, often 2-seeded, more or less torulose; lower joint sometimes seedless.

Hab.: Steppes and fields. C. Negev, Dead Sea area, Moav, Edom.

Var. **puberula** (Boiss.) O. E. Schulz, l.c. 12. *E. aleppica* Gaertn. var. *puberula* Boiss., Fl. 1:365 (1867). Whole plant, including pedicels and fruit, puberulent. Fruit erect with upper joint as long as or shorter than lower.

Hab.: Steppes and deserts. Judean Desert, Lower Jordan Valley.

Var. **horizontalis** (Post) O. E. Schulz, l.c. 12. *E. aleppica* Gaertn. var. *horizontalis* Post, l.c. Fruit glabrous, not winged, with upper joint 2–3-seeded, as long as or shorter than lower; upper fruits spreading to recurved.

Hab.: Steppes, fields and deserts. Judean Desert, Lower Jordan Valley, Dead Sea area, Moav.

Variability in leaves:

Var. **subintegrifolia** (Bornm.) O. E. Schulz, l.c. 12. *E. aleppica* Gaertn. var. *subintegrifolia* Bornm., Verh. Zool.-Bot. Ges. Wien 48:556 (1898) and ssp. *latifolia* Bornm., Mitt. Thür. Bot. Ver. N.F. 30:74 (1913). Lower leaves with large, oblanceolate, coarsely dentate terminal lobe and with lateral narrow, subentire lobes; upper leaves almost entire, oblanceolate.

Hab.: Deserts and fields. Judean Desert, Lower Jordan Valley.

Var. **lyrata** O. E. Schulz, l.c. 12. Leaves lyrate-pinnatifid, terminal lobe obovate, dentate, the lateral small, denticulate.

Hab.: Fields. Judean Mts.

Area of species: Mainly E. Saharo-Arabian, extending into Irano-Turanian territories.

A critical study of *Erucaria hispanica* and *E. boveana* is called for. Schulz's (l.c.) species delimitation and his principles of infraspecific subdivision have been adopted here with some hesitation. Although the geographical ranges of the two species are fairly distinct, the shape and size of the cotyledons are almost the only reliable characters to distinguish between them.

The above species of *Erucaria* are prominent in the landscape. This is especially true for *E. boveana* which lends the "blossoming desert" of rainy years its particular colouring.

Sect. II. ⏻NIA (Boiss.) O. E. Schulz, in Pflznr. 84·13 (1923). Lower joint of fruit straight, upper hooked, tapering into a conical style. Cotyledons oblong, rarely linear, mostly straight, rarely half-circular.

3. Erucaria uncata (Boiss.) Aschers. et Schweinf., Ill. Fl. Eg. 40 (1887); O. E. Schulz, in Pflznr. 84 : 15 (1923). *Hussonia uncata* Boiss., Diagn. ser. 1, 8 : 47 (1849) et Fl. 1 : 367 (1867). [Plate 468]

Annual, subglabrous or slightly puberulent, 15–50 cm. Stems branching below. Branches somewhat flexuous, more or less ascending. Leaves up to 9 (–12) cm., pinnatisect into narrowly linear to oblong, entire, dentate or pinnatifid lobes. Racemes dense and short, elongating in fruit. Pedicels short, pilose. Flowers 5–8 mm. Sepals erect, subequal, sparingly pilose, sometimes coloured. Petals almost twice as long as calyx, white to pink. Fruit 1.5–2.8 × about 0.1 cm., erect to reflexed, terete, recurved or hooked towards apex, glabrous or puberulent, upper joint 2–4-seeded, attenuate, 3–5 times as long as the 2–4-seeded lower one. Seeds small; cotyledons curved. Fl. April.

Var. uncata. Fruit glabrous.
  Hab. : Sandy deserts. W., N. and C. Negev.

Var. dasycarpa O. E. Schulz, l.c. 16. Upper joint of fruit puberulent.
  Hab. : Sandy deserts. W. Negev.
  Area of species : Saharo-Arabian.

59. CAKILE Mill.

Annual glabrous herbs. Leaves succulent, pinnatifid or undivided. Inflorescences racemose. Flowers conspicuous, pedicellate, ebracteate. Calyx erect, slightly 2-gibbous at base. Petals white to pink or violet, short-clawed. Stamens free, without appendages. Style 0; stigma capitate. Fruit indehiscent, more or less tetragonous, consisting of 2 unequal 1–2-seeded joints, the upper somewhat compressed, ovoid-ensiform, separating when mature from the persistent lower one which is shorter, compressed, and often provided with lateral protuberances. Seeds rather large; cotyledons accumbent.

Fifteen species on sea shores of many regions both in the Old and New World.

1. Cakile maritima Scop., Fl. Carn. ed. 2, 2 : 35 (1772); Boiss., Fl. 1 : 365 (1867); O. E. Schulz, in Pflznr. 84 : 19 (1923); E. G. Pobedimova, in Novit. Syst. Pl. Vasc. 116 (1964). [Plate 469]

Annual, succulent, 15–50 cm. Leaves 5–10 cm., petiolate, glabrous, pinnatifid to pinnatisect with more or less linear and obtuse lobes, or almost entire or sinuate-dentate. Inflorescences ebracteate. Flowers 0.7–1.2 cm., lilac to white. Petals twice as long as calyx, clawed; limb obovate. Fruit 1.4–2.5 × 0.4–0.6 cm., on short, stout and thick pedicels, tetragonous, 2-jointed; lower joint mostly top-shaped, with prominent lateral projections above; upper joint distinctly longer than lower, ensiform, ovoid or lanceolate. Seeds solitary in each joint, yellowish-brown, smooth. Fl. March–June.

Hab.: Sandy and shingly sea-shores and further inland on mobile dunes. Coastal Galilee, Acco Plain, Sharon Plain, Philistean Plain.

Area: Mainly Mediterranean and Euro-Siberian.

60. Rapistrum Crantz

Annual or perennial, pubescent or more or less hispid herbs. Leaves pinnately divided or more or less entire. Racemes ebracteate, often paniculate. Flowers conspicuous. Sepals erect-spreading, slightly saccate. Petals yellow, short-clawed. Stamens free, not appendiculate. Style long; stigma capitate or slightly 2-lobed. Fruit a silicle, strongly rugose and ribbed, consisting of 2 indehiscent joints, the lower narrow, cylindrical to obconical, with 1–4 pendulous seeds, the upper globular to pyriform, beaked, with 1 erect seed. Seeds ovoid to spherical; cotyledons conduplicate.

Some 3 species in the Euro-Siberian, Mediterranean and Irano-Turanian regions.

1. Rapistrum rugosum (L.) All., Fl. Ped. 1 : 257 (1785); Boiss., Fl. 1 : 404 (1867); O. E. Schulz, in Pflznr. 70 : 254 (1919). *Myagrum rugosum* L., Sp. Pl. 640 (1753). [Plate 470]

Annual, hispid below, sometimes glabrous, smooth and glaucous above, 30–60 cm. Stems erect, branched. Lower leaves up to 8 cm., petiolate, oblong, lyrate-pinnatifid to subentire; cauline leaves 2–5 cm. or longer, petiolate or subsessile, oblong-linear. Racemes branched, elongating in fruit. Flowers pedicellate. Sepals about 2–4 mm., linear, the inner slightly saccate. Petals 0.6–1 cm., yellow, obovate-cuneate. Fruiting pedicels 2–4 mm., shorter than or as long as lower joint of fruit, erect, somewhat thickened. Fruit 0.7–1.2 cm., hirsute or glabrous; lower joint oblong, cylindrical or narrowly obconical, 2-valved; upper joint ovoid or depressed-globular, longitudinally ribbed or wrinkled, with persistent style, sometimes about as long as upper joint. Seeds ovoid or spherical. Fl. February–May.

Var. **rugosum**. Fruit hirsute or hispid, rarely glabrous, lower joint narrowly obconical, upper joint ovoid-subglobular, longitudinally ribbed or almost smooth, tapering into a long style.

Hab.: Fields and roadsides. Sharon Plain, Philistean Plain, Upper and Lower Galilee, Mt. Carmel, Esdraelon Plain, Samaria, Judean Mts., Hula Plain, Upper Jordan Valley.

Var. **orientale** (L.) Arcang., Comp. Fl. It. 49 (1882); Coss., Comp. Fl. Atl. 2 : 314 (1887); Rech. f., Fl. Aegaea 235 (1943). *R. rugosum* (L.) All. ssp. *orientale* (L.) Rouy et Fouc., Fl. Fr. 2 : 74 (1895). *R. orientale* (L.) Crantz, Class. Cruc. Emend. 106 (1769); Boiss., Fl. 1 : 404 (1867). *Myagrum orientale* L., Sp. Pl. 640 (1753). Fruit glabrous, rarely hirsute; lower joint obconical or short-cylindrical; upper joint larger, depressed, globular-truncate, coarsely rugose, abruptly constricted into a style shorter or longer than upper joint.

Hab.: As above. Acco Plain, Sharon Plain, Philistean Plain, Upper Galilee, Esdraelon Plain, Samaria, Judean Mts., Upper Jordan Valley.

Among the forms observed are plants with densely hirsute fruit, corresponding to var. *hispidum* (Godr.) Coss., Comp. Fl. Atl. 2 : 315 (1887) and subvar. *hirtisiliquum* Thell. in Hegi, Ill. Fl. Mitteleur. 4 : 294 (1918); glabrous plants corresponding to var. *venosum* (Pers.) DC., Syst. 2 : 432 (1821); and others with more or less entire leaves corresponding to var. *indivisum* Vis. et Sacc. (all cited in Schulz, l.c.). The constancy of all these forms is still in need of verification, and also the two varieties recorded above are only tentatively accepted for the local subdivision of the species.

Area of species : Mediterranean and W. Irano-Turanian, extending towards Euro-Siberian provinces.

61. CRAMBE L.

Annual or perennial herbs, glabrous or hairy. Lower leaves large, pinnately divided or lobed, rarely dentate or entire; upper leaves much smaller. Inflorescences ebracteate, racemose, often paniculate. Flowers small, long-pedicelled. Sepals erect-spreading, scarcely subsaccate. Petals longer than sepals, mostly white, short-clawed, sometimes very minute or 0. Inner pair of stamens mostly with appendages. Stigma capitate, mostly sessile. Fruit an indehiscent, 2-jointed silicle; lower joint minute, hardly discernible, rarely 1-seeded, usually sterile; upper joint 1-seeded, globular, wrinkled (4-ribbed) cr smooth, readily separating from lower one. Seeds spherical, rugose or smooth; cotyledons longitudinally conduplicate.

About 20 species, mainly in the Mediterranean, Euro-Siberian and Irano-Turanian regions.

| 1. Perennial herbs. Upper joint of fruit 4-ribbed. | **1. C. orientalis** |
| – Annual herbs. Upper joint of fruit smooth. | **2. C. hispanica** |

Sect. CRAMBE. Sect. *Sarcocrambe* DC., Syst. 2 : 651 (1821). Lower joint of fruit depressed, ovoid. Perennials.

1. Crambe orientalis L., Sp. Pl. 671 (1753); Boiss., Fl. 1 : 407 (1867); O. E. Schulz, in Pflznr. 70 : 237 (1919). *C. orientalis* L. var. *aucheri* Boiss., l.c. [Plate 471]

Perennial, 60–100 cm. Stems many, erect, profusely branching, pilose at base, glabrous above. Lower leaves 15–20 cm., long-petioled, obovate in outline, lyrate, pinnatipartite to pinnatisect, with dentate, more or less pilose lobes, terminal lobe obovate, lateral lobes oblong-ovate; middle leaves about 10 cm., short-petioled, oblong-elliptical, cuneate at base, acute, entire; upper leaves 1–1.5 cm., linear, entire, glabrous. Racemes dense, elongating in fruit. Pedicels 3–5 mm., filiform, erect or recurved. Sepals about 2.5 mm., oblong, obtuse. Petals 3–4 mm., white to cream-coloured; limb obovate.

Fruit on a 0.5–1 cm. long, erect-spreading pedicel; lower joint ovoid, sterile; upper joint 2.5–3.5 mm. in diam., globular, 4-ribbed. Seeds solitary, spherical, brown. Fl. March–May.

Hab. : Steppes. Gilead, Moav, Ammon.

Area : W. Irano-Turanian.

A tumble weed.

Sect. LEPTOCRAMBE DC., Syst. 2 : 655 (1821). Lower joint of fruit cylindrical. Annuals or biennials.

2. Crambe hispanica L., Sp. Pl. 671 (1753); Boiss., Fl. 1 : 408 (1867); O. E. Schulz, in Pflznr. 70 : 241 (1919). [Plate 472]

Annual, hispid at least at base with retrorse or spreading hairs, 40–80 cm. Stems erect, simple or branched above. Radical leaves long-petioled, almost simple or more frequently pinnatisect-lyrate with 1–3 pairs of small, ovate-oblong, acute, dentate or subentire lateral lobes and ovate-orbicular, rather cordate, irregularly crenate-dentate, large terminal lobe; cauline leaves short-petioled with a single pair of lateral lobes, and an ovate terminal lobe; upper leaves sessile, mostly simple, linear. Racemes elongated in fruit. Flowers small, white. Calyx about 2 mm.; sepals glabrous, equal at base. Petals 3–3.5 mm.; limb obovate. Inner stamens with or without tooth-like appendage on filament. Fruit on long, suberect, terete pedicels; lower joint 1.5–2.5 mm., cylindrical, sterile, upper joint 2–3 mm., more or less globular, glabrous, yellowish-green to brown-black, smooth, readily separating from lower joint when mature. Seeds spherical, black at hilum. Fl. March–April.

Var. **hispanica.** Plants hispid throughout. Terminal lobes of radical leaves ovate. Stems rather slender.

Hab. : Batha and fallow fields. Acco Plain, Sharon Plain, Upper and Lower Galilee, Mt. Carmel, Esdraelon Plain, Samaria, Judean Mts., Judean Desert, Dan Valley, Hula Plain, Upper and Lower Jordan Valley, Gilead, Ammon, Moav, Edom.

Var. **glabrata** (DC.) Coss., Comp. Fl. Atl. 2 : 309 (1887); O. E. Schulz, l.c. 242. *C. glabrata* DC., Prodr. 1 : 226 (1824). Stems thick. Leaves large, sparingly hispid below, glabrous above; radical leaves simple or with small to rather abortive lateral lobes, terminal one large, cordate.

Hab. : As above. Sharon Plain, Upper and Lower Galilee, Mt. Carmel, Samaria, Judean Mts., Hula Plain, Upper Jordan Valley, Moav.

Var. **edentula** Boiss., Fl. 1 : 408 (1867); O. E. Schulz, l.c. 242. All filaments without teeth.

Hab. : As above. Samaria, Hula Plain. Not common.

Area or species : N. and E. Mediterranean.

62. CALEPINA Adans.

Annual glabrous herbs. Radical leaves rosulate, lyrate-pinnatifid; cauline amplexicaul-sagittate. Inflorescences many-flowered, ebracteate. Flowers minute, white. Sepals erect, equal at base. Petals unequal, the outer longer. Stamens edentate; filaments dilated at base. Ovary 1-ovuled; stigma capitate, narrower than style. Fruit an indehiscent, 1-celled silicle on a long, terete, erect or incurved pedicel, ovoid to pear-shaped, obsoletely 4-ribbed, tapering into a short blunt beak. Seeds solitary, pendulous; cotyledons conduplicate.

One species in the Mediterranean region and adjacent Euro-Siberian and Irano-Turanian territories.

1. Calepina irregularis (Asso) Thell. in Schinz et Keller, Fl. Schweiz ed. 2, 1 : 218 (1905); O. E. Schulz, in Pflznr. 70 : 225 (1919). *Myagrum irregulare* Asso, Syn. Stirp. Arag. 82 (1779). *C. corvini* (All.) Desv., Journ. Bot. Appl. 3 : 158 (1814); Boiss., Fl. 1 : 409 (1867). [Plate 473]

Annual, glabrous, 10–30 cm. Stems procumbent or ascending, branching from base. Radical leaves rosulate, petiolate, oblong, obovate or oblanceolate, lyrate-pinnatifid with oblong, repand-dentate terminal lobe; cauline leaves 2–6 cm., sessile, oblong-lanceolate, clasping-sagittate, remotely toothed to almost entire. Inflorescences elongating in fruit. Flowers minute, about 2 mm. Sepals erect, equal. Petals longer than calyx, white, the inner somewhat shorter than outer. Fruiting pedicels 0.7–1.2 cm., mostly erect or somewhat incurved. Fruit about 2–4 × 2 mm., erect, ovoid to pyriform, 4-ribbed, tapering into a short blunt beak, wrinkled when dry. Fl. February–April.

Hab.: Fields, mainly on heavy to damp soils in mountainous regions. Upper Galilee, Samaria, Judean Mts., Judean Desert, Ammon, Moav, Edom.

Area: N. Mediterranean and W. Irano-Turanian with extensions into S.W. Euro-Siberian provinces.

63. ZILLA Forssk.

Perennial, spiny, almost leafless shrubs. Stems more or less dichotomously branched. Leaves few, rather fleshy, glabrous, the lower spatulate, sinuate-crenate, the upper oblong, obtuse. Racemes few-flowered, ebracteate, rhachis spinescent. Flowers conspicuous, short-pedicelled, white or pink. Calyx erect, 2-saccate at base. Petals oblong, entire, long-clawed. Stamens free, toothless. Fruit a silicle, indehiscent, 2-celled, crustaceous-woody, ovoid-globular, tapering into a conical beak, ribbed, wrinkled or smooth. Seeds solitary in each cell, pendulous; cotyledons conduplicate.

Some 3 species in the Saharo-Arabian region.

1. Zilla spinosa (L.) Prantl in Engl. et Prantl, Nat. Pflznfam. 3, 2 : 174, f. 112 (1891); O. E. Schulz, in Pflznr. 84 : 30 (1923). *Z. myagrioides* Forssk., Fl. Aeg.-Arab. 121, n. 74, 75 (1775) et Icon. Rer. Nat. ed. Niebuhr 6, t. 17a; 7, t. 18a (1776); Boiss., Fl. 1 : 408 (1867). *Bunias spinosa* L., Mant. 96 (1767). [Plate 474]

Perennial, spiny, glaucous, glabrous, desert shrub, 10–60 cm. Stems erect, longitudinally ribbed, dichotomously branched. Older branches white, leafless, tapering to a sharp spine. Leaves soon decaying or deciduous; the lower ones 2–7.5 cm., subrosulate, spatulate, tapering into a short petiole, repand-crenate or sublyrate-pinnatifid; cauline few, sessile or subsessile, entire or repand-crenate. Racemes few-flowered. Flowers large, short-pedicelled, scattered. Calyx 4–7 mm. Petals 0.6–2 cm., violet or lilac; limb obovate, entire. Fruit 0.6–1 cm. in diam., on a 1–2 mm. long, erect-spreading thick pedicel, globular with truncate base, 6-ribbed with transverse wrinkles between ribs, tapering into a conical, spiny beak; valves and septum thick, hard. Seeds ellipsoidal, dark, smooth. Fl. April–May.

Var. **spinosa.** Fruit 0.8–1 cm. in diam., ribbed and irregularly wrinkled. Flowers 1–2 cm.

Hab.: Mainly sandy deserts. Judean Desert, C. and S. Negev, Lower Jordan Valley, Arava Valley, deserts of Moav and Edom.

Var. **microcarpa** (DC.) Dur. et Schinz, Consp. Fl. Afr. 1, 2 : 149 (1895); O. E. Schulz, l.c. 32. *Z. myagroides* Forssk. var. *microcarpa* DC., Syst. 2 : 647 (1821); Boiss., Fl. 1 : 408 (1867). Low shrubs, 10–20 cm. Branches short, crowded. Flowers 0.6–1 cm. Fruit 6–8 mm. in diam., spherical, smoother than in preceding variety.

Hab.: As above. C. and S. Negev, Dead Sea area, Arava Valley, desert of Edom.

Area of species : Saharo-Arabian.

64. RAPHANUS L.

Annual or biennial herbs, more or less hispid below. Radical leaves lyrate with large terminal lobe; upper linear to oblong. Racemes terminal or axillary, many-flowered, ebracteate. Flowers white, yellow or violet, often purple-veined. Calyx erect; the inner sepals saccate at base. Petals with a long claw and obtuse or retuse limb. Stamens edentate. Stigma depressed, slightly 2-lobed. Fruit an elongated, thick siliqua consisting of 2 joints : the lower joint mostly sterile, narrow, very short and thin, resembling pedicel, top-shaped; upper joint thick, long, indehiscent or breaking into single-seeded portions; beak seedless. Seeds spherical, pendulous, brown; cotyledons conduplicate.

About 8 species, Euro-Siberian and Mediterranean.

1. Flowers yellow, not veined. Pedicels deflexed in fruit.   **3. R. aucheri**
- Flowers white, violet or pink, with prominent purple veins. Fruiting pedicels erect or spreading   2
2. Fruit dagger-shaped. Flowers violet or pink.   **2. R. rostratus**
- Fruit necklace-shaped. Flowers white or yellowish.   **1. R. raphanistrum**

1. Raphanus raphanistrum L., Sp. Pl. 669 (1753); Boiss., Fl. 1 : 401 (1867); O. E. Schulz, in Pflznr. 70 : 196 (1919). [Plate 475]

Annual, glabrous or more or less hispid-scabrous, 30–80 cm. Stems erect, branched. Lower leaves 5–20 cm., long-petioled, lyrate-pinnatisect or pinnatifid, terminal lobe obovate, lateral distant, oblong, obtuse or acute, all cauline leaves 1.5–10 cm., short-petioled, oblong-lanceolate, dentate-serrate. Racemes elongated. Sepals 0.6–1 cm., linear or oblong, obtuse, sometimes purple, the inner ones saccate. Petals 1.5–2 cm., white, lilac or yellow with purple veins, obovate, cuneate. Fruiting pedicels 1.5–2.5 cm., erect to spreading. Fruit 2.5–5.5 × 0.3–0.5 cm., erect; lower joint seedless, very short and scarcely broader than pedicel; upper joint 3–8-seeded, with constrictions between seeds, readily breaking into 1-seeded portions, with a seedless, 1.3–2.5 cm. long, tapering, hispid beak. Seeds globular-ovoid. Fl. January–May.

Hab.: Fallow and cultivated fields. Coast of Galilee, Acco Plain, Sharon Plain, Philistean Plain, Upper Galilee, Mt. Carmel, Judean Mts., Dan Valley, Hula Plain, Upper and Lower Jordan Valley, Edom.

Area: Mediterranean, and Euro-Siberian; also introduced into other regions both of the Old and New World.

In Palestine this species is very common in the coastal plain and rare or uncommon in other districts.

A valuable honey plant.

2. Raphanus rostratus DC., Syst. 2: 666 (1821); O. E. Schulz, in Pflznr. 70: 201 (1919). [Plate 476]

Sparingly scabrous annual, 20–60 cm. Stems erect or ascending, branched. Radical leaves 9–18(–20) cm., lyrate, terminal lobe obovate-oblong, lateral small, linear-triangular, all lobes crenate-dentate; upper leaves subsessile, oblong-linear, acute, more or less crenate-dentate. Flowers large. Sepals 0.8–1.2 cm., erect, linear, obtuse, purplish. Petals 1.5–2.5 cm., violet or pink with purple veins, obovate-cuneate, retuse. Pedicels 1–1.5 cm., erect or spreading. Fruit 3–12.5 × 0.4–0.6 cm.; lower joint minute; upper joint longitudinally septate, dagger-shaped, slightly or not at all constricted between the 2–7 seeds and terminating with a 1.5–7.5 cm. long, conical beak and capitate stigma.

Var. pugioniformis (Boiss.) O. E. Schulz ex Thell. in Hegi, Ill. Fl. Mitteleur. 4: 279 (1918); O. E. Schulz, l.c. 202. *R. pugioniformis* Boiss., Diagn. ser. 1, 8: 46 (1849) et Fl. 1: 400 (1867). Fruit 7–12.5 cm., scabrous, with scattered, appressed hairs directed towards beak. Fl. March–May.

Hab.: Batha and fields. Upper and Lower Galilee, Esdraelon Plain, Mt. Gilboa, Judean Mts., Hula Plain, Upper Jordan Valley, Gilead.

Area: E. Mediterranean.

3. Raphanus aucheri Boiss., Diagn. ser. 1, 8: 45 (1849) et Fl. 1: 401 (1867); O. E. Schulz, in Pflznr. 70: 209 (1919). [Plate 477]

Annual, scabrous-hispid with retrorse hairs, 20–50(–80) cm. Stems erect, much branched. Radical leaves 9–12(–16) cm., almost rosulate, long-petioled, lyrate-pinnatipartite, terminal lobe broadly obovate, irregularly toothed or crenate, lateral lobes much smaller, ovate-oblong, acute; cauline leaves 2–6 cm., subsessile, oblong-linear, acute, rarely lyrate. Flowers conspicuous. Sepals about 5 mm., erect, oblong, obtuse, setulose. Petals 1–1.2 cm., yellow, without dark veins, obovate, clawed. Pedicels 0.5–1.3 cm., deflexed in fruit. Fruit 6–10 × 0.2–0.4 cm., pendulous, terete, linear, strigose with retrorse hairs; lower joint (about 1.5 cm.) seedless, attenuate, the upper sometimes slightly constricted between seeds, terminating with a relatively short beak and a subcapitate stigma. Seeds oblong. Fl. February–May.

Hab.: Batha and rocky places. Upper and Lower Galilee, Esdraelon Plain, Samaria, Hula Plain, Upper and Lower Jordan Valley.

Area: E. Mediterranean and Irano-Turanian (Palestine, Lebanon, Iraq, Iran).

65. ENARTHROCARPUS Labill.

Annual hispid herbs. Radical leaves rather lyrate, pinnately divided and toothed; cauline leaves petiolate or sessile, toothed. Inflorescences bracteate throughout or only at lower flowers, elongated in fruit. Flowers conspicuous, short-pedicelled. Calyx erect; sepals almost equal. Petals longer than calyx, mostly yellow, purple-veined. Stamens edentate. Style long; stigma depressed-capitate, slightly 2-lobed. Fruit a long, indehiscent siliqua, 2-jointed, terete or slightly compressed, hispid or scabrous, the lower joint short, 2-valved, 0–3-seeded; upper joint many-seeded, constricted between seeds, separating into 1-seeded segments, beaked. Seeds in 1 row in each cell; cotyledons longitudinally conduplicate.

Some 5 species in the Mediterranean and Saharo-Arabian regions.

1. Inflorescences ebracteate, except for lower 1–5 flowers. Fruit terete, 3–10 cm. long; lower joint of fruit 1–6 mm., 0–1-seeded. Littoral plants.　**1. E. arcuatus**
– Inflorescences bracteate throughout or almost so. Fruit shorter than 5 cm., more or less compressed. Steppe and desert plants　**2**
2. Flowes 0.9–1.5 cm. long. Fruit necklace-shaped; lower joint about 4 mm., 1–2-seeded, upper 8–12-seeded.　**2. E. strangulatus**
– Flowers 6–7 mm. Fruit obsolètely torulose; lower joint up to 1.4 cm., 2–3-seeded, upper about 3–6-seeded.　**3. E. lyratus**

1. Enarthrocarpus arcuatus Labill., Ic. Pl. Syr. Dec. 5 : 4, t. 2 (1812); Boiss., Fl. 1 : 399 (1867) et Suppl. 67 (1888); O. E. Schulz, in Pflznr. 70 : 211 (1919). [Plate 478]

Annual, appressed-hairy to hispid, 20–45(–60) cm. Stems ascending to decumbent, branching from base. Radical leaves up to 12 cm., petiolate, lyrate-pinnatisect or pinnatipartite into about 4 pairs of alternate, oblong, acute or obtuse, irregularly dentate lobes; cauline leaves short-petioled or sessile, oblong-ovate, laciniate or dentate. Inflorescences dense, the lowermost 1–5 flowers bracteate. Flowers 0.9–1.5 cm. Calyx 5–7 mm.; sepals erect, hispid, sometimes violet, equal. Petals about twice as long as sepals, yellow, purple-veined, clawed; limb oblong-obovate. Fruit 3–10 cm., on long (up to 1 cm.), erect-spreading pedicel, terete, torulose, recurved-arcuate, hispid; lower joint 0–1-seeded; upper joint many-seeded, separating into 1-seeded segments and with a 0.5–1.2 cm. long, conical beak. Seeds in 1 row, small. Fl. March–May.

Hab.: Littoral sands and rocks. Coastal Galilee, Acco Plain, Sharon Plain.
Area: E. Mediterranean.

2. Enarthrocarpus strangulatus Boiss., Diagn. ser. 1, 8 : 44 (1849) et Fl. 1 : 399 (1867); O. E. Schulz, in Pflznr. 70 : 214 (1919). [Plate 479]

Annual, densely hispid-hirsute, 15–45(–65) cm. Stems many, procumbent to ascending, branched from base. Radical leaves varying in size and shape, 5–15 cm., lyrate, pinnatisect or pinnatipartite with (3–)4–7(–9) pairs of opposite or alternate, sometimes more or less runcinate, obtuse or acute, densely or sparingly dentate, lateral lobes; cauline leaves subsessile, obovate-oblong, pinnatifid or dentate. Inflorescences bracteate throughout, compact. Calyx 5–6 mm.; sepals almost erect, hispidulous, violet. Petals 0.9–1.5 cm., cream-coloured or yellow, purple-veined, clawed; limb obovate. Fruit 2.5–4.5 cm., on about 1 cm. long, curved pedicels, more or less compressed, lanceolate-linear, curved, hispid, longitudinally striate; lower joint about 4 mm., 1–2-seeded; upper 8–12-seeded, with swollen nodes alternating with oblong 1-seeded segments, separating at maturity; beak about 5 mm., conical-flat. Seeds about 1 mm. in diam., brown. Fl. January–April.

Hab.: Steppes, deserts and fields, mostly on sandy soil. Judean Desert, W., N. and C. Negev, Lower Jordan Valley, Dead Sea area, Arava Valley.

Area: Saharo-Arabian.

3. Enarthrocarpus lyratus (Forssk.) DC., Syst. 2 : 661 (1821); Boiss., Fl. 1 : 399 (1867); O. E. Schulz, in Pflznr. 70 : 216 (1919). *Raphanus lyratus* Forssk., Fl. Aeg.-Arab. 119 (1775).

Annual, hispid with scattered simple hairs, 10–45 cm. Stems ascending to erect, branching from base. Radical leaves petiolate, lyrate, pinnately divided, terminal lobe obovate, more or less 3-lobed, lateral lobes oblong, dentate; cauline leaves short-petioled, dentate or pinnately incised. Inflorescences bracteate almost throughout, compact. Flowers up to 7 mm. Sepals about 4 mm. Petals pale yellow, purple-veined. Fruit 2–4 cm., on short (up to 6 mm.) pedicel, compressed, obsoletely torulose, recurved, hispid with stiff hairs at margin; lower joint 0.7–1.4 cm., 2–3-seeded; upper joint 3–6-seeded, with 5–7 longitudinal ridges; beak 5 mm. or shorter, attenuate. Seeds brown. Fl. February–April.

Hab.: Steppes and fields. Recorded by Dinsmore (in Post ed. 2, 1 : 126, 1932) from Philistean Plain, Negev and Lower Jordan Valley. Not observed by the present author.

Area: E. Mediterranean.

66. CORDYLOCARPUS Desf.

Annual herbs with retrorse simple hairs. Leaves from almost entire to pinnatifid. Inflorescences ebracteate, elongating in fruit. Flowers large, short-pedicelled. Sepals more or less erect, the inner ones slightly saccate. Petals pale yellow, clawed, limb entire. Stamens edentate. Fruit indehiscent, 2-jointed; lower joint cylindrical, with a few 1-seeded transversal cells; upper joint 1-seeded, ovoid-globular, long-rostrate, muricate, with 4 rows of prickles and 2–4 entire or repand wings. Seeds slightly compressed, oblong; cotyledons longitudinally conduplicate.

One species, mainly Saharo-Arabian.

1. Cordylocarpus muricatus Desf., Fl. Atl. 2 : 79 (1798); Coss., Comp. Fl. Atl. 2 : 319 (1887); O. E. Schulz, in Pflznr. 70 : 223 (1919). [Plate 480]

Annual, hispid-hirsute, 15–60 cm. Stems erect, branched, tetragonous. Radical leaves petiolate, oblong-ovate, entire or irregularly dentate or pinnatifid towards base;

cauline leaves subsessile, oblong, acute. Inflorescences elongating in fruit. Sepals about 4 mm., glabrous or hispid. Petals about twice as long as sepals, pale yellow. Fruit 1.8–2.5 cm., on a thick, spreading or horizontal pedicel; lower joint 1–4-seeded, subglabrous or pubescent; upper joint short and muricate. Seeds subcompressed.

Var. **palaestinus** Eig ex Zoh., Palest. Journ. Bot. Jerusalem ser. 2 : 163 (1941). Plant with long, retrorse hairs. Pedicels and calyx densely hirsute-pubescent. Fl. April.

Hab. : Sandy soils. Coastal Negev. Very rare.

Area of species : Saharo-Arabian.

46. RESEDACEAE

Annual or perennial herbs, rarely shrubs. Leaves alternate or fasciculate, entire or variously cut or divided. Flowers hermaphrodite or rarely unisexual, in bracteate racemes or spikes, zygomorphic. Sepals (2–)4–8. Petals (2–)4–8, rarely 0, the 2 posterior ones usually larger and often more dissected than others. Stamens 3–40, usually inserted on a unilaterally expanded disk (rarely disk 0). Ovary superior, sessile or stipitate, consisting either of 2–6 free, 1-ovuled carpels or of 3–7 connate carpels with numerous ovules on parietal (or basal) placentae; style very short or 0; stigmas or stigmatic lobes as many as carpels. Fruit either a 1-celled capsule with 3–4 teeth, open or closed at the top, or a 6(–7)-celled capsule, rarely a berry. Seeds numerous, subspherical or reniform with a curved embryo and without endosperm; cotyledons incumbent.

About 6 genera and 70 species in the Mediterranean, Saharo-Arabian, Irano-Turanian and Sudanian and also in the Euro-Siberian, Macaronesian, S. African and N. W. Pacific regions.

Literature : F. Bolle, Resedaceae, in : Engl. u. Prantl, *Nat. Pflznfam.* ed. 2, 17b : 659–692 (1936). J. Müller-Argoviensis, *Monographie de la famille des Résédacées.* Leipzig (1857).

1. Petals 0. Fruit a berry. Shrubs with linear, deciduous leaves.   **1. Ochradenus**
– Petals 2–8. Fruit a dry capsule. Annual or perennial herbs   2
2. Petals 2. Stamens mostly 3. Disk 0. Herbs with entire, narrowly linear leaves.   **2. Oligomeris**
– Petals 4–8. Stamens mostly more than 8. Disk one-sided   3
3. Leaves mostly divided, rarely simple. Gynophore 0 or much shorter than calyx. Sepals and petals 4–8. Capsule syncarpous, 1-celled, with many seeds on parietal placentae. Stigmas 3–4.   **3. Reseda**
– Leaves simple. Gynophore longer than calyx. Sepals and petals 5. Capsule almost apocarpous, usually with 6, almost separate carpels, each with 1(–2) seeds on basal placentae. Stigmas 6.   **4. Caylusea**

Trib. RESEDEAE. Placentae parietal; ovules pendulous; carpels 2–5, almost entirely connate.

1. Ochradenus Del.

Glabrous, sometimes dioecious, much branching shrubs. Stems green. Branches some-what thorny when dry. Leaves simple, soon deciduous. Flowers minute, unisexual (plants then polygamous or dioecious) or hermaphrodite, in terminal, more or less spike-like, bracteate racemes. Sepals 5–6. Petals absent. Disk dilated unilaterally. Stamens 10–15. Ovary sessile, ovoid, 3-carpelled, 3-dentate; placentae parietal with pendulous ovules. Capsule closed, dry and membranous or fleshy, baccate. Seeds few, yellow to brownish-red or almost black.

Four species, mainly Sudanian.

1. Ochradenus baccatus Del., Fl. Eg. 15 (1813); Müll., Monogr. Res. 94 (1857); Boiss., Fl. 1 : 422 (1867). [Plate 481]

Glabrous, dioecious or polygamous shrub, 50–100 cm. Stems erect, woody, with divaricate, twiggy or simple branches. Leaves soon deciduous, frequently fasciculate, narrowly linear, sometimes subulate or oblanceolate or spatulate, acuminate at apex, glabrous. Flowering racemes more or less spike-like, rigid. Bracts minute, lanceolate, frequently somewhat concave. Flowers mostly unisexual, short-pedicelled. Sepals 4–6 mm., more or less equalling pedicel in length, persistent, linear-lanceolate. Petals 0. Stamens 10–13; filaments longer than sepals, soon deciduous. Fruit a berry, 4–6 mm., ovoid or obovoid-globular, with rounded apex, 3-denticulate, glabrous, whitish. Seeds 1–1.5 mm., reniform, tuberculate, yellow to brown. Fl. Mainly December–May.

Hab.: Hot desert wadis and depressions. Judean Desert, W., N. and C. Negev, Lower Jordan Valley, Dead Sea area, Arava Valley, deserts of Moav and Edom.

Area: Sudanian, extending into adjacent Saharo-Arabian territories.

Fruit sweet, edible; plant browsed by camels.

2. Oligomeris Cambess.

Annual or perennial herbs or dwarf-shrubs with entire, scattered or fasciculate leaves. Flowers minute, hermaphrodite or unisexual (plants then monoecious), sessile in slender spike-like racemes, bracteolate. Sepals 2–6, persistent. Petals 2, white, posterior, without claw, entire or lobed, sometimes connate. Floral disk 0. Stamens 3 or 4 or up to 10 on the posterior part of the flower or all around. Ovary sessile, 3–5-carpelled; placentae parietal, with numerous pendulous ovules; stigmas as many as carpels. Capsule char-taceous, 1-celled, open at apex. Seeds smooth.

Eight species, most of them S. African, one in the Sudanian and Saharo-Arabian regions.

1. Oligomeris subulata (Del.) Boiss., Fl. 1 : 435 (1867). *Reseda subulata* Del., Fl. Eg. 15 (1813). *O. linifolia* (Vahl ex Hornem.) Macbride, Contr. Gray Herb. ser. 2, 53 : 13 (1918). *R. linifolia* Vahl ex Hornem., Hort. Hafn. 2 : 501 (1815). [Plate 482]

Annual or biennial, somewhat glaucous, glabrous, 10–40 cm. Stems decumbent to erect, leafy, rather branched. Leaves 2–7 × 0.1–0.3 cm., fasciculate, simple, narrowly linear to almost subulate or narrowly oblanceolate, acuminate, entire. Racemes spike-like, erect. Bracts 1–2 mm., persistent, lanceolate-subulate, hyaline and somewhat crenate at margin. Flowers 1–2 mm., almost sessile. Calyx often persistent, with 2–5 lanceolate sepals, similar to bracts but a little shorter. Petals white, longer than sepals, simple or often lobed. Stamens 3, unilateral. Fruiting racemes elongated, rather loose. Capsule 3–4 mm., sessile, 4-carpelled, depressed-globular, 4-dentate, 8-sulcate, papillose. Seeds smaller than 1 mm., reniform, rather smooth, shining, yellow to brown. Fl. March–May.

Hab.: Hot deserts. Samaria (desert), Judean Desert, C. and S. Negev, Lower Jordan Valley, Dead Sea area, Arava Valley, deserts of Moav and Edom.

Area: Sudanian, extending into adjacent Saharo-Arabian territories.

3. RESEDA L.

Annual or perennial herbs or half-shrubs. Leaves pinnately or ternately divided or dissected, rarely undivided; stipules gland-like. Flowers usually hermaphrodite, in spike-like racemes. Sepals 4–8, persistent or caducous. Petals 4–8, white, yellow or yellowish-green, unequal, the upper and often the lateral ones (rarely all petals) more or less dissected. Stamens (7–)10–40, inserted on the eccentric disk, on one side of flower. Ovary often short-stipitate, 3–4-carpelled; ovules numerous, on parietal placentae; carpels united to apex or up to two thirds their length. Capsule 1-celled, open at top; stigmas 3–4. Seeds many, reniform to subglobular.

About 50 species, mainly in the Mediterranean, Saharo-Arabian, Irano-Turanian and Sudanian regions.

1. Leaves all entire, lanceolate-linear, sometimes undulate at margin. Petals light yellow. Flowers 4-merous, almost sessile. Filaments mostly persistent. Capsule erect.
                                                                    13. R. luteola
– Leaves all or only upper ones ternately or pinnately divided   2
2. Calyx and corolla 5(–6)-merous. Leaves pinnatisect. Capsule 4-toothed   3
– Calyx and corolla (4–)6–8-merous. All or only some of upper leaves ternately divided or lobed. Capsule usually 3-toothed   4
3. Herbs 30–80 cm. high. Petals much longer than sepals. Capsule 0.8–1.3 cm., generally oblong-ellipsoidal, frequently constricted at base of teeth.   **1. R. alba**
– Herbs 10–30 cm. high. Petals a little longer than calyx. Capsule 4–6(–8) mm., obovoid, not constricted above, with truncate teeth. Desert plants.   **2. R. decursiva**
4 (2). Capsule distinctly 6-lobed, depressed-subglobular or disciform. Seeds smooth. Filaments persistent.   **8. R. globulosa**
– Capsule 3-lobed, obovoid, globular or ovoid to cylindrical   5
5. Calyx usually deciduous. Seeds smooth to somewhat tuberculate   6
– Calyx persistent   7
6. Stems and leaves muricate. Capsule obovoid-ellipsoidal, sometimes pyriform or subglobular. Filaments persistent.   **10. R. muricata**
– Stems and leaves subglabrous to sparingly papillose. Capsule club-shaped, oblong or obconical, truncate, 3–4-toothed. Filaments deciduous.   **9. R. stenostachya**
7 (5). Flowers white   8
– Flowers yellow or cream-coloured   11
8. Lateral segments of upper (posterior) petal crenate or only slightly lobed. Seeds smooth.

Bracts usually deciduous.   **11. R. boissieri**
- Lateral segments of upper (posterior) petals deeply dissected into long lobes. Seeds
  smooth or tuberculate-wrinkled. Bracts usually persistent   9
9. Capsule erect.   **12. R. maris-mortui**
- Capsule pendulous   10
10. Filaments persistent. Ovary sessile. Stems sparingly pruinose. Desert plants.
                                                                         3. R. arabica
— Filaments deciduous. Ovary stipitate. Stems densely scabrous-papillose. Fragrant Mediterranean plants.   **5. R. orientalis**
11 (7). Capsule 1–1.5 cm. across. Stems covered with dense and long papillae.
                                                                         4. R. alopecuros
— Capsule much narrower than above   12
12. Seeds usually smooth. Flowers yellow. All leaves ternately divided. Ruderal and segetal
    plants.   **7. R. lutea**
— Seeds wrinkled-tuberculate. Flowers cream-coloured. Upper leaves entire and 3-fid.
   Fragrant plants cultivated and escaped from cultivation.   **6. R. odorata**

Sect. LEUCORESEDA DC. in Duby, Bot. Gall. 1 : 67 (1828). Calyx and filaments
persistent. Perianth 5–6-merous. Capsule erect, 4-toothed. Petals white, upper frequently
3-lobed. Seeds tuberculate. Leaves 1–2-pinnatisect.

1. Reseda alba L., Sp. Pl. 449 (1753); Müll., Monogr. Res. 100 (1857); Boiss., Fl.
1 : 425 (1867). [Plate 483]

Annual or biennial, glabrous or sparingly papillose, 30–80 cm. Stems erect, sometimes woody at base, simple or branching from base. Leaves pinnatisect into 6 or more
linear to oblong-lanceolate, entire or sparingly denticulate and undulate segments;
upper leaves smaller, the lowermost crowded. Flowering racemes spike-like, rather dense.
Bracts small, persistent, acicular-linear. Flowers 3–5 mm., short-pedicelled. Calyx a
little shorter than pedicel and much shorter than corolla, persistent, with 5–6 sepals.
Petals 5–6, white, 3-partite; the 2 upper with somewhat lobulate or entire lateral
lobes. Stamens 10–13, with long, mostly persistent filaments. Fruiting racemes rather
dense. Fruiting pedicels longer than calyx. Capsule 0.8–1.3 cm., erect or deflexed,
oblong-ellipsoidal, sometimes ovoid to cylindrical, with 4 teeth, frequently constricted at
base of teeth. Seeds spherical-reniform, rough-tuberculate, yellow to brown. 2n=20.
Fl. February–July (–November).

Hab.: Waste places, roadsides and abandoned gardens. Coastal Galilee, Acco Plain,
Sharon Plain, Philistean Plain, Upper and Lower Galilee, Mt. Carmel, Esdraelon
Plain, Samaria, Judean Mts., Judean Desert, N. Negev, Upper and Lower Jordan
Valley, Gilead, Edom. Common.

Area: Mainly Mediterranean and W. Irano-Turanian, extending into Euro-Siberian territories.

2. Reseda decursiva Forssk., Fl. Aeg.-Arab. LXVI (1775). *R. propinqua* Boiss., Fl.
1 : 425 (1867) non R. Br. (1826). *R. eremophila* Boiss., Diagn. ser. 1, 8 : 54 (1849);
Müll., Monogr. Res. 109 (1857). [Plate 484]

Annual, glabrous, 10–30 cm. Stems procumbent to erect, simple or branching from base. Leaves mostly crowded at base, deeply pinnatisect with variable number of narrowly linear to lanceolate, subdenticulate or undulate segments. Flowering racemes slender. Bracts small, persistent, acicular. Flowers 2–3 mm., short-pedicelled to almost sessile. Sepals 5, a little shorter than petals, persistent. Petals 5, white, 3-partite; the upper with shortly 2-lobulate lateral lobes. Stamens about 10, with short, persistent filaments. Fruiting racemes rather dense with pedicels shorter than calyx. Capsule 4–6(–8) mm., subsessile, obovoid to oblong, with 4 truncate teeth, mostly not constricted below teeth. Seeds smaller than 1 mm., reniform-globular, tuberculate, yellow-brown. Fl. March–May.

Hab.: Steppes and deserts. Judean Desert, W. and C. Negev, Lower Jordan Valley, Dead Sea area, Arava Valley, deserts of Ammon, Moav and Edom.

Area: Saharo-Arabian.

Intermediate forms between this species and *R. alba* in border areas between Mediterranean and Saharo-Arabian territory (e.g. Sharon Plain, Philistean Plain, Judean Desert, N. Negev) may perhaps be looked upon as hybrids.

Contrary to Bornmueller, Verh. Zool.-Bot. Ges. Wien 48:558 (1898) we find it impossible to differentiate *R. alba* L. var. *hookeri* (Guss.) Heldr., Bull. Herb. Boiss. 6:236 (1898) from these intermediate forms.

Var. *foliosa* (Post) Hand.-Mazzetti, Ann. Naturh. Mus. Wien 26:56 (1912), recorded by Dinsmore (in Post, Fl. Syr. Pal. Sin. ed. 2, 1:137, 1932) in the Judean Mts. and the Lower Jordan Valley, is in our opinion one of the few forms around *R. alba* and *R. decursiva* calling for an intensive study.

Sect. RESEDA. Sect. *Resedastrum* Duby, Bot. Gall. 1:66 (1828). Calyx and filaments soon deciduous or sometimes persistent. Perianth 6–8-merous. Capsule 3-toothed. Petals white or yellow, the upper 3-partite with crenate, lobed or dissected lateral lobes. Seeds tuberculate or smooth. Leaves simple or 1–2-ternate.

3. Reseda arabica Boiss., Diagn. ser. 1, 1:6 (1843) et Fl. 1:426 (1867); Müll., Monogr. Res. 124 (1857). [Plate 485]

Annual, sparingly pruinose, 8–40 cm. Stems erect, often branching from base, subglabrous or sparingly muricate above. Leaves tapering at base, the lowest entire, oblong-lanceolate, the upper ternate with linear to oblanceolate, obtuse, sometimes wavy lobes of which the middle one is longer than the lateral. Flowering racemes more or less spike-like, rather loose. Bracts 2–3 mm., mostly persistent, linear-lanceolate. Pedicels longer than calyx. Flowers 2–3 mm., white. Calyx persistent, with 6 oblong sepals. Petals 6, somewhat shorter than sepals; upper petals 3-lobed with 6–8-partite lateral lobes and an entire, shorter, lingulate middle lobe; lower petals entire or 2-lobed. Stamens about 20, with persistent filaments. Ovary sessile. Fruiting pedicels a little longer than calyx. Capsule 0.6–1 cm., pendulous, globular-ellipsoidal, 3-toothed. Seeds over 1 mm., reniform, tuberculate-wrinkled, light yellow to grey-brown. Fl. March–April.

Hab.: Deserts. Negev, Arava Valley, Edom.

Area: Saharo-Arabian.

4. Reseda alopecuros Boiss., Diagn. ser. 1, 8 : 55 (1849) et Fl. 1 : 426 (1867); Müll., Monogr. Res. 121 (1857). [Plate 486]

Annual, hispid-papillose, frequently woody at base, 30–70 cm. Stems erect, branching. Leaves tapering to a cuneate base, frequently ciliate and wavy at margin; the lowest entire, oblanceolate, obtuse; the upper ternate or 2-ternate with oblong-lanceolate, rather acuminate lobes. Racemes long, dense. Bracts 3–5 mm., mostly persistent, lanceolate-linear. Pedicels twice as long as calyx. Flowers 4–5 mm. Calyx persistent, with 6–7, linear-oblong to spatulate sepals, equalling petals in length. Petals 6, yellowish-white; the upper ones 3-sect with lateral lobes 7–9-partite into lingulate segments and an entire middle lobe; the lower petals mostly entire, ovate, ciliate. Stamens 18–20; filaments deciduous. Fruiting pedicels scabrous, about as long as capsule. Capsule 1–1.5 cm. long and broad, deflexed, ovoid to globular, 3-toothed, scabrous at ribs. Seeds 1.5–2.5 mm., reniform, wrinkled-tuberculate, light yellow to grey-brown. Fl. March–May.

Hab. : Fields and semisteppe batha. Coastal Galilee, Sharon Plain, Philistean Plain, Upper and Lower Galilee, Mt. Carmel, Esdraelon Plain, Samaria, Judean Mts., Judean Desert, Hula Plain, Upper and Lower Jordan Valley, Gilead.

Area : Mediterranean (endemic in Palestine).

5. Reseda orientalis (Müll.) Boiss., Fl. 1 : 427 (1867). *R. macrosperma* Reichb. var. *orientalis* Müll., Monogr. Res. 135 (1857). [Plate 487]

Annual, papillose-scabrous, 10–50 cm. Stems ascending to decumbent, branching, especially from base. Leaves with a tapering base and an obtuse or acuminate apex, papillose at margin, some oblanceolate-spatulate, entire, others 3-partite into oblong-linear to spatulate lobes. Flowering racemes loose. Bracts 3–5 mm., persistent, lanceolate-subulate or linear. Pedicels distinctly longer than calyx. Flowers 3–6 mm., white, fragrant. Calyx longer than corolla, persistent, with 6 linear to spatulate sepals scabrous at margin. Petals 6, 3–4 mm.; the upper ones cut into 3 lobes, the lateral lobes deeply 4–6-partite, the middle one entire; lower petals entire. Stamens 15–20, with deciduous filaments. Fruiting racemes loose. Fruiting pedicels a little longer than capsule. Capsule 0.8–1.5(–1.8) cm., pendulous, oblong-obovoid, 3-toothed, constricted at opening, scabrous. Seeds 2–3 mm., reniform, tuberculate-wrinkled, light yellow to grey-brown. Fl. January–February.

Hab. : Sandy soils. Acco Plain, Sharon Plain, Philistean Plain, N. Negev.

Area : E. Mediterranean.

6. Reseda odorata L., Syst. ed. 10, 2 : 1046 (1759); Müll., Monogr. Res. 128 (1857); Boiss., Fl. 1 : 428 (1867).

Annual, subglabrous, 20–30 cm. Stems procumbent to ascending, diffusely branching. Leaves tapering at base, lanceolate to oblanceolate-spatulate, entire; some of the upper leaves 3-fid. Racemes short. Bracts 2–3 mm., persistent, linear-lanceolate, obtuse. Pedicels twice as long as calyx. Flowers 3–5 mm., cream-coloured, sweet-scented. Calyx as long as corolla, persistent, with 6 linear or oblong to spatulate sepals. Petals 6; the upper 3-partite with lateral segments deeply incised into 4–6 obovate-spatulate lobes; lower petals entire. Stamens about 20, with deciduous filaments. Fruiting racemes loose. Fruiting pedicels much longer than calyx and somewhat longer than capsule. Capsule 0.9–1.1 cm., pendulous, globular-ovoid to obovoid, 3-toothed, glabrous. Seeds 1.5–2 mm., reniform, tuberculate-wrinkled, cream-coloured to yellow. Fl. March–April.

Hab.: Walls. Upper Galilee. Very rare. Probably escaped from cultivation.

Origin not known.

The record in Post, Fl. Syr. Pal. Sin. ed. 2, 1 : 138 (1932) from Wady Zuwayrah (Dead Sea area) is very doubtful.

7. Reseda lutea L., Sp. Pl. 449 (1753); Müll., Monogr. Res. 183 (1857); Boiss., Fl. 1 : 429 (1867). [Plate 488]

Annual or biennial, subglabrous to papillose, 15–70 cm. Stems leafy, ascending to erect, sometimes decumbent, branching, especially below. Leaves cuneate at base, acuminate or somewhat apiculate, glabrous, frequently somewhat papillose at margin and on median nerve, ternately or 2-ternately divided into linear, oblanceolate or spatulate lobes; some of the lower leaves entire, spatulate. Racemes rather dense. Pedicels slightly papillose, longer than calyx. Bracts deciduous, linear-lanceolate, almost hyaline. Flowers 2–4 mm., yellow. Calyx persistent, with 6 linear-oblanceolate sepals, glabrous or somewhat papillose at margin. Petals slightly longer than calyx; the upper ones 3-partite with crenate or lobulate lateral lobes and shorter, narrow, entire middle lobe; the lower petals 3-partite, with lobes more or less equal in length. Stamens 15–20, with deciduous filaments. Fruiting racemes loose. Fruiting pedicels much longer than calyx, about as long as or shorter than capsule. Capsule 0.6–1.8 cm., erect or deflexed, sometimes pendulous, ellipsoidal, cylindrical or oblong-obovoid, angular, 3-toothed, glabrous to almost papillose. Seeds about 2 mm., reniform, smooth or slightly tuberculate, shining, yellow to dark brown. Fl. March–July (rarely in other seasons).

Var. **lutea**. *R. lutea* L. var. *vulgaris* Müll., l.c. 185. Capsule oblong or oblanceolate-ellipsoidal, erect or ascending.

Hab.: Roadsides, gardens and fields. Acco Plain, Sharon Plain, Philistean Plain, Upper and Lower Galilee, Judean Mts., Judean Desert, N. Negev, Upper Jordan Valley, Gilead, Ammon, Edom.

Var. **nutans** Boiss., Fl. 1 : 430 (1867). *R. clausa* Reichb. ex Müll., l.c. 192. Plants smaller than in preceding variety. Capsule ovoid-oblong, deflexed or pendulous.

Hab.: Roadsides. Samaria, Judean Mts., Edom.

Area of species : Mediterranean, Irano-Turanian and Euro-Siberian.

8. Reseda globulosa Fisch. et Mey., Ind. Sem. Hort. Petrop. 4 : 45 (1837) et Linnaea 12 : 167 (1838); Müll., Monogr. Res. 194 (1857); Boiss., Fl. 1 : 430 (1867). [Plate 489]

Annual, nearly glabrous, 30–50 cm. Stems often decumbent, branching. Leaves tapering at base, with somewhat serrulate margin, ternately, rarely pinnately divided into linear-oblanceolate lobes. Flowering racemes rather loose. Bracts 1.5–2 mm., mostly persistent. Flowers very short-pedicelled. Calyx persistent, with 6 oblong-lanceolate sepals, membranous and often denticulate at margin. Petals 6, cream-coloured; the upper longer than sepals, 2–3-partite; the lower shorter, entire, linear. Stamens 10–12, with persistent filaments. Fruiting racemes loose. Fruiting pedicels a little longer than capsule. Capsule broader than long, 3.5–4.5 × 5–7.5 mm., erect or deflexed, disciform, depressed-subglobular, prominently 6-lobed, 3-toothed. Seeds about 1 mm., reniform, smooth, yellow to brown. Fl. April–May.

Hab. : Fields and steppes. Judean Mts., N. Negev, Moav. Rare.

Area : W. Irano-Turanian.

Our specimens fit var. *brevipes* Post, Fl. Syr. Pal. Sin. 112 (1883–1896), which seems to deserve the rank of a species.

9. Reseda stenostachya Boiss., Diagn. ser. 1, 1 : 5 (1843) et Fl. 1 : 431 (1867); Müll., Monogr. Res. 156, t. 8, f. 108 (1857). [Plate 490]

Annual or biennial, sparingly papillose to rather glabrous, sometimes pruinose, 10–60 cm. Stems ascending to erect, rather rigid, leafy, simple to branched. Leaves somewhat fleshy, some entire, others ternately, rarely pinnately divided into linear to lanceolate, entire or crenate, subglabrous lobes. Racemes erect, rather dense, spike-like. Bracts 1–3 mm., usually deciduous, linear. Pedicels as long as calyx, about 1–2 mm., subglabrous to papillose. Flowers 1–2 mm., white. Calyx mostly deciduous, consisting of 6 linear-lanceolate, somewhat papillose-squamose sepals, membranous at margin. Petals a little longer than sepals; the upper ones 5–7-partite into linear to spatulate lobes; lower petals entire or 2–3-partite. Stamens 15–20; filaments mostly deciduous. Fruiting racemes rather dense. Fruiting pedicels much shorter than capsule. Capsule 0.8–1.3 cm., erect, club-shaped, oblong or obconical, 3 (–4)-toothed, glabrous to papillose-squamose especially when young. Seeds less than 1 mm., reniform, somewhat tuberculate, often shining, light brown to black-brown. Fl. (February–) April–September (–December).

Var. stenostachya. Sparingly papillose, slightly pruinose. Bracts and sepals deciduous.

Hab. : Deserts. C. and S. Negev, Lower Jordan Valley, Dead Sea area, Arava Valley.

Var. eilatensis Plitm. et Zoh. * Rather green, glabrous. Bracts and sepals rarely deciduous.

Hab. : Deserts. Arava Valley.

Area of species : E. Saharo-Arabian, extending into adjacent Sudanian territories.

R. stenostachya is very close to *R. pruinosa* Del., Fl. Aeg. Ill. 15 (1813) and is distinguished from it mainly by indumentum characters. The relations between these two species and their varieties are not sufficiently clear as yet.

10. Reseda muricata C. Presl, Bot. Bemerk. 8 (1844); Müll., Monogr. Res. 159 (1857); Boiss., Fl. 1 : 431 (1867). [Plate 491]

Perennial, scabrous-muricate herbs, woody at base, 10–50 cm. Stems ascending to erect, rigid, almost simple or branching from lower parts. Leaves glabrous to sparingly papillose-scabrous, especially at margin, mostly ternately or 2-ternately divided into linear to oblong-lanceolate lobes; lower leaves sometimes entire, spatulate or lanceolate or linear. Racemes more or less dense. Bracts 2.5–4 mm., deciduous, linear-oblanceolate, membranous at margin. Pedicels shorter than calyx, glabrous or some-

* See Appendix at end of this volume.

what scabrous-papillose. Flowers 3–4 mm., white. Calyx shorter than corolla, mostly deciduous, of 6 lanceolate, rather glabrous sepals. Petals 6, 2–3 mm.; the upper divided into 7 linear to spatulate lobes; the lower entire. Stamens 14–17, with persistent, rarely deciduous filaments. Fruiting racemes elongating. Fruiting pedicels usually shorter than capsule. Capsule 6–8 mm., erect, obovoid-ellipsoidal, sometimes pyriform or subglobular, with rather truncate, 3-denticulate apex, glabrous to papillose-squamose or muricate. Seeds less than 1 mm., reniform, smooth to somewhat tuberculate, shining, yellow to brown, rarely greyish-brown. Fl. March–April.

Hab.: Steppes and deserts, mainly on gypsaceous soil. Judean Desert, C. Negev, Lower Jordan Valley, Arava Valley.

Area: E. Saharo-Arabian.

R. muricata is rather polymorphic. There are specimens with very long racemes, pedicels and capsules (var. *hierochuntica* Zoh., Palest. Journ. Bot. Jerusalem ser. 2 : 164, 1941). Others differ from the type by their dense, scale-like papillae on the fruit (var. *lepidocarpa* Zoh., l.c.). A third form (var. *macrocarpa* Zoh., l.c.) is distinguished by its pyriform capsule.

All these forms, as well as var. *undulata* Post, Fl. Syr. Pal. Sin. 113 (1883–1896), should be examined for constancy.

11. Reseda boissieri Müll., Bot. Zeit. 14 : 37 (1856) et Monogr. Res. 175 (1857). *R. cahirana* Müll., Monogr. Res. 176 (1857). *R. kahirina* Müll. var. *boissieri* (Müll.) Boiss., Fl. 1 : 430 (1867). [Plate 492]

Annual, glabrous to sparingly papillose, 10–40 cm. Stems ascending to erect, mostly branching. Leaves tapering to a cuneate base; the lowest mostly entire, oblanceolate-spatulate; the upper mostly divided into 3 oblong-lanceolate to linear lobes. Flowering racemes rather loose. Bracts about 2 mm., usually deciduous, sometimes persistent, linear, almost hyaline. Pedicels papillose, twice as long as calyx. Flowers 2–4 mm., white. Calyx about as long as corolla, persistent, with 6 linear-oblong to obovate sepals, scabrous. Petals 6, 3-partite; the upper with crenate or very slightly lobed lateral segments and an entire, linear, much longer middle one; lower petals with 3 linear or linear-spatulate, entire lobes. Stamens 13–17, usually with deciduous filaments. Fruiting racemes loose. Fruiting pedicels much longer than calyx and somewhat longer than capsule. Capsule up to 1 cm., conspicuously stipitate, erect or pendulous, oblong, ovoid or obovoid, 3-denticulate. Seeds about 1 mm., reniform, smooth, light yellow to grey-brown. Fl. March–April.

Var. **boissieri.** Capsule erect, oblong-ovoid.

Hab.: Deserts. Fairly common. Judean Desert, C. and S. Negev, Lower Jordan Valley, Dead Sea area, Arava Valley.

Var. **pendula** Plitm. et Zoh. * Capsule pendulous, obovoid to oblong-obovoid.

Hab.: Deserts. Arava Valley.

Area of species: W. Sudanian.

* See Appendix at end of this volume.

12. Reseda maris-mortui Eig, Palest. Journ. Bot. Jerusalem ser. 4 : 171 (1948). [Plate 492a]

Annual, slightly glabrous, 10–25 cm. Stems procumbent to ascending, branching from base. Leaves tapering below, somewhat papillose at margin; lower and intermediate leaves entire, lanceolate-spatulate; upper leaves ternately divided with middle lobe much larger than lateral ones. Flowering racemes rather loose. Bracts 2 mm., mostly persistent, linear, somewhat hyaline. Pedicels papillose, longer than calyx. Flowers 2–3 mm., white. Calyx persistent, with 6 oblanceolate or linear-spatulate sepals. Petals 6, 3-partite; lateral segments of the upper ones deeply dissected into linear, filiform lobes, middle segment entire, linear, equal to or somewhat longer than lateral ones; the lower petals divided into 3 equal, linear to filiform lobes. Stamens 14–20, with deciduous, white filaments. Fruiting racemes loose. Fruiting pedicels a little longer than capsule. Capsule 5–8 mm., erect, ovoid or oblong, 3-dentate. Seeds about 1 mm. or less, smooth, shining, yellow to dark brown. Fl. March–April.

Hab. : Deserts. Judean Desert, S. Negev, Lower Jordan Valley, Arava Valley.

Area : Saharo-Arabian (endemic in Palestine).

Despite the differences between *R. maris-mortui* and *R. boissieri* in the petals, leaves and sepals, the specific rank of the former is still questionable.

Sect. LUTEOLA DC. in Duby, Bot. Gall. 1 : 67 (1828). Calyx and filaments persistent. Perianth 4-merous. Capsule 3-toothed. Petals light yellow, upper 3-partite with 2–4-fid lobes. Seeds smooth. Leaves all entire.

13. Reseda luteola L., Sp. Pl. 448 (1753); Müll., Monogr. Res. 202 (1857); Boiss., Fl. 1 : 434 (1867). [Plate 493]

Annual or biennial, woody at base, 10–100 cm. Stems erect, frequently almost simple, sometimes branching. Leaves entire, linear to oblong-lanceolate, tapering towards base, mucronulate or obtuse at apex, sometimes denticulate or undulate at margin. Racemes 10–50 cm., spike-like, sometimes branching and capitate, rather dense. Bracts 2–4 mm., mostly persistent, linear-subulate, green or somewhat hyaline. Flowers 3–4 mm., very short-pedicelled to subsessile. Calyx persistent, with 4 ovate-lanceolate sepals, shorter than corolla. Petals 4, 3–4 mm. long, twice as long as calyx, light yellow; the upper ones usually with a 4–8-lobed limb, the rest entire or 2–4-partite. Stamens about 20; filaments persistent. Fruiting racemes elongated; pedicels shorter than capsule. Capsule 3–6 mm., obovoid-truncate, knobbed by transverse constrictions, acutely 3-toothed, half-closed at top, glabrous. Seeds usually 1 mm., reniform, smooth, brown. Fl. April–May.

Var. **luteola.** Stems 50–100 cm. Leaves linear to oblong-lanceolate, entire or very slightly denticulate or undulate at margin.

Hab. : Fallow fields. Esdraelon Plain, Ammon.

Var. **crispata** (Link) Müll., l.c. 206. *R. crispata* Link, Enum. Hort. Berol. Alt. 2 : 8 (1822). Stems lower, 10–40 cm. Leaves obtuse at apex, distinctly undulate at margin.

Hab. : As above. W. and N. Negev, Ammon, Edom.

More common than preceding variety.

Var. **glomerata** Zoh., Palest Journ. Bot. Jerusalem ser. 2 : 165 (1941). Much branching herbs, 10–40 cm. Leaves with undulate margin. Flowers and fruit clustered in capitate racemes on upper part of stems.

Hab. : As above. Judean Mts., N. Negev.

Area of species : Mainly Mediterranean, extending into the Euro-Siberian and Irano-Turanian regions.

Trib. CAYLUSEAE. Placentae basal; ovules erect; carpels 5–6, connate only at base so as to form an almost apocarpous gynoecium.

4. CAYLUSEA St. Hil.

Annual or perennial herbs with entire leaves. Flowers hermaphrodite, subsessile in spike-like, bracteate racemes. Sepals 5, persistent, the upper smaller. Petals 5, the upper larger with many-parted limb. Stamens 10–18; filaments deciduous. Disk one-sided, upper part much dilated. Gynophore elongated, slender. Carpels 5–6, almost free, whorled at apex of gynophore, 1-celled, many-ovuled, 1 (–2)-seeded, boat-shaped, tapering, with ciliate margin; placentae at base of carpels. Seeds almost globular to reniform.

Two species, mainly E. Sudanian.

1. **Caylusea hexagyna** (Forssk.) Green, Prop. Brit. Bot. 102 (1929). *Reseda hexagyna* Forssk., Fl. Aeg.-Arab. 92 (1775); Taylor, Kew Bull. 13 : 285 (1958). *C. canescens* (L. ed. Murray) Walp., Repert. 2 : 754 (1843); Webb in Hook. f., Niger Fl. 101 (1849); Müll., Monogr. Res. 226 (1857); Boiss., Fl. 1 : 436 (1867). *R. canescens* L., Syst. ed. 12, 2 : 33 (1767) non L. (1753) nec L. (1767). [Plate 494]

Annual, hirsute-papillose herb, 10–70 cm. Stems decumbent to erect, sometimes woody at base. Leaves 2–5 cm., subsessile, simple, linear-lanceolate or oblanceolate or elliptical, obtuse to acuminate at apex, often with undulate margin. Racemes dense, spike-like. Bracts small, mostly deciduous, lanceolate or ovate-oblong, subglabrous. Sepals 5, somewhat longer than pedicels. Petals 5, longer than calyx, white; the upper ones 5–7-partite into linear to spatulate lobes; the rest often 3-partite. Stamens 10–15; filaments sparingly hispidulous. Gynophore longer than calyx. Carpels 6. Fruiting racemes elongated, loose, with pedicels longer than calyx. Capsule depressed-globular, consisting of (5–)6 (–7) boat-shaped, 1 (–2)-seeded, almost separate carpels, hirsute only at margin. Seeds about 1 mm., reniform, somewhat tuberculate, yellow to brown. Fl. March–May.

Hab. : Desert wadis. Philistean Plain (very rare), Judean Desert, N. and C. Negev, Lower Jordan Valley, Dead Sea area, deserts of Ammon, Moav and Edom.

Area : E. Sudanian, extending into adjacent Saharo-Arabian territories.

Varies considerably in stature and in size of leaves.

47. MORINGACEAE

Deciduous low trees or shrubs. Leaves alternate, 2–3-imparipinnate with alternate or opposite pinnae; stipules sometimes present, mostly 0 or reduced to basal glands. Inflorescence an axillary, cymose, many-flowered, hairy panicle. Flowers hermaphrodite, zygomorphic, with perianth and stamens borne on the margin of a hypanthial cupule filled by a disk. Sepals 5. Petals 5, unequal. Stamens 5, anther-bearing, unequal, alternating with the 3–5 staminodes; filaments free; anthers 1-celled, dehiscing longitudinally. Ovary superior, 1-celled, borne on a short gynophore; carpels 3; placentation parietal; ovules numerous, anatropous, pendulous in 2 rows on each of the 3 placentae; style 1; stigma truncate. Fruit a many-seeded, 1-celled, beaked capsule, opening by 3 valves. Seeds many, large, 3-winged or wingless; endosperm 0; embryo straight.

One genus and 10 species in the tropics of the Old World.

1. MORINGA Adans.

Description as for family.

1. Moringa peregrina (Forssk.) Fiori, Agricolt. Colon. 5 : 59 (1911). *Hyperanthera peregrina* Forssk., Fl. Aeg.-Arab. 67 (1775). *M. aptera* Gaertn., Fruct. 2 : 315 (1791); Boiss., Fl. 2 : 23 (1872). *M. arabica* (Lam.) Pers., Syn. 1 : 460 (1805). [Plate 495]

Green-glaucous, sparingly leafy tree, 4–10 m. Branches divaricate or ascending, forming an ovoid or obovoid crown. Leaves 30 cm. or more; primary rhachis elongated, jointed, with 3 pairs of opposite slender ribs bearing a few remote, 1–2 × 0.1–0.2 cm., early deciduous, obovate-oblanceolate, obtuse leaflets. Panicles up to 30 cm., loose. Flowers 1.5 cm. or more across, pedicellate, white-pinkish to pale yellow. Hypanthial cupule very short, cup-shaped. Sepals oblong-lanceolate, acuminate. Petals up to 1.5 × 0.5 cm., oblong-obovate, obtuse. Capsule about 10–30 × 1.5 cm., pendulous, ribbed, brown. Seeds up to 1 cm., trigonous, ovoid, not winged, white. Fl. March–May.

Hab.: Tropical oases. Judean Desert (hot ravines), S. Negev, Lower Jordan Valley, Dead Sea area, Arava Valley, Edom.

Area: E. Sudanian.

DIAGNOSES PLANTARUM NOVARUM

Polygonum palaestinum Zoh. sp. nov. var. **palaestinum**. *P. equisetiforme* Sibth. et Sm. var. *arenarium* Eig et Feinbr., Palest. Journ. Bot. Jerusalem ser.2 : 99 (1940). [Tab. 56]

Planta perennis, erecta, glabra, elata (usque ad 150 cm. alta). Caules numerosi, plus minusve herbacei, rigidi, striati, internodiis elongatis. Folia 1–6 ×0.1–0.8 cm., cito decidua, subsessilia vel breviter petiolata, basi articulata, anguste elliptico-oblanceolata vel linearia, integra, utrinque cum nervis prominentibus; ochreae usque ad 2.5 cm. longae, membranaceae, basi brunneae, apice hyalinae, laceratae. Racemi simplices vel ramosi, elongati, valde interrupti, fasciculis axillaribus 2–3-floris distantibus compositi. Bracteae parvae, fasciculis breviores; ochreolae basi brunneo-virides, apice hyalinae, laceratae. Pedicelli floribus paulo breviores. Flores 3–4 × 3.5 mm. Tepala late obovata, superne alba, inferne virescente-rosea, fructifera parte superiore patentia, marginem aliformem circum achaenium formantia. Stylus brevissimus, tripartitus. Achaenium 3 mm. longum, perigonio conspicue exsertum, ovoideum, trigonum cum faciebus laeviter concavis, brunneum, nitidum. Fl. Martio–Octobri.

Holotype : Sharon Plain, 23 km. on the Nathania–Tel-Aviv road, sand dunes, 1952 *Orshan* 1020 (HUJ).

Var. negevense Zoh. et Waisel var. nov.

Planta elata, erecta. Caules graciliores quam in typo. Folia angustiora brevioraque. Flores et achaenia quam in typo minores, ca. 2 mm. longi. Tepala obtusa, truncata, nunquam apiculata.

Holotype : N. Negev, near Oron, wadi in desert, 1966 *Zohary* 1010 (HUJ).

Emex spinosa (L.) Campd., Monogr. Rum. 58 (1819) var. **minor** Zoh. et Waisel var. nov.

Folia 1–5 × 1–2.5 cm., flavido-viridia; petiolus lamina saepe longior. Fructus parvus, 4–6 mm. longus, multo spinosior quam in typo. Cotyledones lineares.

Holotype : C. Negev, Revivim, 1949 *D'Angelis* 191 (HUJ).

Silene telavivensis Zoh. et Plitm. sp. nov. [Tab. 117]

Annua, plus minusve pubescens, ad 35 cm. alta. Caules numerosi, a basi ramosi, erecti, tenues. Folia inferne conferta, oblongo-spatulata vel oblanceolato-linearia, margine ciliata, in petiolum attenuata, basi paulo dilatata et breviter connata. Inflorescentia unilateralis, spiciformis. Bracteae calyce multo breviores, lanceolato-lineares, ad marginem albidae. Pedicelli brevissimi. Calyx 8–9 mm. longus, anguste cylindricus, umbilicatus, cum nervis papilloso-hirsutis prominentibus, fructiferus aliquanto clavatus; dentes calycini breves, lineari-lanceolati, in fructu paulo recurvi. Petala alba vel rosea; unguis plus minusve exsertus; lamina in lacinias lineares bi-partita; coronae laciniae squamiformes. Capsula inclusa, cylindrica, carpophoro longior. Semina 0.5–1 mm., seriatim tuberculata. Fl. Aprili–Maio.

Holotype : Philistean Plain, Tel-Aviv, 1928 *Feinbrun* 03207 (HUJ).

Silene arabica Boiss., Fl. 1:593 (1867) var. **moabitica** Zoh. var. nov. *S. moabitica* Eig (in herb.). [Tab. 123]

Annua, hirsuto-glandulosa, 20–40 cm. alta, pilis simplicibus longissimis patentibus et pilis glandulosis longis brevibusve dense obsita. Folia oblongo-lanceolata, caulina late linearia, non plicata. Pedicelli inferiores calyce multo longiores. Petala alba vel rosea. Capsula carpophoro hirsuto aequilonga vel brevior.

Holotype: Moav, E. slopes facing Wadi Zerka Main, 1930 *Eig* and *Zohary* 1237 (HUJ).

Paronychia sinaica Fresen., Mus. Senckenb. 1 : 180 (1834) var. **negevensis** Zoh. var. nov.

Bracteae parvae, ca. 3 mm. longae. Capitula minora quam in typo.

Holotype: Negev, Asluj–Hafir, 20 km. before Hafir, 1929 *Naftolsky* 02592 (HUJ).

Suaeda vera Forssk. ex J. F. Gmel., Syst. ed. 13, 2 : 503 (1791) var. **deserti** Zoh. et Baum. var. nov.

Perianthii segmenta lanceolata, acuta. Bracteolae cum nervo medio inconspicuo.

Holotype: Negev, descent to Wadi Baqqara, small wadi bed, 1946 *D. Zohary* 911 (HUJ).

Haloxylon persicum Bge. in Buhse, Nouv. Mém. Soc. Nat. Mosc. 12 : 189 (1860).

Var. **idumaeum** Zoh. var. nov. [Tab. 242]

Folia acutiuscula, brevissime cuspidata.

Holotype: Arava Valley, Wadi Ghadian, sandy wadi, 1942 *D. Zohary* 50150 (HUJ).

Var. **maris-mortui** Zoh. var. nov.

Folia ut in varietate praecedente sed alae perianthii plus minusve regulariter crenatae.

Holotype: Dead Sea area, outlet of Wadi Jeib into the Ghor, 1942 *M.* and *D. Zohary* 101 (HUJ).

Salsola baryosma (Roem. et Schult.) Dandy in Andrews, Fl. Pl. Angl.-Eg. Sudan 1 : 111 (1950) var. **viridis** Zoh. var. nov. [Tab. 256]

Planta viridis, plus minusve regulariter ramosa, non foetida. Flores et perianthium fructiferum quam in typo minores. Antherae non appendiculatae. Ovarium subglobosum.

Holotype: Dead Sea area, Arnon River, 1936 *Eig, Feinbrun* and *Grizi* 121 (HUJ).

Nigella arvensis L., Sp. Pl. 534 (1753) ssp. **arvensis.**

Var. **beershevensis** Zoh. var. nov.

Planta gracilis, 15–25 cm. alta. Folia in segmenta longa, filiformia, rigida, tenuiter dissecta. Petali labium exterius lobis valde incurvis. Antherae plus minusve aristatae.

Holotype: N. Negev, env. of Beersheva, loess soil, 1957 *Zohary* and *Waisel* 875 (HUJ).

Nigella arvensis L. ssp. **tuberculata** (Griseb.) Bornm., Verh. Zool.-Bot. Ges. Wien 48 : 457 (1898).

Var. **negevensis** Zoh. var. nov.

Caules parce ramosi. Flores parvi (usque ad 1.5 cm. diam.). Petala 3 mm. lata; petali labium exterius lobis linearibus, 0.3–0.5 mm. latis, apice subclavatis. Antherae muticae.

Holotype : N.E. Negev, Mishor Yemin, sandy soil, 1957 *Zohary* and *Waisel* 482 (HUJ).

Var. **multicaulis** Zoh. var. nov.

Planta valde ramosa. Flores majusculi (2–2.5 cm. diam.). Sepala albido-viridia. Petali labium exterius lobis 2 mm. longis et circa 1 mm. latis, apice breviter ovato-clavatis. Carpella sparse tuberculata.

Holotype : W. Negev, env. of Wadi Sheneq, sandy-loess soils, 1949 *Zohary* and *Orshan* 837 (HUJ).

Nigella arvensis L. ssp. **divaricata** (Beaupré ex DC.) Zoh. stat. nov.

Var. **palaestina** Zoh. var. nov.

Folia parce vel vix divisa; divisiones oblongo-ellipticae, media 3–8 mm. vel ultra lata; folia superiora non divisa, oblongo-elliptica.

Holotype : Sharon Plain, sandy coast, 1956 *Zohary* and *Waisel* 898 (HUJ).

Var. **daucifolia** Zoh. var. nov.

Planta usque ad 20(–30) cm. alta. Folia usque ad 6 cm. vel ultra longa, 1–2-pin-natifida, inferiora dense rosulata, superiora in lobos lineares dissecta.

Holotype : Coast of Upper Galilee, near Rosh-Haniqra, 1956 *Zohary* and *Waisel* 911 (HUJ).

Consolida scleroclada (Boiss.) Schrödgr., Ann. Naturh. Mus. Wien 27 : 44 (1913) var. **moabitica** Zoh. var. nov.

Sinus inter corollae lobos superiores et laterales laevissimus; lobi laterales et inferiores brevissimi.

Holotype : Moav, Mt. Nebo, N. slope, grey stony soil, 1942 *Zohary* and *Feinbrun* 711 (HUJ).

Reseda stenostachya Boiss., Diagn. ser. 1, 1 : 5 (1843) var. **eilatensis** Plitm. et Zoh. var. nov.

Planta viridis, glabra. Bracteae et sepala raro decidui.

Holotype : Env. of Aqaba, Ras-el-Naqb, wadi, 1943 *Glober* 924 (HUJ).

Reseda boissieri Müll., Bot. Zeit. 14 : 37 (1856) var. **pendula** Plitm. et Zoh. var. nov. [Tab. 492]

Capsula pendula, obovoidea vel oblongo-obovoidea.

Holotype : Arava Valley, Wadi Fuqra, 1938 *Evenari* 111 (HUJ).

ADDENDA AND ERRATA

p. 6, line 6 from top, read (L. f.); add *Adiantum fragrans* L. f., Suppl. 447 (1781).

p. 6, line 18 from bottom, add *Acrostichum catanense* Cosent., Atti Accad. Gioenia ser. 1, 2 : 2107/218 (1827).

p. 139, line 10 from bottom, after L. add ssp. *maritima*

p. 147, line 12 from top, after *turcomanica* add (Moq.)

p. 200, after line 17 from top, add **Myosurus minimus** L., Sp. Pl. 284 (1753); damp rocks, Upper Galilee.

p. 324, line 4 from bottom, read *Z. myagrioides* Forssk.

Plate 404, for (Ait.) read (Sol.)

INDEX

(Synonyms in italics)

Achyranthes L.   188
  alternifolia L.   *187*
  argentea Lam.   *188*
  aspera L.   188
    var. *argentea* (Lam.) Boiss.   *188*
    var. sicula L.   188
  muricata L.   *187*
Acrostichum catanense Cosent.   *344*
  velleum Ait.   *6*
ADIANTACEAE   7
Adiantum L.   7
  capillus-veneris L.   7
  fragrans L. f.   *344*
Adonis L.   211
  aestivalis L.   212
    var. aestivalis   212
    var. *cupaniana* (Guss.) Huth   *212*
    var. palaestina (Boiss.) Zoh.   212
  aleppica Boiss.   211
  annua L.   213
  autumnalis L.   *213*
  cupaniana Guss.   212
  dentata Del.   213
    var. dentata   213
    var. subinermis Boiss.   213
  flammea Jacq.   212
  microcarpa Boiss.   *212*
  palaestina Boiss.   *212*
Aellenia Ulbrich   167
  autrani (Post) Zoh.   168
    var. *hierochuntica* (Bornm.) Zoh.   *168*
  hierochuntica (Bornm.) Aellen   *168*
  lancifolia (Boiss.) Ulbrich   168
Aerva Forssk.   186
  javanica auct. non (Burm. f.) Juss.   *186*
  javanica (Burm. f.) Juss. var. *bovei*
    Webb   *187*
  persica (Burm. f.) Merr.   186
    var. bovei (Webb) Chiov.   187
    var. persica   187
  tomentosa Forssk.   *186*
    var. *bovei* (Webb) C. B. Clarke   *187*
Aethionema R. Br.   296
  carneum (Banks et Sol.) Fedtsch.   297

Aethionema (cont.)
  cristatum DC.   *297*
  gileadense Post   296
  heterocarpum J. Gay   297
Agathophora alopecuroides (Del.) Bge.
  179
Agrostemma L.   80
  githago L.   81
  gracile Boiss.   81
AIZOACEAE   74
Aizoon L.   74
  canariense L.   75
  hispanicum L.   74
Albersia blitum auct. non Kunth   *186*
Alsine bocconi Scheele   *125*
  decipiens Fenzl   *116*
  formosa Fenzl   *116*
  globulosa (Labill.) C. A. Mey.   *115*
  media L.   *117*
  mediterranea (Ledeb.) J. Maly   *114*
  meyeri Boiss.   *115*
  mucronata Sibth. et Sm.   *114*
  picta (Sibth. et Sm.) Fenzl   *116*
  smithii Fenzl   *115*
  succulenta Del.   *127*
  tenuifolia (L.) Crantz   *114*
    var. *maritima* Boiss. et Heldr.   *114*
    var. *mucronata* (Sibth. et Sm.) Boiss.
      114
Alternanthera Forssk.   188
  pungens Kunth   189
  sessilis (L.) DC.   189
Alyssum L.   283
  aureum (Fenzl) Boiss.   *285*
  baumgartnerianum Bornm.   288
  campestre L.   *287*
  clypeatum L.   *283*
  damascenum Boiss. et Gaill.   285
  dasycarpum Steph. ex Willd.   286
  dimorphosepalum Eig   *286*
  homalocarpum (Fisch. et Mey.) Boiss.
    286
  iranicum Hausskn. ex Baumg.   286
  linifolium Steph. ex Willd.   285

Alyssum (cont.)
  marginatum Steud. ex Boiss.   287
    var. macrocarpum Zoh.   287
  meniocoides Boiss.   284
    var. aureum (Fenzl) Zoh.   285
  minus (L.) Rothm.   287
    var. minus   287
    var. strigosum (Banks et Sol.) Zoh.
      287
  parviflorum M. B.   *287*
  strigosum Banks et Sol.   *287*
  subspinosum Dudl.   288
  tetrastemon Boiss. var. *latifolium* Boiss.
    288
AMARANTHACEAE   180
Amaranthus L.   180
  albus L.   185
  angustifolius Lam.   *185*
    var. *graecizans* (L.) Thell.   *185*
  arenicola J. M. Johnston   183
  blitoides S. Wats.   184
  blitum L.   *185*
  caudatus L.   184
  chlorostachys Willd.   *181*
  delilei Richter et Loret   *182*
  gracilis Desf.   186
  graecizans L.   185
    var. graecizans   185
    var. sylvestris (Vill.) Aschers.   185
  hybridus L.   181
    ssp. *hypochondriacus* (L.) Thell. var.
      chlorostachys (Willd.) Thell.   *181*
  lividus auct. non L.   *186*
  muricatus Gillies ex Moq.   184
  palmeri S. Wats.   183
  retroflexus L.   182
    var. delilei (Richter et Loret) Thell.
      182
    var. retroflexus   182
  silvestris Vill.   *185*
  spinosus L.   183
  sylvestris Desf. var. *graecizans* (L.) Boiss.
    185
  tricolor L.   184
Anabasis L.   176
  aphylla auct. non L.   *177*
  articulata (Forssk.) Moq.   177
  haussknechtii auct. non Bge. ex Boiss.
    177

Anabasis (cont.)
    var. *longispicata* Eig   *177*
  setifera Moq.   178
  syriaca Iljin   177
    var. syriaca   177
    var. zoharyi (Iljin) Zoh.   177
  zoharyi Iljin   *177*
Anastatica L.   276
  hierochuntica L.   276
Anemone L.   198
  coronaria L.   198
    var. *albiflora* Rouy et Fouc.   *199*
    var. *coccinea* (Jord.) Rouy et Fouc.
      199
    var. *cyanea* (Risso) Ard.   *199*
    var. *incisa* Boiss.   *199*
    var. *parviflora* Boiss.   *199*
    var. *rosea* (Hanry) Rouy et Fouc.
      199
Ankyropetalum Fenzl   102
  coelesyriacum Boiss.   *103*
  gypsophiloides Fenzl   103
    var. coelesyriacum (Boiss.) Barkoudah
      103
    var. gypsophiloides   103
Anogramma Link   8
  leptophylla (L.) Link   8
Arabidopsis (DC.) Heynh.   257
  pumila (Steph. ex Willd.) Busch   257
Arabis L.   278
  albida Stev.   *280*
  aucheri Boiss.   280
  auriculata Lam.   *279*
  caucasica Schlecht.   280
  cremocarpa Boiss. et Bal.   *281*
  laxa Sibth. et Sm.   *281*
    var. *cremocarpa* (Boiss. et Bal.) Boiss.
      281
  nova Vill.   279
  turrita L.   280
  verna (L.) R. Br.   279
Arenaria L.   112
  bocconi Sol.   *125*
  deflexa Decne.   112
  diandra Guss.   *124*
  flaccida Roxb.   *122*
  globulosa Labill.   *115*
  glutinosa M. B.   *119*
  graveolens auct. non Schreb.   *112*

Arenaria (cont.)
 graveolens Schreb. var. *minuta* Post *112*
 hybrida Vill. *114*
 leptoclados (Reichb.) Boiss. *112*
 leptoclados (Reichb.) Guss. 112
 marginata DC. *123*
 media L. *123*
 mediterranea Ledeb. *114*
 mucronata Sibth. et Sm. *114*
 picta Sibth. et Sm. *116*
 rubra L. *124*
  var. *campestris* L. *124*
  var. *marina* L. *124*
 serpyllifolia sensu Guss. *112*
 serpyllifolia L. var. *leptoclados* Reichb. *113*
 tenuifolia L. *114*
 tremula Boiss. 113
Argemone mexicana L. 229
Aristolochia L. 47
 altissima Desf. *48*
 billardieri Jaub. et Sp. 48
  var. galilaea Zoh. 49
 maurorum L. 49
  var. *latifolia* Boiss. *49*
 paecilantha Boiss. 49
 parvifolia Sibth. et Sm. 48
 sempervirens L. 48
ARISTOLOCHIACEAE 47
Arthrocnemum Moq. 155
 fruticosum (L.) Moq. 156
 glaucum (Del.) Ung.-Sternb. *156*
 macrostachyum (Moric.) Moris et Delponte 156
 perenne (Mill.) Moss 156
Asarum hypocistis L. *50*
ASPIDIACEAE 12
Aspidium rigidum Swartz *12*
  var. *australe* Ten. *12*
ASPLENIACEAE 10
Asplenium L. 10
 adiantum-nigrum L. 10
  ssp. onopteris (L.) Heufl. 10
  var. *virgilii* (Bory) Boiss. *10*
 ceterach L. *11*
 hemionitis L. *11*
 onopteris L. *10*
Atraphaxis L. 68

Atraphaxis (cont.)
 spinosa L. 68
  var. sinaica (Jaub. et Sp.) Boiss. 69
Atriplex L. 143
 alexandrinum Boiss. *146*
 dimorphostegia Kar. et Kir. 147
 halimus L. 145
  var. halimus 145
  var. schweinfurthii Boiss. 145
 hastata L. 149
  var. microtheea Schum. 149
 inamoena Aellen *147*
 laciniata L. var. *turcomanica* Moq. *147*
 lasiantha Boiss. 149
 leucoclada Boiss. 146
  ssp. *eu-leucoclada* Aellen *147*
  var. inamoena (Aellen) Zoh. 147
  var. leucoclada 147
  ssp. *turcomanica* (Moq.) Aellen 344
  var. turcomanica (Moq.) Zoh. 147
 microcarpa Benth. *150*
 nitens Schkuhr 147
 palaestinum Boiss. *146*
 parvifolium Lowe var. *alexandrinum* (Boiss.) Eig *146*
  var. *confertum* Eig *146*
  var. *palaestinum* (Boiss.) Eig *146*
 patulum L. var. *palaestinum* Eig *149*
 portulacoides L. *151*
 rosea L. 148
 semibaccata R. Br. 150
  var. melanocarpa Aellen 150
  var. microcarpa (Benth.) Aellen 150
  var. semibaccata 150
 stylosa Viv. 146
  var. alexandrina (Boiss.) Zoh. 146
  var. conferta (Eig) Zoh. 146
  var. stylosa 146
 tatarica L. 148
  var. desertorum Eig 149
 tataricum L. var. *hierosolymitanum* Eig *149*
  var. *virgatum* Boiss. *149*

Bassia All. 152
 eriophora (Schrad.) Aschers. 153
  var. *rosea* Eig *153*
 joppensis Bornm. et Dinsmore *153*

348   INDEX

Bassia (cont.)
   muricata (L.) Aschers.   152
      var. *brevispina* Bornm.   *153*
Behen vulgaris Moench   *85*
BERBERIDACEAE   214
Beta L.   138
  maritima L.   *139*
  vulgaris L.   139
    var. *foliosa* Aellen   *139*
    ssp. *foliosa* auct.   *139*
    var. *glabra* (Del.) Aellen   *139*
    ssp. macrocarpa (Guss.) Thell.   139
    var. maritima   139
    ssp. maritima (L.) Arcang.   139
    var. *maritima* (L.) Boiss.   *139*
    var. orientalis Moq.   139
    var. pilosa (Del. ex Moq.) Aellen
      139
Biscutella L.   298
  ciliata DC.   *299*
  columnae Ten.   *299*
  didyma L.   298
    var. ciliata (DC.) Hal.   299
    var. columnae (Ten.) Hal.   299
    var. leiocarpa (DC.) Hal.   299
  leiocarpa DC.   *299*
Boerhavia L.   71
  africana Lour.   *70*
  plumbaginea Cav.   *70*
    var. *glabrata* Boiss.   *71*
    var. *viscosa* Ehrenb. ex Aschers. et
      Schweinf.   *71*
  repens L.   72
  verticillata Poir.   *71*
Bolanthus (Ser.) Reichb.   102
  filicaulis (Boiss.) Barkoudah   102
Bongardia C. A. Mey.   215
  chrysogonum (L.) Sp.   215
Boreava Jaub. et Sp.   258
  aptera Boiss. et Heldr.   258
Brassica L.   308
  acris O. Ktze. f. *brevisiliquosa* O. Ktze.
    307
  bracteolata Fisch. et Mey.   *310*
  erysimoides Sieb. ex Spreng.   *314*
  nigra (L.) Koch   309
    var. bracteolata (Fisch. et Mey.) Sp.
      ex Coss.   310
    var. nigra   310

Brassica (cont.)
  orientalis L.   *303*
  purpurascens Banks et Sol.   *276*
  sinaica Boiss.   *305*
  tournefortii Gouan   309
Bufonia L.   117
  virgata Boiss.   117
Bunias aegyptiaca L.   *257*
  myagroides L.   *317*
  spinosa L.   *324*

Cakile Mill.   320
  maritima Scop.   320
Calepina Adans.   324
  corvini (All.) Desv.   *324*
  irregularis (Asso) Thell.   324
Calligonum L.   67
  comosum L'Hér.   68
Camelina Crantz   292
  hispida Boiss.   292
  lasiocarpa Boiss. et Bl.   *293*
  persistens Rech. f.   *292*
CAPPARACEAE   240
Capparis L.   241
  aegyptia Lam.   *242*
  cartilaginea Decne.   244
  decidua (Forssk.) Edgew.   244
  galeata Fresen.   *244*
  herbacea Willd. var. *microphylla* Ledeb.
    244
  ovata Desf.   243
    var. microphylla (Ledeb.) Zoh.   244
    var. palaestina Zoh.   243
  sodada R. Br.   *244*
  spinosa L.   242
    var. aegyptia (Lam.) Boiss.   242
    var. aravensis Zoh.   242
    var. *canescens* Coss.   *243*
    var. spinosa   243
Capsella Medik.   294
  bursa-pastoris (L.) Medik.   294
    var. bursa-pastoris   294
    var. minuta Post   294
  procumbens (L.) Fries   *295*
  rubella Reut.   294
Cardamine L.   278
  hirsuta L.   278
  lunaria L.   *282*
Cardaria Desv.   302

Cardaria (cont.)
  draba (L.) Desv. 302
Caroxylon salicornicum Moq. *164*
Carrichtera Adans. 314
  annua (L.) DC. 314
  vellae DC. *314*
CARYOPHYLLACEAE 78
Caylusea St. Hil. 339
  canescens (L. ed Murray) Walp. *339*
  hexagyna (Forssk.) Green 339
Celtis australis L. 36
Cerastium L. 119
  anomalum Waldst. et Kit. ex Willd. *120*
  dichotomum L. 120
  dubium (Bast.) O. Schwarz 120
  glomeratum Thuill. 121
  inflatum Link ex Sweet 121
  viscosum L. *121*
Ceratocapnos Dur. 236
  palaestinus Boiss. 237
Ceratocephala Moench 210
  falcata (L.) Pers. 210
    var. exscapa Boiss. 211
    var. incurva (Stev.) Boiss. 211
    var. vulgaris Boiss. 211
CERATOPHYLLACEAE 218
Ceratophyllum L. 219
  demersum L. 219
  submersum L. 219
Ceterach DC. 10
  officinarum DC. 11
Cheilanthes Swartz 5
  catanensis (Cosent.) H. P. Fuchs 6
  fragrans (L.) Webb et Berth. *6*
  fragrans (L. f.) Swartz 6
  pteridioides (Reichard) Christens. *6*
Cheiranthus chius L. *265*
  bicornis Sibth. et Sm. *273*
  farsetia L. *281*
  lividus Del. *273*
  longipetalus Vent. *272*
  parviflorus Schousb. *272*
  tricuspidatus L. *271*
Chelidonium corniculatum L. *231*
  hybridum L. *229*
Chenolea Thunb. 152
  arabica Boiss. 152
CHENOPODIACEAE 136

Chenopodina asphaltica Boiss. *158*
  setigera (DC.) Moq. *161*
Chenopodium L. 140
  aegyptiacum Hasselq. *161*
  album L. 142
  ambrosioides L. 140
  baryosmon Roem. et Schult. *174*
  fruticosum L. *159*
  murale L. 142
    var. *humile* Peterm. *143*
    var. *microphyllum* Boiss. *143*
  opulifolium Schrad. ex Koch et Ziz 142
  polyspermum L. 141
  rubrum L. 143
  vulvaria L. 141
Chorispora R. Br. 275
  purpurascens (Banks et Sol.) Eig 276
  syriaca Boiss. *276*
Clematis L. 199
  cirrhosa L. 199
  flammula L. 200
Cleome L. 245
  africana Botsch. *246*
  arabica Botsch. *246*
  arabica L. 246
  droserifolia (Forssk.) Del. 245
  trinervia Fresen. 246
Clypeola L. 289
  aspera (Grauer) Turrill 290
  echinata DC. *290*
  jonthlaspi L. 290
    var. glabriuscula Grun. 290
    var. jonthlaspi 290
    var. lasiocarpa Guss. 290
  lappacea Boiss. 291
  microcarpa Moris *290*
  minor L. *287*
Cocculus DC. 216
  laeba (Del.) DC. *216*
  pendulus (J. R. et G. Forst.) Diels 216
Cometes L. 135
  abyssinica R. Br. 135
Commicarpus Standl. 70
  africanus (Lour.) Dandy 70
    var. africanus 71
    var. viscosus (Ehrenb. ex Aschers. et Schweinf.) Baum 71
  verticillatus (Poir.) Standl. 71

Conringia Adans.   303
  orientalis (L.) Andrz.   303
Consolida (DC.) S. F. Gray   196
  flava (DC.) Schrödgr.   197
  rigida (DC.) Bornm. 197
  scleroclada (Boiss.) Schrödgr.   198
    var. moabitica Zoh.   198, 343
Cordylocarpus Desf.   328
  muricatus Desf.   328
    var. palaestinus Eig ex Zoh.   329
Coronopus Hall.   302
  squamatus (Forssk.) Aschers.   303
Corrigiola L.   129
  litoralis L.   129
    ssp. *eulittoralis* Briq.   *129*
    ssp. litoralis   129
    ssp. telephiifolia (Pourr.) Briq.   129
  repens Forssk.   *127*
  telephiifolia Pourr.   *129*
Crambe L.   322
  glabrata DC.   *323*
  hispanica L.   323
    var. edentula Boiss.   323
    var. glabrata (DC.) Coss.   323
    var. hispanica   323
  orientalis L.   322
    var. *aucheri* Boiss.   *322*
CRUCIFERAE   246
Cryophytum crystallinum (L.) N.E. Br.
  76
  nodiflorum (L.) L. Bol.   *76*
Cucubalus aegyptiacus L.   *89*
  behen L.   *85*
  inflatus Salisb.   *85*
  italicus L.   *84*
  venosus Gilib.   *85*
CUPRESSACEAE   18
Cupressus L.   19
  horizontalis Mill.   *19*
  sempervirens L.   19
    var. horizontalis (Mill.) Gordon   19
    var. pyramidalis (Targ.-Tozz.) Nym.
    19
CYNOMORIACEAE   50
Cynomorium L.   51
  coccineum L.   51
CYTINACEAE   50
Cytinus L.   50
  hypocistis (L.) L.   50

Cytinus (cont.)
  ssp. orientalis Wettst.   50

Delphinium L.   195
  anthoroideum Boiss. var. *sclerocladum*
    (Boiss.) Boiss.   *198*
  bovei Decne.   *196*
  chodati Opphr.   *196*
  deserti Boiss.   *197*
  eriocarpum (Boiss.) Hal.   *196*
  flavum DC.   *197*
  halteratum Sibth. et Sm.   *196*
  ithaburense Boiss.   195
  peregrinum L.   196
  rigidum DC.   *197*
  sclerocladum Boiss.   *198*
  virgatum Poir.   *196*
Descurainia Webb et Berth.   256
  sophia (L.) Webb ex Prantl   256
    var. brachycarpa (Boiss.) O. E. Schulz
    256
    var. sophia   256
Dianthus L.   106
  cyri Fisch. et Mey.   109
    var. corymbosus Opphr.   110
  deserti Post   *108*
  judaicus Boiss.   110
  libanotis Labill.   110
  multipunctatus Ser.   *108*
    var. *velutinus* Boiss.   *108*
  palaestinus Freyn   *108*
  pallens Sibth. et Sm. var. *oxylepis* Boiss.
    110
  pendulus Boiss. et Bl.   107
  polycladus Boiss.   109
  sinaicus Boiss.   111
  strictus Banks et Sol.   108
    var. *axilliflorus* (Fenzl) Eig   *109*
    var. strictus   108
    var. velutinus (Boiss.) Eig   108
  strictus Sibth. et Sm.   *108*
  tripunctatus Sibth. et Sm.   109
  velutinus Guss.   *106*
Didesmus rostratus Boiss.   318
Digera Forssk.   187
  alternifolia (L.) Aschers.   *187*
  arvensis Forssk.   *187*
  muricata (L.) Mart.   187
Diplotaxis DC.   305

Diplotaxis (cont.)
 acris (Forssk.) Boiss.   306
 biloba C. Koch   *254*
 erucoides (L.) DC.   307
  var. erucoides   307
  var. valentina (Pau) O. E. Schulz
   307
 harra (Forssk.) Boiss.   306
  var. brevisiliquosa (O. Ktze.) O. E.
   Schulz   307
  var. *glabra* Post   *307*
  var. harra   307
  var. subglabra (DC.) O. E. Schulz
   307
 hispida DC. var. *subglabra* DC.   *307*
 valentina Pau   *307*
 viminea (L.) DC.   308
  var. integrifolia Guss.   308
  var. subcaulescens Lojacono-Poj.   308
  var. viminea   308
Draba verna L.   *291*
Dryopteris Adans.   12
 rigida (Swartz) A. Gray   *12*
 thelypteris (L.) A. Gray   *9*
 villarii (Bellardi) H. Woynar ex Schinz
  et Thell.   12
  var. australis (Ten.) Maire   12

Emex Campd.   66
 spinosa (L.) Campd.   67
  var. minor Zoh. et Waisel   67, 341
  var. pusilla Bég. et Vaccari   67
  var. spinosa   67
Enarthrocarpus Labill.   327
 arcuatus Labill.   327
 lyratus (Forssk.) DC.   328
 strangulatus Boiss.   327
Ephedra L.   21
 alata Decne.   22
  var. decaisnei Stapf   22
 alte C. A. Mey.   22
 campylopoda C. A. Mey.   22
 foliata Boiss. et Ky. ex Boiss.   23
 peduncularis Boiss.   23
Ephedraceae   21
Epibaterium pendulum J. R. et G. Forst.
 216
Equisetaceae   3
Equisetum L.   3

maximum auct. non Lam.   *3*
 ramosissimum Desf.   4
  var. altissimum A. Br. ex Milde   4
  var. procerum (Poll.) Aschers.   4
  var. simplex (Döll.) Milde   4
 ramosum Schl. ex DC.   *4*
 telmateia Ehrh.   3
Eremobium Boiss.   266
 aegyptiacum (Spreng.) Aschers. et
  Schweinf. ex Boiss.   266
  var. aegyptiacum   267
  var. lineare (Del.) Zoh.   267
 diffusum (Decne.) Botsch.   *267*
 lineare (Del.) Aschers. et Schweinf. ex
  Boiss.   *267*
 pyramidatum (Presl) Botsch.   *267*
Erophila DC.   291
 minima C. A. Mey.   292
 verna (L.) Bess.   291
  ssp. praecox (Stev.) Walters   292
Eruca Mill.   313
 cappadocica Reut. var. *eriocarpa* Boiss.
  314
 sativa Mill.   313
  var. eriocarpa (Boiss.) Post   314
  var. erysimoides (Sieb. et Spreng.)
   Fiori   314
  prol. *longirostris* (v. Uechtritz) Rouy
   314
  var. sativa   313
Erucaria Gaertn.   316
 aleppica Gaertn.   *317*
  var. *horizontalis* Post   *319*
  ssp. *latifolia* Bornm.   *319*
  var. *polysperma* Boiss.   *318*
  var. *puberula* Boiss.   *319*
  var. *subintegrifolia* Bornm.   *319*
 boveana Coss.   318
  var. alata Zoh.   319
  var. boveana   319
  var. horizontalis (Post) O. E. Schulz
   319
  var. lyrata O. E. Schulz   319
  var. maris-mortui Zoh.   319
  var. patula Zoh.   319
  var. puberula (Boiss.) O. E. Schulz
   319
  var. subintegrifolia (Bornm.) O. E.
   Schulz   319

Erucaria (cont.)
    var. torulosa Zoh. 319
  hispanica (L.) Druce 317
    var. hispanica 317
    var. patula Zoh. 318
    var. polysperma (Boiss.) Zoh. 318
    var. tanacetifolia Opphr. 318
  lineariloba Post *318*
  microcarpa Boiss. *315*
  myagroides (L.) Hal. *317*
    var. *patula* Zoh. *318*
    var. *polysperma* (Boiss.) O. E. Schulz
      318
    var. *tel-avivensis* Zoh. *317*
  uncata (Boiss.) Aschers. et Schweinf.
    320
    var. dasycarpa O. E. Schulz 320
    var. uncata 320
Erysimum L. 261
  bicorne Sol. *275*
  crassipes Fisch. et Mey. 262
  officinale L. *255*
  oleifolium J. Gay 262
  repandum L. 262
  rigidum DC. *262*
  scabrum DC. 262
Exolus muricatus Moq. *184*

FAGACEAE 29
Farsetia Turra 281
  aegyptiaca Turra 281
    var. aegyptiaca 282
    var. ovalis (Boiss.) Post 282
  eriocarpa DC. *283*
  ovalis Boiss. *282*
  rostrata Schenk *283*
Fibigia Medik. 283
  clypeata (L.) Medik. 283
    var. clypeata 283
    var. eriocarpa (DC.) Thiéb. 283
  eriocarpa (DC.) Boiss. *283*
  rostrata (Schenk) Boiss. *283*
Ficus L. 37
  carica L. 37
    var. caprificus Risso 38
  johannis Boiss. 38
  palmata Schweinf. *38*
  palmata Forssk. 38
  persica Boiss. 38

Ficus (cont.)
  pseudo-sycomorus Decne. 38
  sycomorus L. 38
Forsskaolea L. 42
  tenacissima L. 43
Fumaria L. 237
  agraria Lag. 240
  anatolica Boiss. *239*
  asepala Boiss. 240
  bracteosa Pomel *239*
  capreolata L. 238
  densiflora DC. 239
    var. densiflora 239
    var. parlatoriana (Kral. ex Boiss.) Zoh.
      239
  gaillardotii Boiss. 240
  judaica Boiss. 238
    f. bipunctata Aarons. ex Opphr. et
      Evenari 238
  kralikii Jord. 239
  macrocarpa Parl. 238
    var. *oxyloba* (Boiss.) Hamm. *238*
  megalocarpa Boiss. et Sprunn. *238*
  micrantha Lag. *239*
    var. *parlatoriana* Kral. ex Boiss. *239*
  parviflora Lam. 240
  thuretii Boiss. 239
FUMARIACEAE 233

Gamosepalum alyssoides (L.) Hausskn.
  288
Garidella unguicularis Poir. *192*
Girgensohnia Bge. 175
  oppositiflora (Pall.) Fenzl 176
Gisekia L. 73
  pharnacioides L. 73
Githago gracilis (Boiss.) Boiss. *81*
  segetum Desf. *81*
Glaucium Mill. 230
  aleppicum Boiss. et Hausskn. ex Boiss.
    232
  arabicum Fresen. 231
    var. *gracilescens* Fedde *232*
  corniculatum (L.) J. H. Rud. 231
  flavum Crantz 233
  grandiflorum Boiss. et Huet 232
    var. grandiflorum 232
    var. judaicum (Bornm.) Sam. ex
      Rech. f. 232

Glaucium (cont.)
judaicum Bornm. *232*
luteum Scop. *233*
Glinus L. 72
dictamnoides Burm. f. *73*
lotoides L. 73
var. dictamnoides (Burm. f.) Maire 73
var. lotoides 73
Gomphrena sessilis L. *189*
Gymnocarpos Forssk. 129
decandrum Forssk. 130
fruticosum (Vahl) Pers. *130*
GYMNOGRAMMACEAE 8
Gymnogramme leptophylla (L.) Desv. *8*
Gypsophila L. 100
arabica Barkoudah 100
capillaris (Forssk.) Christens. *101*
filicaulis (Boiss.) Bornm. *102*
hirsuta (Labill.) Spreng. var. *filicaulis* Boiss. *102*
pilosa Huds. 101
porrigens (L.) Boiss. *101*
rokejeka auct. non Del. *100*
rokejeka Del. 101
viscosa Murr. 101

Halimione Aellen 150
portulacoides (L.) Aellen 151
Halimocnemis pilosa Moq. *178*
Halocnemum M. B. 154
strobilaceum (Pall.) M. B. 155
Halogeton C. A. Mey. 179
alopecuroides (Del.) Moq. 179
var. alopecuroides 179
var. papillosus Maire 179
Halopeplis Bge. ex Ung.-Sternb. 154
amplexicaulis (Vahl) Ung.-Sternb. ex Ces., Passer. et Gib. 154
Halotis Bge. 178
pilosa (Moq.) Iljin 178
Haloxylon Bge. 166
articulatum (Cav.) Bge. *164*
ssp. *ramosissimum* (Benth. et Hook. f.) Eig *166*
persicum Bge. 166
var. idumaeum Zoh. 167, 342
var. maris-mortui Zoh. 167, 342
ramosissimum Benth. et Hook. f. *166*

Haloxylon (cont.)
salicornicum (Moq.) Bge. ex Boiss. *164*
schmittianum Pomel *165*
schweinfurthii Aschers. *165*
scoparium Pomel *164*
Hammada Iljin 163
eigii Iljin 166
elegans (Bge.) Botsch. *165*
hispanica Botsch. 164
negevensis Iljin et Zoh. 164
ramosissima (Benth. et Hook. f.) Iljin *166*
salicornica (Moq.) Iljin 164
schmittiana (Pomel) Botsch. 165
scoparia (Pomel) Iljin 164
syriaca Botsch. *166*
Herniaria L. 133
alsines-folia Mill. *126*
cinerea DC. *133*
diandra Bge. *134*
glabra L. 133
hemistemon J. Gay 134
hirsuta L. 133
lenticulata Forssk. *132*
Hesperis L. 263
acris Forssk. *306*
africana L. *265*
crenulata DC. *265*
diffusa Decne. *267*
nitens Viv. *305*
pendula DC. 263
pulchella DC. *269*
pygmaea Del. *269*
verna L. *279*
Hirschfeldia Moench 312
adpressa Moench *312*
incana (L.) Lagrèze-Fossat 312
var. *geniculata* (Desf.) Bonn. *313*
Holosteum L. 119
glutinosum (M. B.) Fisch. et Mey. 119
linifolium Fisch. et Mey. *119*
umbellatum L. 119
Hussonia uncata Boiss. *320*
Hutchinsia procumbens (L.) Desv. *295*
Hymenolobus Nutt. ex Torr. et A. Gray 295
procumbens (L.) Nutt. ex Torr. et A. Gray 295
Hypecoum L. 234

Hypecoum (cont.)
  aegyptiacum (Forssk.) Aschers. et
    Schweinf.   235
  deuteroparviflorum Fedde   235
  geslinii Coss. et Kral.   236
  grandiflorum Benth.   *235*
  imberbe Sm.   235
  parviflorum Barb.   *235*
  pendulum L.   236
  procumbens L.   234
Hyperanthera peregrina Forssk.   *340*
HYPERICACEAE   220
Hypericum L.   221
  acutum Moench   223
    var. anagallidioides (Jaub. et Sp.)
      Boiss.   223
  anagallidioides Jaub. et Sp.   *223*
  crispum L.   *223*
  hircinum L.   221
  hyssopifolium Chaix   222
    var. latifolium Boiss.   222
  lanuginosum Lam.   223
  nanum Poir.   222
  serpyllifolium Lam.   222
  tetrapterum Fries   *223*
  triquetrifolium Turra   223

Iberis L.   297
  odorata L.   298
Illecebrum arabicum L.   *132*
Iresine persica Burm. f.   *186*
Isatis L.   259
  aegyptica L.   *260*
  aleppica Scop.   *260*
  lusitanica L.   260
    var. *pamphylica* Boiss.   *260*
  microcarpa J. Gay ex Boiss.   260
    var. blepharocarpa Aschers.   260
    var. microcarpa   260

Juniperus L.   19
  oxycedrus L.   20
  phoenica L.   20

Kochia Roth   153
  eriophora Schrad.   *153*
  indica Wight   153
  latifolia Fresen.   *153*
  muricata Schrad.   *152*

Kohlrauschia velutina (Guss.) Reichb.
  106
Koniga arabica Boiss.   *289*
  libyca (Viv.) R. Br.   *289*

LAURACEAE   189
Laurus L.   190
  nobilis L.   190
Leontice L.   214
  chrysogonum L.   *215*
  leontopetalum L.   214
    var. *minor* Ky. ap. Sam. ex Rech. f.
      215
  minor Boiss.   *215*
Lepidium L.   299
  aucheri Boiss.   301
  cornutum Sm.   *300*
  draba L.   *302*
  latifolium L.   300
  procumbens L.   *295*
  ruderale L.   302
  sativum L.   301
  spinescens DC.   300
  spinosum Ard.   300
  squamatum Forssk.   *303*
Leptaleum DC.   269
  filifolium (Willd.) DC.   270
Lobularia Desv.   288
  arabica (Boiss.) Muschl.   289
  libyca (Viv.) Webb et Berth.   289
Loeflingia L.   128
  hispanica L.   128
LORANTHACEAE   45
Loranthus L.   46
  acaciae Zucc.   46
Lunaria libyca Viv.   *289*
  parviflora Del.   *315*

Maerua Forssk.   241
  crassifolia Forssk.   241
    var. maris-mortui Zoh.   241
  uniflora Vahl   *241*
Malcolmia R. Br.   264
  aegyptiaca Spreng.   *266*
    var. *aegyptiaca* Coss.   *267*
    var. *linearis* (Del.) Coss.   *267*
  africana (L.) R. Br.   265
  chia (L.) DC.   265
  confusa Boiss.   *268*

Malcolmia (cont.)
  conringioides Boiss.   *266*
  crenulata (DC.) Boiss.   265
    var. albiflora Eig   266
    var. crenulata   266
  exacoides (DC.) Spreng.   266
  nana (DC.) Boiss.   *268*
  pulchella (DC.) Boiss.   *269*
  pygmea Boiss.   *269*
  torulosa (Desf.) Boiss.   *267*
    var. *leiocarpa* Boiss.   *268*
Maresia Pomel   268
  nana (DC.) Batt.   268
    var. nana   268
  pulchella (DC.) O. E. Schulz   269
  pygmaea (Del.) O. E. Schulz   269
Marsilea L.   13
  diffusa auct. non Lepr. ex A. Br.   *14*
  minuta L.   14
MARSILEACEAE   13
Matthiola R. Br.   270
  arabica Boiss.   271
  aspera Boiss.   271
    var. aspera   271
    var. leiocarpa Bornm.   271
  bicornis (Sibth. et Sm.) DC.   *273*
  linearis Del.   *267*
  livida (Del.) DC.   273
    var. leiocarpa Zoh.   273
    var. livida   273
  longipetala (Vent.) DC.   272
    var. bicornis (Sibth. et Sm.) Zoh.
      273
    var. brachyloma Zoh.   273
    var. contorta Zoh.   273
    var. longipetala   273
    var. *macrocarpa* (Eig) Zoh.   *273*
    var. *oxyceras* (DC.) Zoh.   *273*
  oxyceras DC.   *273*
    var. *forcipifera* Boiss.   *273*
    var. *lunata* Boiss.   *273*
  parviflora (Schousb.) R. Br.   272
    var. maris-mortui Zoh.   272
  tricuspidata (L.) R. Br.   271
Meniocus aureus Fenzl   *285*
MENISPERMACEAE   215
Mesembryanthemum L.   76
  crystallinum L.   76
  forskahlei Hochst.   *77*

Mesembryanthemum (cont.)
  forsskalii Hochst. ex Boiss.   77
  nodiflorum L.   76
Minuartia L.   113
  decipiens (Fenzl) Bornm.   116
  formosa (Fenzl) Mattf.   116
    var. glabra Opphr.   117
  globulosa (Labill.) Schinz et Thell.   115
  hybrida (Vill.) Schischk.   114
    ssp. hybrida var. hybrida   115
    var. palaestina McNeill   115
  mediterranea (Ledeb.) K. Maly   114
  mesogitana (Boiss.) Hand.-Mazzetti   115
    ssp. mesogitana   115
  meyeri (Boiss.) Bornm.   115
  picta (Sibth. et Sm.) Bornm.   116
    var. albiflora (Eig) Zoh.   116
    var. brachypetala (Boiss.) Zoh.   116
  tenuifolia (L.) Hiern.   *114*
Mnemosilla aegyptiaca Forssk.   *235*
MOLLUGINACEAE   72
Mollugo tetraphylla L.   *126*
MORACEAE   37
Morettia DC.   274
  canescens Boiss.   274
  parviflora Boiss.   275
  philaeana (Del.) DC.   274
Moricandia DC.   304
  clavata Boiss. et Reut.   *304*
  dumosa Boiss.   *305*
  nitens (Viv.) Dur. et Barr.   305
  sinaica (Boiss.) Boiss.   305
Moringa Adans.   340
  aptera Gaertn.   *340*
  arabica (Lam.) Pers.   *340*
  peregrina (Forssk.) Fiori   340
MORINGACEAE   340
Myagrum L.   258
  irregulare Asso   *324*
  orientale L.   *322*
  perfoliatum L.   258
  rugosum L.   *321*
Myosurus minimus L.   344

Nasturtiopsis Boiss.   263
  arabica Boiss.   263
  coronopifolia (Desf.) Boiss. var. *arabica*
    (Boiss.) O. E. Schulz   *263*
Nasturtium R. Br.   277

Nasturtium (cont.)
  amphibium (L.) R. Br.   *277*
  officinale R. Br.   277
Nephrodium rigidum (Swartz) Desv.   *12*
  thelypteris (L.) Stremp.   *9*
Neslia Desv.   293
  apiculata Fisch., Mey. et Avé-Lall.   293
  paniculata (L.) Boiss.   *293*
Nigella L.   191
  arvensis L.   192
    var. arabica (Boiss.) Zoh.   193
    ssp. arvensis   193
    var. arvensis   193
    var. beershevensis Zoh.   193, 342
    var. daucifolia Zoh.   194, 343
    var. divaricata   194
    ssp. divaricata (Beaupré ex DC.) Zoh.
      194
    var. *glauca* Boiss.   *193*
    var. multicaulis Zoh.   194, 343
    var. negevensis Zoh.   194, 343
    var. palaestina Zoh.   194, 343
    var. submutica Bornm.   194
    var. tuberculata   193
    ssp. tuberculata (Griseb.) Bornm.
      193
  ciliaris DC.   194
  deserti Boiss.   *193*
    var. *arabica* Boiss.   *193*
  divaricata Beaupré ex DC.   *194*
  sativa L.   195
  tuberculata Griseb.   *193*
  unguicularis (Poir.) Spenn.   192
Noaea Moq.   175
  mucronata (Forssk.) Aschers. et
    Schweinf.   175
  spinosissima (L. f.) Moq.   *175*
Notholaena lanuginosa (Desf.) Desv. ex
  Poir.   *6*
  vellea (Ait.) R. Br. emend. Desv.   *6*
Notoceras R. Br.   275
  bicorne (Sol.) Caruel   275
  canariense R. Br.   *275*
Nuphar Sm.   218
  lutea (L.) Sm.   218
Nyctaginaceae   70
Nymphaea L.   217
  alba L.   217
  caerulea Savigny   217

Nymphaea (cont.)
  lutea L.   *218*
Nymphaeaceae   216

Obione portulacoides (L.) Moq.   *151*
Ochradenus Del.   330
  baccatus Del.   330
Ochthodium DC.   257
  aegyptiacum (L.) DC.   257
Oligomeris Cambess.   330
  linifolia (Vahl ex Hornem.) Macbride
    330
  subulata (Del.) Boiss.   330
Ophioglossaceae   5
Ophioglossum L.   5
  lusitanicum L.   5
Opophytum forskahlii (Hochst. ex Boiss.)
  N. E. Br.   *77*
Osyris L.   44
  alba L.   44

Paeonia L.   220
  corallina Retz.   *220*
  mascula (L.) Mill.   220
  officinalis L. var. *mascula* L.   *220*
Paeoniaceae   219
Panderia Fisch. et Mey.   151
  pilosa Fisch. et Mey.   151
Papaver L.   224
  apulum Ten. var. *gracillimum* Fedde
    228
  argemone L.   228
    var. argemone   228
    var. glabratum (Coss. et Germ.) Rouy
      et Fouc.   228
  belangeri Boiss.   *228*
  carmeli Feinbr.   226
  divergens Fedde et Bornm.   229
  humile Fedde   227
    ssp. humile   227
    ssp. sharonense Feinbr.   227
  hybridum L.   228
    var. grandiflorum Boiss.   229
    var. *siculum* (Guss.) Arcang.   *229*
  obtusifolium Desf. var. *barbeyi* Fedde
    226
  polytrichum Boiss. et Ky.   227
    f. subintegrum (O. Ktze.) Fedde   228
  rhoeas L. var. *oblongatum* Boiss.   *226*

Papaver (cont.)
var. *syriacum* Boiss.   *226*
strigosum (Bönningh.) Schur var. *gaillardotii* Fedde   *227*
subpiriforme Fedde   226
syriacum Boiss. et Bl.   226
Papaveraceae   224
Parietaria L.   41
alsinifolia Del.   42
diffusa Mert. et Koch   41
judaica sensu Boiss.   *41*
lusitanica L.   42
Paronychia Mill.   130
arabica (L.) DC.   132
var. breviseta Aschers. et Schweinf.   132
var. longiseta (Bertol.) Aschers. et Schweinf.   132
argentea Lam.   131
capitata (L.) Lam.   131
desertorum Boiss.   132
echinata auct. non (Desf.) Lam.   *132*
echinulata Chater   132
lenticulata (Forssk.) Aschers. et Schweinf.   *132*
palaestina Eig   130
sclerocephala Decne.   *135*
sinaica Fresen.   131
var. flavescens Boiss.   131
var. negevensis Zoh.   131, 342
var. sinaica   131
Peltaria aspera Grauer   *290*
glastifolia DC.   *259*
Petrorhagia (Ser.) Link   105
arabica (Boiss.) P. W. Ball et Heywood   106
cretica (L.) P. W. Ball. et Heywood   105
velutina (Guss.) P. W. Ball et Heywood   106
Phyllitis Hill   11
hemionitis (Swartz) O. Ktze.   *11*
sagittata (DC.) Guinea et Heywood   11
Phytolacca L.   69
americana L.   69
decandra L.   *69*
Phytolaccaceae   69
Pinaceae   17
Pinus L.   17

halepensis Mill.   18
Polycarpaea Lam.   127
fragilis Del.   *127*
repens (Forssk.) Aschers. et Schweinf.   127
Polycarpon L.   126
alsinefolium (Biv.) DC.   *126*
arabicum Boiss.   *127*
succulentum (Del.) J. Gay   127
tetraphyllum (L.) L.   126
var. alsinifolium (Mill.) Arcang.   126
var. tetraphyllum   126
var. *verticillatum* Fenzl   *126*
Polygonaceae   51
Polygonum L.   52
acuminatum Kunth   57
arenastrum Bor.   55
aviculare auct. non L.   *55*
bellardi auct. non All.   *55*
equisetiforme Sibth. et Sm.   53
var. *arenarium* Eig et Feinbr.   *54*
lanigerum R. Br.   57
lapathifolium L.   56
maritimum L.   54
palaestinum Zoh.   54, 341
var. negevense Zoh. et Waisel   54, 341
var. palaestinum   54
patulum M. B.   55
salicifolium Brouss. ex Willd.   56
var. glabrescens (Feinbr.) Zoh.   57
var. salicifolium   57
var. *salicifolium* (Del. ex Meissn.) Maire et Weiller   *57*
var. *serrulatum* (Lag.) Maire et Weiller   *57*
scabrum Poir.   *56*
var. *glabrescens* Feinbr.   *57*
senegalense Meissn.   58
serrulatum Lag.   *56*
Polypodiaceae   12
Polypodium L.   13
fragrans L.   *6*
leptophyllum L.   *8*
pteridioides Reichard   *6*
villarii Bellardi   *12*
vulgare L.   13
var. serratum Willd.   13
Populus L.   29

Populus (cont.)
  euphratica Oliv.   29
Portulaca L.   77
  oleracea L.   78
    var. sativa L.   78
PORTULACACEAE   77
Pseuderucaria (Boiss.) O. E. Schulz   304
  clavata (Boiss. et Reut.) O. E. Schulz
    304
Psilonema homalocarpum Fisch. et Mey.
  286
Pteranthus Forssk.   135
  dichotomus Forssk.   136
  echinatus Desf.   *136*
PTERIDACEAE   7
Pteris L.   8
  longifolia auct. non L.   *8*
  vittata L.   8

Quercus L.   30
  aegilops L. var. *ithaburensis* (Decne.)
    Boiss.   *32*
    ssp. *ithaburensis* (Decne.) Eig   *32*
  boissieri Reut.   31
    var. boissieri   31
    ssp. *latifolia* (Boiss.) O. Schwarz   *31*
    var. latifolia (Boiss.) Zoh.   31
  brachybalanos Ky.   *35*
  calliprinos Webb   33
    var. *arcuata* (Ky.) A. DC.   *35*
    var. brachybalanos (Ky.) A. DC.   35
    var. calliprinos   34
    var. *dispar* Ky.   *34*
    var. eigii A. Camus   35
    var. *eucalliprinos* A. DC.   *34*
    var. fenzlii (Ky.) A. Camus   35
    var. palaestina (Ky.) Zoh.   35
    var. puberula Zoh.   35
    var. subglobosa Zoh.   35
  coccifera L. var. *calliprinos* (Webb)
    Boiss.   *33*
    var. *palaestina* (Ky.) Boiss.   *35*
  fenzlii Ky.   *35*
  infectoria Oliv. ssp. *boissieri* (Reut.)
    Gürke   *31*
    var. *latifolia* (Boiss.) A. Camus   *31*
  ithaburensis Decne.   32
    var. calliprinoides Zoh.   33
    var. dolicholepis Zoh.   33

Quercus (cont.)
    var. ithaburensis   32
    var. subcalva Zoh.   33
    var. subinclusa Zoh.   33
  palaestina Ky.   *35*
  pseudococcifera Labill.   *34*
  pyrami Ky.   *32*
  ungeri Ky.   *32*

RAFFLESIACEAE   50
RANUNCULACEAE   190
Ranunculus L.   200
  aquatilis L. ssp. heleophilus (Arv.-Touv.)
    Rikli   209
    var. microcarpus Meikle   209
    var. *sphaerospermus* (Boiss. et Bl.)
      Boiss.   *209*
    var. *submersus* Gren. et Godr.   *209*
  aquatilis sensu Post   *209*
  arvensis L.   207
    var. arvensis   207
    var. longispinus Post   207
    var. tuberculatus (DC.) Fiori   208
  asiaticus L.   204
    var. asiaticus   204
    var. tenuilobus Boiss.   204
  calthaefolius (Guss.) Jord.   *202*
  chaerophyllus auct. non L.   *203*
  chius DC.   207
  circinatus Dinsmore   *209*
  constantinopolitanus (DC.) Urv.   204
    var. *palaestinus* (Boiss.) Boiss.   *204*
  cornutus DC.   205
  damascenus Boiss. et Gaill.   202
  dasycarpus (Stev.) Boiss.   *203*
    var. *edumeus* Zoh.   *203*
    var. *leiocarpus* Zoh.   *203*
    var. *macrorhynchus* (Boiss.) Zoh.
      203
  falcatus L.   *210*
  ficaria L. ssp. ficariiformis (F. W.
    Schultz) Rouy et Fouc.   202
  ficariaeformis F. W. Schultz   *202*
  flabellatus Desf.   *203*
  guilelmi-jordani Aschers.   206
  heleophilus Arv.-Touv.   *209*
  hierosolymitanus Boiss.   *203*
  lanuginosus L. var. *constantinopolitanus*
    DC.   *204*

Ranunculus (cont.)
  lomatocarpus Fisch. et Mey. *205*
  macrorhynchus Boiss. 203
    ssp. *trigonocarpus* (Boiss.) P. H. Davis *203*
  marginatus Urv. 205
    var. marginatus 206
    var. scandicinus (Boiss.) Zoh. 206
  millefolius Banks et Sol. 203
  muricatus L. 206
  myriophyllus Russ. ex Schrad. *203*
  ophioglossifolius Vill. 208
  orientalis L. *203*
    ssp. *hierosolymitanus* (Boiss.) P. H. Davis *203*
  oxyspermus M. B. ssp. *damascenus* (Boiss. et Gaill.) P. H. Davis *202*
    var. *damascenus* (Boiss. et Gaill.) Post *202*
  paludosus Poir. 203
  peltatus sensu Meikle *209*
  peltatus Schrank ssp. *sphaerospermus* (Boiss. et Bl.) Meikle *209*
  saniculifolius Viv. 209
  sceleratus L. 208
  sphaerospermus Boiss. et Bl. 209
  trachycarpus Fisch. et Mey. *205*
    var. *minor* Zoh. *206*
    var. *scandicinus* Boiss. *206*
  trichophyllus Chaix 210
  trigonocarpus Boiss. *203*
  tuberculatus DC. var. *prostratus* Post *208*
Raphanus L. 325
  aucheri Boiss. 326
  lyratus Forssk. *328*
  pinnatus Viv. *315*
  pugioniformis Boiss. *326*
  raphanistrum L. 325
  rostratus DC. 326
    var. pugioniformis (Boiss.) O. E. Schulz ex Thell. 326
Rapistrum Crantz 321
  orientale (L.) Crantz *322*
  rugosum (L.) All. 321
    subvar. hirtisiliquum Thell. 322
    var. hispidum (Godr.) Coss. 322
    var. indivisum Vis. et Sacc. 322
    var. orientale (L.) Arcang. 321

Rapistrum (cont.)
  ssp. *orientale* (L.) Rouy et Fouc. *321*
  var. rugosum 321
  var. venosum (Pers.) DC. 322
Reboudia Coss. et Dur. 315
  pinnata (Viv.) O. E. Schulz 315
    var. absalomii Opphr. 316
    var. *boissieri* O. E. Schulz *316*
    var. liocarpa O. E. Schulz 316
    var. *major* (Post) O. E. Schulz *316*
    var. pinnata 316
Reseda L. 331
  alba L. 332
    var. *hookeri* (Guss.) Heldr. *333*
  alopecuros Boiss. 334
  arabica Boiss. 333
  boissieri Müll. 337
    var. boissieri 337
    var. pendula Plitm. et Zoh. 337, 343
  cahirana Müll. *337*
  canescens L. *339*
  clausa Reichb. ex Müll. *335*
  crispata Link *338*
  decursiva Forssk. 332
    var. foliosa (Post) Hand.-Mazzetti 333
  eremophila Boiss. *332*
  globulosa Fisch. et Mey. 335
    var. brevipes Post 336
  hexagyna Forssk. *339*
  kahirina Müll. var. *boissieri* (Müll.) Boiss. *337*
  linifolia Vahl ex Hornem. *330*
  lutea L. 335
    var. lutea 335
    var. nutans Boiss. 335
    var. *vulgaris* Müll. *335*
  luteola L. 338
    var. crispata (Link) Müll. 338
    var. glomerata Zoh. 339
    var. luteola 338
  macrosperma Reichb. var. *orientalis* Müll. *334*
  maris-mortui Eig 338
  muricata C. Presl 336
    var. hierochuntica Zoh. 337
    var. lepidocarpa Zoh. 337
    var. macrocarpa Zoh. 337
    var. undulata Post 337

Reseda (cont.)
odorata L. 334
orientalis (Müll.) Boiss. 334
propinqua Boiss. *332*
pruinosa Del. 336
stenostachya Boiss. 336
var. eilatensis Plitm. et Zoh. 336, 343
var. stenostachya 336
subulata Del. *330*
RESEDACEAE 329
Rheum L. 58
palaestinum Feinbr. 58
Ricotia L. 282
lunaria (L.) DC. 282
Robbairea Boiss. 127
delileana Milne-Redhead 128
prostrata (Forssk.) Boiss. *128*
Roemeria Medik. 229
dodecandra (Forssk.) Stapf *229*
hybrida (L.) DC. 229
var. velutina DC. 230
var. velutino-eriocarpa Fedde 230
procumbens Aarons. et Opphr. 230
Rokejeka capillaris Forssk. *101*
Roridula droserifolia Forssk. *245*
Rorippa Scop. 277
amphibia (L.) Bess. 277
nasturtium-aquaticum (L.) Hay. *277*
Rumex L. 59
bucephalophorus L. 66
var. bucephalophorus 66
var. papillosus Feinbr. 66
callosissimus Meissn. *65*
cassius Boiss. 64
conglomeratus Murr. 63
crispus L. 63
var. unicallosus Peterm. 63
cyprius Murb. 61
ssp. *disciformis* Sam. *61*
dentatus L. 65
ssp. callosissimus (Meissn.) Rech. f. 65
ssp. mesopotamicus Rech. f. 65
var. *pleiodon* Boiss. *65*
divaricatus L. *64*
lacerus Balbis *62*
var. *macrocarpus* Boiss. *62*
maritimus L. 65

Rumex (cont.)
occultans Sam. 62
pictus Forssk. 62
pulcher L. 64
ssp. anodontus (Hausskn.) Rech. f. 64
ssp. *cassius* (Boiss.) Rech. f. *64*
ssp. divaricatus (L.) Murb. 64
var. *palaestinus* Feinbr. *64*
roseus L. *60, 61*
rothschildianus Aarons. ex Evenari 62
spinosus L. *67*
tingitanus L. 60
tuberosus L. 60
vesicarius L. 61

Sagina L. 121
apetala Ard. 121
SALICACEAE 24
Salicornia L. 157
amplexicaulis Vahl *154*
europaea L. 157
var. *fruticosa* L. *156*
fruticosa (L.) L. *156*
glauca Del. *156*
herbacea (L.) L. *157*
macrostachya Moric. *156*
mucronata Lag. *156*
perennis Mill. *156*
radicans Sm. *156*
strobilacea Pall. *155*
Salix L. 25
acmophylla Boiss. 26
× alba 28
× babylonica 28
var. *pseudo-safsaf* (A. Camus et Gomb.) Thiéb. *27*
alba L. 26
× babylonica 28
var. alba 26
var. micans (Anderss.) Anderss. 27
australior Anderss. 28
babylonica L. 27
dinsmorei Enander ex Dinsmore 28
micans Anderss. *27*
pseudo-safsaf A. Camus et Gomb. 27
subserrata Willd. 28
triandra L. 28
× pseudo-safsaf 28

Salsola L. 168
  alopecuroides Del. *179*
  articulata Cav. *164*
  articulata Forssk. *177*
  autrani Post *168*
  baryosma (Roem. et Schult.) Dandy
    174
    var. baryosma 174
    var. viridis Zoh. 175, 342
  crassa M. B. 171
  delileana Botsch. *174*
  foetida Del. ex Spreng. *174*
  inermis Forssk. 171
  jordanicola Eig 171
  kali L. 170
    var. crassifolia Reichb. 170
    var. tenuifolia Tausch 170
  lancifolia Boiss. *168*
  longifolia Forssk. 173
  mucronata Forssk. *175*
  muricata L. *152*
  oppositiflora Pall. *176*
  oppositifolia Desf. *173*
  pachoi Volkens et Aschers. *172*
  rigida Pall. var. *tenuifolia* Boiss. *174*
  rosmarinus (Bge. ex Boiss.) Solms-Laub.
    167
  schweinfurthii Solms-Laub. 173
  sieberi C. Presl *173*
  soda L. 170
  splendens Pourr. *161*
  tetragona Del. 172
    var. *tetrandra* (Forssk.) Boiss. *172*
  tetrandra Forssk. 172
  vermiculata L. 174
    ssp. *villosa* (Del.) Eig *174*
    var. villosa (Del. ex Roem. et Schult.)
      Moq. 174
  villosa Del. ex Roem. et Schult. *174*
  volkensii Schweinf. et Aschers. 171
SANTALACEAE 43
Saponaria L. 103
  cretica L. *105*
  filicaulis Boiss. *102*
  mesogitana Boiss. 103
  oxyodonta Boiss. *104*
  vaccaria L. *104*
Savignya DC. 315
  aegyptiaca DC. *315*

Savignya (cont.)
  parviflora (Del.) Webb 315
    var. *oblonga* (Boiss.) O. E. Schulz
      315
Schanginia baccata (Forssk.) Moq. *161*
  hortensis (Forssk.) Moq. *162*
Schimpera Hochst. et Steud. 260
  arabica Hochst. et Steud. ex Boiss. 261
    var. arabica 261
    var. lasiocarpa Boiss. 261
Sclerocephalus Boiss. 134
  arabicus Boiss. 135
Scolopendrium hemionitis Swartz *11*
  sagittatum DC. *11*
Seidlitzia Bge. ex Boiss. 167
  rosmarinus Bge. ex Boiss. 167
Senebiera coronopus (L.) Poir. *303*
Silene L. 81
  aegyptiaca (L.) L. f. 89
    var. alba Aarons. et Evenari ex
      Opphr. et Evenari 89
    var. umbrosa Nab. 89
  affinis Boiss. *94*
  apetala Willd. 98
    var. alexandrina Aschers. 99
    var. apetala 99
    var. grandiflora Boiss. 98
  arabica Boiss. 94
    var. arabica 95
    var. moabitica Zoh. 95, 342
    var. nabathaea (Gomb. et A. Camus)
      Zoh. 95
    var. viscida (Boiss.) Zoh. 95
  atocion Murr. *89*
  behen L. *85*
  behen L. 90
  bipartita Desf. *97*
  calycina Salzm. ex Rohrb. *97*
  chaetodonta Boiss. var. *modesta* (Boiss.
    et Bl.) Boiss. *88*
  chloraefolia Sm. var. *swertiaefolia*
    (Boiss.) Rohrb. *85*
  colorata Poir. 97
    ssp. colorata 98
    ssp. oliveriana (Otth) Rohrb. 98
  coniflora Nees ex Otth 99
  conoidea L. 99
  crassipes Fenzl 91
  damascena Boiss. et Gaill. 93

Silene (cont.)

dichotoma Ehrh. var. *racemosa* (Otth)
    Rohrb. *92*
fuscata Link ex Brot.   89
gallica L.   96
grisea Boiss.   86
hussonii Boiss.   87
inflata (Salisb.) Sm.   *85*
italica (L.) Pers.   84
juncea Sibth. et Sm.   *88*
linearis Decne.   88
longipetala Vent.   84
    var. purpurascens Boiss.   84
macrodonta Boiss.   99
moabitica Eig   *95*
modesta Boiss. et Bl.   88
muscipula L.   91
nabathaea Gomb. et A. Camus   *95*
nocturna L.   96
    var. brachypetala (Rob. et Cast.)
        Rohrb. 96
    var. genuina Rohrb.   96
    var. prostrata Post   96
oliveriana Otth   *98*
oxyodonta Barb.   97
palaestina Boiss.   94
    var. *damascena* (Boiss. et Gaill.)
        Rohrb. *93*
papillosa Boiss.   92
physalodes Boiss.   86
picta Pers.   *88*
racemosa Otth   *92*
reinwardtii Roth   88
rubella L.   90
sedoides Poir.   90
setacea Viv.   *95*
    var. *viscida* Boiss.   *95*
sibthorpiana Reichb.   *92*
succulenta Forssk.   87
    var. *eliezeri* Eig   *87*
swertiifolia Boiss.   85
telavivensis Zoh. et Plitm.   92, 341
tridentata Desf.   97
trinervis Banks et Sol.   92
venosa (Gilib.) Aschers.   *86*
    var. *commutata* (Guss.) Dinsmore   *86*
villosa Forssk.   93
    var. ismaelitica Schweinf.   93
    var. villosa   93

Silene (cont.)

vivianii Steud.   95
vulgaris (Moench) Garcke   85
    var. rubriflora Boiss.   86
Sinapis L.   310
alba L.   311
    var. alba   311
    var. melanosperma Alef.   311
arvensis L.   311
    var. arvensis   312
    var. *divaricata* O. E. Schulz ex Thell.
        312
    var. orientalis (L.) Koch et Ziz   312
    var. *schkuhriana* (Reichb.) Hag.   *312*
    var. *stricta* Čelak.   *312*
erucoides L.   *307*
harra Forssk.   *306*
hispanica L.   *317*
incana L.   *312*
nigra L.   *309*
philaeana Del.   *274*
SINOPTERIDACEAE   5
Sisymbrium L.   252
amphibium L.   *277*
bilobum (C. Koch) Grossh.   254
columnae Jacq.   *254*
damascenum Boiss. et Gaill.   253
erysimoides Desf.   255
exacoides DC.   *266*
filifolium Willd.   *270*
hirsutum Lag.   *253*
irio L.   253
nanum DC.   *268*
nasturtium-aquaticum L.   *277*
officinale (L.) Scop.   255
    var. leiocarpum DC.   256
    var. officinale   256
orientale L.   254
pumilum Steph. ex Willd.   *257*
runcinatum Lag.   252
    var. hirsutum (Lag.) Coss.   253
    var. runcinatum   253
scorpiuroides Boiss.   *268*
septulatum DC. prol. *bilobum* (C. Koch)
    O. E. Schulz   *254*
sophia L.   *256*
    var. *brachycarpum* Boiss.   *256*
torulosum Desf.   *267*
vimineum L.   *308*

Sodada decidua Forssk. *244*
Spergula L. 122
 arvensis L. 122
 fallax (Lowe) Krause 122
 flaccida (Roxb.) Aschers. *122*
 pentandra L. var. *intermedia* Boiss.
  122
Spergularia (Pers.) J. et C. Presl 123
 bocconii (Sol. ex Scheele) Aschers. et
  Graebn. 125
 campestris (L.) Aschers. *124*
 diandra (Guss.) Heldr. et Sart. 124
 fallax Lowe *122*
 marginata (DC.) Kitt. *123*
 marina (L.) Griseb. *124*
 media Boiss. *124*
 media (L.) C. Presl 123
 rubra (L.) J. et C. Presl 124
 salina J. et C. Presl 123, 124
Stellaria L. 117
 apetala auct. non Ucria *118*
 dubia Bast. *120*
 media (L.) Vill. 117
  var. *apetala* (Ucria) Ledeb. *118*
  var. *eliezeri* Eig *118*
  ssp. eliezeri (Eig) Zoh. 118
  var. *major* Koch *118*
  var. *neglecta* (Weihe) Mert. et Koch
   118
  ssp. postii Holmboe 118
 neglecta Weihe *118*
 pallida (Dumort.) Piré 118
Stigmatella Eig 264
 longistyla Eig 264
Suaeda Forssk. ex Scop. 157
 aegyptiaca (Hasselq.) Zoh. 161
 asphaltica (Boiss.) Boiss. 158
 baccata Forssk. ex J. F. Gmel. *161*
 fruticosa Forssk. ex J. F. Gmel. 159
 fruticosa (L.) Forssk. *159*
 hortensis Forssk. ex J. F. Gmel. 162
 linifolia Pall. 162
 maris-mortui Post 162
 maritima (L.) Dumort. 162
 monoica Forssk. ex J. F. Gmel. 160
 palaestina Eig et Zoh. 160
 rosmarinus Ehrenb. ex Boiss. *167*
 salsa Pall. 162
 setigera (DC.) Moq. *161*

Suaeda (cont.)
 splendens (Pourr.) Gren. et Godr. 161
 vera Forssk. ex J. F. Gmel. 159
  var. deserti Zoh. et Baum 159, 342
  var. vera 159
 vermiculata Forssk. ex J. F. Gmel. 160

Telephium L. 125
 sphaerospermum Boiss. 125
Tetragonia expansa Murr. *75*
Tetragonia tetragonoides (Pall.) O. Ktze.
 75
Texiera Jaub. et Sp. 259
 glastifolia (DC.) Jaub. et Sp. 259
THELYPTERIDACEAE 9
Thelypteris Schmidel 9
 palustris Schott 9
Thesium L. 44
 bergeri Zucc. 45
 humile Vahl 45
Thlaspi L. 295
 arvense L. 295
 bursa-pastoris L. *294*
 carneum Banks et Sol. *297*
 perfoliatum L. 296
Torularia (Coss.) O. E. Schulz 267
 torulosa (Desf.) O. E. Schulz 267
  var. scorpiuroides (Boiss.) O. E.
   Schulz 268
  var. torulosa 268
Traganum Del. 162
 nudatum Del. 162
Trianthema L. 75
 pentandra L. 75
Tunica arabica Boiss. *106*
 pachygona Fisch. et Mey. *105*
 velutina (Guss.) Fisch. et Mey. *106*
Turritis L. 281
 laxa (Sibth. et Sm.) Hay. 281

ULMACEAE 36
Ulmus L. 36
 canescens Melv. 36
Urtica L. 39
 caudata Vahl *40*
 dioica L. 41
 dubia Forssk. 40
 hulensis Feinbr. 40
 membranacea Poir. *40*

Urtica (cont.)
 pilulifera L.   40
 urens L.   40
URTICACEAE   39

Vaccaria Moench   104
 oxyodonta Boiss.   *104*
 parviflora Moench   *104*
 pyramidata Medik.   104
  var. oxyodonta (Boiss.) Zoh.   104
  var. pyramidata   104
 segetalis (Neck.) Garcke ex Aschers.
  104
Velezia L.   111
 fasciculata Boiss.   *112*
 rigida L.   111
  var. fasciculata (Boiss.) Post   112

Velezia (cont.)
  var. rigida   111
Vella annua L.   *314*
Viscum L.   46
 cruciatum Sieb. ex Boiss.   46
Vogelia apiculata (Fisch., Mey. et Avé-
 Lall.) Vierh.   *293*

Zilla Forssk.   324
 myagrioides Forssk.   *344*
 myagroides Forssk. var. *microcarpa* DC.
  325
 spinosa (L.) Prantl   324
  var. microcarpa (DC.) Dur. et Schinz
   325
  var. spinosa   325

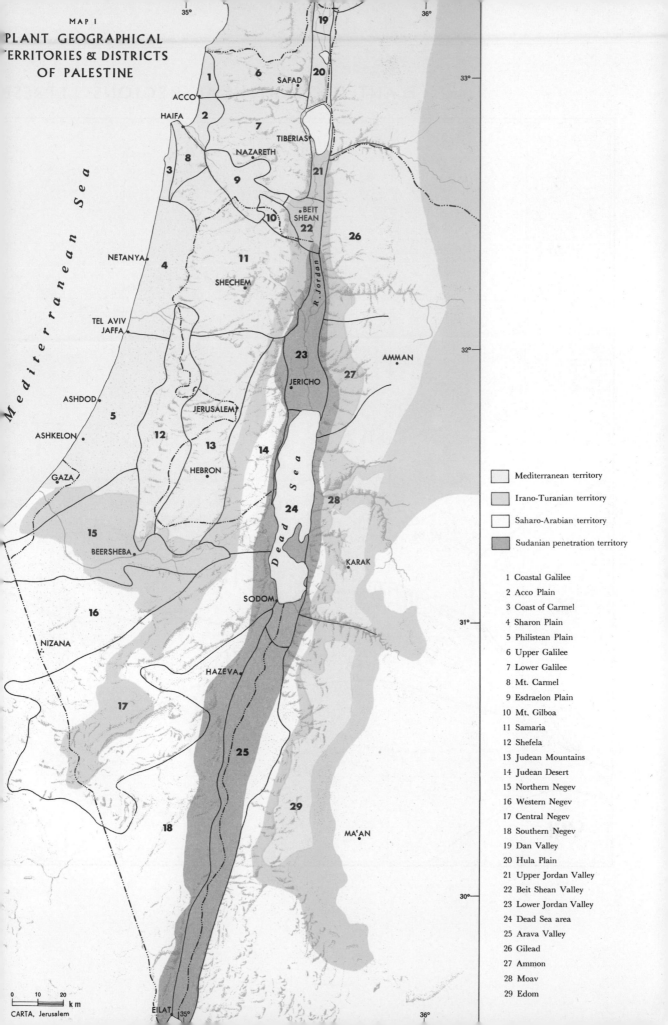

MAP 1

PLANT GEOGRAPHICAL
TERRITORIES & DISTRICTS
OF PALESTINE

Mediterranean Sea

R. Jordan

Dead Sea

Legend:

☐ Mediterranean territory

☐ Irano-Turanian territory

☐ Saharo-Arabian territory

☐ Sudanian penetration territory

1 Coastal Galilee
2 Acco Plain
3 Coast of Carmel
4 Sharon Plain
5 Philistean Plain
6 Upper Galilee
7 Lower Galilee
8 Mt. Carmel
9 Esdraelon Plain
10 Mt. Gilboa
11 Samaria
12 Shefela
13 Judean Mountains
14 Judean Desert
15 Northern Negev
16 Western Negev
17 Central Negev
18 Southern Negev
19 Dan Valley
20 Hula Plain
21 Upper Jordan Valley
22 Beit Shean Valley
23 Lower Jordan Valley
24 Dead Sea area
25 Arava Valley
26 Gilead
27 Ammon
28 Moav
29 Edom

Place names: ACCO, HAIFA, SAFAD, NAZARETH, TIBERIAS, BEIT SHEAN, NETANYA, SHECHEM, TEL AVIV JAFFA, ASHDOD, JERUSALEM, ASHKELON, GAZA, HEBRON, JERICHO, AMMAN, BEERSHEBA, KARAK, SODOM, NIZANA, HAZEVA, MA'AN, EILAT

Scale: 0 10 20 km

CARTA, Jerusalem

MAP 2

PLANT GEOGRAPHICAL REGIONS REPRESE

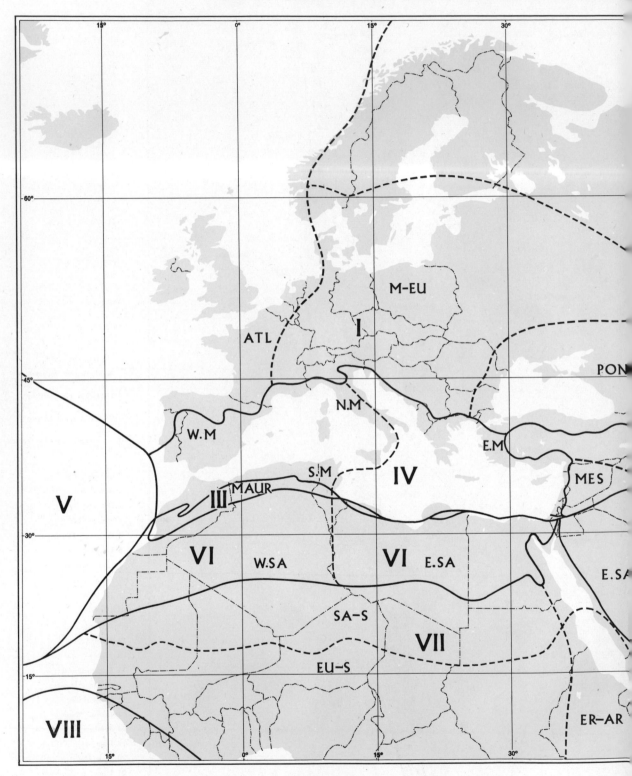

IN FLORA PALAESTINA

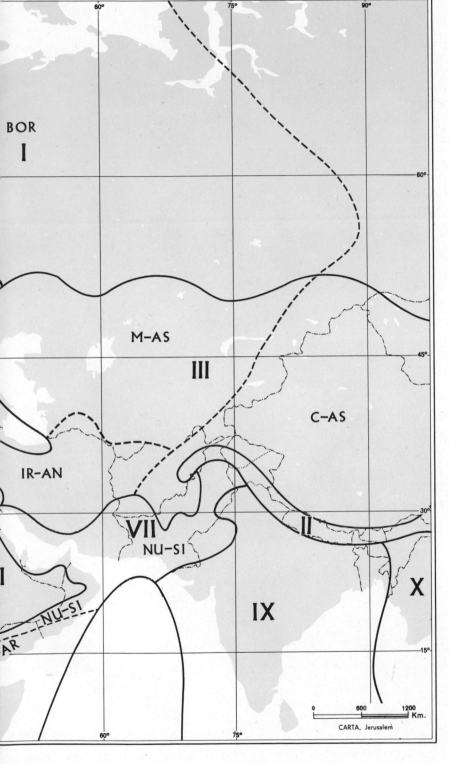

I   Euro-Siberian region

               West Euro-Siberian subregion
ATL   *Atlantic province*
BOR   *Boreal province*
M-EU   *Medio-European province*
PON   *Pontic province*

II   Sino-Japanese region

III   Irano-Turanian region

               West Irano-Turanian subregion
MAUR   *Mauritanian steppes province*
MES   *Mesopotamian province*
IR-AN   *Irano-Anatolian province*
M-AS   *Medio-Asiatic province*

               East Irano-Turanian subregion
C-AS   *Centro-Asiatic province*

IV   Mediterranean region

W.M   *West Mediterranean subregion*
N.M   *North Mediterranean part*
S.M   *South Mediterranean part*
E.M   *East Mediterranean subregion*

V   Macaronesian region

VI   Saharo-Arabian region

W.SA   *West Saharo-Arabian subregion*
E.SA   *East Saharo-Arabian subregion*

VII   Sudanian region

               West Sudanian subregion
SA-S   *Sahelo-Sudanian province*
EU-S   *Eu-Sudanian province*
E.S   *East Sudanian subregion*
NU-SI   *Nubo-Sindian province*
ER-AR   *Eritreo-Arabian province*

VIII   Guineo-Congolese region
IX   Indian region
X   Malaysian region